# Fundamentals of Mathematical Analysis

# Fundamentals of
# Mathematical Analysis

ADEL N. BOULES

# OXFORD
## UNIVERSITY PRESS

Great Clarendon Street, Oxford, OX2 6DP,
United Kingdom

Oxford University Press is a department of the University of Oxford.
It furthers the University's objective of excellence in research, scholarship,
and education by publishing worldwide. Oxford is a registered trade mark of
Oxford University Press in the UK and in certain other countries

First Edition published in 2021

Impression: 2

Published in the United States of America by Oxford University Press
198 Madison Avenue, New York, NY 10016, United States of America

British Library Cataloguing in Publication Data

Data available

Library of Congress Control Number: 2020952673

ISBN 978-0-19-886878-1 (hbk.)
ISBN 978-0-19-886879-8 (pbk.)

DOI: 10.1093/oso/9780198868781.001.0001

Printed and bound by
CPI Group (UK) Ltd, Croydon, CR0 4YY

*This work is a tribute*
*To all my teachers*
*The ones I met and the ones I did not*

*Dedication*
*To all my children*

# Preface

This is a beginning graduate book on real and functional analysis, with a significant component on topology. The prerequisites include a solid understanding of undergraduate real analysis and linear algebra, and a good degree of mathematical maturity. Rudimentary knowledge of metric spaces, although not required, is a huge asset. With the singular exception of Liouville's theorem (stated without proof), and a passing reference to Laurent series, knowledge of complex analysis is neither assumed nor needed.

It is possible for students with high mathematical aptitude to study this book independently. However, the book is designed as a textbook for well-prepared students of mathematics, to be taught under the able guidance of an instructor. I like to think of this book as an accessible classical introduction to the subject. The goal is to provide a springboard from which students can dive into greater depths in the sea of mathematics.

The book is neither encyclopedic nor a shallow introduction. The aim is to achieve excellent breadth and depth. The topics are organized logically but not rigidly, in order to maximize utility and the potential readership. The careful sequencing of the sections is designed to allow instructors to select topics that suit their course goals, student backgrounds, and time limitations. Although the proofs are detailed, I hope the reader will find the writing style clear and concise. The section exercises constitute an important complement to the results in the main body of the section. Indeed, some of the exercises provide alternative approaches to some topics, and generalizations of some of the results in the main text are considered in the exercises. The book synopsis included after the preface furnishes more details on the structure of the book and brief chapter descriptions.

I deliberately avoided making specific bibliographic citations within the body of the text. There are two main reasons for this. First, all the results in this book are well established and can be found in multiple sources. Second, the book contains no original results. Therefore, the lack of bibliographic citations or the absence of any specific source must not be conflated with claims of originality. I did not number the definitions, in order to prevent item numbers within sections from escalating to an annoying level. Definitions are seldom referenced far from where they first appear, and the extensive index and the glossary of symbols should help the reader locate items easily. Examples are locally and manually numbered within each section.

Almost all of the historical information contained in this book is abridged, with large excerpts included without quotation marks, from J. J. O'Connor and E. F. Robertson's articles in the *MacTutor History of Mathematics* archive, School of Mathematics and Statistics, University of St Andrews, Scotland (see http://www-history.mcs.st-andrews.ac.uk/index.html).

Sir Isaac Newton once said that if he had seen further than others, it was by standing on the shoulders of giants. I am no giant, but this book is the shoulder I have to offer. Perhaps a few students will climb and will be able to see farther than I have.

# Acknowledgments

I would like to express my sincere appreciation to a long succession of upper academic administrators at the University of North Florida for their support throughout my tenure there. I must expressly mention Lewis Radonovich, Mark Workman, and Barbara Hetrick. I also owe a debt to the College of Arts and Sciences for granting me a sabbatical for the academic year 2017–18, during which a significant bulk of writing was achieved.

I thank the anonymous reviewers of this book for their insightful criticism. Their suggestions greatly contributed to the richness of the book and the cohesion of its topics. I also thank the commissioning editor Dan Taber and the assistant commissioning editor Katherine Ward for their prompt and professional assistance during the review and production stages, and my son Youssef for his assistance with the typesetting and formatting of the graphics.

Finally but foremost, my deep gratitude goes to my wife and life companion. Her kind, patient, and trusting nature touched many lives and transformed mine.

**Jacksonville, Florida**
**July 2020**

# The Book in Synopsis

The book in its entirety contains enough material for a two-semester course. The core of the book can be used for an easy paced two-semester course. If a definition of the core contents of the book is desirable, I define the core to consist of the following sections, in addition to the very basic ideas in sections 1.1, 1.2, and 3.1:

▶ Sections 2.1 and 2.2
▶ Sections 3.2–3.4, 3.6, and 3.7
▶ Sections 4.1–4.10
▶ Sections 5.1–5.4 and 5.6–5.8
▶ Sections 6.1–6.4
▶ Sections 7.1 and 7.2
▶ Sections 8.1–8.4

## Part I. Background Material

Instructors can choose material from this part as their students' background warrants. The most basic results in the first three chapters are stated without proof.

**Chapter 1.** This chapter furnishes a brief refresher of basic concepts. The natural, rational, and real number systems are taken for granted, although we develop the completeness of the real line and the Bolzano-Weierstrass theorem at length, as well as the complex number field, including its completeness. Embryonic manifestations of completeness and compactness can be seen in this chapter. Examples include the nested interval theorem and the uniform continuity of continuous functions on compact intervals, and our proof of the Heine-Borel theorem in chapter 4 is squarely based on the Bolzano-Weierstrass property of bounded sets.

**Chapter 2.** This chapter fills in any potential gaps that may exist in the student's knowledge of set theory. Sections 2.1 and 2.2 are essential for a proper understanding of the rest of the book. In particular, a thorough understanding of countability and Zorn's lemma is indispensable. Some of section 2.3 may be included, but only an intuitive understanding of cardinal numbers is sufficient. Studying section 2.3 up to theorem 2.3.4, together with theorem 2.3.13, is sufficient to follow the discussion on the existence of a vector space of arbitrary (infinite) dimension, and the existence of inseparable Hilbert spaces. Cardinal arithmetic can be omitted. Indeed, the results on cardinal arithmetic are applied only once in order to prove the invariance of the cardinality of a linear basis of a vector space. Ordinal numbers have been carefully avoided.

**Chapter 3.** It is this author's observation that the undergraduate linear algebra curriculum has settled into a matrix theory mode without enough exposure to vector space theory. This chapter aims to provide a solid but brief account of the theory of vector spaces. The reader is assumed to have good knowledge of the basic definitions, which are briefly summarized in section 3.1. The aim of sections 3.2 and 3.3 is to provide a thorough presentation of the concepts of basis and dimension, especially for infinite-dimensional vector spaces, as these are topics that are not normally developed rigorously in the undergraduate curriculum. The approach is unified in the sense that we do not treat finite and infinite-dimensional spaces separately. Important concepts make their first debut in section 3.4. These include algebraic complements, quotient spaces, direct sums, projections, linear functionals, and invariant subspaces. Section 3.5 provides a brief refresher of matrix representations and diagonalization. Section 3.6 introduces normed linear spaces and is followed by an extensive study of inner product spaces in section 3.7. The presentation of inner product spaces in this section and in section 4.10 is not limited to finite-dimensional spaces but rather to many of the properties of inner products that do not require completeness. The chapter concludes with the finite-dimensional spectral theory.

# Part II. Topology

A respectable one-semester course on topology can be based on chapters 4 and 5. It is my belief that an adequate mastery of the basics of topology is a necessary prerequisite for an organized study of higher mathematics. This is a focal point of the book philosophy. It is fair to say that the book, generally speaking, has a mild topological flavor. Chapters 4 and 5 provide a solid launch pad into the last three chapters of the book. It is possible for the instructor, with a moderate amount of maneuvering, to navigate most of the rest of book while avoiding chapter 5. This chapter, however, contributes richly to the depth of the book.

**Chapter 4.** This chapter provides an extensive account of the metric topology and is a prerequisite for all the subsequent chapters. The leading sections furnish basic concepts such as closure, continuity, separation properties, product spaces, and countability axioms. This is followed by a detailed study of completeness, compactness, and function spaces. Chapter applications include contraction mappings, nowhere differentiable functions, and space-filling curves. The chapter concludes with a detailed section on Fourier series and orthogonal polynomials, which, together with section 3.7, provides an excellent background for Hilbert spaces. Our study of sequence and function spaces in this chapter leads up gently into Banach spaces.

**Chapter 5.** This chapter emphasizes the nonmetric properties of topology. Sections 5.1–5.8 constitute the core of the chapter. Section 5.5 is terminal and may be omitted. The remaining sections are more advanced and can be omitted. Section 5.9 (locally compact spaces) is the transitional section between the core of the chapter and the last three sections. At various points in the book, I point out how results stated for the metric case can be extended to topological spaces, especially locally compact spaces. Some such results are developed in the exercises. Sections 5.10–5.12 are optional, and little subsequent material is based on them. I provided a specialized proof of Urysohn's lemma for $\mathbb{R}^n$ in section 8.4 in order to help instructors avoid section 5.11, if they so choose. Tychonoff's theorem appears twice: once in section 5.8, for the product of finitely many topological spaces, and again in section 5.12, for the product of infinitely many spaces. The proofs are different, and both are worthy of inclusion, if an instructor decides to include section 5.12.

## Part III. Functional Analysis

An introductory course on functional analysis can be based on the instructor's choice of the background material and chapters 4, 6, and 7.

**Chapter 6.** This chapter introduces Banach spaces. Sections 6.1–6.4 form the core of the chapter. It would be accurate to characterize sections 6.1–6.4 as quite classical. Section 6.5 is needed for sections 7.3 and 7.4. Section 6.6 can be omitted if a brief introduction is the goal. In this case, section 7.5 must also be omitted. Section 6.7 is terminal and may be omitted without consequence. I have enriched the chapter by including such topics as Gelfand's theorem, Schauder bases, and complemented subspaces. Chapters 6 and 7 include a good number of applications of the four fundamental theorems of functional analysis.

**Chapter 7.** This chapter introduces Hilbert spaces and the elements of operator theory. Sections 7.3 and 7.4 contain a good set of results on self-adjoint and compact operators. The section exercises contain problems that suggest alternative approaches and hence allow the instructor to shorten these two sections while preserving good depth. For example, the Fredholm theory can be bypassed if the instructor wishes to limit the discussion to compact, self-adjoint operators on

Hilbert spaces. Sections 7.3 and 7.4 are written in such a way to facilitate extending the results to compact operators on Banach spaces (section 7.5). For example, we used Riesz's lemma instead of the projection theorem in order to keep the proofs adaptable for extension to Banach spaces. Sections 7.3–7.5 contain more results than are typically found in an introductory course.

## Part IV. Integration Theory

Together with chapter 4, this chapter constitutes the general/real analysis component of the book, and a good course on real analysis can be built on the background material and those two chapters.

**Chapter 8.** Section 8.1 furnishes a brief but rigorous introduction to the Riemann integral of continuous functions on compact boxes in $\mathbb{R}^n$. Although it has intrinsic value, the section is included for the express purpose of developing section 8.4. Section 8.4 develops the Lebesgue measure on $\mathbb{R}^n$, and the approach is to extend the positive linear functional provided by the Riemann integral on the space of continuous, compactly supported functions on $\mathbb{R}^n$. This very nearly amounts to developing the Radon measure theory on locally compact Hausdorff spaces. However, I chose to limit the discussion to Lebesgue measure on $\mathbb{R}^n$ because I did not wish to base the presentation heavily on chapter 5. I did, nonetheless, include an excursion into Radon measures as an optional topic. The rest of the chapter is largely independent of sections 8.1 and 8.4 and constitutes a decent introduction to general measure and integration theories. The section on complex measures has intrinsic value but is also included in order to facilitate the study of the duals of $\mathfrak{L}^p$ spaces. In particular, I limited the discussion of signed measures to real measures, this is, signed measures that are not allowed to assume infinite values. This turned out to be sufficient for our purposes. The selection of topics and the approach in sections 8.6 and 8.8 are quite classical and cover the basics of $\mathfrak{L}^p$ spaces and product measures. Section 8.7 contains an excellent collection of approximations theorems, including approximations by $\mathcal{C}^\infty$ functions. The title of the last section accurately captures its contents: a mere glimpse of the subject. However, the section finally settles questions started in sections 3.7 and 4.10 and concludes with the unraveling of the mystery about the completeness of orthogonal polynomials.

# Appendices

**Appendix A**. This appendix contains the proof of the equivalence of the axiom of choice, Zorn's lemma, and the well-ordering principle. I created this appendix in order to avoid distraction if instructors decide not to include the proof in their course.

**Appendix B**. This appendix is rather elementary in nature. It develops matrix factorizations and is used for deriving the change of variables formula in the exercises on section 8.8. Reference to this appendix is also made in section 3.5.

# Contents

1. Preliminaries                                                                   1
    1.1   Sets, Functions, and Relations                                     2
    1.2   The Real and Complex Number Fields                                  9

2. Set Theory                                                                      25
    2.1   Finite, Countable, and Uncountable Sets                             26
    2.2   Zorn's Lemma and the Axiom of Choice                                33
    2.3   Cardinal Numbers                                                    39

3. Vector Spaces                                                                   47
    3.1   Definitions and Basic Properties                                    49
    3.2   Independent Sets and Bases                                          53
    3.3   The Dimension of a Vector Space                                     57
    3.4   Linear Mappings, Quotient Spaces, and Direct Sums                   61
    3.5   Matrix Representation and Diagonalization                           70
    3.6   Normed Linear Spaces                                                75
    3.7   Inner Product Spaces                                                85

4. The Metric Topology                                                             103
    4.1   Definitions and Basic Properties                                    105
    4.2   Interior, Closure, and Boundary                                     110
    4.3   Continuity and Equivalent Metrics                                   119
    4.4   Product Spaces                                                      129
    4.5   Separable Spaces                                                    133
    4.6   Completeness                                                        136
    4.7   Compactness                                                         149
    4.8   Function Spaces                                                     160
    4.9   The Stone-Weierstrass Theorem                                       171
    4.10  Fourier Series and Orthogonal Polynomials                           175

5. Essentials of General Topology                                                  191
    5.1   Definitions and Basic Properties                                    192
    5.2   Bases and Subbases                                                  197
    5.3   Continuity                                                          200
    5.4   The Product Topology: The Finite Case                               205
    5.5   Connected Spaces                                                    208
    5.6   Separation by Open Sets                                             213
    5.7   Second Countable Spaces                                             217
    5.8   Compact Spaces                                                      221
    5.9   Locally Compact Spaces                                              226

5.10  Compactification                                   229
5.11  Metrization                                        233
5.12  The Product of Infinitely Many Spaces              238

6.  Banach Spaces                                        245
  6.1  Finite vs. Infinite-Dimensional Spaces            247
  6.2  Bounded Linear Mappings                           253
  6.3  Three Fundamental Theorems                        260
  6.4  The Hahn-Banach Theorem                           266
  6.5  The Spectrum of an Operator                       272
  6.6  Adjoint Operators and Quotient Spaces             278
  6.7  Weak Topologies                                   284

7.  Hilbert Spaces                                       291
  7.1  Definitions and Basic Properties                  292
  7.2  Orthonormal Bases and Fourier Series              300
  7.3  Self-Adjoint Operators                            308
  7.4  Compact Operators                                 319
  7.5  Compact Operators on Banach Spaces                336

8.  Integration Theory                                   341
  8.1  The Riemann Integral                              342
  8.2  Measure Spaces                                    349
  8.3  Abstract Integration                              364
  8.4  Lebesgue Measure on $\mathbb{R}^n$                373
  8.5  Complex Measures                                  393
  8.6  $\mathcal{L}^p$ Spaces                            402
  8.7  Approximation                                     408
  8.8  Product Measures                                  418
  8.9  A Glimpse of Fourier Analysis                     430

Appendix A: The Equivalence of Zorn's Lemma, the Axiom of Choice,
  and the Well Ordering Principle                        445

Appendix B: Matrix Factorizations                        449

Bibliography                                             453
Glossary of Symbols                                      455
Index                                                    457

# 1
# Preliminaries

*We are justified in calling numbers a free creation of the human mind.*

Richard Dedekind

Richard Dedekind. 1831–1916

In 1848, at the age of 16, Dedekind entered The Collegium Carolinum, an educational institution between a high school and a university. He then attended the University of Göttingen in 1850, and in 1852 completed his doctoral work in four semesters under Gauss's supervision. Dedekind was to be Gauss's last pupil. Dedekind spent the following two years in Berlin for further training, returning to Göttingen in 1855, the year Gauss died.

Dirichlet was appointed to fill Gauss's chair at Göttingen, soon became Dedekind's friend and mentor, and had a strong influence in shaping his mathematical interests. While at Göttingen, Dedekind studied the work of Galois and was the first to lecture on Galois theory.

Dedekind was later appointed to the Polytechnic of Zürich and began teaching there in 1858. By the 1860s, The Collegium Carolinum in Brunswick had been upgraded to the Brunswick Polytechnic, and Dedekind was appointed to it in

*Fundamentals of Mathematical Analysis.* Adel N. Boules, Oxford University Press (2021). © Adel N. Boules.
DOI: 10.1093/oso/9780198868781.003.0001

1862. With this appointment he returned to his hometown and remained there for the rest of his life.

Dedekind made a number of highly significant contributions to mathematics, including his definition of finite and infinite sets, and his construction of the real numbers as cuts in the set of rational numbers. Dedekind's definitions are accepted today as the standard definitions.

Among Dedekind's other notable contributions to mathematics were his editions of the collected works of Dirichlet, Gauss, and Riemann. His study of Dirichlet's work led him to study algebraic number fields, where he realized the importance of rings and ideals. The general term *ring* did not appear in Dedekind's work; it was introduced later by David Hilbert, and Dedekind's notion of an ideal was taken up and extended by Hilbert and then later by Emmy Noether.

Dedekind retired in 1894. His life was long, healthy, and contented. He never married and instead lived with one of his sisters, who also remained unmarried, for most of his adult life. "He did not feel pressed to have a more marked effect in the outside world: such confirmation of himself was unnecessary."[1]

"Dedekind's legacy ... consisted not only of important theorems, examples, and concepts, but a whole style of mathematics that has been an inspiration to each succeeding generation."[2]

## 1.1 Sets, Functions, and Relations

The reader is expected to be familiar with basic set theoretic concepts such as containment, unions, and intersections and should be comfortable with set notation. Most of the essential definitions will be stated in this section. A number of basic facts will be stated as theorems, without proof.

We use the symbols $\mathbb{N}, \mathbb{Z}, \mathbb{Q}, \mathbb{R}$, and $\mathbb{C}$ to denote, respectively, the natural numbers, the integers, rational numbers, real numbers, and complex numbers. The symbol $\emptyset$ denotes the empty set.

[1] J. J. O'Connor and E. F. Robertson, "Julius Wilhelm Richard Dedekind," in *MacTutor History of Mathematics*, (St Andrews: University of St Andrews, 1998), http://mathshistory.st-andrews. ac.uk/Biographies/Dedekind/, accessed Oct. 31, 2020.
[2] O'Connor and Robertson, "Julius Wilhelm Richard Dedekind."

**Notation.** If X is a suitable universal set for a particular problem and $A \subseteq X$, we use the notation $X - A$ to denote the complement of $A$ in $X$:

$$X - A = \{x \in X : x \notin A\}.$$

We use the same notation for relative differences (the complement of $B$ in $A$):

$$A - B = \{x \in A : x \notin B\} = A \cap (X - B).$$

**Theorem 1.1.1 (distributive laws).** *Let $A$ and $B_1, B_2, \ldots, B_n$ be subsets of a set $X$. Then*

(a) $A \cup (\cap_{i=1}^n B_i) = \cap_{i=1}^n (A \cup B_i)$,
(b) $A \cap (\cup_{i=1}^n B_i) = \cup_{i=1}^n (A \cap B_i)$. ∎

**Theorem 1.1.2 (De Morgan's laws).** *Let $A_1, A_2, \ldots, A_n$ be subsets of a set $X$. Then*

(a) $X - \cup_{i=1}^n A_i = \cap_{i=1}^n (X - A_i)$,
(b) $X - \cap_{i=1}^n A_i = \cup_{i=1}^n (X - A_i)$. ∎

**Definition.** If $x$ and $y$ are objects (e.g., numbers, functions, sets), the **ordered pair** $(x, y)$ is defined by $(x, y) = \{x, \{x, y\}\}$. The reader can verify that the definition guarantees that $(x, y) = (a, b)$ if and only if $x = a$, and $y = b$.

**Definition.** Let $X$ and $Y$ be nonempty sets. The **Cartesian product** of $X$ and $Y$ is the set of all ordered pairs:

$$X \times Y = \{(x, y) : x \in X, y \in Y\}.$$

**Definitions.** Let $X$ and $Y$ be nonempty sets. A function $f$ from $X$ to $Y$ is a subset of $X \times Y$ such that for any $x \in X$, there is a unique $y \in Y$ such that $(x, y) \in f$. We use the more common notation $y = f(x)$ instead of the cumbersome $(x, y) \in f$. We use the notation $f : X \to Y$ to indicate that $f$ is a function from $X$ to $Y$; $X$ is called the domain of $f$, denoted $Dom(f)$, and the range of $f$, denoted $\mathfrak{R}(f)$, is the set of all function values,

$$\mathfrak{R}(f) = \{f(x) : x \in X\}.$$

If $A \subseteq X$, the image of $A$ under $f$ is the set $f(A) = \{f(a) : a \in A\}$. The inverse image of a set $B \subseteq Y$ is the set $f^{-1}(B) = \{x \in X : f(x) \in B\}$.

**Definitions.** A function $f : X \to Y$ is **onto** (or **surjective**) if $\mathfrak{R}(f) = Y$. Thus, every $y \in Y$ is the image of some $x \in X$. A function $f$ is called **one-to-one** (or **injective**) if, for $x_1, x_2 \in X, x_1 \neq x_2$ implies $f(x_1) \neq f(x_2)$. Finally, f is called a **one-to-one correspondence** (or a **bijection**) if $f$ is one-to-one and onto.

**Definition.** The **identity function** $I_X$ on a set $X$ is the function $I_X(x) = x$ for all $x \in X$.

**Definition.** Let $f : X \to Y$ and $g : Y \to Z$. The **composition** of $f$ and $g$ is the function $gof : X \to Z$ defined by $(gof)(x) = g(f(x))$.

We sometimes use the notation $gf$ if there is no danger of confusing the composition of $f$ and $g$ with the product of $f$ and $g$.

**Definition.** Let $f : X \to Y$ and $g : Y \to X$ be functions. We say that $g$ is the inverse of $f$ if $gof = I_X$ and $fog = I_Y$.

We write $g = f^{-1}$ to indicate that $g$ is the inverse of $f$. Notice that a function $f$ has an inverse if and only if it is bijective. Also, if $g = f^{-1}$, then $f = g^{-1}$.

**Definition.** A **finite sequence** in a set $A$ is a function $a : \{1, 2, \dots, n\} \to A$. The element $a(i)$ is often denoted by $a_i$. It is sometimes the case that a distinction must be made between the sequence (as a function) and its range (as a set). We denote a sequence by the notation $(a_i)_{i=1}^n$, and its range by $\{a_1, a_2, \dots, a_n\}$. An **infinite sequence** in $A$ in is a function $a : \mathbb{N} \to A$. An infinite sequence is often given the notation $(a_n)$.

## Indexed Sets

Let $I$ be a set (the indexing set) and let $\mathfrak{A}$ be a collection of sets. An indexing of $\mathfrak{A}$ by $I$ is a bijection $A : I \to \mathfrak{A}$. The image of an element $\alpha \in I$ is denoted by $A_\alpha$ instead of $A(\alpha)$. Thus $\mathfrak{A} = \{A_\alpha : \alpha \in I\}$. Indexing is, of course, not limited to sets; one can index, for example, a set of numbers, or functions. If there is no danger of ambiguity, we sometimes omit reference to the indexing set $I$ and write $\{A_\alpha\}_\alpha$. Indexing is clearly a generalization of sequencing, as illustrated by the examples below.

**Example 1.** For each $n \in \mathbb{Z}$, let $A_n = (n, n+1)$. This is a collection of intervals indexed by $\mathbb{Z}$. ♦

**Example 2.** Let $I = \mathbb{R}$ and, for $\alpha \in I$, let $B_\alpha = (\alpha, \infty)$. ♦

**Example 3.** We can index the set of linear homogeneous functions in one real variable as $\{f_\alpha : \alpha \in \mathbb{R}\}$, where, for $x \in \mathbb{R}, f_\alpha(x) = \alpha x.$ ♦

**Definition.** Let $\mathfrak{A} = \{A_\alpha : \alpha \in I\}$ be an indexed family of subsets of a set $X$. We define the union and intersection of $\mathfrak{A}$ as follows:

$$\cup_{\alpha \in I} A_\alpha = \{x \in X : x \in A_\alpha \text{ for some } \alpha \in I\},$$
$$\cap_{\alpha \in I} A_\alpha = \{x \in X : x \in A_\alpha \text{ for all } \alpha \in I\}.$$

**Example 4.** In example 1, $\cup_{n \in \mathbb{Z}} A_n = \mathbb{R} - \mathbb{Z}, \cap_{n \in \mathbb{Z}} A_n = \emptyset.$ ♦

**Example 5.** In example 2, $\cup_{\alpha \in \mathbb{R}} B_\alpha = \mathbb{R}, \cap_{\alpha \in \mathbb{R}} B_\alpha = \emptyset.$ ♦

**Definition.** A family of sets $\{A_\alpha\}_\alpha$ is said to be **disjoint** if $A_\alpha \cap A_\beta = \emptyset$ whenever $\alpha \neq \beta$.

For example, the family $\{A_n\}$ in example 1 is a disjoint family.
The following theorem will be used frequently in this book.

**Theorem 1.1.3.** *Let $(A_n)$ be a sequence of subsets of a given set $X$. Then*

(a) *There exists a sequence of sets $(B_n)$ such that $B_1 \subseteq B_2 \subseteq \ldots$ and $\cup_{n=1}^\infty A_n = \cup_{n=1}^\infty B_n$. We simply define $B_n = \cup_{i=1}^n A_i$.*
(b) *There exists a disjoint sequence of sets $(C_n)$ such that $\cup_{n=1}^\infty A_n = \cup_{n=1}^\infty C_n$. The sequence we seek is $C_1 = A_1$ and, for $n \geq 2$, $C_n = A_n - \cup_{i=1}^{n-1} A_i$.* ■

**Definition.** A sequence $(B_n)$ of sets is said to be **ascending** if $B_1 \subseteq B_2 \subseteq \ldots$. A sequence $(B_n)$ of sets is said to be **descending** if $B_1 \supseteq B_2 \supseteq \ldots$.

The following two theorems generalize theorems 1.1.1 and 1.1.2.

**Theorem 1.1.4 (distributive laws).** *Let $\{B_\alpha\}_\alpha$ be an indexed family of subsets of a set $X$, and let $A$ be a subset of $X$. Then*

(a) $A \cup (\cap_\alpha B_\alpha) = \cap_\alpha (A \cup B_\alpha),$
(b) $A \cap (\cup_\alpha B_\alpha) = \cup_\alpha (A \cap B_\alpha).$ ■

**Theorem 1.1.5 (De Morgan's laws).** *Let $\{A_\alpha\}_\alpha$ be an indexed family of subsets of a set $X$. Then*

(a) $X - \cap_\alpha A_\alpha = \cup_\alpha (X - A_\alpha)$,
(b) $X - \cup_\alpha A_\alpha = \cap_\alpha (X - A_\alpha)$. ■

**Theorem 1.1.6.** *Let $f : X \to Y$, let $\{A_\alpha\}_\alpha$ be a collection of subsets of $X$, and let $\{B_\beta\}_\beta$ be a collection of subsets of $Y$. Then*

(a) $f(\cup_\alpha A_\alpha) = \cup_\alpha f(A_\alpha)$,
(b) $f(\cap_\alpha A_\alpha) \subseteq \cap_\alpha f(A_\alpha)$,
(c) $f^{-1}(\cup_\beta B_\beta) = \cup_\beta f^{-1}(B_\beta)$,
(d) $f^{-1}(\cap_\beta B_\beta) = \cap_\beta f^{-1}(B_\beta)$. ■

**Definition (Cartesian products).** Let $\{X_\alpha\}_{\alpha \in I}$ be a nonempty collection of nonempty sets. The product $\prod_{\alpha \in I} X_\alpha$ is the collection of all functions

$$x : I \to \cup_{\alpha \in I} X_\alpha$$

such that $x(\alpha) \in X_\alpha$ for all $\alpha \in I$. We write $x_\alpha$ for $x(\alpha)$.

We will denote the function $x$ in the above definition by $(x_\alpha)_{\alpha \in I}$, or simply $(x_\alpha)$. The above definition generalizes the definition of the Cartesian product of a finite number of sets. Indeed, for sets $X_1, X_2, \ldots, X_n$, the Cartesian product $\prod_{i=1}^{n} X_i$ is the set of all sequences $(x_1, x_2, \ldots, x_n)$ such that $x_i \in X_i$ for all $1 \leq i \leq n$. A sequence is nothing but a function $x : \{1, 2, \ldots, n\} \to \cup_{i=1}^{n} X_i$ such that $x_i = x(i) \in X_i$ for all $1 \leq i \leq n$.

**Example 6.** $\mathbb{R}^n = \mathbb{R} \times \ldots \times \mathbb{R}$ ($n$ factors) is the **Euclidean $n$-space**. The complex n-space $\mathbb{C}^n$ is defined similarly. ♦

**Example 7.** Let $\mathbb{R}^{\mathbb{N}}$ be the set of all infinite sequences in $\mathbb{R}$. This is also the product $\prod_{i=1}^{\infty} X_i$, where each $X_i = \mathbb{R}$. ♦

**Example 8.** Let $A$ be a set, and let $2^A$ denote the set of all functions from $A$ to the set $\{0, 1\}$. Indeed, $2^A$ is a product because if we define $X_\alpha = \{0, 1\}$ for all $\alpha \in A$, then $2^A = \prod_{\alpha \in A} X_\alpha$. As a special case, the set $2^{\mathbb{N}}$ is the set of all **binary sequences**. ♦

**Definition (set exponentiation).** Let $A$ and $B$ be nonempty sets. Define $A^B$ to be the set of all functions $f : B \to A$. We leave it to the reader to interpret $A^B$ as a product.

**Definition.** Let $A$ be a nonempty set. The collection of all the subsets of $A$, including the empty set, is known as the **power set** of $A$ and is denoted by $\mathcal{P}(A)$.

**Definition.** For a subset $S$ of a set $A$, we define the **characteristic function** of $S$ by

$$\chi_S(x) = \begin{cases} 1 & \text{if } x \in S, \\ 0 & \text{if } x \notin S. \end{cases}$$

Clearly, $\chi_S \in 2^A$. Moreover, the correspondence $\chi : \mathcal{P}(A) \to 2^A$ that assigns to each element $S \in \mathcal{P}(A)$ (i.e., $S \subseteq A$) its characteristic function $\chi_S$ is a bijection from $\mathcal{P}(A)$ to $2^A$. We leave it to the reader to verify the details.

**Definition.** Let $\{X_\alpha\}_{\alpha \in I}$ be a collection of sets, and let $X = \prod_{\alpha \in I} X_\alpha$. For each $\alpha \in I$, define the **projection** $\pi_\alpha : X \to X_\alpha$ by $\pi_\alpha(x) = x_\alpha$. Here $x = (x_\alpha)_{\alpha \in I}$ is an element of $X$.

**Example 9.** Let $X = \mathbb{R}^n$. Then, $\pi_1 : \mathbb{R}^n \to \mathbb{R}$ is indeed what we think of as the projection of $\mathbb{R}^n$ onto the $x_1$-axis: $\pi_1(x_1, \dots, x_n) = x_1$. ◆

**Example 10.** Consider the set $D$ of all functions $f : [0,1] \to \mathbb{R}$. This set can be thought of as $\prod_{\alpha \in [0,1]} X_\alpha$, where each $X_\alpha = \mathbb{R}$. If $f \in D$ and $a \in [0,1]$, then $\pi_a(f) = f(a)$. Fix an element $a \in [0,1]$ and an interval $U \subseteq \mathbb{R}$. It makes sense to ask what $\pi_a^{-1}(U)$ is. This is simply the set of all functions $f \in D$ such that $\pi_a(f) \in U$ or simply $f(a) \in U$. Thus $\pi_a^{-1}(U)$ is the set of all the functions on the closed unit interval whose graphs cross the line segment $\{a\} \times U$. ◆

**Definition.** A **relation** $R$ on a set $A$ is a subset of $A \times A$. Thus $R$ is a set of ordered pairs $(x, y)$, where $x, y \in A$. Instead of writing $(x, y) \in R$, we write $xRy$. If $xRy$, we say that $x$ is related to $y$.

**Definition.** A relation $R$ on a set $A$ is said to be

(a) **reflexive** if, for all $x \in A$, $xRx$;
(b) **symmetric** if $yRx$ whenever $xRy$;
(c) **transitive** if $xRy$ and $yRz$ imply $xRz$; and
(d) an **equivalence relation** if it is reflexive, symmetric, and transitive.

**Definition.** Let $R$ be an equivalence relation on a set $A$, and let $x \in A$. The equivalence class of $x$, denoted $[x]$, is $[x] = \{y \in A : yRx\}$.

**Theorem 1.1.7.** *Let R be an equivalence relation on a set A, and let $x, y \in A$. Then*

*(a) $[x] = [y]$ if and only if $xRy$.*
*(b) If $[x] \neq [y]$, then $[x] \cap [y] = \emptyset$.* ∎

Thus the union of the equivalence classes is $A$, and distinct equivalence classes are disjoint. The common terminology is that the equivalence classes partition $A$.

## Exercises

In the exercises below, $A, A_n, B, C$, and so on are subsets of a nonempty set $X$.

1.  Prove that $A \cap (B - C) = (A \cap B) - (A \cap C)$.
    Is it true that $A \cup (B - C) = (A \cup B) - (A \cup C)$?
2.  Find $\cup_{n \in \mathbb{N}} [\frac{1}{n}, n]$, $\cap_{\alpha \in (0,1)} [\alpha - 1, \alpha + 1]$, and $\cup_{\alpha \in (0,1)} [\alpha - 1, \alpha + 1]$.
3.  (a) Show that $A \subseteq B$ if and only if $\mathcal{P}(A) \subseteq \mathcal{P}(B)$.
    (b) Show that $\mathcal{P}(A) \cup \mathcal{P}(B) \subseteq \mathcal{P}(A \cup B)$.
    (c) Show that $\mathcal{P}(A) \cap \mathcal{P}(B) = \mathcal{P}(A \cap B)$.
4.  For $r \in \mathbb{R}, r > 0$, let $B_r = \{(x, y) \in \mathbb{R}^2 : x^2 + y^2 \leq r^2\}$. Find $\cap_{r > 0} B_r$ and $\cup_{r > 0} B_r$.
5.  Describe the following sets in words: $\cup_{n=1}^{\infty} \cap_{k=n}^{\infty} A_k$ and $\cap_{n=1}^{\infty} \cup_{k=n}^{\infty} A_k$
6.  Let $\{A_\alpha\}_\alpha$ be an indexed family of sets, and let $B$ be a set. Show that
    (a) $(\cup_\alpha A_\alpha) \times B = \cup_\alpha (A_\alpha \times B)$, and
    (b) $(\cap_\alpha A_\alpha) \times B = \cap_\alpha (A_\alpha \times B)$.
7.  Let $A = \{x \in \mathbb{R} : |x| \leq 1\}$, $B = \{x \in \mathbb{R} : |x| \geq 1\}$. Give a geometric interpretation of $A \times B$.
8.  Consider the product $\prod_{\alpha \in I} X_\alpha$ of a collection of sets. Suppose that $\{\alpha_1, \alpha_2, ..., \alpha_n\}$ is a finite subset of $I$ and that $U_{\alpha_i} \subseteq X_{\alpha_i}, 1 \leq i \leq n$. Describe the set $\cap_{i=1}^{n} \pi_{\alpha_i}^{-1}(U_{\alpha_i})$.
9.  Prove theorem 1.1.6.
10. Let $f : X \to Y$. Show that the following are equivalent:
    (a) $f$ is one-to-one.
    (b) $f(A_1 \cap A_2) = f(A_1) \cap f(A_2)$ for every $A_1, A_2 \subseteq X$.
    (c) $f^{-1}(f(A)) = A$ for every $A \subseteq X$.
    (d) $f(X - A) \subseteq Y - f(A)$ for every $A \subseteq X$.
11. Let $f : X \to Y$. Show that the following are equivalent:
    (a) $f$ is onto.
    (b) $f(f^{-1}(B)) = B$ for every $B \subseteq Y$.
    (c) $f(X - A) \supseteq Y - f(A)$ for every $A \subseteq X$.

12. Show that the composition of two injective (respectively, surjective, bijective) functions is injective (respectively, surjective, bijective).

13. Let $f : A \to B$ and $g : B \to C$ be bijections. Show that $(g \circ f)^{-1} = f^{-1} \circ g^{-1}$.

14. (a) Show that if $f : A \to B$ is injective, then there exists a function $g : B \to A$ such that $g \circ f = I_A$.
    (b) Show that if $f : A \to B$ is surjective, then there exists a function $g : B \to A$ such that $f \circ g = I_B$.

15. Show that the function $f : \mathbb{N} \times \mathbb{N} \to \mathbb{N}$ given by $f(m, n) = 2^{m-1}(2n - 1)$ is a one-to-one correspondence.

16. Verify the one-to-one correspondence between $2^A$ and $\mathcal{P}(A)$.

17. Show that if $A$ has $n$ elements and $B$ has $m$ elements, then $A^B$ has $n^m$ elements. Conclude that $\mathcal{P}(A)$ has $2^n$ elements.

18. Let $S$ and $T$ be subsets of a set $A$. Show that
    (a) $\chi_{S \cap T} = \chi_S \cdot \chi_T$; and
    (b) $\chi_{S \cup T} = \chi_S + \chi_T - \chi_{S \cap T}$.

19. Prove theorem 1.1.7.

20. Fix an integer $n > 1$, and define a relation $R$ on $\mathbb{Z}$ as follows: $xRy$ if $x - y$ is a multiple of $n$. Show that $R$ is an equivalence relation, and describe the equivalence classes.

21. Define a relation $R$ on $\mathbb{R}$ as follows: $xRy$ if and only if $x - y \in \mathbb{Q}$. Show that $R$ is an equivalence relation.

22. Define a relation $R$ on $\mathbb{Z}$ by $xRy$ if and only if $x^2 + y^2$ is even. Show that $R$ is an equivalence relation.

23. Define a relation $R$ on $\mathbb{R}$ by $xRy$ if and only if $xy \geq 0$. Is $R$ an equivalence relation?

## 1.2  The Real and Complex Number Fields

An organized study of mathematics must be rooted in a proper understanding of number systems. Authors of textbooks such as this one are often divided between two extremes: either they provide an extensive development of number systems from scratch or they ignore the entire matter and consider knowledge of the real numbers to be a prerequisite. This presentation is a compromise between the two extremes. It is assumed that the reader has a thorough knowledge of integers and the rational number field, including such topics as divisibility, prime factorizations, the infinitude of the set of prime numbers, and the construction of $\mathbb{Q}$ in the usual manner as the quotient field of $\mathbb{Z}$. We basically accept the completeness of real numbers as an axiom, then prove the Cauchy criterion and the Bolzano-Weierstrass property, which we decided to develop at length since it is a cornerstone theorem. The section concludes with the definition of complex numbers and a study of their basic properties, including completeness. Although the section is not totally self-contained, there is value in its inclusion because it illustrates a number of important proof techniques and provides a succinct summary of the properties of real and complex number fields.

**Definition.** Let $F$ be a nonempty set endowed with two binary operations, $+$ (addition) and $\times$ (multiplication). The triple $(F, +, \times)$ is said to be a **field** if the following conditions are satisfied for all $a, b, c \in F$:

(a) $a + b = b + a$.
(b) $a + (b + c) = (a + b) + c$.
(c) There is an element $0 \in F$ such that $a + 0 = a$.
(d) For every $a \in F$, there is an element $-a \in F$ such that $a + (-a) = 0$.
(e) $a \times b = b \times a$.
(f) $a \times (b \times c) = (a \times b) \times c$.
(g) There is an element $1 \in F$ such that $a \times 1 = a$.
(h) For every $a \neq 0$, there is an element $a^{-1}$ such that $a \times a^{-1} = 1$.
(i) $a \times (b + c) = a \times b + a \times c$.

We often omit the symbol for multiplication and write $ab$ or $a.b$ for $a \times b$. The element 0 is called the additive identity, and 1 is called the multiplicative identity of the filed. A field must clearly contain at least two elements.

With the usual operations of addition and multiplication of numbers, the rational numbers, $\mathbb{Q}$, and the real numbers, $\mathbb{R}$, are fields. We will see later in this section that complex numbers also form a field.

**Example 1.** Let $p$ be a prime number. Define an equivalence relation $\equiv$ on $\mathbb{Z}$ by $a \equiv b$ if $a - b$ is divisible by $p$. Since the remainder upon dividing a whole number by $p$ is an integer between 0 and $p - 1$, the equivalence classes containing $0, 1, \ldots, p - 1$ are all the equivalence classes of $\equiv$. The field of $p$ elements (also called the integers modulo $p$) consists of the equivalence classes of $0, 1, \ldots, p - 1$ and is often given the symbol $\mathbb{Z}_p$. The equivalence class of an integer $n$ is denoted $\bar{n}$. Addition and multiplication in $\mathbb{Z}_p$ are defined as follows: $\bar{n} + \bar{m} = \overline{n + m}$, and $\bar{n}.\bar{m} = \overline{nm}$. This simply means that we add or multiply integers representing the class, then reduce the result modulo $p$. For example, let $p = 7$. Then $\bar{6} + \bar{5} = \overline{11} = \bar{4}$. ◆

## Real Numbers

**Definition.** A subset $A$ of $\mathbb{R}$ is said to be **bounded above** if there is a real number $M$ such that, for all $x \in A, x \leq M$. The number $M$ is called an **upper bound** of $A$. $A$ is said to be **bounded below** if there is a real number $m$ such that, for all

$x \in A, x \geq m$. The number $m$ is called a **lower bound** of $A$. Finally, $A$ is **bounded** if it is bounded above and below.

It is clear that if $M$ is an upper bound of $A$, then every real number greater than $M$ is also an upper bound of $A$. This leads to the following definition.

**Definition.** The **least upper bound** of a set $A \subseteq \mathbb{R}$ is the number $M$ such that

(a) $M$ is an upper bound of $A$, and
(b) for all $\epsilon > 0$, $M - \epsilon$ is not an upper bound of $A$. Thus there is an element $x \in A$ such that $x > M - \epsilon$.

The least upper bound of $A$ is also called the **supremum** of $A$ and is denoted by $supA$. If $A$ is not bounded above, we set $supA = \infty$.

**Definition.** The **greatest lower bound** of a set $A \subseteq \mathbb{R}$ is the number $m$ such that

(a) $m$ is a lower bound of $A$, and
(b) for all $\epsilon > 0$, $m + \epsilon$ is not a lower bound of $A$. Thus there is an element $x \in A$ such that $x < m + \epsilon$.

The greatest lower bound of $A$ is also called the **infimum** of $A$ and is given the notation $inf\, A$. If $A$ is not bounded below, we set $infA = -\infty$.

**Example 2.** Let $A_1 = (-\infty, 1), A_2 = \{\frac{1}{n} : n \in \mathbb{N}\}$, and $A_3 = \{\frac{n}{n+1} : n \in \mathbb{N}\}$. Then, $supA_1 = supA_2 = supA_3 = 1, infA_1 = -\infty$, $infA_2 = 0$, and $infA_3 = 1/2$. ♦

The **completeness** of $\mathbb{R}$. We accept the following fact as true:

Let $A \subset \mathbb{R}$ be bounded above. Then $A$ has a least upper bound.

The above fact is not trivial; its establishment requires delving deep into the very definition of the real numbers, which we will not do here. The following example shows that $\mathbb{Q}$ is not complete and illustrates that the completeness of $\mathbb{R}$ is not to be mistaken for a simple fact.

**Example 3.** Let $A = \{x \in \mathbb{Q} : x^2 < 2\}$. Clearly, $A$ is bounded above and below. However, $supA$ and $infA$ are not in $\mathbb{Q}$. ♦

**Definition.** A sequence $(a_n)$ of real numbers is said to **converge** to $a \in \mathbb{R}$ if, for every $\epsilon > 0$, there is a natural number $N$ such that $|a_n - a| < \epsilon$ for all $n > N$. In this case, we say the limit of $(a_n)$ is $a$, and we write $\lim_{n \to \infty} a_n = a$ or simply $\lim_n a_n = a$.

**Definition.** A sequence $(a_n)$ of real numbers is said to diverge to $\infty$ if, for every $M \in \mathbb{R}$, there is a natural number $N$ such that $a_n > M$ for all $n > N$. In this case, we also say that $a_n$ has limit $\infty$, and we write $\lim_n a_n = \infty$. The sequence $(a_n)$ is said to diverge to $-\infty$ if, for every $m \in \mathbb{R}$, there is a natural number $N$ such that $a_n < m$ for all $n > N$. In this case, we also say that $a_n$ has limit $-\infty$, and we write $\lim_n a_n = -\infty$.

**Example 4.** Let $a_n = n + \frac{1}{n}, b_n = 1 + (-1)^n, c_n = e^{-n}$.
The sequence $(a_n)$ diverges to $\infty$, while $(b_n)$ does not converge, nor does it diverge to $\pm\infty$. Finally, $\lim_n c_n = 0.$◆

**Example 5.** If, for every $n \in \mathbb{N}, a_n \leq b_n$ and $\lim_n a_n = a, \lim_n b_n = b$, then $a \leq b$.
Suppose for a contradiction that $b < a$. Let $\epsilon = (a - b)/3$. Observe that $b - \epsilon < b + \epsilon < a - \epsilon < a + \epsilon$. There exist integers $N_1$ and $N_2$ such that, for $n > N_1, b_n \in (b - \epsilon, b + \epsilon)$, and for $n > N_2, a_n \in (a - \epsilon, a + \epsilon)$. Now, for any $n > max\{N_1, N_2\}$, $b_n < a_n$, which is a contradiction. ◆

**Theorem 1.2.1.** *If $a_n \leq c_n \leq b_n$ and $\lim_n a_n = \lim_n b_n = a$, then $\lim_n c_n = a$.* ∎

**Definition.** A sequence $(a_n)$ is **bounded** if its range $\{a_1, a_2, ...\}$ is a bounded set. Thus there is a positive number $M$ such that, for all $n \in \mathbb{N}, |a_n| \leq M$.

**Theorem 1.2.2.** *A convergent sequence is bounded.* ∎

**Definitions.** A sequence $(a_n)$ is **non-decreasing** if $a_1 \leq a_2 \leq ....$
A sequence $(a_n)$ is **(strictly) increasing** if $a_1 < a_2 < ....$
A sequence $(a_n)$ is **non-increasing** if $a_1 \geq a_2 \geq ....$
A sequence $(a_n)$ is **(strictly) decreasing** if $a_1 > a_2 > ....$
A sequence $(a_n)$ is **monotonic** if it is non-decreasing or non-increasing.

**Example 6.** $a_n = 1/n$ is decreasing, but $b_n = (-1)^n/n$ is not monotonic.◆

**Theorem 1.2.3.** *A monotonic sequence is convergent if and only if it is bounded.*

*Proof. Without loss of generality, let $(a_n)$ be a bounded, non-decreasing sequence, and let $A = \{a_n : n \in \mathbb{N}\}$. By assumption, $A$ is bounded, so, by the completeness of $\mathbb{R}$, $a = supA$ exists. We show that $\lim_n a_n = a$. Let $\epsilon > 0$. There is an integer $N > 0$ such that $a - \epsilon < a_N$. Because $(a_n)$ is non-decreasing, for any $n > N$, $a - \epsilon < a_N \leq a_n \leq a < a + \epsilon$; hence $\lim_n a_n = a$. The converse is a special case of theorem 1.2.2.* ∎

**Definition.** A sequence $a_n$ is said to be a **Cauchy sequence** if, for every $\epsilon > 0$, there is a natural number $N$ such that, for all $m, n > N, |a_n - a_m| < \epsilon$

**Theorem 1.2.4.** *A convergent sequence is a Cauchy sequence.* ∎

**Theorem 1.2.5.** *A Cauchy sequence is bounded.*

*Proof.* Let $\epsilon = 1$. There is a positive integer $N$ such that, for $m, n \geq N$, $|a_n - a_m| < 1$. In particular, taking $m = N, |a_n - a_N| < 1$ for all $n \geq N$ Thus, by the triangle inequality, for every $n \geq N$, $|a_n| = |(a_n - a_N) + a_N| \leq |a_n - a_N| + |a_N| \leq 1 + |a_N|$. Let $M = max\{|a_1|, \dots, |a_{N-1}|, 1 + |a_N|\}$. Clearly, $|a_n| \leq M$ for all $n$. ∎

**Definition.** Let $(a_n)$ be a sequence, and let $(n_1, n_2, \dots)$ be a strictly increasing sequence of natural numbers. We say that $(a_{n_k})_{k=1}^{\infty}$ is a **subsequence** of $(a_n)$.

**Theorem 1.2.6.** *A subsequence of a convergent sequence is convergent to the same limit. Thus if* $\lim_n a_n = a$ *and* $(a_{n_k})$ *is a subsequence of* $(a_n)$, *then* $\lim_{k \to \infty} a_{n_k} = a$.

*Proof.* Let $\epsilon > 0$. Since $\lim_n a_n = a$, there is a positive integer $N$ such that, for $n > N$, $|a_n - a| < \epsilon$. Since $(n_k)$ is an increasing sequence of natural numbers, $n_k \geq k$ for every $k \in \mathbb{N}$. Thus, for $k > N, n_k > N$ and $|a_{n_k} - a| < \epsilon$. ∎

**Theorem 1.2.7.** *Every sequence* $(a_n)$ *contains a monotonic subsequence.*

*Proof.* Define a term $a_n$ of the sequence to be a peak if, for every $i \geq n, a_n \geq a_i$. There are two cases:

Case 1. The sequence $(a_n)$ has finitely many peaks. Suppose $k_0$ is the largest positive integer for which $a_{k_0}$ is a peak, and let $n_1 = k_0 + 1$. Since $a_{n_1}$ is not a peak, there is an integer $n_2 > n_1$ such that $a_{n_2} > a_{n_1}$. Continuing inductively, one can construct a strictly increasing sequence of positive integers $n_1 < n_2 < n_3, \dots$ such that $a_{n_1} < a_{n_2} < a_{n_3} \dots$. The sequence $(a_{n_k})$ is an increasing subsequence of $(a_n)$.
Case 2. The sequence, $(a_n)$ contains infinitely many peaks, $a_{n_1} \geq a_{n_2} \geq a_{n_3} \geq \dots$, where $n_k$ is an increasing sequence of positive integers. The sequence $(a_{n_k})$ is a non-increasing subsequence of $(a_n)$. ∎

**Theorem 1.2.8 (the Bolzano-Weierstrass theorem).** *Every bounded sequence contains a convergent subsequence.*

*Proof.* Let $(a_n)$ be a bounded sequence. By the previous theorem, $(a_n)$ contains a monotonic subsequence, $(a_{n_k})$, which is convergent by theorem 1.2.3. ∎

The following two examples are fixtures in undergraduate real analysis books. The proof technique is quite common and is valid for general compact sets. See chapter 4. First we remind the reader of the definition of continuity.

**Definition.** Let $X$ be a subset of $\mathbb{R}$, and let $f : X \to \mathbb{R}$. We say that $f$ is continuous at a point $x \in X$ if, for every $\epsilon > 0$, there exists $\delta > 0$ such that $|f(y) - f(x)| < \epsilon$ whenever $y \in X$ and $|y - x| < \delta$. The function $f$ is said to be continuous on $X$ if it is continuous at every point $x \in X$.

It is easy to see that if $f$ is continuous at $x$, and $x_n \in X$ is such that $\lim_n x_n = x$, then $\lim_n f(x_n) = f(x)$.

**Example 7.** A continuous real-valued function $f$ on a closed bounded interval $[a, b]$ is bounded and attains its supremum and infimum values.

Suppose, for a contradiction, that $f$ is unbounded. Without loss of generality, assume that $\sup_{x \in [a,b]} f(x) = \infty$. There exists a sequence $(x_n)$ in $[a, b]$ such that $\lim_n f(x_n) = \infty$. By the Bolzano-Weierstrass theorem, $(x_n)$ contains a convergent subsequence $x_{n_k}$. Because $[a, b]$ is closed, $x = \lim_k x_{n_k} \in [a, b]$. Now we have the following contradiction: $f(x) = \lim_k f(x_{n_k}) = \infty$.

The proof that $f$ attains its supremum and infimum values replicates the above argument. ♦

**Definition.** Let $X$ be a subset of $\mathbb{R}$, and let $f : X \to \mathbb{R}$. We say that $f$ is **uniformly continuous** on $X$, if for every $\epsilon > 0$, there exists $\delta > 0$ such that $|f(y) - f(x)| < \epsilon$ whenever $x, y \in X$ and $|y - x| < \delta$.

The number $\delta$ in the above definition depends on $\epsilon$ only and not on any particular $x \in X$. For example, the function $f : (0, 1) \to \mathbb{R}$ defined by $f(x) = 1/x$ is continuous but not uniformly continuous.

**Example 8.** A continuous real-valued function $f$ on a closed bounded interval $[a, b]$ is uniformly continuous.

Suppose that $f$ is not uniformly continuous. Then there exists a positive number $\epsilon$ such that, for every $n \in \mathbb{N}$, $[a, b]$ contains a pair of points $x_n$ and $y_n$ such that $|x_n - y_n| < \frac{1}{n}$ and $|f(y_n) - f(x_n)| \geq \epsilon$. By the Bolzano-Weierstrass theorem, $(x_n)$ contains a convergent subsequence $x_{n_k}$. Let $x = \lim_k x_{n_k}$. Observe that $x \in [a, b]$ and $\lim_k y_{n_k} = x$. Now, for all $k \in \mathbb{N}$, $|f(y_{n_k}) - f(x_{n_k})| \geq \epsilon$. This is a contradiction because if we take the limit as $k \to \infty$ of the left-hand side of the last inequality and use the continuity of $f$, we obtain $0 = |f(x) - f(x)| = \lim_k |f(y_{n_k}) - f(x_{n_k})| \geq \epsilon$. ♦

**Theorem 1.2.9 (the Cauchy criterion).** *A sequence in $\mathbb{R}$ is a Cauchy sequence if and only if it is convergent.*

*Proof.* *By theorem 1.2.4, every convergent sequence is a Cauchy sequence. To prove the converse, let $(a_n)$ be a Cauchy sequence. By theorem 1.2.5, $(a_n)$ is bounded; hence, by theorem 1.2.8, $(a_n)$ contains a convergent subsequence, $(a_{n_k})$. Let $\lim_k a_{n_k} = a$. We show that $\lim_n a_n = a$. Let $\epsilon > 0$. Since $(a_n)$ is Cauchy, there is a positive integer $N$ such that, for $n, m > N, |a_n - a_m| < \epsilon/2$. Since $\lim_k a_{n_k} = a$, there is an integer $K$ such that for $k \geq K, |a_{n_k} - a| < \epsilon/2$. Without loss of generality, we may assume that $K > N$; thus $n_K \geq K > N$. Taking $m = n_K$ and using the triangle inequality, for all $n > N, |a_n - a| \leq |a_n - a_{n_K}| + |a_{n_K} - a| < \epsilon/2 + \epsilon/2 = \epsilon.$* ∎

**Example 9.** The rational field $\mathbb{Q}$ does not satisfy the Cauchy criterion. For example, the sequence $\sum_{i=0}^{n} \frac{1}{i!}$ is a Cauchy sequence in $\mathbb{Q}$, but its limit, $e$, is not in $\mathbb{Q}$. ◆

**Remark.** The completeness of $\mathbb{R}$ is, in fact, equivalent to the Cauchy criterion. See example 10 below. This is why the Cauchy criterion is sometimes used as a definition of the completeness of $\mathbb{R}$.

**Definition.** Let $A$ be a subset of $\mathbb{R}$. A real number $x$ is called a **limit point** of $A$ if, for every $\delta > 0, (x - \delta, x + \delta) \cap A$ contains a point other than $x$.

**Theorem 1.2.10 (the Bolzano-Weierstrass property of bounded sets).** *Every bounded infinite subset $A$ of $\mathbb{R}$ has a limit point.*

*Proof.* *Let $I_1 = [a, b]$ be a closed bounded interval that contains $A$. Bisect $I_1$ into two congruent closed subintervals. One of the two subintervals contains infinitely many points of $A$. Denote that interval by $I_2$. Continuing this process produces a sequence of subintervals $I_1 \supseteq I_2 \supseteq ...$ such that $A \cap I_n$ is infinite for all $n \in \mathbb{N}$, and the length of $I_n, l(I_n) = \frac{b-a}{2^{n-1}}$. For each $n \in \mathbb{N}$, pick a point $a_n \in A \cap I_n$. If $m > n, I_n \supseteq I_m$, and $a_m, a_n \in I_n$; hence $|a_n - a_m| < \frac{b-a}{2^{n-1}}$. Since $\lim_n \frac{b-a}{2^{n-1}} = 0, (a_n)$ is a Cauchy sequence. Let $a = \lim_n a_n$. Since $a_i \in I_n$ for all $i \geq n, a \in I_n$ for all $n$ (see exercise 9 at the end of this section). Now let $\delta > 0$. Since $\lim_n l(I_n) = 0$, and $a \in \cap_{n=1}^{\infty} I_n, I_n \subseteq (a - \delta, a + \delta)$ for sufficiently large $n$. Thus $(a - \delta, a + \delta)$ contains infinitely many points of $A$ because $I_n$ does. In particular, $(a - \delta, a + \delta) \cap A$ contains a point other than $a$.* ∎

**Example 10.** The completeness of $\mathbb{R}$ is equivalent to the Cauchy criterion. The fact that the completeness of $\mathbb{R}$ implies the Cauchy criterion has been established in theorem 1.2.9. Observe that the proof of theorem 1.2.9 depends heavily (through the intervening theorems) on theorem 1.2.3, where the completeness of $\mathbb{R}$ was crucial. We now prove that the Cauchy criterion implies the completeness of $\mathbb{R}$.

Let $A$ be a subset of $\mathbb{R}$ that is bounded above, pick an element $a_0 \in A$, and let $b_0$ be an upper bound of $A$. We construct two sequences $(a_n)$ and $(b_n)$ such that

(1) each $a_n \in A$, and each $b_n$ is an upper bound of $A$;
(2) $[a_{n+1}, b_{n+1}] \subseteq [a_n, b_n]$; and
(3) $b_n - a_n \leq \frac{b_0 - a_0}{2^n}$.

Consequently, $a_{n+1} - a_n \leq \frac{b_0 - a_0}{2^n}$ and $b_n - b_{n+1} \leq \frac{b_0 - a_0}{2^n}$.

Suppose $a_1, ..., a_n$ and $b_1, ..., b_n$ have been found. We define $a_{n+1}$ and $b_{n+1}$ as follows. Let $m = \frac{a_n + b_n}{2}$. If $m$ is an upper bound of $A$, let $a_{n+1} = a_n$, and let $b_{n+1} = m$. If $m$ is not an upper bound of $A$, choose an element $a_{n+1} \in A$ such that $m < a_{n+1} < b_n$, and define $b_{n+1} = b_n$.[3]

We now show that $(b_n)$ is a Cauchy sequence. Let $\epsilon > 0$, and choose an integer $N$ such that $(b_0 - a_0)/2^{N-1} < \epsilon$. For $m > n > N$, we have

$$|b_n - b_m| = b_n - b_m = (b_n - b_{n+1}) + (b_{n+1} - b_{n+2}) + ... + (b_{m-1} - b_m)$$
$$\leq (b_0 - a_0)[1/2^n + ... + 1/2^{m-1}] < (b_0 - a_0)[1 + 1/2 + ...]/2^n$$
$$= (b_0 - a_0)/2^{n-1} < (b_0 - a_0)/2^{N-1} < \epsilon$$

By the Cauchy criterion, the sequence $(b_n)$ has a limit, say, $b$. An argument identical to the one above shows that $(a_n)$ is convergent, and since $b_n - a_n \leq (b_0 - a_0)/2^n$, $\lim_n a_n = b$.

Finally, we prove that $b = \sup A$. If $a > b$ for an element $a \in A$, then $a > b_n$ for some $n$, which contradicts the fact that $b_n$ is an upper bound of $A$. Thus $b$ is an upper bound of $A$. For any number $c < b$, let $\epsilon = b - c$. Since $\lim_n a_n = b$, there exists an integer $n$ such that $a_n \in (b - \epsilon, b + \epsilon)$. In particular, $a_n > c$; hence $c$ is not an upper bound of $A$. ◆

[3] Observe that if $a_{n+1} = b_n$, the process terminates and $b_n$ is the least upper bound (in fact, the maximum) of $A$. Otherwise, the process continues ad infinitum, and each $b_n$ is a strict upper bound of $A$.

**Definition.** The **extended real line** is $\overline{\mathbb{R}} = \mathbb{R} \cup \{-\infty, \infty\}$. We need this extension of $\mathbb{R}$ because the limits of some sequences are infinite and because it is sometimes convenient to allow functions to take infinite values. We retain the usual ordering on $\mathbb{R}$, and, for $x \in \mathbb{R}$, we define $-\infty < x < \infty$. The following rules of arithmetic in $\overline{\mathbb{R}}$ are convenient and widely accepted:

(a) For a real number $a, a + \infty = \infty, a - \infty = -\infty$.
(b) If $a > 0$, then $a.\infty = \infty$, and $a.(-\infty) = -\infty$, while if $a < 0$, then $a.\infty = -\infty$, and $a.(-\infty) = \infty$.
(c) For any real number $a$, $a/\pm\infty = 0$.
(d) In chapter 8, we adopt the convention that $0.\infty = 0$. However, this definition is specific to integration theory.

We do not define the operation $\infty - \infty$.

**Definition.** A point $a \in \mathbb{R}$ is a **limit point of a sequence** $(a_n)$ if, for every $\epsilon > 0, |a_n - a| < \epsilon$ for infinitely many $n$. We say that $\infty$ is a limit point of $a_n$ if, for every $M \in \mathbb{R}, a_n > M$ for infinitely many n. Likewise, $-\infty$ is a limit point of $a_n$ if, for every $m \in \mathbb{R}, a_n < m$ for infinitely many $n$.

**Remark.** Not every limit point of a sequence is a limit point of its range. For example, the sequence $a_n = (-1)^n$ has two limit points, $\pm 1$, while its range, the set $\{-1, 1\}$, has no limit points.

**Theorem 1.2.11.** *An extended real number $a$ is a limit point of $(a_n)$ if and only if there exists a subsequence $(a_{n_k})$ of $(a_n)$ such that $\lim_k a_{n_k} = a$.*

*Proof.* Suppose $a \in \mathbb{R}$ is a limit point of $(a_n)$. There exists $n_1 \in \mathbb{N}$ such that $|a_{n_1} - a| < 1$. Now we can find an integer $n_2 > n_1$ such that $|a_{n_2} - a| < 1/2$. Continuing this construction produces a sequence $n_1 < n_2 < n_3 < \dots$ of integers such that $|a_{n_k} - a| < 1/k$. Thus $\lim_k a_{n_k} = a$. The converse is trivial. We leave it to the reader to provide the details when $a = \pm\infty$. ∎

**Definition.** Let $(a_n)$ be a sequence, and consider the sequences

$$\alpha_n = sup_{k \geq n} a_k, \text{ and } \beta_n = inf_{k \geq n} a_k.$$

Clearly, $\alpha_n$ is non-increasing, and $\beta_n$ is non-decreasing. Therefore $\lim_n \alpha_n$ and $\lim_n \beta_n$ exist. We define the **limit superior**, or the **upper limit** and the **limit inferior**, or the **lower limit** of $(a_n)$, respectively, as follows:

$$\limsup_n a_n = \lim_n \alpha_n = \lim_n sup_{k \geq n} a_k = inf_{n \in \mathbb{N}} \alpha_n,$$
$$\liminf_n a_n = \lim_n \beta_n = \lim_n inf_{k \geq n} a_k = sup_{n \in \mathbb{N}} \beta_n.$$

**Theorem 1.2.12.** $\alpha = \limsup_n a_n$ if and only if, for every $\epsilon > 0$,

(a) there is a positive integer $N$ such that $a_n < \alpha + \epsilon$ for all $n > N$, and
(b) $a_n > \alpha - \epsilon$ for infinitely many $n \in \mathbb{N}$.

*Proof.* Suppose $\alpha = \limsup_n a_n$. Since $\alpha = \inf_n \alpha_n$, there is a positive integer $N$ such that $\alpha_N < \alpha + \epsilon$. Now, because $a_n \le \alpha_n$ and $\alpha_n$ is non-increasing, $a_n \le \alpha_n \le \alpha_N < \alpha + \epsilon$, for all $n \ge N$. This proves (a).

To prove (b), note that $\alpha - \epsilon < \alpha \le \alpha_1 = \sup\{a_1, a_2, ...\}$. Thus there is a positive integer $n_1$ such that $a_{n_1} > \alpha - \epsilon$. Now $\alpha - \epsilon < \alpha \le \alpha_{n_1+1} = \sup\{a_{n_1+1}, a_{n_1+2}, ...\}$. Thus there is a positive integer $n_2 > n_1$ such that $a_{n_2} > \alpha - \epsilon$. This process produces a subsequence $a_{n_k}$ of $a_n$ such that $a_{n_k} > \alpha - \epsilon$.

To prove the converse, suppose $\alpha \in \mathbb{R}$ satisfies conditions (a) and (b), and let $\epsilon > 0$. By condition (b), for every $n \in \mathbb{N}$, there exists an integer $k \ge n$ such that $a_k > \alpha - \epsilon$. Thus $\alpha_n = \sup_{k \ge n} a_k \ge \alpha - \epsilon$. Taking the limit as $n \to \infty$, produces $\limsup_n a_n \ge \alpha - \epsilon$. Since $\epsilon$ is arbitrary, $\limsup_n a_n \ge \alpha$.

By condition (a), there exists an integer $N$ such that $a_k < \alpha + \epsilon$ for every $k > N$. Thus, for every $n > N$, $\alpha_n = \sup_{k \ge n} \alpha_k \le \alpha + \epsilon$. Taking the limit as $n \to \infty$, we obtain $\limsup_n a_n \le \alpha + \epsilon$. Because $\epsilon$ is arbitrary, $\limsup_n a_n \le \alpha$. ∎

**Theorem 1.2.13.** *The upper limit of a sequence $(a_n)$ is the largest limit point of $(a_n)$.*

*Proof.* Let $\epsilon > 0$. By the previous theorem (and its proof), there is a positive integer $N$ such that, for all $n > N$, $a_n < \alpha + \epsilon$, and a subsequence $(a_{n_k})$ such that $a_{n_k} > \alpha - \epsilon$. Since $\lim_k n_k = \infty$, there is a positive integer $K$ such that $n_k > N$ for all $k > K$. Therefore $\alpha - \epsilon < a_{n_k} < \alpha + \epsilon$ for all $k > K$. By theorem 1.2.11, $\alpha$ is a limit point of $(a_n)$. If $t$ is a limit point of $(a_n)$, then, for infinitely many positive integers $n$, $t - \epsilon < a_n$. By theorem 1.2.12, there is an integer $N$ such that, for all $n > N$, $a_n < \alpha + \epsilon$. Choosing $n$ large enough for the last two inequalities to be simultaneously satisfied, we have $t - \epsilon < a_n < \alpha + \epsilon$. Therefore $t < \alpha + 2\epsilon$. Since $\epsilon$ is arbitrary, $t \le \alpha$. ∎

**Theorem 1.2.14.** *A sequence $(a_n)$ converges to $a$ if and only if $\limsup_n a_n = \liminf_n a_n = a$.*

*Proof.* Let $\alpha = \limsup_n a_n$, $\beta = \liminf_n a_n$, and suppose that $\alpha = \beta$. By theorems 1.2.12 and problem 17 at the end of this section, there is a positive integer $N$ such that, for $n > N$, $a_n < \alpha + \epsilon$ and $\alpha - \epsilon = \beta - \epsilon < a_n$. Thus, for $n > N$, $\alpha - \epsilon < a_n < \alpha + \epsilon$; hence $\lim_n a_n = \alpha$. Conversely, if $\lim_n a_n = a$, then it is easy to verify that the conditions of theorem 1.2.12 and those of problem 17 are met with $\alpha = a$ and $\beta = a$, respectively. Hence $\alpha = \beta$. ∎

## Complex Numbers

**Definition.** A **complex number** $z$ is an ordered pair $(x, y)$ of real numbers. We use the symbol $\mathbb{C}$ for the set of complex numbers.

The definition so far makes $\mathbb{C}$ nothing more than the Euclidean plane $\mathbb{R}^2$. This is why the set of complex numbers is also called the complex plane. What sets $\mathbb{C}$ apart from $\mathbb{R}^2$ is the following pair of binary operations.

**Definition.** For complex numbers $z = (x, y)$ and $w = (a, b)$, we define the sum $z + w = (x + a, y + b)$ and the product $zw = (ax - by, ay + bx)$. The real field $\mathbb{R}$ is embedded into the complex plane in a natural way: we identify a real number $x$ with the complex number $(x, 0)$. Under the operations of complex addition and multiplication, the subset $\tilde{\mathbb{R}} = \{(x, 0) \in \mathbb{C} : x \in \mathbb{R}\}$ is closed in the sense that if $z$ and $w$ are in $\tilde{\mathbb{R}}$, then $z + w$ and $zw$ are in $\tilde{\mathbb{R}}$. Indeed $z + w = (x + a, 0)$ and $zw = (ax, 0)$. From now on, we make no distinction between $\mathbb{R}$ and $\tilde{\mathbb{R}}$ and simply write $x$ for $(x, 0)$. With this understanding, we see that if $x \in \mathbb{R}$, then $xw = (x, 0)(a, b) = (xa, xb)$. It is also straightforward to verify that the elements $0 = (0, 0)$ and $1 = (1, 0)$ satisfy $z + 0 = z$ and $z.1 = z$ for all $z \in \mathbb{C}$. Thus 0 and 1 are the identity elements for complex addition and multiplication, respectively.

**Definition.** The complex number $i = (0, 1)$ is called the imaginary number. Now $i^2 = (0, 1).(0, 1) = (-1, 0) = -1$. We therefore think of $i$ as the square root of $-1$.

Armed with the imaginary number $i$, we now have a convenient and notationally simple way to represent complex numbers. An arbitrary complex number $z$ can be written as $z = (x, y) = (x, 0) + (0, y) = x + y(0, 1) = x + iy$. With this way of representing complex numbers, we can restate the definitions of complex addition and multiplication as follows. For complex numbers $z = x + iy$ and $w = a + ib, z + w = (x + a) + i(y + b)$ and $zw = (ax - by) + i(ay + bx)$. Note that the complex operations obey the same rules as the addition and multiplication of linear polynomials, taking into account that $i^2 = -1$. Indeed, if we *multiply out* the product of the binomials $x + iy$ and $a + ib$ according to the usual rules of algebra, we obtain $(x + iy)(a + ib) = ax + iay + ibx + i^2 by = (ax - by) + (ay + bx)i$, which is consistent with the original definition of complex multiplication. Now that we have a convenient way of manipulating complex numbers, we can prove the following theorem.

**Theorem 1.2.15.** *With the operations of complex addition and multiplication, $\mathbb{C}$ is a field.*

*Proof.* *Most of the defining properties of a field are easy to verify. As a sample of the calculations, we verify the following two properties:*

(a) *Complex multiplication is associative. Let $z = x + iy, w = a + ib$, and let $t = r + is$ be complex numbers. Then $(zw)t = [(ax - by) + i(ay + bx)](r + is) = r(ax - by) - s(ay + bx) + i[r(ay + bx) + s(ax - by)]$. The reader can calculate $z(wt)$ and reconcile the result with the above expression for $(zw)t$.*

(b) *A slightly less obvious fact is the inversion formula of a nonzero complex number. If $z = x + iy \neq 0$, then $z^{-1} = \frac{x}{x^2+y^2} - i\frac{y}{x^2+y^2}$. It is easy to verify that $zz^{-1} = 1$.* ∎

**Definition.** For a complex number $z = x + iy$, $x$ is called the real part of $z$, and $y$ is the imaginary part of $z$. We use the notation $x = Re(z)$, and $y = Im(z)$.

**Definition.** The complex conjugate of a complex number $z = x + iy$ is the number $\bar{z} = x - iy$, and the absolute value (or modulus) of $z$ is $|z| = \sqrt{x^2 + y^2}$.

**Theorem 1.2.16.** *If $z$ and $w$ are complex numbers, then*

(a) *$\overline{z + w} = \bar{z} + \bar{w}$ and $\overline{zw} = \bar{z}\,\bar{w}$;*
(b) *$z + \bar{z} = 2Re(z)$ and $z - \bar{z} = 2iIm(z)$;*
(c) *$|\bar{z}| = |z|$ and $z\bar{z} = |z|^2$;*
(d) *$|Re(z)| \leq |z|$, $|Im(z)| \leq |z|$, and $|z| \leq |Re(z)| + |Im(z)|$;*
(e) *$z^{-1} = \dfrac{\bar{z}}{|z|^2}$; and*
(f) *the triangle inequality $|z + w| \leq |z| + |w|$.*

*Proof.* *The proofs are mostly computational and are left to the reader to check. We prove the triangle inequality below.*

*Note that $\overline{zw}$ is the conjugate of $\bar{z}w$; hence $\bar{z}w + z\bar{w} = 2Re(z\bar{w}) \leq 2|z\bar{w}| = 2|z||\bar{w}| = 2|z||w|$. Using this, we have $|z + w|^2 = (z + w)(\bar{z} + \bar{w}) = z\bar{z} + \bar{z}w + z\bar{w} + w\bar{w} \leq |z|^2 + 2|z||w| + |w|^2 = (|z| + |w|)^2$. The result follows by taking square roots of the extreme sides of the above string of inequalities.* ∎

Now that we have a measure of the length of a complex number, we have a measure of the distance between two points in the complex plane. For complex numbers $z_1$ and $z_2$, the quantity $|z_1 - z_2|$ is exactly the Euclidean distance between $z_1$ and $z_2$. Now we can generalize many of the properties of subsets of the real line to the complex plane. For example, a bounded subset of $\mathbb{C}$ is a set $A$ of complex numbers such that $sup\{|z| : z \in A\} < \infty$. For a complex number $a$, and a positive real number $\delta$, the set $\{z \in \mathbb{C} : |z - a| < \delta\}$ is an open disk of radius $\delta$ and centered at $a$. A point $z \in \mathbb{C}$ is a **limit point** of a set $A$ of complex numbers if every open disk centered at $z$ contains points of $A$ other than $z$. We urge the reader to examine the rest of the concepts we studied for real numbers and generalize them to the complex field, whenever possible. One important distinction between $\mathbb{R}$ and $\mathbb{C}$ is that there is no natural (or useful) way to order the complex field. We conclude the section with the following theorem.

**Theorem 1.2.17 (completeness of the complex field).** *A complex sequence $(z_n)$ is a Cauchy sequence if and only if it is convergent.*

*Proof.* Let $z_n = x_n + iy_n$, where $(x_n)$ and $(y_n)$ are real sequences. If $(z_n)$ is Cauchy, then, given $\epsilon > 0$, there exists a positive integer $N$ such that, for $n, m > N$, $|z_n - z_m| < \epsilon$. It follows that the real sequences $(x_n)$ and $(y_n)$ are Cauchy sequences, since $|x_n - x_m| \leq |z_n - z_m|$ and $|y_n - y_m| \leq |z_n - z_m|$. By the completeness of $\mathbb{R}$, $(x_n)$ and $(y_n)$ converge to real numbers $x$ and $y$, respectively. Clearly, $(z_n)$ converges to $z = x + iy$ because $|z_n - z| \leq |x_n - x| + |y_n - y|$. We leave the proof of the converse to the reader. ∎

The following example establishes the Bolzano-Weierstrass theorem for complex sequences.

**Example 11.** Every bounded complex sequence contains a convergent subsequence. Let $z_n = x_n + iy_n$ be a bounded sequence in $\mathbb{C}$. Since $|x_n| \leq |z_n|$, $(x_n)$ is bounded. By theorem 1.2.8, $(x_n)$ has a convergent subsequence $(x_{n_k})$. Let $x = \lim_k x_{n_k}$. Now the sequence $(y_{n_k})$ is bounded, so it contains a convergent subsequence $(y_{n_{k_p}})$. Let $y = \lim_p (y_{n_{k_p}})$. The subsequence $z_{n_{k_p}} = x_{n_{k_p}} + iy_{n_{k_p}}$ of $(z_n)$ clearly converges to $x + iy$ as $p \to \infty$. ♦

## Exercises

1. Prove that the finite union of bounded subsets of $\mathbb{R}$ is bounded, and give an example to show that the conclusion is false for an infinite union of bounded sets.

2. Prove that if $A \subseteq \mathbb{R}$ is bounded below, then $A$ has a greatest lower bound. Hint: Define $-A = \{-x : x \in A\}$. Show that $\inf A = -\sup\{-A\}$.

3. Prove that if $\lim_n a_n = a, \lim_n b_n = b$, then $\lim_n (a_n \pm b_n) = a \pm b$ and that $\lim_n (a_n b_n) = ab$.

4. Let $\lim_n b_n = b \neq 0$. Prove that there is a natural number $N$ such that, for all $n > N, |b_n| \geq |b|/2$. Hence prove that if, in addition, $\lim_n a_n = a$, then $\lim_n \frac{a_n}{b_n} = \frac{a}{b}$.

5. Prove theorem 1.2.1.

6. Prove theorem 1.2.2.

7. Prove theorem 1.2.4.

8. Show that the limit of a convergent sequence is unique.

9. Suppose all the terms of a sequence $(a_n)$ are in a closed interval $I$. Prove that if $\lim_n a_n = a$, then $a \in I$.

10. Show that $\lim_n a_n = a$ if and only if every interval centered at $a$ contains all but finitely many terms of the sequence.

11. Let $(\delta_n)$ be a positive sequence such that $\sum_{n=1}^{\infty} \delta_n < \infty$. Show that if $(a_n)$ is such that $|a_{n+1} - a_n| \leq \delta_n$, then $(a_n)$ is convergent. Hint: Examine the proof in example 10.

12. Prove that a point $x$ is a limit point of a subset $A$ of $\mathbb{R}$ if and only if every interval centered at $x$ contains infinitely many points of $A$.

13. Show that if $A$ is bounded above and $a = \sup A$, then there is a sequence $(a_n)$ in $A$ such that $\lim_n a_n = a$. Also prove that if $a \notin A$, the terms of the sequence $(a_n)$ can be chosen to be distinct.

14. Prove that if $\{I_n\}_{n \in \mathbb{N}}$ is a descending sequence of closed bounded intervals, then $\cap_{n \in \mathbb{N}} I_n$ is a closed interval or a point. Give examples to show that the result is false if either of the conditions *closed* or *bounded* is omitted.

15. Let $A$ and $B$ be nonempty subsets of $\mathbb{Q}$ such that $A \cup B = \mathbb{Q}$ and, for every $a \in A$ and every $b \in B, a < b$.[4] Prove that exactly one of the following alternatives holds:
    (a) there exists a number $\alpha \in \mathbb{Q}$ such that $A = \mathbb{Q} \cap (-\infty, \alpha]$,
    (b) there exists a number $\alpha \in \mathbb{Q}$ such that $A = \mathbb{Q} \cap (-\infty, \alpha)$, or
    (c) there exists a number $\alpha \in \mathbb{R} - \mathbb{Q}$ such that $A = \mathbb{Q} \cap (-\infty, \alpha)$.

16. Suppose $(a_n)$ and $(b_n)$ are real sequences. Prove that
    (a) $\liminf_n a_n \leq \limsup_n a_n$;
    (b) $\liminf_n(-a_n) = -\limsup_n a_n$ and $\limsup_n(-a_n) = -\liminf_n a_n$;
    (c) $\liminf_n a_n + \liminf_n b_n \leq \liminf_n(a_n + b_n)$;
    (d) $\limsup_n(a_n + b_n) \leq \limsup_n a_n + \limsup_n b_n$; and
    (e) if $a_n \leq b_n$, then $\limsup_n a_n \leq \limsup_n b_n$.

17. Prove that $\beta = \liminf_n a_n$ if and only if, for every $\epsilon > 0$,
    (a) there is a positive integer $N$ such that $a_n > \beta - \epsilon$ for all $n > N$, and
    (b) $a_n < \beta + \epsilon$ for infinitely many $n \in \mathbb{N}$.

18. Show that $\liminf_n a_n$ is the smallest limit point of $(a_n)$.

19. Let $a_n$ be a positive sequence. Prove that $\limsup_n \dfrac{1}{a_n} = \dfrac{1}{\liminf_n a_n}$.

20. Verify the details of the proof of theorem 1.2.15.

21. Verify the details of the proof of theorem 1.2.16.

22. Verify the details of the proof of theorem 1.2.17.

23. Prove that every bounded infinite subset of $\mathbb{C}$ has a limit point.

---

[4] Such a partition of $\mathbb{Q}$ is called a Dedekind cut.

24. (a) Show that the series $\sum_{n=0}^{\infty} \frac{z^n}{n!}$ is absolutely convergent for all $z \in \mathbb{C}$.

(b) Define $e^z = \sum_{n=0}^{\infty} \frac{z^n}{n!}$. Show that, for all $z, w \in \mathbb{C}$, $e^z e^w = e^{z+w}$. Conclude that, for $n \in \mathbb{N}$, and $z \in \mathbb{C}$, $(e^z)^n = e^{nz}$. Hint: Recall that absolutely convergent series can be multiplied term by term. The reader will recognize $e^z$ as the complex exponential function.

25. (a) Show that, for $\theta \in \mathbb{R}$, $e^{i\theta} = \cos\theta + i\sin\theta$. Hint: Recall that the terms of an absolutely convergent series can be rearranged without affecting the sum of the series.

(b) Show that if $z$ is a nonzero complex number, then there is a unique positive number $r$ and a unique real number $\theta \in [0, 2\pi)$ such that $z = re^{i\theta}$. Hint: Write $z = |z|w$, where $w = \frac{z}{|z|}$. Note that $|w| = 1$.

26. Show that, for $\theta \in \mathbb{R}$, $(\cos\theta + i\sin\theta)^n = \cos(n\theta) + i\sin(n\theta)$.

27. Let $z$ be a nonzero complex number, and write $z = re^{i\theta}$. Show that, for $n \geq 2$, each of the numbers $\xi_k = r^{1/n} e^{i(\theta + 2\pi k)/n}$, $0 \leq k \leq n-1$, satisfies $\xi_k^n = z$. The numbers $\xi_k$ are the $n^{th}$ roots of $z$.

# 2
# Set Theory

*A false conclusion once arrived at and widely accepted is not easily dislodged and the less it is understood the more tenaciously it is held.*

Georg Cantor

Georg Cantor. 1845–1918

Georg Cantor entered the Polytechnic of Zürich in 1862 to study engineering. He later moved to the University of Berlin, where he attended lectures by Weierstrass, Kummer, and Kronecker, completing his dissertation on number theory in 1867.

In 1873 Cantor proved the countability of the set of rational numbers. He then proved that the real numbers were uncountable and published the result in 1874. It is in that paper that the idea of a one-to-one correspondence appeared for the first time. He next pondered the question of whether the unit interval could be put in a one-to-one correspondence with the unit square. He initially dismissed the possibility and wrote that "the answer seems so clearly to be 'no' that proof appears almost unnecessary." When he did prove the result, he wrote to Dedekind in 1877, "I see it, but I don't believe it!" In a paper published in 1878, he made the concept of one-to-one correspondence precise and discussed sets of equal power, that is, sets which have equal cardinality.

*Fundamentals of Mathematical Analysis.* Adel N. Boules, Oxford University Press (2021). © Adel N. Boules.
DOI: 10.1093/oso/9780198868781.003.0002

Between 1879 and 1884, Cantor published a series of six papers designed to provide a basic introduction to set theory, and this is when he realized that his work was not finding the acceptance that he had hoped. In fact, Cantor's ideas earned him the strong antagonism of Kronecker, among other mathematicians and philosophers. Dedekind was sympathetic to Cantor's work and in 1888 wrote his article *Was sind und was sollen die Zahlen* [What are numbers and what should they be], partially in defense of Cantor's work.

Cantor's last major papers on set theory appeared in 1895 and 1897, where he had hoped, without success, to include a proof of the continuum hypothesis. He did, however, succeed in formulating his theory of well-ordered sets and ordinal numbers. It was also during those years that Cantor discovered the first paradoxes of set theory.

Hilbert described Cantor's work as "the finest product of mathematical genius and one of the supreme achievements of purely intellectual human activity."

Cantor's personal life was not entirely a happy one. For more than thirty years, Cantor was troubled with bouts of depression, and, in 1899, he suffered the death of his youngest son. He spent the last year of his life confined to a sanatorium, where he died of a heart attack.

## 2.1   Finite, Countable, and Uncountable Sets

Cantor's revolutionary ideas were initially focused on understanding infinite sets. His starting point was, as is ours in this chapter, set equivalence. The title of the section accurately captures its objectives: to formulate clear definitions of finite and infinite sets, and to study their properties in good detail. Among the results we establish are Dedekind's definition of an infinite set, the countability of $\mathbb{Q}$, and, in general, the countability of a countable union of countable sets. We conclude the section by showing the existence of uncountable sets through the establishment of the fact that $2^{\mathbb{N}}$ and $\mathbb{R}$ are uncountable.

**Definition.  Two sets $A$ and $B$ are equivalent** if there is a bijection from $A$ to $B$.[1]
We use the notation $A \approx B$ to indicate the equivalence of $A$ and $B$.

**Example 1.** The set $2\mathbb{N}$ of even positive integers is equivalent to $\mathbb{N}$. The function $f : \mathbb{N} \to 2\mathbb{N}$ defined by $f(n) = 2n$ is bijective. ◆

---

[1] The term *equipotent* is also used.

**Example 2.** The closed interval $[0,1]$ is equivalent to an arbitrary closed interval $[a,b]$ $(a < b)$. The function $f(x) = \frac{x-a}{b-a}$ is a bijection from $[a,b]$ to $[0,1]$. ♦

**Example 3.** The closed interval $[0,1]$ is equivalent to the open interval $(0,1)$. Define a function $f : [0,1] \to (0,1)$ as follows:

$$f(x) = \begin{cases} 1/2 & \text{if } x = 0, \\ 1/(n+2) & \text{if } x = 1/n, n \in \mathbb{N}, \\ x & \text{otherwise.} \end{cases}$$

It is easy to verify that $f$ is a bijection. ♦

**Example 4.** Let $A = (-\pi/2, \pi/2), B = \mathbb{R}$. The function $f(x) = tan(x)$ is a bijection from $A$ to $B$. Thus $A \approx B$. ♦

**Theorem 2.1.1.** *Let $A, B$ and $C$ be sets. Then*

*(a) $A \approx A$.*
*(b) If $A \approx B$, then $B \approx A$.*
*(c) If $A \approx B$ and $B \approx C$, then $A \approx C$.*

*Proof. (a) The identity function $I_A : A \to A$ is a bijection.*
*(b) If $f : A \to B$ is a bijection, then $f^{-1} : B \to A$ is a bijection.*
*(c) If $f : A \to B$ and $g : B \to C$ are bijections, then $g \circ f : A \to C$ is a bijection.* ∎

**Definitions.** For $n \in \mathbb{N}$, let $\mathbb{N}_n = \{1, 2, ..., n\}$. A set $A$ is **finite** if $A \approx \mathbb{N}_n$ for some $n \in \mathbb{N}$. A set is said to be **infinite** if it is not finite. If $A \approx \mathbb{N}_n$, we say that the cardinality of $A$ is $n$, and we write $Card(A) = n$. The cardinality of a finite set is simply the number of elements in it. We also define $Card(\emptyset) = 0$.

**Theorem 2.1.2.**
*(a) A proper subset $B$ of $\mathbb{N}_n$ is finite, and $Card(B) = m$ for some $m < n$.*
*(b) A proper subset $B$ of a finite, set $A$ is finite, and $Card(B) < Card(A)$.*
*(c) If $m, n \in \mathbb{N}$ and $m < n$, then there is no injection from $\mathbb{N}_n$ to $\mathbb{N}_m$.*
*(d) A finite set is not equivalent to any of its proper subsets.*
*(e) $\mathbb{N}$ is infinite.*

*Proof. (a) We proceed by induction. The only proper subset of $\mathbb{N}_1$ is $\emptyset$, and $Card(\emptyset) = 0 < 1 = Card(\mathbb{N}_1)$. Suppose the statement is true for some integer $n \geq 1$, and let $B$ be a proper subset of $\mathbb{N}_{n+1}$. If $n + 1 \notin B$, then $B \subseteq \mathbb{N}_n$, and, by the inductive hypothesis, $B$ is finite and $Card(B) \leq n < n+1$. Otherwise,*

$B = \{n+1\} \cup C$, where $C$ is a proper subset of $\mathbb{N}_n$. By the inductive hypothesis, $C$ is finite, and $Card(C) = m < n$. Let $g$ be a bijection from $\mathbb{N}_m$ to $C$. Define $f : \mathbb{N}_{m+1} \to B$ by

$$f(x) = \begin{cases} g(x) & \text{if } x \in \mathbb{N}_m, \\ n+1 & \text{if } x = m+1. \end{cases}$$

Clearly, $f$ is a bijection; hence $Card(B) = m + 1 < n + 1$.

(b) Suppose $Card(A) = n$, and let $f$ be a bijection from $A$ to $\mathbb{N}_n$. The restriction of $f$ to $B$ is a bijection from $B$ onto $f(B)$. By part (a), $f(B)$ is finite, and $Card(f(B)) < n$; thus $Card(B) = Card(f(B)) < n$.

(c) If $f$ is an injection from $\mathbb{N}_n$ to $\mathbb{N}_m$, then $B = f(\mathbb{N}_n)$ is a subset of $\mathbb{N}_m$. By part (a), $n = Card(\mathbb{N}_n) = Card(B) \leq m$. This contradiction shows that no such $f$ exists.

(d) Suppose $B$ is a proper subset of a finite set $A$, and let $n = Card(A)$, and $m = Card(B)$. By part (b), $m < n$. Let $g : B \to \mathbb{N}_m$ and $h : \mathbb{N}_n \to A$ be bijections. If there is a bijection $f : A \to B$, then $g \circ f \circ h$ would be an injection from $\mathbb{N}_n$ to $\mathbb{N}_m$. This contradicts part (c).

(e) If, for some positive integer $m$, there exists a bijection $f : \mathbb{N} \to \mathbb{N}_m$, then, for any integer $n > m$, the restriction of $f$ to $\mathbb{N}_n$ would be an injection from $\mathbb{N}_n$ into $\mathbb{N}_m$. This contradicts part (c) and completes the proof. ∎

**Corollary 2.1.3.** *If $A$ is infinite and $A \subseteq B$, then $B$ is infinite.*

*Proof.* If $B$ were finite, $A$ would also be finite by theorem 2.1.2. ∎

**Theorem 2.1.4.** *A set $A$ is infinite if and only if it contains a sequence of distinct elements.*

*Proof.* Suppose $A$ is infinite. First we show that, for $n \in \mathbb{N}$, $A$ contains a set of exactly $n$ elements. The proof is inductive, and here is the inductive step: Having found a subset $\{a_1, ..., a_n\}$ of $A$ containing exactly $n$ elements, we pick an element $a_{n+1} \in A - \{a_1, ..., a_n\}$. Such an $a_{n+1}$ exists because otherwise $A$ would be be equal to $\{a_1, ..., a_n\}$, which is finite. The set $\{a_1, ..., a_{n+1}\}$ has exactly $n + 1$ elements.

For $n = 0, 1, 2, ...,$ let $B_n$ be a subset of $A$ of exactly $2^n$ elements. Define $C_0 = B_0$, and, for $n \in \mathbb{N}$, let $C_n = B_n - \cup_{i=0}^{n-1} B_i$. Now $Card(\cup_{i=0}^{n-1} B_i) \leq \sum_{i=0}^{n-1} 2^i = 2^n - 1$. Hence $Card(C_n) \geq 2^n - (2^n - 1) = 1$. Thus the sets $C_n$ are disjoint and nonempty. We choose an element $c_n$ from each $C_n$, and we obtain a sequence of distinct elements of $A$. The converse is true because $\mathbb{N}$ is infinite. ∎

**Theorem 2.1.5.** *A set A is infinite if and only if it is equivalent to one of its proper subsets.*

*Proof.* *If A is equivalent to one of its proper subsets, A is infinite by theorem 2.1.2(d). Conversely, if A is infinite, by theorem 2.1.4, A contains a sequence of distinct elements $(b_1, b_2, \ldots)$. Let $B = \{b_1, b_2, \ldots\}$ and define a function $f : A \to A - \{b_1\}$ as follows: for $n \in \mathbb{N}, f(b_n) = b_{n+1}$, and $f(x) = x$ if $x \notin B$. As f is clearly a bijection, $A \approx A - \{b_1\}$.* ∎

**Example 5.** The closed interval $[0, 1]$ is infinite because it is equivalent to its subset $(0, 1)$. (See example 3.)

**Definition.** A set $A$ is **countable** if it is equivalent to $\mathbb{N}$. A bijection $f : \mathbb{N} \to A$ is called an enumeration (sequencing) of $A$. A set is said to be **at most countable** if it is finite or countable. If an infinite set is not countable, we say it is **uncountable**.

**Theorem 2.1.6.** $\mathbb{N} \times \mathbb{N}$ *is countable.*

*Proof.* *We enumerate $\mathbb{N} \times \mathbb{N}$ recursively as follows: $f(1) = (1, 1)$, and once $f(n)$ has been defined, say, $f(n) = (a, b)$, define*

$$f(n+1) = \begin{cases} (a-1, b+1) & \text{if } a > 1, \\ (b+1, 1) & \text{if } a = 1. \end{cases}$$

*We pictorially think of $\mathbb{N} \times \mathbb{N}$ as the integer points in the open first quadrant of the plane. The above enumeration sequences the integer points in the open first quadrant along each diagonal $a + b = \text{constant}, a, b \in \mathbb{N}$, from bottom to top. Once the top of a diagonal has been reached, we start at the bottom of the next diagonal. See figure 2.1. It is clear that f is a bijection from $\mathbb{N}$ to $\mathbb{N} \times \mathbb{N}$. The enumeration trick in this proof is attributed to Cantor. Another proof of this theorem is provided by exercise 15 on section 1.1.* ∎

**Theorem 2.1.7.** *An infinite subset B of a countable set A is countable.*

*Proof.* *Let $A = \{a_1, a_2, \ldots\}$ be an enumeration of A. Let $n_1$ be the least positive integer such that $a_{n_1} \in B$. Suppose we have found integers $n_1 < n_2 < \ldots < n_k$ such that, for $1 \leq i \leq k - 1$, $n_{i+1}$ is the least positive integer greater than $n_i$ for which $a_{n_{i+1}} \in B$. Define $n_{k+1}$ to be the least integer greater than $n_k$ such that $a_{n_{k+1}} \in B$.*

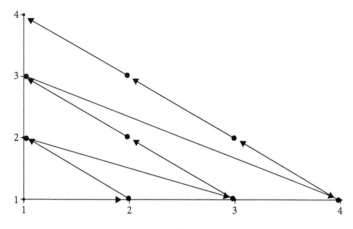

**Figure 2.1** Cantor's trick

We claim that $B = \{a_{n_k} : k \in \mathbb{N}\}$. If not, then there exists an element $b \in B$ such that $b \neq a_{n_k}$ for all $k \in \mathbb{N}$. Now $b \in A$, so $b = a_n$ for some positive integer $n$. By assumption, $n \neq n_k$ for all $k \in \mathbb{N}$. Because $n_k$ is a strictly increasing sequence of positive integers, there are two possibilities: either $n < n_1$ or there is a unique $k \in \mathbb{N}$ such that $n_k < n < n_{k+1}$. The former possibility would contradict the definition of $n_1$, and the latter possibility contradicts the definition of $n_{k+1}$. This shows that $B = \{a_{n_k} : k \in \mathbb{N}\}$. ∎

**Theorem 2.1.8.** *If there exists an injection $f$ from a set $A$ to $\mathbb{N}$, then $A$ is at most countable.*

*Proof. Without loss of generality, assume $A$ is infinite. Since $f$ is one-to-one, $\mathfrak{R}(f)$ (the range of $f$) is an infinite subset of $\mathbb{N}$. By theorem 2.1.7, $\mathfrak{R}(f)$ is countable. Therefore $A$ is countable since it is in one-to-one correspondence with $\mathfrak{R}(f)$. ∎*

**Theorem 2.1.9.** *A set $A$ is at most countable if and only if there is a surjection $f : \mathbb{N} \to A$.*

*Proof. If $A$ is countable, a bijection exists from $\mathbb{N}$ onto $A$. If $A$ is finite and $\text{Card}(A) = n$, then there exists a bijection $f : \{1, 2, \ldots, n\} \to A$. Extend $f$ to a surjection $f : \mathbb{N} \to A$ by defining $f(m) = f(n)$ for all $m > n$.*

*Conversely, if there exists a surjection $f : \mathbb{N} \to A$, the sets $S_a = f^{-1}(a), a \in A$, are mutually disjoint and nonempty. Choose an element $n_a$ from each of the sets $S_a$ and define a function $g : A \to \mathbb{N}$ by $g(a) = n_a$. Clearly, $g$ is one-to-one. Now $A$ is at most countable by theorem 2.1.8. ∎*

**Theorem 2.1.10.** *A countable union of countable sets is countable.*

*Proof.* Let $\{A_n\}$ be a countable collection of countable sets, and let $A = \cup_{n=1}^{\infty} A_n$. Write $A_n = \{a_{n1}, a_{n2}, \ldots\}$. Define $f : \mathbb{N} \times \mathbb{N} \to \cup_{n=1}^{\infty} A_n$ by $f(m,n) = a_{mn}$. Clearly, $f$ is onto. By theorem 2.1.6, there exists a bijection $g : \mathbb{N} \to \mathbb{N} \times \mathbb{N}$. The composition $f \circ g$ maps $\mathbb{N}$ onto $A$. By theorem 2.1.9, $A$ is countable. ∎

**Corollary 2.1.11.** $\mathbb{Z}$ *and* $\mathbb{Q}$ *are countable.*

*Proof.* Use theorem 2.1.10 and the facts that $\mathbb{Z} = \mathbb{N} \cup \{0\} \cup -\mathbb{N}$ and $\mathbb{Q} = \cup_{n=1}^{\infty} \{\frac{m}{n} : m \in \mathbb{Z}\}$. ∎

**Theorem 2.1.12.** *The set* $\mathfrak{F}$ *of finite sequences in a countable set $A$ is countable.*

*Proof.* For each $n \in \mathbb{N}$, let $A^n$ be the family of sequences in $A$ of exact length $n$. As a consequence of theorem 2.1.6 (see problem 9 at the end of this section), $A^n$ is countable. Since $\mathfrak{F} = \cup_{n=1}^{\infty} A^n$, $\mathfrak{F}$ is countable by theorem 2.1.10. ∎

**Theorem 2.1.13.** $2^{\mathbb{N}}$ *is uncountable.*

*Proof.* Recall that $2^{\mathbb{N}}$ is the set of all sequences from $\mathbb{N}$ in $\{0,1\}$ (binary sequences). Suppose, for a contradiction, that $2^{\mathbb{N}}$ is countable. Then $2^{\mathbb{N}} = \{x_1, x_2, \ldots\}$, where each $x_i$ is a binary sequence, say, $x_i = (x_{i1}, x_{i2}, \ldots)$ and each $x_{ij}$ is 0 or 1. The binary sequence $y = (y_1, y_2, \ldots)$, where

$$y_i = \begin{cases} 0 & \text{if } x_{ii} = 1, \\ 1 & \text{if } x_{ii} = 0. \end{cases}$$

is clearly not equal to any $x_i$. This contradiction establishes the theorem. ∎

**Corollary 2.1.14.** *If a set $A$ contains at least two elements, then $A^{\mathbb{N}}$ is uncountable.*

*Proof.* Pick two distinct elements $a_0$ and $a_1$ in $A$. By the previous theorem, $\{a_0, a_1\}^{\mathbb{N}}$ is uncountable. Consequently, $A^{\mathbb{N}}$ is uncountable because it contains $\{a_0, a_1\}^{\mathbb{N}}$. See problem 4 at the end of this section. ∎

**Theorem 2.1.15.** *The interval $(0,1]$ is uncountable. Consequently, $\mathbb{R}$ is uncountable.*

*Proof.* Let $T$ be the set of binary sequences which contain only a finite number of nonzero terms. Each of the sets $T_n$ of binary sequences of length $n$ is finite, and $T$ is equivalent to $\cup_{n=1}^{\infty} T_n$. Therefore $T$ is countable. It follows that

$A = 2^{\mathbb{N}} - T$ is uncountable, by problem 6 at the end of this section. We construct an injection $f$ from $A$ to $(0,1]$ as follows: for a sequence $a = (a_1, a_2, \ldots) \in A$, define $f(a) = \sum_{i=1}^{\infty} \frac{a_i}{2^i}$. To prove that $f$ is one-to-one, suppose $a = (a_1, a_2, \ldots)$ and $b = (b_1, b_2, \ldots)$ are distinct sequences in $A$. Let $n$ be the least positive integer for which $a_n \neq b_n$, and assume that $a_n = 1, b_n = 0$. If $t = \sum_{i=1}^{n-1} \frac{a_i}{2^i} = \sum_{i=1}^{n-1} \frac{b_i}{2^i}$, then $f(b) \leq t + \sum_{i=n+1}^{\infty} 1/2^i = t + 1/2^n$. Since $(a_n)$ contains an infinite number of terms that are equal to 1, let $m$ be such that $m > n$ and $a_m = 1$. Now $f(a) > t + 1/2^n + 1/2^m > f(b)$. Thus $\mathfrak{R}(f)$ is uncountable. Since $\mathfrak{R}(f) \subseteq (0,1] \subseteq \mathbb{R}$, both $(0,1]$ and $\mathbb{R}$ are uncountable. See problem 4 at the end of this section. ∎

**Remark.** It is easy to see that the function $f$ in the above proof is onto and hence $A \approx (0,1]$. See problem 11 at the end of this section.

## Exercises

1. Show that any two (bounded or unbounded) intervals in $\mathbb{R}$ are equivalent.
2. Show that $A \times B \approx B \times A$.
3. Show that if $A \approx B$ and $C \approx D$, then $A \times C \approx B \times D$ and $A^C \approx B^D$.
4. Prove that if $A \subseteq B$ and $A$ is uncountable then $B$ is uncountable.
5. Show that, for any two sets $A$ and $B$, there exist disjoint sets $C$ and $D$ such that $A \approx C$ and $B \approx D$.
6. Prove that if $A$ is a countable subset of an uncountable set $B$, then $B - A$ is uncountable and $B - A \approx B$.
7. Let $f : A \to B$.
   (a) Show that if $f$ is onto and $B$ is uncountable, then $A$ is uncountable.
   (b) Show that if $f$ is one-to-one and $A$ is uncountable, then $B$ is uncountable.
8. Let $g : \mathbb{N} \times \mathbb{N} \to \mathbb{N}$ be the inverse of the function $f$ in the proof of theorem 2.1.6. Show that, for $(a,b) \in \mathbb{N} \times \mathbb{N}$, $g(a,b) = \frac{1}{2}(a+b-2)(a+b-1)+b$.
9. Show that if $A$ is a countable set, then $A^n$ is countable.
10. Let $A$ be a countable set. Show that the collection of finite subsets of $A$ is countable.
11. In connection with the proof of theorem 2.1.15, show that $A \approx (0,1] \approx \mathbb{R}$.
12. A real number $x$ is said to be algebraic if it is a root of some polynomial equation with integer coefficients. For example, $\sqrt{2}$ is algebraic. Prove that the set of algebraic numbers is countable. If a real number is not algebraic, it is said to be transcendental. It can be shown, for example, that $e$ and $\pi$ are transcendental numbers. Conclude that the set of transcendental numbers is uncountable.

## 2.2  Zorn's Lemma and the Axiom of Choice

The axiom of choice is one of the most useful tools in set theory. Although it is easy to state and widely accepted, the axiom of choice has also generated much controversy among mathematicians. In this section, we study the axiom of choice and its most famous and widely applicable equivalent: Zorn's lemma, which is an indispensable tool in this book. The section and the section exercises contain typical but illuminating illustrations of how Zorn's lemma is applied. In this section, we also study partially ordered, linearly ordered, and well-ordered sets and establish results such as the Schröder-Bernstein theorem, which will help us study cardinal numbers in the next section. Although ordinal numbers have been avoided in this book, the section exercises are largely focused on well-ordered sets.

**Definition.**  Let $A$ be a nonempty set. A **partial ordering** on $A$ is a relation $\leq$ on $A$ such that, for all $x, y$, and $z \in A$,

(a)  $x \leq x$,
(b)  if $x \leq y$ and $y \leq z$, then $x \leq z$, and
(c)  if $x \leq y$ and $y \leq x$, then $x = y$.

If $x \leq y$ and $x \neq y$, we write $x < y$.

A relation satisfying condition (c) is called **antisymmetric**.

**Definition.**  Let $A$ be a nonempty set. A partial ordering $\leq$ on $A$ is said to be a **linear (or total) ordering** if it also satisfies the condition that, for $x, y \in A$, either $x \leq y$ or $y \leq x$. In this case, we say that $A$ is linearly ordered by $\leq$. A linearly ordered set is commonly called **a chain**.

**Example 1.**  Let $A = \mathcal{P}(\mathbb{N})$. Order $A$ by set inclusion. Thus if $S$ and $T$ are subsets of $\mathbb{N}$, then $S \leq T$ means that $S \subseteq T$. Set inclusion is a partial ordering of A. It is not total because if $S$ and $T$ are subsets of $\mathbb{N}$, it need not be the case that $T \subseteq S$ or $S \subseteq T$. The set $\{\mathbb{N}_n : n \in \mathbb{N}\}$ is a chain in $A$. ◆

**Definitions.**  Let $(A, \leq)$ be a partially ordered set and let $S \subseteq A$.

(a)  An element $s \in S$ is **the greatest element** of $S$ if, for all $t \in S, t \leq s$. Thus $s$ exceeds every other element of $S$. The greatest element of a set, if one exists, is unique.
(b)  An element $s \in S$ is **maximal** if, for all $t \in S, t \geq s$ implies that $t = s$. Thus $s$ is not exceeded by any other element of $S$. A maximal element need not be unique.

(c) An element $a \in A$ is an **upper bound** of $S$ if $s \leq a$ for all $s \in S$. Notice that an upper bound of $S$ need not be in $S$.

(d) An element $a \in A$ is the **least upper bound** of $S$ if $a$ is an upper bound of $S$ and, if $b < a$, then $b$ is not an upper bound of $S$.

**Example 2.** Let $A = \mathbb{R}^2$ and define $\leq$ on $A$ as follows: $P = (a,b) \leq Q = (x,y)$ if $a \leq x$ and $b \leq y$. Thus P is below and to the left of Q. This ordering of $\mathbb{R}^2$ is not linear because the points $(1,2)$ and $(2,1)$ are not comparable. Let $S$ (the shaded region in fig. 2.2) be the closed subset of the third quadrant below the line $x + y + 1 = 0$. The fact that $S$ is closed means that $S$ contains its three straight boundaries. The set S has no greatest element, since no point of $S$ is strictly above and to the right of every other point of $S$. Every point on the line segment $x + y + 1 = 0, -1 \leq x \leq 0$, is a maximal element of $S$. The set of upper bounds of $S$ is the closed first quadrant, and the least upper bound of $S$ is $(0,0)$. ◆

**Definition.** A linearly ordered set $A$ is said to be **well ordered** if every nonempty subset of $A$ contains a least element. Thus if $S$ is a subset of $A$, then there is an element $s \in S$ such that $s \leq t$ for every $t \in S$. The least element of $S$ is also called **the first element of** $S$.

**Example 3.** $\mathbb{N}$ is well ordered with the usual ordering of the real numbers. We often use the well ordering of $\mathbb{N}$ without explicit mention; see, for example, the proof of theorem 2.1.7. ◆

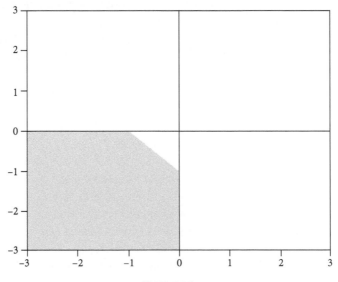

Figure 2.2

**Example 4.** Let $A = \mathbb{N} \cup \{\omega\}$, where $\omega$ is any object not in $\mathbb{N}$. The ordering on the subset $\mathbb{N}$ of $A$ is the natural ordering of the integers. We define $n < \omega$ for all $n \in \mathbb{N}$. Thus we simply define $\omega$ to be the largest element of A. The set $(A, \leq)$ is well ordered. ♦

**Example 5.** Let $B = \{x_1, x_2, \ldots\}$ be a countable set of distinct elements such that $B \cap \mathbb{N} = \varnothing$. Define an ordering $\leq$ on $A = \mathbb{N} \cup B$ as follows: the restriction of $\leq$ to $\mathbb{N}$ is the usual ordering on $\mathbb{N}$ and if $x_n$ and $x_m$ are in $B$, $x_n \leq x_m$ if $n \leq m$. Finally, if $n \in \mathbb{N}, x \in B$, we define $n < x$. The set $(A, \leq)$ is well ordered. ♦

We now state the well ordering principle, which is really an axiom. It simply states that any set can be well ordered.

**The well ordering principle:** given a nonempty set $A$, there exists a well ordering on $A$.

It should be clear that a countable set can be well ordered. If $\{a_1, a_2, \ldots\}$ is an enumeration of a countable set $A$, then we can well order $A$ in a natural way: define $a_n \leq a_m$ if $n \leq m$. Therefore the challenge is when $A$ is uncountable. Notice that an arbitrary uncountable set contains an abundance of well-ordered subsets, namely, all the countable subsets of $A$. In order to make the terminology we use in this section unambiguous, when we speak of a well-ordered subset of $A$, we mean a subset that can be well ordered.

**The axiom of choice:** if $\{X_\alpha\}_{\alpha \in I}$ is a nonempty collection of nonempty sets, then $\prod_\alpha X_\alpha$ is nonempty.

Recall that an element of $\prod_\alpha X_\alpha$ is a function $x : I \to \cup_{\alpha \in I} X_\alpha$ such that $x_\alpha \in X_\alpha$ for all $\alpha \in I$. Such a function $x$ is called a **choice function** because it is constructed by *choosing* an element $x_\alpha$ from each of the sets $X_\alpha$, hence the name "Axiom of Choice," which we can restate as follows: choice functions exist. Notice that the axiom of choice is not needed when the product $\prod_\alpha X_\alpha$ involves a finite number of factor sets. Also when each of the factor sets $X_\alpha$ contains a distinguishable element, then one does not need the axiom of choice to assert that $\prod_\alpha X_\alpha$ is nonempty. For example, in $\mathbb{N}$, we can pick a distinguishable element, say, 1, and $\mathbb{N}^\mathbb{N}$ is not empty because it contains the constant sequence $(1, 1, 1, \ldots)$.

The axiom of choice is often applied in the following equivalent form: if $X$ is a nonempty set, it is possible to choose an element from each of the nonempty subsets of $X$.

The axiom of choice is perhaps the most believable axiom of set theory. However, it is neither a simple fact nor obvious. In fact, the axiom of choice is equivalent to the well ordering principle and to the following axiom, which is less intuitive than the well ordering principle or the axiom of choice:

**Zorn's lemma:** if $A$ is a partially ordered set such that every chain in $A$ has an upper bound, then $A$ contains a maximal element.

**Theorem 2.2.1.** *The axiom of choice, Zorn's lemma, and the well ordering principle are all equivalent.* ∎

We will include the lengthy proof of the above theorem in Appendix A.

The rest of the results in this section are well-known theorems and lay the foundation for studying the cardinality of sets. The theorem below is a typical example of how Zorn's lemma is applied.

**Theorem 2.2.2.** *Let $A$ and $B$ be nonempty sets. Then there is an injection from $A$ to $B$ or an injection from $B$ to $A$.*

*Proof. Let $\mathfrak{B}$ be the collection of all injective functions $f$ such that $\mathrm{Dom}(f) \subseteq A$ and $\mathfrak{R}(f) \subseteq B$. Let $\{f_\alpha : \alpha \in I\}$ be an indexing of $\mathfrak{B}$ and, for $\alpha \in I$, write $A_\alpha = \mathrm{Dom}(f_\alpha)$, and $B_\alpha = \mathfrak{R}(f_\alpha)$. Partially order $\mathfrak{B}$ as follows: $f_\alpha \leq f_\beta$ iff $f_\beta$ extends $f_\alpha$. More explicitly, $f_\alpha \leq f_\beta$ means that $A_\alpha \subseteq A_\beta$, $B_\alpha \subseteq B_\beta$, and the restriction of $f_\beta$ to $A_\alpha$ is $f_\alpha$. Clearly, $\leq$ is a partial ordering of $\mathfrak{B}$. Now let $\mathfrak{C}$ be a chain in $\mathfrak{B}$, and index $\mathfrak{C}$ by a subset $J$ of $I$; $\mathfrak{C} = \{f_\alpha : \alpha \in J\}$. We show that $\mathfrak{C}$ has an upper bound: let $A_\alpha = \mathrm{Dom}(f_\alpha)$, $B_\alpha = \mathfrak{R}(f_\alpha)$, $S = \cup_{\alpha \in J} A_\alpha$, and $T = \cup_{\alpha \in J} B_\alpha$. Define $f : S \to T$ as follows: if $x \in S$, choose a set $A_\alpha$ that contains $x$, and let $f(x) = f_\alpha(x)$. The function $f$ is well defined because $\mathfrak{C}$ is a chain. Specifically, if $x \in A_\alpha \cap A_\beta$ $(\alpha, \beta \in J)$, then $f_\alpha \leq f_\beta$ or $f_\beta \leq f_\alpha$; say, the former. Since $f_\beta$ extends $f_\alpha$, $f_\alpha(x) = f_\beta(x)$. We leave it to the reader to verify that $f$ is an injection. Clearly, $f$ is an upper bound of $\mathfrak{C}$. By Zorn's lemma, $\mathfrak{B}$ has a maximal element, say, $f_1$. Write $A_1 = \mathrm{Dom}(f_1)$ and $B_1 = \mathfrak{R}(f_1)$. If $A_1 = A$, then $f_1$ is an injection from $A$ into $B$. If $B_1 = B$, then $f_1^{-1}$ is an injection from $B$ to $A$, and the proof is complete. We now show that $A_1 \neq A$ and $B_1 \neq B$ cannot occur simultaneously. If that were the case, pick elements $a \in A - A_1, b \in B - B_1$, and extend $f_1$ to a function $f : A_1 \cup \{a\} \to B_1 \cup \{b\}$ by defining $f(a) = b$ and $f|_{A_1} = f_1$. Clearly, $f$ is a strict extension of $f_1$, and this contradicts the maximality of $f_1$.* ∎

**Lemma 2.2.3.** *Let $B$ be a nonempty subset of a set $A$. If there is an injection $f : A \to B$, then $A \approx B$.*

*Proof.* Assume, without loss of generality, that $\mathfrak{R}(f) \subset B \subset A$ (strict inclusions). Define the powers of $f$ as follows: $f^{(1)}(x) = f(x)$, $f^{(n+1)}(x) = f(f^{(n)}(x))$, $n \geq 1$. Let $B' = \{f^{(n)}(x) : x \in A - B, n \in \mathbb{N}\}$. Note that $B' \subseteq \mathfrak{R}(f)$ and that $f(B') \subseteq B'$. Let $C = (A - B) \cup B'$, and let $f_1$ be the restriction of $f$ to $C$. Thus $f_1$ is an injection from $C$ to $B'$. We now show that $f_1$ is a bijection. If $y \in B'$, then $y = f^{(n)}(x)$ for some $x \in A - B$. If $n = 1$, then $y = f(x), x \in A - B \subseteq C$. If $n > 1$, then $y = f(z), z = f^{(n-1)}(x) \in B' \subseteq C$. Now let $D = B - B'$. Since $B' \subseteq \mathfrak{R}(f)$, $D = B - B' \supseteq B - \mathfrak{R}(f) \neq \emptyset$. The reader can check that $B'$ and $D$ partition $B$ and that the three sets $B', D$, and $A - B$ partition $A$; a simple Venn diagram makes this abundantly clear. Thus $C$ and $D$ partition $A$, and $B'$ and $D$ partitions $B$. The function $h : A \to B$ defined below is a bijection from $A$ to $B$:

$$h(x) = \begin{cases} g(x) & \text{if } x \in C, \\ x & \text{if } x \in D. \end{cases} \blacksquare$$

**Example 6.** Any two open disks in the plane are equivalent. Let $0 < r_1 < r_2$, and consider the disks $D_1 = \{(x,y) \in \mathbb{R}^2 : x^2 + y^2 < r_1^2\}$ and $D_2 = \{(x,y) \in \mathbb{R}^2 : x^2 + y^2 < r_2^2\}$. Choose a number $a$ such that $0 < a < r_1$. The function $f : D_2 \to D_1$ defined by $f(x,y) = \frac{a}{r_2}(x,y)$ is an injection, as the reader can easily verify. By the previous lemma, $D_1 \approx D_2$. In a similar manner, one can prove that any two open squares are equivalent. ♦

**Theorem 2.2.4 (the Schröder-Bernstein theorem).** *Let $A$ and $B$ be nonempty sets. If there exist injections $f : A \to B$ and $g : B \to A$, then $A \approx B$.*

*Proof.* Let $A_1 = \mathfrak{R}(g) \subseteq A$. The function $g \circ f : A \to A_1$ is an injection. By lemma 2.2.3, there exists a bijection $h : A \to A_1$. Now $g^{-1}$ is a bijection from $A_1$ onto $B$, and the composition $g^{-1} \circ h$ is the desired bijection from $A$ to $B$. $\blacksquare$

**Example 7.** An open square is equivalent to an open disk. Let $S = \{(x,y) : |x| < 2, |y| < 2\}$, let $D_3 = \{(x,y) : x^2 + y^2 < 9\}$, and let $D_1 = \{(x,y) : x^2 + y^2 < 1\}$. Observe that $D_1 \subseteq S \subseteq D_3$. Set inclusion is clearly an injection from $S$ into $D_3$. By example 6, there is an injection from $D_3$ into $D_1$, and hence from $D_3$ into $S$. By the Schröder-Bernstein theorem, $S \approx D_3$. ♦

## Exercises

1. Let $A$ be a partially ordered set and let $S \subseteq A$. State the definition of each of the following terms: the least element of $S$, a minimal element of $S$, a lower bound of $S$, and the greatest lower bound of $S$.

2. Prove that a nonempty subset of a well-ordered set is well ordered.
3. Let $A$ be a partially ordered set such that every nonempty subset of $A$ has a least element. Prove that $A$ is linearly ordered and hence well-ordered.
4. Let $(A, \leq)$ be a linearly ordered set. Prove that $\leq$ is a well ordering if and only if $A$ does not contain a strictly decreasing sequence $a_1 > a_2 > a_3 > \ldots$
5. Prove that if every countable subset of a linearly ordered set $A$ is well ordered, then $A$ is well ordered.

**Definition.** Let $A$ be a linearly ordered set and let $a \in A$. The **initial segment** of $A$ determined by $a$ is the set $S(a) = \{x \in A : x < a\}$.

6. Let $A$ be linearly ordered and let $a, b \in A$. Show that $S(a) = S(b)$ if and only if $a = b$.
7. Prove that if every segment of a linearly ordered set $A$ is well ordered, then $A$ is well ordered.
8. Let $A$ be a well-ordered set, and let $B$ be a proper subset of $A$ with the property that the conditions $b \in B$ and $c < b$ imply that $c \in B$. Prove that $B$ is a segment of $A$.
9. Let $A$ be a well-ordered set, and let $B$ be a proper subset of $A$ such that, for every $b \in B$ and for every $a \in A - B$, $b < a$. Prove that $B$ is a segment of $A$.
10. **The principle of transfinite induction.** Suppose that $A$ is a well-ordered set and that $\emptyset \neq B \subseteq A$ is such that whenever $S(x) \subseteq B$, $x \in B$. Prove that $B = A$.
11. Suppose that $A$ is a well-ordered set and that $B \subseteq A$. Prove that either $\cup_{x \in B} S(x) = A$, or $\cup_{x \in B} S(x)$ is an initial segment of $A$.
12. Prove that there exists an uncountable, well-ordered set $\Omega$ such that every initial segment of $\Omega$ is countable.
13. Let $\Omega$ be as in the previous problem. Prove that every countable subset of $\Omega$ has an upper bound.
14. Give a direct proof of the fact that Zorn's lemma implies the axiom of choice. Hint: Let $\{X_\alpha\}_{\alpha \in I}$ be a nonempty collection of nonempty sets, and let $\mathcal{B} = \{(J, g) : J \subseteq I, g \in \prod_{\alpha \in J} X_\alpha\}$, that is, $g$ is a choice function on $\{X_\alpha\}_{\alpha \in J}$. The set $\mathcal{B} \neq \emptyset$ because finite subsets $J$ of $I$ generate such functions. Partially order $\mathcal{B}$ as follows: $(J_1, g_1) \leq (J_2, g_2)$ if $J_1 \subseteq J_2$ and $g_2$ extends $g_1$.
15. Let $A$ be a linearly ordered set. Is the union of a collection of well-ordered subsets of $A$ necessarily a well-ordered set?
16. **The Hausdorff maximal principle.** Every partially ordered set contains a maximal chain, that is, a chain that is not properly contained in any other chain.

   Prove that the Hausdorff maximal principle is equivalent to Zorn's lemma.

Hint: To prove that the Hausdorff maximal principle implies Zorn's lemma, let $(X, \leq)$ be a partially ordered set that satisfies the conditions of Zorn's lemma. Let $C$ be a maximal chain, and let $x$ be an upper bound of $C$. To prove the converse, let $\mathfrak{C}$ be the collection of all chains in $X$, and order $\mathfrak{C}$ by set inclusion. Verify that the conditions of Zorn's lemma are met and hence $\mathfrak{C}$ contains a maximal member, that is, a maximal chain in $X$.

17. Prove that any open disk is equivalent to any closed disk.

18. Prove that any open square is equivalent to any closed square.

## 2.3  Cardinal Numbers

In section 2.1 we took a small step toward showing that infinite sets are not created equal. In this section, we show that there are *infinitely many types of infinities*, in the sense that there is a whole cascade (loosely speaking) of infinite sets of unequal sizes, or cardinalities. This is the first result in the section. Our approach to infinite cardinals is intuitive rather than axiomatic. We proceed to show that the set of integers is the *smallest infinite set*, then we prove that a set of infinite sets is well ordered by size, or cardinality. Only an intuitive understanding of cardinal numbers is essential for subsequent material that make reference to cardinality. Thus the discussion of cardinal arithmetic and sums of infinitely many cardinals can be omitted on the first reading if the goal is to take the fastest route to chapter 4.

**Definition.** Let $A$ and $B$ be nonempty sets. We say that $A$ and $B$ have **the same cardinality** if $A \approx B$. We also say that $A$ and $B$ define **the same cardinal number**, and we write $Card(A) = Card(B)$.

**Definition.** Let $a = Card(A)$ and $b = Card(B)$. We say that $a \leq b$ if there is an injection from $A$ to $B$. We also write $a < b$ to mean that there is a injection from $A$ to $B$ but that $A$ and $B$ are not equivalent. By the Schröder-Bernstein theorem, this is equivalent to saying that there is no injection from $B$ to $A$.

**Theorem 2.3.1.** *For any set $A$, $Card(A) < Card(\mathcal{P}(A))$.*

*Proof. Recall that the notation $\mathcal{P}(A)$ stands for the power set of $A$. Define a function $f : A \to \mathcal{P}(A)$ by $f(x) = \{x\}$. Clearly, $f$ is one-to-one; therefore, $Card(A) \leq Card(\mathcal{P}(A))$. If $Card(A) = Card(\mathcal{P}(A))$, then there exists a bijection $g : A \to \mathcal{P}(A)$. Define $S = \{x \in A : x \notin g(x)\}$. Since $g$ is onto, let $a$ be such that $g(a) = S$. If $a \in S$, then, by the definition of $S$, $a \notin S$. If $a \notin S$, then again, by the definition of $S$, $a \in S$. This contradiction completes the proof.* ∎

The reader may have observed that while our definition of what it means for two sets to have the same cardinality is unambiguous, we have not really defined what a *cardinal number* is. We can give a slightly more tangible definition of *a set of cardinal numbers* as follows: Let $\mathfrak{S}$ be a set of sets. By theorem 2.1.1, set equivalence is an equivalence relation on $\mathfrak{S}$. We can define the cardinal numbers in $\mathfrak{S}$ to be equivalence classes of set equivalence in $\mathfrak{S}$. This does not define *all cardinal numbers* because if $A$ is a set that is not equivalent to any set in $\mathfrak{S}$,[2] then $Card(A) \neq Card(S)$ for all $S \in \mathfrak{S}$.

One might be tempted to generalize the idea of the last paragraph by considering *the set of all sets*, instead of a fixed set of sets. However, within the limitations of naïve set theory, this is paradoxical for the following reason: If we were allowed to use terms such as *the set of all sets*, $\mathfrak{S}$, let $U = \cup\{S : S \in \mathfrak{S}\}$. Since $U$ contains every $S \in \mathfrak{S}$, $Card(S) \leq Card(U)$. Since $\mathfrak{S}$ contains *all sets*, $Card(U)$ would be the largest cardinal number. This is a paradox because, by theorem 2.3.1, $Card(\mathcal{P}(U)) > Card(U)$. Such paradoxes can be avoided in an axiomatic treatment of set theory. Such a treatment is hardly essential for our purposes because we will never refer to cardinal numbers as an absolute concept. We will be content to think of cardinal numbers as a comparative measure of the size of sets in the sense of the opening definition of this section.

Some common cardinals are:

$$n = Card(\mathbb{N}_n)$$
$$\aleph_0 = Card(\mathbb{N})$$
$$\mathfrak{c} = Card(\mathbb{R})$$

The natural numbers are the finite cardinals, and all other cardinals are infinite.

**Theorem 2.3.2.** $\aleph_0$ *is the smallest infinite cardinal number.*

*Proof.* Let $A$ be an infinite set. By theorem 2.1.4, $A$ contains a countable set of distinct elements, $B = \{b_1, b_2, \ldots\}$. Since the inclusion $B \subseteq A$ is an injection, $\aleph_0 = Card(B) \leq Card(A)$. ∎

**Theorem 2.3.3.** *Let $\mathfrak{S}$ be a set of cardinal numbers. Then $\mathfrak{S}$ is linearly ordered.*

*Proof.* Let $a, b \in \mathfrak{S}$ and let $A$ and $B$ be sets such that $a = Card(A)$, and $b = Card(B)$. By theorem 2.2.2, there is an injection from $A$ to $B$ or one from $B$ to $A$. Thus $a \leq b$ or $b \leq a$. To check antisymmetry, suppose that $a \leq b \leq a$. Then is an injection from $A$ to $B$ and one from $B$ to $A$. By the Schröder-Bernstein theorem, $A \approx B$ and $a = b$. ∎

---

[2] Such a set $A$ exists. One can take $A$ to be the power set of $\cup\{S : S \in \mathfrak{S}\}$.

The following theorem establishes the fact that any set of cardinal numbers is well ordered.

**Theorem 2.3.4.** *If* $\mathfrak{S} = \{\xi_\alpha\}_{\alpha \in I}$ *is a set of cardinal numbers, then there is an element* $\alpha_0 \in I$ *such that* $\xi_{\alpha_0} \le \xi_\alpha$ *for all* $\alpha \in I$.

*Proof.* Let $\{X_\alpha\}_{\alpha \in I}$ be sets such that $\xi_\alpha = Card(X_\alpha)$. If $\mathfrak{S}$ contains any integers, the smallest of these integers is the least cardinal in $\mathfrak{S}$. Otherwise, all the sets $X_\alpha$ are infinite. We prove that there is $\alpha_0 \in I$ such that, for every $\alpha \in I$, there is an injection $f_\alpha : X_{\alpha_0} \to X_\alpha$.

Let $X = \prod_{\alpha \in I} X_\alpha$, and let $\mathfrak{B}$ be the collection of subsets $B$ of $X$ with the property that if $x = (x_\alpha)$ and $y = (y_\alpha)$ are distinct elements of $B$, then $x_\alpha \ne y_\alpha$ for all $\alpha \in I$. Order $\mathfrak{B}$ by set inclusion. It is clear that if $\mathfrak{C}$ is a chain in $\mathfrak{B}$, then $\mathfrak{C}$ has an upper bound, namely, $\cup\{C : C \in \mathfrak{C}\}$. By Zorn's lemma, $\mathfrak{B}$ has a maximal member, $B$. We claim that, for some $\alpha_0 \in I$, $\pi_{\alpha_0}(B) = X_{\alpha_0}$. If this is not the case, then, for each $\alpha \in I$, choose an element $a_\alpha \in X_\alpha - \pi_\alpha(B)$ and let $a = (a_\alpha)$. The set $B \cup \{a\}$ is clearly in $\mathfrak{B}$, which contradicts the maximality of $B$ and shows that, for some $\alpha_0 \in I$, $\pi_{\alpha_0}(B) = X_{\alpha_0}$.

Now, for each $a \in X_{\alpha_0}$, there is a unique element $x \in B$ such that $\pi_{\alpha_0}(x) = a$. Such $x$ exists because $\pi_{\alpha_0}$ is onto, and it is unique by the definition of $\mathfrak{B}$ and the fact that $B \in \mathfrak{B}$. Now define $f_\alpha : X_{\alpha_0} \to X_\alpha$ as follows: $f_\alpha(a) = \pi_\alpha(x)$, where $x$ is the element of $B$ constructed above.[3] By construction, $f_\alpha$ is an injection. ∎

## Cardinal Arithmetic

**Definition.** Let $A$ and $B$ be disjoint sets, and write $a = Card(A), b = Card(B)$. By definition:

the sum of $a$ and $b$: $a + b = Card(A \cup B)$
the product of $a$ and $b$: $ab = Card(A \times B)$
exponentiation: $a^b = Card(A^B)$; recall that $A^B$ is the set of all functions $B \to A$

The above operations are well defined in the sense that they are independent of the particular sets $A$ and $B$ chosen to represent $a$ and $b$. For example, if $A \approx C$ and $B \approx D$, then $A \times B \approx C \times D$ and $A^B \approx C^D$. See the exercises on section 2.1.

**Example 1.** For any cardinal number $a$, $a < 2^a$.

---

[3] In fact, $f_\alpha(a) = \pi_\alpha(\pi_{\alpha_0}^{-1}(a))$.

By theorem 2.3.1 and problem 16 on section 1.1, $a < Card(\mathcal{P}(A)) = Card(2^A) = 2^a$. ◆

**Example 2.** Let $a$ and $b$ be cardinal numbers. Then $a \leq b$ if and only if there is a cardinal number $c$ such that $a + c = b$.

Suppose $b = a + c$ and let $A$, $B$, and $C$ be sets such that $Card(A) = a$, $Card(B) = b$, and $Card(C) = c$, and suppose that $A \cap C = \varnothing$. By assumption, there is a bijection $f : A \cup C \to B$. The restriction of $f$ to $A$ injects $A$ into $B$. Hence $a \leq b$. Conversely, if $a \leq b$, let $A$ and $B$ be as above, let $f : A \to B$ be an injection and assume that $A \cap B = \varnothing$. Define $C = B - f(A)$. The function $g : A \cup C \to B$ defined below is easily seen to be a bijection:

$$g(x) = \begin{cases} f(x) & \text{if } x \in A, \\ x & \text{if } x \in C. \end{cases}$$

If $c = Card(C)$ then, by definition, $a + c = b$. ◆

**Theorem 2.3.5.** *Let $a, b,$ and $c$ be cardinal numbers. Then*

1. $a + b = b + a$ *and* $ab = ba$.
2. $a + (b + c) = (a + b) + c$ *and* $a(bc) = (ab)c$.
3. $a(b + c) = ab + ac$.
4. $a^b a^c = a^{b+c}$.
5. $a^c b^c = (ab)^c$.
6. $(a^b)^c = a^{bc}$.
7. *If* $a \leq b$, *then* $a + c \leq b + c$.
8. *If* $a \leq b$, *and* $c \geq 1$, *then* $ac \leq bc$.

*Proof. Most of the rules of cardinal arithmetic are obvious. We prove property 6, as an example. Let $A, B,$ and $C$ be such that $a = Card(A), b = Card(B)$ and $c = Card(C)$. We need to show that $(A^B)^C$ is equivalent to $A^{B \times C}$. Let $f \in (A^B)^C$. Then for $c \in C, f(c)$ is a function from $B$ to $A$. We write $f_c$ instead of $f(c)$. For such an $f$, define a function $\phi(f) = g : B \times C \to A$ by $g(b, c) = f_c(b)$. The assignment $\phi : f \mapsto g$ maps $(A^B)^C$ to $A^{B \times C}$.*

*$\phi$ is onto: If $g : B \times C \to A$, define $f : C \to A^B$ by $f_c(b) = g(b, c)$. Clearly $g = \phi(f)$.*

*$\phi$ is one-to-one: Let $f$ and $f'$ be in $(A^B)^C$ be such that $f \neq f'$. Then there is $c \in C$ such that $f_c \neq f'_c$. Thus there is $b \in B$ such that $f_c(b) \neq f'_c(b)$. Now if $g = \phi(f), g' = \phi(f')$, then $g(b, c) = f_c(b) \neq f'_c(b) = g'(b, c)$.* ∎

**Example 3.** Let $a, b, c$, and $d$ be cardinal numbers such that $a \le b$ and $c \le d$. Then $ac \le bd$.

By example 2, there are cardinal numbers $r$ and $s$ such that $a + r = b$, and $c + s = d$. Now $bd = (a + r)(c + s) = ac + (as + rc + rs)$. Again by example 2, $ac \le bd$. ◆

**Example 4.** If $a, b$, and $c$ are cardinal numbers and $a \le b$, then $a^c \le b^c$.

Let $A, B$, and $C$ be such that $Card(A) = a$, $Card(B) = b$, and $Card(C) = c$. By assumption, there is an injection $g : A \to B$. Define a function $\phi : A^C \to B^C$ by $\phi(f) = gof$. The function $\phi$ is an injection; hence $a^c \le b^c$. ◆

**Theorem 2.3.6.** *Let $a$ be an infinite cardinal. Then $a.a = a$.*

*Proof. Let $A$ be such that $Card(A) = a$, and let $\mathfrak{B} = \{(A_\alpha, f_\alpha)\}_{\alpha \in I}$ be the collection of all bijections $f_\alpha : A_\alpha \to A_\alpha \times A_\alpha$, where $A_\alpha \subseteq A$. To see that $\mathfrak{B} \ne \varnothing$, pick a countable subset $G$ of $A$. By theorem 2.1.6, $G \approx G \times G$; hence $\mathfrak{B} \ne \varnothing$.*

*Order $\mathfrak{B}$ as follows: for $\alpha$ and $\beta \in I$, $(A_\alpha, f_\alpha) \le (A_\beta, f_\beta)$ if $A_\alpha \subseteq A_\beta$, and $f_\beta$ extends $f_\alpha$. If $\mathfrak{C} = \{(A_\alpha, f_\alpha)\}_{\alpha \in J}$ is a chain in $\mathfrak{B}$, let $C = \cup_{\alpha \in J} A_\alpha$ and define a function $f : C \to C \times C$ by $f(x) = f_\alpha(x)$, where $\alpha \in J$ is such that $x \in A_\alpha$. The function $f$ is a well-defined bijection from $C \to C \times C$, as the reader can verify. Clearly, $(C, f)$ extends every member in $\mathfrak{C}$ and hence is an upper bound of $\mathfrak{C}$. By Zorn's lemma, $\mathfrak{B}$ contains a maximal member, $(C, g)$. We claim that $Card(C) = a$. Suppose for a contradiction that $Card(C) = b < a$. First observe that $b \le b + b \le b.b = Card(C \times C) = Card(C) = b$, and hence $b + b = b.b = b$. Now let $d = Card(A - C)$. If $d \le b$, then $a = b + d \le b + b = b$, which contradicts the supposition that $b < a$. Therefore $b < d$, and $A - C$ contains a subset $E$ such that $Card(E) = b$.*

*Now $(C \cup E) \times (C \cup E) = (C \times C) \cup K$, where $K = (C \times E) \cup (E \times C) \cup (E \times E)$. Since $K$ is the disjoint union of three sets each of cardinality $b$, $Card(K) = b + b + b = (b + b) + b = b + b = b$. Therefore there is an bijection $h : E \to K$. Now define a function $f : C \cup E \to (C \times C) \cup K = (C \cup E) \times (C \cup E)$ by*

$$f(x) = \begin{cases} g(x) & \text{if } x \in C, \\ h(x) & \text{if } x \in E. \end{cases}$$

*Clearly, the pair $(C \cup E, f) \in \mathfrak{B}$ is a strict extension of $(C, g)$, which contradicts the maximality of $(C, g)$. This shows that the supposition $b < a$ is false; hence $a = b$. This concludes the proof because $a.a = Card(C \times C) = Card(C) = a$. ∎*

**Corollary 2.3.7.** *If $a$ is an infinite cardinal number and $1 \le b \le a$, then $ab = a$.*

*Proof.* $a \leq ab \leq a.a = a.$ ∎

**Theorem 2.3.8.** *Let $a$ be an infinite cardinal. Then $a + a = a$.*

*Proof.* $a \leq a + a \leq a.a = a$. Therefore $a + a = a$. ∎

**Corollary 2.3.9.** *If $a$ is an infinite cardinal and $1 \leq b \leq a$, then $a + b = a$*

*Proof.* $a \leq a + b \leq a + a = a$.∎

**Example 5.** Let $b$ be an infinite cardinal number, and suppose that $1 < a \leq b$. Then $a^b = 2^b$.

Since $a < 2^a$, $a^b \leq (2^a)^b = 2^{ab} = 2^b$. Because $2 \leq a$, $2^b \leq a^b$, and $2^b = a^b$. ◆

We conclude our study of cardinal arithmetic with a brief exploration of sums of infinitely many cardinals.

**Definition.** Let $\{a_\alpha\}_{\alpha \in I}$ be a set of cardinal numbers, and let $\{A_\alpha\}$ be a collection of disjoint sets such that $Card(A_\alpha) = a_\alpha$. Define $\sum_{\alpha \in I} a_\alpha = Card(\cup_{\alpha \in I} A_\alpha)$.

**Theorem 2.3.10.** *Let $\{a_\alpha\}_{\alpha \in I}$ be a collection of equal cardinal numbers, say, $a_\alpha = a$ and let $b = Card(I)$. Then $\sum_{\alpha \in I} a_\alpha = ab$.*

*Proof.* Let $A$ be such that $Card(A) = a$, and let $\{A_\alpha\}$ be a collection of disjoint sets such that $Card(A_\alpha) = a$. Then there are bijections $f_\alpha : A \to A_\alpha$. Define a function $f : A \times I \to \cup_{\alpha \in I} A_\alpha$ by $f(x, \alpha) = f_\alpha(x)$. Verifying that $f$ is a bijection is straightforward. Therefore $\sum_{\alpha \in I} a_\alpha = Card(\cup_{\alpha \in I} A_\alpha) = Card(A \times I) = ab$. ∎

**Theorem 2.3.11.** *If $a_\alpha \leq b_\alpha$ for every $\alpha \in I$, then $\sum_{\alpha \in I} a_\alpha \leq \sum_{\alpha \in I} b_\alpha$.*

*Proof.* Let $\{A_\alpha\}$ be a collection of disjoint sets such that $Card(A_\alpha) = a_\alpha$, and let $\{B_\alpha\}$ be a collection of disjoint sets such that $Card(B_\alpha) = b_\alpha$. By assumption, there exist injections $f_\alpha : A_\alpha \to B_\alpha$. Define a function $f : \cup_{\alpha \in I} A_\alpha \to \cup_{\alpha \in I} B_\alpha$ by $f(x) = f_\alpha(x)$ if $x \in A_\alpha$. The function $f$ is well defined because $\{A_\alpha\}$ is a disjoint family. Clearly, $f$ is an injection from $\cup_{\alpha \in I} A_\alpha$ into $\cup_{\alpha \in I} B_\alpha$. ∎

The following theorem is a far-reaching generalization of theorem 2.1.12:

**Theorem 2.3.12.** *Let $I$ be an infinite set, and let $b = Card(I)$. Then the family $\mathfrak{F}$ of finite sequences in $I$ has cardinality $b$.*

*Proof. Let $I^n$ be the family of sequences in $I$ of length exactly $n$. Since $I^n = I^{\{1,2,\ldots,n\}}$, $Card(I^n) = b^n = b$. Now $\mathfrak{F} = \cup_{n=1}^{\infty} I^n$, so $Card(\mathfrak{F}) = \sum_{n=1}^{\infty} Card(I^n)$. By theorem 2.3.10, $\sum_{n=1}^{\infty} Card(I^n) = \aleph_0 b = b$.* ∎

We conclude the section with a well-known theorem and a famous conjecture.

**Theorem 2.3.13.** $2^{\aleph_0} = \mathfrak{c}$.

*Proof. By definition, $2^{\aleph_0} = Card(2^{\mathbb{N}})$. Let $T$ be the set of all binary sequences that contain only a finite number of nonzero terms. By problem 6 on section 2.1, $Card(2^{\mathbb{N}} - T) = Card(2^{\mathbb{N}}) = 2^{\aleph_0}$. By the proof of theorem 2.1.15 (see also problem 11 on section 2.1), $2^{\mathbb{N}} - T \approx (0,1] \approx \mathbb{R}$. Thus $\mathfrak{c} = Card(\mathbb{R}) = Card((0,1]) = Card(2^{\mathbb{N}} - T) = 2^{\aleph_0}$.* ∎

**Example 6.** $\mathfrak{c}^{\aleph_0} = \mathfrak{c}$.

Using theorems 2.3.6 and 2.3.13, $\mathfrak{c}^{\aleph_0} = (2^{\aleph_0})^{\aleph_0} = 2^{\aleph_0 \aleph_0} = 2^{\aleph_0} = \mathfrak{c}$. ♦

**Example 7.** There exits a sequence of cardinal numbers $a_1, a_2, \ldots$ such that $a_1 < a_2 < \ldots$ and $a_n^{\aleph_0} = a_n$.

Take $a_1 = \mathfrak{c}$. By the previous example, $a_1^{\aleph_0} = a_1$. For $n \geq 1$, define $a_{n+1} = 2^{a_n}$. ♦

## The Continuum Hypothesis

Take a sufficiently large infinite cardinal such as $\mathfrak{a} = 2^{2^{2^{\aleph_0}}}$. There are several infinite cardinals between $\aleph_0$ and $\mathfrak{a}$, such as $2^{\aleph_0}$, and $2^{2^{\aleph_0}}$. Consider the set of cardinals strictly between $\aleph_0$ and $\mathfrak{a}$. By theorem 2.3.4, there is a smallest cardinal in that set. We call it $\aleph_1$. Thus $\aleph_1$ is the immediate successor of $\aleph_0$ in the sense that there are no cardinals strictly between $\aleph_0$ and $\aleph_1$.

We know that $2^{\aleph_0} > \aleph_0$. By the above paragraph, $\aleph_1 \leq 2^{\aleph_0}$. A famous conjecture of set theory is **the continuum hypothesis**, which states that $2^{\aleph_0} = \aleph_1$. In other words, there are no cardinals strictly between $\aleph_0$ and $2^{\aleph_0} = \mathfrak{c}$.

The **generalized continuum hypothesis** states that, for any infinite cardinal $a$, there is no cardinal number $b$ such that $a < b < 2^a$, that is, $2^a$ is the immediate successor of $a$.

## Exercises

1. Provide proofs of the statements of theorem 2.3.5.

2. Prove that if $a \geq 2$ is a cardinal number, then $a + a \leq a.a$. This result was used in the proof of theorems 2.3.6 and 2.3.8.

3. Let $a$ and $b$ be infinite cardinal numbers. Prove that if $a + a = a + b$, then $a \geq b$.

4. Let $a, b$, and $c$ be infinite cardinal numbers. Prove that if $a + b < a + c$, then $b < c$.

5. What is $\aleph_0^{\aleph_0}$ ?

6. Prove that $\sum_{n=1}^{\infty} n = \aleph_0$.

7. Let $A$ and $B$ be infinite sets and let $f : A \to B$ be a surjection.
   (a) Prove that $Card(A) \geq Card(B)$.
   (b) Prove that if $f^{-1}(b)$ is at most countable for each $b \in B$, then $A \approx B$.

8. Let $\{A_\alpha\}_{\alpha \in I}$ be a family of nonempty sets. Prove that $Card(\cup_{\alpha \in I} A_\alpha) \leq \sum_{\alpha \in I} Card(A_\alpha)$.

# 3

# Vector Spaces

*Questions that pertain to the foundations of mathematics, although treated by many in recent times, still lack a satisfactory solution. Ambiguity of language is philosophy's main source of problems. That is why it is of the utmost importance to examine attentively the very words we use.*

Giuseppe Peano

Giuseppe Peano. 1858–1932

Peano was born in a farmhouse about 5 km from Cuneo, where he received his early education. One of Peano's uncles was a priest and a lawyer in Turin, and he realized the child's talent. He took him to Turin in 1870 for his secondary schooling. Peano entered the University of Turin in 1876, graduated in 1880 doctor of mathematics, and was appointed to the university the same year. He received his qualification to be a university professor in 1884.

In 1886 Peano proved the existence of the solution of the differential equation $dy/dx = f(x, y)$ under the mere assumption that $f$ is continuous in the neighborhood of the initial point $(x_0, y_0)$. In 1888 he published the book *Geometrical Calculus*, which begins with a chapter on mathematical logic. A significant feature of the book is that, in it, Peano sets out with great clarity the ideas of Grassmann, who made the first attempt to define a vector space, albeit in a rather obscure way.

*Fundamentals of Mathematical Analysis.* Adel N. Boules, Oxford University Press (2021). © Adel N. Boules.
DOI: 10.1093/oso/9780198868781.003.0003

This book contains the first definition of a vector space given with a remarkably modern notation and style. This was, without a doubt, a big development in the history of mathematics. In 1889 Peano published an axiomatic approach to the definition of the natural numbers that was based on the notion of the successor function. In 1890 he made the stunning discovery that there are continuous surjective mappings from [0,1] onto the unit square, which came to be known as space-filling curves.

Peano's career was strangely divided into two periods. The period up to 1900 is one where he showed great originality and a remarkable feel for topics that would be important in the development of mathematics. His achievements were outstanding, and he had a modern style quite ahead of his own time. However, this feel for what was important seemed to leave him, and, after 1900, he worked with great enthusiasm on two projects of great difficulty, which were enormous undertakings but proved quite unimportant in the development of mathematics.

From around 1892, Peano embarked on a new and extremely ambitious project, namely, the *Formulario Mathematico*. As he explained:[1]

> of the greatest usefulness would be the publication of collections of all the theorems now known that refer to given branches of the mathematical sciences... Such a collection, which would be long and difficult in ordinary language, is made noticeably easier by using the notation of mathematical logic

Even before the *Formulario Mathematico* project was completed, Peano took up the project of finding an international, artificial language, "Latino sine flexione," which was based on Latin but stripped of all grammar. He compiled the vocabulary by taking words from English, French, German, and Latin. In fact, the final edition of the *Formulario Mathematico* was written in Latino sine flexione, which is another reason the work was so little used.

## The Evolution of the Concept of a Vector Space[2]

The emergence of the modern definition of a vector space was delayed for a considerable length of time because of several reasons. It appears that early attempts to define what we know now as a vector space were hindered by the insistence on incorporating axioms for determinants. The lack of awareness of the importance of axiomatics and abstract thinking was also a major obstacle. Grassmann's

---

[1] J. J. O'Connor and E. F. Robertson, "Giuseppe Peano," in *MacTutor History of Mathematics*, (St Andrews: University of St Andrews, 1998), http://mathshistory.st-andrews.ac.uk/Biographies/Peano/, accessed Nov. 3, 2020.

[2] All the historical information in this article can be found in Jean-Luc Dorier, "A general outline of the genesis of vector space theory." *Historia Mathematica* 22, no. 3 (1995): 227–61.

pioneering ideas were obscured by philosophical language, and although Peano's definition was a long step toward axiomatization, it did not produce the modern definition. The founders of functional analysis were instrumental in framing the modern definition. In 1916 Riesz studied spaces of continuous functions and defined linear transformations and even the concept of bounded linear operators. Decisive steps toward axiomatization were taken independently by Banach in 1920 and by Hahn, in two papers published in 1922 and 1927. In 1920 Banach took Riesz's ideas one step further and defined what is known in modern terminology as a Banach space. The function spaces Banach and Riesz studied are infinite dimensional, and this makes the use of an axiomatic approach compulsory. Banach's approach was confined to function spaces, and his axioms did not coincide with the modern definition in that some axioms are redundant and some are missing. Modern algebra finally paved the way toward the modern definition of a vector space: determinants were dropped from the axiomatic approach, and this unified the definition of finite and infinite-dimensional spaces. The definition was made accessible to beginning students in books that were published in the 1940s by Birkhoff and MacLane, Halmos, and Bourbaki.

## 3.1  Definitions and Basic Properties

This section is a summary of the most basic concepts of vector space theory. The main reason for including this section is to establish terminology and provide a collection of important examples. The reader should pay particular attention to the examples, because the sequence and function spaces we introduce here are of fundamental importance for the rest of the book. The theorems are stated without proof.

**Definition.** Let $\mathbb{K}$ be a field, and suppose $U$ is a nonempty set equipped with a binary operation, $+$ (vector addition). Suppose also that there is a function $\times : \mathbb{K} \times U \to U$ (scalar multiplication) that assigns to each pair $(a, u) \in \mathbb{K} \times U$ an element $a \times u$ (or simply $au$) in $U$. The triple $(U, +, x)$ is called a **vector space** over the field $\mathbb{K}$ if the following conditions are satisfied by all elements $a, b \in \mathbb{K}$ and all elements $u, v, w \in U$:

(a) $u + v = v + u$.
(b) $u + (v + w) = (u + v) + w$.
(c) There is an element $0 \in V$ (the zero vector) such that $u + 0 = u$.
(d) For every $u \in U$, there is an element $-u \in U$ such that $u + (-u) = 0$.
(e) $a(u + v) = au + av$.
(f) $(a + b)u = au + bu$.
(g) $(ab)u = a(bu)$.
(h) $1.u = u$.

The field $\mathbb{K}$ is called the base field, the elements of $\mathbb{K}$ are referred to as scalars, and the elements in U are called vectors. The only two fields we will use in this book are the real field, $\mathbb{R}$, and the complex field, $\mathbb{C}$. Either of these two fields will be denoted by $\mathbb{K}$. Most of the results we will obtain apply equally whether the underlying field is $\mathbb{R}$ or $\mathbb{C}$. When a given result applies to only one field but not the other, we will explicitly state the base field.

**Example 1.** For each $n \in \mathbb{N}$, let $\mathbb{K}^n$ be the set of sequences in $\mathbb{K}$ of length $n$. The set $\mathbb{K}^n$ is a vector space with the operations

$$(x_1, \ldots, x_n) + (y_1, \ldots, y_n) = (x_1 + y_1, \ldots, x_n + y_n), a(x_1, \ldots, x_n) = (ax_1, \ldots, ax_n). \blacklozenge$$

**Example 2.** Let $I$ be a nonempty set, and let $\mathbb{K}^I$ be the space of all functions from $I$ to $\mathbb{K}$. For functions $x = (x_\alpha)_{\alpha \in I}, y = (y_\alpha)_{\alpha \in I}$, and for $a \in \mathbb{K}$, define

$$x + y = (x_\alpha + y_\alpha)_{\alpha \in I}, ax = (ax_\alpha)_{\alpha \in I}. \blacklozenge$$

**Example 3.** The space $\mathbb{K}(I)$ is the space of all functions $x : I \to \mathbb{K}$ such that $x_\alpha = 0$ for all but a finite number of elements $\alpha \in I$. Addition and scalar multiplication are defined as in example 2. $\blacklozenge$

**Example 4.** Important special cases of examples 2 and 3 are obtained when $I = \mathbb{N}$; $\mathbb{K}^\mathbb{N}$ is the space of all sequences in $\mathbb{K}$, and $\mathbb{K}(\mathbb{N})$ is the space of sequences that have a finite number of terms different from 0. $\blacklozenge$

**Example 5.** Let $\mathbb{P}$ be the set of all polynomials with coefficients in $\mathbb{K}$. We add polynomials by adding the coefficients of equal powers of $x$, and scalar multiplication is defined by $a(\sum_{i=0}^n a_i x^i) = \sum_{i=0}^n (aa_i)x^i$. Let $\mathbb{P}_n$ be the space of polynomials of degree $\leq n$. Clearly, $\mathbb{P}_n$ is contained in $\mathbb{P}$ for all $n \in \mathbb{N}$. In fact, $\mathbb{P} = \cup_{n=0}^\infty \mathbb{P}_n. \blacklozenge$

**Example 6.** Let $\mathbb{K}_{m \times n}$ be the space of all $m \times n$ matrices. Addition and scalar multiplication are defined entrywise, in the usual manner. $\blacklozenge$

**Example 7.** For real numbers $a < b$, define $X = \mathcal{B}[a,b]$ as the space of all bounded (real or complex) functions on the interval $[a,b]$. For $f, g \in X, x \in [a,b]$, and $c \in \mathbb{K}$, define vector addition and scalar multiplication in $X$, respectively, by

$$(f+g)(x) = f(x) + g(x),$$
$$(cf)(x) = cf(x). \blacklozenge$$

**Example 8.** Another important example is the set $\mathcal{C}[a,b]$ of continuous functions on the closed bounded interval $[a,b]$. Vector addition and scalar multiplication are defined as in the previous example. Because continuous functions are closed under addition and scalar multiplication (the sum of two continuous functions is continuous, etc.), $\mathcal{C}[a,b]$ is a vector space. ◆

**Example 9.** The space $\mathcal{C}^{\infty}(\mathbb{R})$ consists of all real-valued functions on $\mathbb{R}$ that have derivatives of all orders. Vector addition and scalar multiplication are defined as in example 7. ◆

**Theorem 3.1.1.**
   *(a) The zero vector is unique.*
   *(b) For $u \in U$ and $a \in \mathbb{K}$, $0.u = O$ and $a.O = O$ ($0$ is the scalar zero and $O$ is the zero vector).*
   *(c) $(-a)u = a(-u) = -(au)$.*
   *(d) $(\sum_{i=1}^{n} a_i)u = \sum_{i=1}^{n} a_i u$.*
   *(e) $a(\sum_{i=1}^{n} u_i) = \sum_{i=1}^{n} a u_i$.* ∎

**Definition.** A subset $V$ of a vector space $U$ is called a **subspace** of $U$ if $V$ is closed under vector addition and scalar multiplication. Thus $V$ is a subspace if, for all $v, w \in V$, and all $a \in \mathbb{K}$, $v + w \in V$, and $av \in V$. It is clear that $V$ is a vector space in its own right.

**Example 10.** The set $V = \{(x_1, x_2, 0) : x_1, x_2 \in \mathbb{K}\}$ is a subspace of $\mathbb{K}^3$. ◆

**Example 11.** More generally, for $n < m$, $\mathbb{K}^n$ can be viewed as a subspace of $\mathbb{K}^m$ if we identify an element $(x_1, \ldots, x_n) \in \mathbb{K}^n$ with the element $(x_1, \ldots, x_n, 0, \ldots, 0) \in \mathbb{K}^m$. ◆

**Example 12.** For every $n \in \mathbb{N}$, $\mathbb{P}_n$ is a subspace of $\mathbb{P}$. ◆

**Example 13.** For an arbitrary nonempty set $I$, $\mathbb{K}(I)$ is a subspace of $\mathbb{K}^I$. In particular, $\mathbb{K}(\mathbb{N})$ is a subspace of $\mathbb{K}^{\mathbb{N}}$. A particularly important subspace of $\mathbb{K}^{\mathbb{N}}$ is the **space of bounded sequences**,

$$l^{\infty} = \{(x_1, x_2, \ldots) : sup_n |x_n| < \infty\}. ◆$$

**Example 14.** Two well-known subspaces of $l^{\infty}$ are the **space $c$ of convergent sequences**, and the space $c_0$ of all sequences that converge to 0. We also call $c_0$ the **space of null sequences**. ◆

**Example 15.** The space $\mathcal{C}[a,b]$ is a subspace of $\mathcal{B}[a,b]$ because a continuous function on a closed bounded interval is bounded. ◆

**Theorem 3.1.2.** *A subset $V$ of a vector space $U$ is a subspace of $U$ if and only if for all $v, w \in V$, and all $a, b \in \mathbb{K}$, $av + bw \in V$.* ■

**Definitions. The Canonical Vectors:**

(a) The canonical vectors in $\mathbb{K}^n$ are the $n$ vectors

$$e_1 = (1, 0, \ldots, 0), e_2 = (0, 1, 0, \ldots, 0), \ldots, \text{ and } e_n = (0, 0, \ldots, 1).$$

(b) The canonical vectors in $\mathbb{K}(\mathbb{N})$ are the sequences $e_n (n \in \mathbb{N})$, where the $n$th term of $e_n$ is one, and all the other terms are zero.

(c) The canonical vectors in $\mathbb{K}(I)$ are the functions $e_\alpha : I \to \mathbb{K}$, defined by $e_\alpha(\beta) = \delta_{\alpha, \beta}$. Here $\delta_{\alpha, \beta}$ is the **Kronecker delta**:

$$\delta_{\alpha, \beta} = \begin{cases} 1 & \text{if } \alpha = \beta, \\ 0 & \text{if } \alpha \neq \beta. \end{cases}$$

**Definition.** Let $\{u_1, u_2, \ldots, u_n\}$ be a finite subset of a vector space $U$. **A linear combination** of $u_1, u_2, \ldots, u_n$ is an element of $U$ of the form $\sum_{i=1}^{n} a_i u_i$ for some scalars $a_1, \ldots a_n$.

**Example 16.** Every vector in $\mathbb{K}^n$ is a linear combination of the canonical vectors in $\mathbb{K}^n$. Indeed, if $x = (x_1, \ldots, x_n) \in \mathbb{K}^n$, then $x = \sum_{i=1}^{n} x_i e_i$. ◆

**Example 17.** In $\mathbb{K}(\mathbb{N})$, every vector is a linear combination of a finite number of the canonical vectors, because if $f \in \mathbb{K}(\mathbb{N})$, and $a_1 = f(k_1), \ldots, a_n = f(k_n)$ are all the nonzero terms of $f$, then $f = \sum_{i=1}^{n} a_i e_{k_i}$. ◆

**Example 18.** Every polynomial in $\mathbb{P}_n$ is a linear combination of the $n+1$ vectors $1, x, x^2, \ldots, x^n$. ◆

**Definition.** Let $S$ be a subset of a vector space $U$. The **span** of $S$, written $Span(S)$, is the collection of all linear combinations of finite subsets of $S$ (common terminology: finite linear combinations of $S$). To reiterate, a finite linear combination of $S$ is a vector in $U$ of the form $\sum_{i=1}^{n} a_i u_i$, where $u_i \in S$ and $a_i \in \mathbb{K}$.

**Example 19.** In $\mathbb{K}^n$, $Span(\{e_1, e_2, \ldots, e_n\}) = \mathbb{K}^n$. In $\mathbb{K}(\mathbb{N})$, $Span(\{e_1, e_2\})$ is the set of all sequences where only the first two terms may be nonzero. ◆

It is easy of see that $Span(S)$ is a subspace of $U$. If $V$ is a subspace of $U$ that contains $S$, then $V$ contains all finite linear combinations of $S$. Thus $Span(S) \subseteq V$, hence the following result.

**Theorem 3.1.3.** *Span(S) is the smallest subspace of U containing S.* ■

**Theorem 3.1.4.** *If $\{V_\alpha\}$ is a collection of subspaces of U, then $\cap_\alpha V_\alpha$ is a subspace of U.* ■

**Theorem 3.1.5.** *Span(S) is the intersection of all the subspaces containing S.* ■

## Exercises

1. Prove theorem 3.1.1.
2. Prove theorem 3.1.2.
3. Prove theorem 3.1.4.
4. Prove theorem 3.1.5.
5. Let $S_1$ and $S_2$ be subsets of a vector space $U$. Prove that $Span(S_1 \cup S_2) = Span(S_1)$ if and only if $S_2 \subseteq Span(S_1)$.

## 3.2 Independent Sets and Bases

This section is focused on the concepts on linear independence and bases. Our approach to studying bases is unified in the sense that we do not treat finite-dimensional and infinite-dimensional spaces separately. We use Zorn's lemma to prove the existence of a basis. A number of important equivalent characterizations of a basis are also discussed, both in the body of the section as well as in the section exercises.

**Definition.** A finite subset $\{u_1, u_2, \ldots, u_n\}$ of a vector space $U$ is **dependent** if there exist scalars $a_1, a_2, \ldots, a_n$, not all zero, such that $\sum_{i=1}^n a_i u_i = 0$.

Terminology. A vector of the form $\sum_{i=1}^n a_i u_i$, where at least one $a_i \neq 0$, is called a **nontrivial linear combination** of $u_1, u_2, \ldots, u_n$. The above definition can be restated as follows: $\{u_1, u_2, \ldots, u_n\}$ is dependent if some nontrivial linear combination of $u_1, u_2, \ldots, u_n$ is zero.

**Theorem 3.2.1.** *A subset $S = \{u_1, u_2, \ldots, u_n\}$ of a vector space U is dependent if and only if one of the vectors in S is a linear combination of the remaining vectors.*

*Proof.* Suppose $\{u_1, u_2, \ldots, u_n\}$ is dependent. Then $\sum_{i=1}^n a_i u_i = 0$ for scalars $a_1, a_2, \ldots, a_n$, not all zero. Say $a_i \neq 0$. Then $u_i = \frac{-1}{a_i} \sum_{j \neq i}^n a_j u_j$. Conversely, if $u_i = \sum_{j \neq i}^n a_j u_j$, then $1.u_i - \sum_{j \neq i}^n a_j u_j = 0$, and $\{u_1, u_2, \ldots, u_n\}$ is dependent. ■

**Definition.** A finite subset $\{u_1, u_2, \ldots, u_n\}$ of a vector space $U$ is **independent** if it is not dependent. Equivalently, $\{u_1, u_2, \ldots, u_n\}$ is independent if a linear combination $\sum_{i=1}^{n} a_i u_i$ is equal to zero if and only if each $a_i = 0$.

**Example 1.** The set $\{e_1, \ldots, e_n\}$ is independent in $\mathbb{K}^n$. ◆

**Example 2.** Any finite subset of $\{e_n : n \in \mathbb{N}\}$ is independent in $\mathbb{K}(\mathbb{N})$. ◆

**Example 3.** Any finite subset of the monomials $1, x, x^2, \ldots$ is independent in $\mathbb{P}$. ◆

**Example 4.** Any finite subset of the canonical vectors $e_\alpha$ in $\mathbb{K}(I)$ is independent. ◆

**Definitions.** An infinite subset $S$ of a vector space $U$ is **independent** if every finite subset of $S$ is independent. An infinite subset $S$ of vectors is dependent if some finite subset of $S$ is **dependent**.

The following follow immediately from the previous set of examples.

**Example 5.** The set of canonical vectors $\{e_n : n \in \mathbb{N}\}$ is independent in $\mathbb{K}(\mathbb{N})$. ◆

**Example 6.** The set of all monomials $\{1, x, x^2, \ldots\}$ is independent in $\mathbb{P}$. ◆

**Example 7.** The set of canonical vectors $\{e_\alpha : \alpha \in I\}$ is independent in $\mathbb{K}(I)$. ◆

**Example 8.** The functions $f_\alpha(x) = e^{\alpha x}, \alpha \in \mathbb{R}$, are independent in $\mathcal{C}^\infty(\mathbb{R})$.

We show that, for any finite set $\{\alpha_1, \ldots, \alpha_n\}$ of distinct real numbers, the functions $e^{\alpha_1 x}, \ldots, e^{\alpha_n x}$ are independent. Suppose that, for constants $c_1, \ldots, c_n$, $\sum_{i=1}^{n} c_i e^{\alpha_i x} = 0$. Repeated differentiation of the above identity ($n - 1$ times) yields $\sum_{i=1}^{n} \alpha_i^j c_i e^{\alpha_i x} = 0, j = 0, \ldots, n - 1$. Evaluating each of the last identities at $x = 0$ yields the system of linear equations

$$\sum_{i=1}^{n} \alpha_i^j c_i = 0, \, j = 0, \ldots, n - 1.$$

The matrix of the system is the famous Vandermonde matrix,

$$J = \begin{pmatrix} 1 & 1 & \cdots & 1 \\ \alpha_1 & \alpha_2 & \cdots & \alpha_n \\ \vdots & & & \vdots \\ \alpha_1^{n-1} & \alpha_2^{n-1} & \cdots & \alpha_n^{n-1} \end{pmatrix}.$$

Since $det(J) = \prod_{1 \leq i < j \leq n}(\alpha_j - \alpha_i) \neq 0$, we must have $c_1 = ... = c_n = 0$, establishing the independence of the set $\{ f_\alpha(x) = e^{\alpha x} : \alpha \in \mathbb{R} \}$. ♦

**Definition.** Let $U$ be a vector space. **A basis** for $U$ is a maximal independent subset of $U$. To rephrase, $S$ is a basis if $S$ is independent and any subset of $U$ properly containing $S$ is dependent. A basis for a vector space is sometimes called a **linear basis** or a **Hamel basis**.

**Theorem 3.2.2.** *Every vector space $U$ has a basis.*

*Proof. Let $\mathfrak{B}$ be the collection of all independent subsets of $U$. Order $\mathfrak{B}$ by set inclusion, and let $\mathfrak{C}$ be a chain in $\mathfrak{B}$. We show that $\cup\{C : C \in \mathfrak{C}\}$ is independent. Let $\{u_1, u_2, ..., u_n\}$ be a subset of $\cup\{C : C \in \mathfrak{C}\}$. Then, for each $1 \leq i \leq n$, there is a member $C_i \in \mathfrak{C}$ such that $u_i \in C_i$. Because $\mathfrak{C}$ is a chain, one set $C_i$ contains all the other sets $C_j, 1 \leq j \leq n$. Therefore $\{u_1, u_2, ..., u_n\}$ is a subset of $C_i$ and hence is independent. Thus $\mathfrak{C}$ has an upper bound, namely, $\cup\{C : C \in \mathfrak{C}\}$. By Zorn's lemma, $\mathfrak{B}$ has a maximal member, that is, a maximal independent subset of $U$, that is, a basis for $U$.* ∎

The corollary below says that an independent subset of a vector space can be augmented to a basis.

**Corollary 3.2.3.** *Let $S_1$ be an independent subset of a vector space $U$. Then there is a basis $S$ for $U$ containing $S_1$.*

*Proof. The proof parallels that of theorem 3.2.2, except that $\mathfrak{B}$ is defined to be the family of all independent subsets of $U$ that contain $S_1$; $\mathfrak{B} \neq \varnothing$ because $S_1 \in \mathfrak{B}$.* ∎

**Theorem 3.2.4.** *Let $S$ be a subset of a vector space $U$. The following are equivalent:*

*(a) $S$ is a basis.*

*(b) $S$ is independent and spans $U$ (meaning that $Span(S) = U$).*

*(c) Every nonzero element of $U$ can be written uniquely as a finite linear combination of vectors in $S$. Specifically, if $u \neq 0$, then there exists a unique subset $\{u_1, ..., u_n\}$ of $S$ and a unique set of nonzero scalars $\{a_1, ..., a_n\}$ such that $u = \sum_{i=1}^{n} a_i u_i$.*

*Proof. (a) implies (b). Since a basis is independent, we only need to show that $Span(S) = U$. Let $u \in U$ and, without loss of generality, assume that $u \notin S$. Then $S_1 = S \cup \{u\}$ is dependent, so a finite subset $S_2$ of $S_1$ is dependent. $S_2$ must contain* *u* because the other elements of $S_2$ are independent. Write $S_2 = \{u, u_1, ..., u_n\}$. Then there are scalars $a, a_1, ..., a_n$, not all zero, such that $au + a_1 u_1 + ... + a_n u_n = 0$; $a \neq 0$ because otherwise, $\{u_1, ..., u_n\}$ would be dependent. Hence $u = \frac{1}{a}(a_1 u_1 + ... + a_n u_n)$. Thus $S$ spans $U$.

*(b) implies (c).* We only need to show the uniqueness of the representation of a nonzero element $u \in U$ as a finite linear combination of $S$. Suppose there are finite subsets $E$ and $F$ of $S$ such that $u$ can be written as a linear combination of the elements of both $E$ and $F$. We will show that $E \neq F$ leads to a contradiction. We adopt the notation $E \cap F = \{u_1, ..., u_r\}$, $E - F = \{u_{r+1}, ..., u_s\}$, and $F - E = \{u_{s+1}, ..., u_n\}$. The assumption is that there are nonzero scalars $a_1, ..., b_1, ...,$ such that

$$u = \sum_{i=1}^{r} a_i u_i + \sum_{i=r+1}^{s} a_i u_i = \sum_{i=1}^{r} b_i u_i + \sum_{i=s+1}^{n} b_i u_i.$$

Rearranging the above equation, we have $\sum_{i=1}^{r}(a_i - b_i)u_i + \sum_{i=r+1}^{s} a_i u_i - \sum_{i=s+1}^{n} b_i u_i = 0$. This would contradict the independence of $E \cup F$ unless $E - F = \varnothing = F - E$. Now $\sum_{i=1}^{r}(a_i - b_i)u_i = 0$, and the independence of $E$ forces $a_i = b_i$ for all $1 \leq i \leq r$.

*(c) implies (a)* First observe that the zero vector is not in $S$ because otherwise the uniqueness of representation of any finite linear combination of $S$ would be violated by adding 1.0 to it. To show the independence of $S$, suppose a linear combination of some finite subset of $S$ is equal to zero, say, $\sum_{i=1}^{n} a_i u_i = 0$. By the previous observation, at least two of the coefficients are nonzero, say, $a_1 \neq 0 \neq a_2$. In this case, $a_1 u_1 = -\sum_{i=2}^{n} a_i u_i$. This contradicts the uniqueness of the representation of $a_1 u_1$ and proves the independence of $S$. To show that $S$ is maximal, let $u \in U - S$. Then $u$ is a finite linear combination of elements $u_1, ..., u_n$ of $S$. This implies that $\{u, u_1, ..., u_n\}$ is dependent, and hence $\{u\} \cup S$ is dependent. This establishes the maximality of $S$. ∎

**Example 9.** The canonical vectors $e_1, ..., e_n$ are independent and span $\mathbb{K}^n$ and therefore form a basis for $\mathbb{K}^n$. ♦

**Example 10.** The set $S = \{1, x, x^2, ...\}$ is independent and spans $\mathbb{P}$ and is therefore a basis for $\mathbb{P}$. Naturally, we call $S$ the canonical basis for $\mathbb{P}$. ♦

**Example 11.** For the same reason, $\{e_n : n \in \mathbb{N}\}$ is a basis for $\mathbb{K}(\mathbb{N})$. ♦

**Theorem 3.2.5.** *If $S$ is a basis for a vector space $U$, then $S$ is **a minimal spanning set** for $U$ in the sense that $Span(S) = U$ and no proper subset of $S$ spans $U$.*

*Proof.* *If S is a basis for U, then, by the previous theorem, S is independent and*
*Span(S) = U. If $S_1$ is a proper subset of S that also spans U, then, again by the*
*previous theorem, $S_1$ would also be a basis for U. This contradicts the maximality*
*of $S_1$ and hence the very definition of a basis, because $S_1$ is a proper subset*
*of S.* ∎

## Exercises

1. Prove that a subset of an independent set is independent.
2. Prove that a set containing a dependent set is dependent.
3. Prove that a minimal spanning subset of a vector space $U$ is a basis for $U$.
   This is the converse of theorem 3.2.5.
4. Prove that every spanning subset of a vector space $U$ contains a basis for $U$.
5. Find a basis for $\mathbb{K}_{m \times n}$.

## 3.3 The Dimension of a Vector Space

In this section, we discuss the definition of dimension and prove the invariance
of the cardinality of the basis. Some results on cardinal arithmetic are needed in
the infinite-dimensional case. We also prove the existence of a vector space of any
given dimension.

**Definition.** A vector space $U$ is said to be **finite dimensional** if it contains a finite
basis.

**Example 1.** $\mathbb{K}^n$ and $\mathbb{P}_n$ are finite dimensional. ♦

**Lemma 3.3.1.** *Consider the following system of linear equations with coefficients*
*in $\mathbb{K}$:*

$$a_{11}x_1 + a_{12}x_2 + \ldots + a_{1m}x_m = 0,$$
$$a_{21}x_1 + a_{22}x_2 + \ldots + a_{2m}x_m = 0,$$
$$\vdots$$
$$\vdots$$
$$a_{n1}x_1 + a_{n2}x_2 + \ldots + a_{nm}x_n = 0.$$

*If $m > n$, then the system has a nontrivial (i.e., nonzero) solution*
*$(x_1, \ldots, x_m) \in \mathbb{K}^m$.*

*Proof.* *Without loss of generality, assume that $m = n + 1$, because we can augment the system by adding $m - n - 1$ equations with zero coefficients to the system.*

*Since at least one of the coefficients is different from zero, we may assume, by reordering the equations and renumbering the variables, that $a_{11} \neq 0$. We prove the theorem by induction on $n$. Subtracting $\frac{a_{i,1}}{a_{11}}$ times the top equation from equation $i, 2 \leq i \leq n$ yields the equivalent system*

$$a_{11}x_1 + a_{12}x_2 + \ldots + a_{1,n+1}x_{n+1} = 0,$$
$$b_{22}x_2 + \ldots + b_{2,n+1}x_{n+1} = 0,$$
$$\vdots$$
$$\vdots$$
$$b_{n2}x_2 + \ldots + b_{n,n+1}x_{n+1} = 0,$$

*where $b_{ij} = a_{ij} - a_{i1}a_{1j}/a_{11}, 2 \leq i \leq n, 2 \leq j \leq n + 1$. The bottom $n - 1$ equations of the above system have a nontrivial solution $(x_2, \ldots, x_{n+1})$, by the inductive hypothesis. Defining $x_1 = \frac{-1}{a_{11}} \sum_{j=2}^{n+1} a_{1j}x_j$ yields a nontrivial solution $(x_1, \ldots, x_{n+1})$ of the original system.* ∎

**Lemma 3.3.2.** *If a finite dimensional space $U$ has a basis $S = \{u_1, \ldots, u_n\}$ of $n$ vectors, then any subset of $U$ containing more than $n$ elements is dependent.*

*Proof.* *Let $\{v_1, \ldots, v_m\}$ be a subset of $U$ with $m > n$. Each $v_j$ is a linear combination of $S$, say, $v_j = \sum_{i=1}^{n} a_{ij}u_i, (1 \leq j \leq m)$. By the previous theorem, there exists a nontrivial solution $(x_1, \ldots, x_m)$ of the system $\sum_{j=1}^{m} a_{ij}x_j = 0, i = 1, 2, \ldots, n$. Now*
$$\sum_{j=1}^{m} x_j v_j = \sum_{j=1}^{m} x_j \sum_{i=1}^{n} a_{ij}u_i = \sum_{i=1}^{n} (\sum_{j=1}^{m} a_{ij}x_j)u_i = \sum_{i=1}^{n} 0.u_i = 0.$$
*Thus $\{v_1, \ldots, v_m\}$ are dependent.* ∎

We now prove **the invariance of the number of vectors in a basis** for a finite-dimensional space.

**Theorem 3.3.3.** *If $S = \{u_1, \ldots, u_n\}$ and $T = \{v_1, \ldots, v_m\}$ are bases for a finite-dimensional vector space $U$, then $n = m$.*

*Proof.* *Since $S$ is independent and $T$ is a basis, $n \leq m$, by the previous lemma. For the same reason, $m \leq n$.* ∎

**Definition.** The **dimension of a finite-dimensional vector space** $U$ is the number of elements in a basis for $U$. This number is independent of the basis by the previous theorem.

**Example 2.** The dimension of $\mathbb{K}^n$ is $n$. The dimension of $\mathbb{P}_n$ is $n + 1$. The dimension of $\mathbb{K}_{m \times n}$ is $mn$. ◆

**Definition.** A vector space $U$ is said to be **infinite dimensional** if it is not finite dimensional. Thus $U$ is infinite dimensional if every basis for $U$ is infinite.

As in the finite-dimensional case, the cardinality of a basis for an infinite-dimensional space is an invariant of the space, as the following theorem shows.

**Theorem 3.3.4.** *Let $\{u_\alpha\}_{\alpha \in I}$ and $\{v_\beta\}_{\beta \in J}$ be bases for an infinite-dimensional space $U$. Then $Card(I) = Card(J)$.*

*Proof. For each $\beta \in J$, there is a finite subset $I_\beta \subseteq I$ such that $v_\beta$ is a linear combination of the finite set $\{u_\alpha : \alpha \in I_\beta\}$. Therefore*

$$U = Span(\{v_\beta : \beta \in J\}) \subseteq Span(\{u_\alpha : \alpha \in \cup_{\beta \in J} I_\beta\}) \subseteq U.$$

*Since no proper subset of $\{u_\alpha\}_{\alpha \in I}$ spans $U$ (theorem 3.2.5), $I = \cup_{\beta \in J} I_\beta$. Using theorems 2.3.11, and 2.3.10 (also see problem 8 on section 2.3),*

$$Card(I) = Card(\cup_{\beta \in J} I_\beta) \leq \sum_{\beta \in J} Card(I_\beta) \leq \sum_{\beta \in J} \aleph_0 = \aleph_0 Card(J) = Card(J).$$

*Likewise, $Card(J) \leq Card(I)$, and the proof is complete.* ∎

Now that we proved the **invariance of the cardinality of a basis in an infinite-dimensional space**, we define the **dimension** of such a space to be the cardinality of any basis for the space.

**Notation.** We use the notation $dim_\mathbb{K}(U)$ to denote the dimension of a vector space $U$ over the field $\mathbb{K}$. If the base field is understood, we simply write $dim(U)$.

**Example 3.** $dim(\mathbb{K}(\mathbb{N})) = \aleph_0 = dim(\mathbb{P})$. ◆

**Example 4.** In example 8 in section 3.2, we proved that the set of functions $\{f_\alpha(x) = e^{\alpha x} : \alpha \in \mathbb{R}\}$ is independent in $\mathcal{C}^\infty(\mathbb{R})$. This shows that $dim(\mathcal{C}^\infty(\mathbb{R})) \geq \mathfrak{c}$. ◆

We now show the existence of a vector space of any given dimension. The essential uniqueness of such a space will be discussed in section 3.4.

**Theorem 3.3.5.** *Let $\aleph$ be a cardinal number. Then there is a vector space of dimension $\aleph$.*

*Proof. If $\aleph$ is a finite cardinal, $n$, then $\mathbb{K}^n$ has dimension $n$. So assume that $\aleph$ is infinite, and let $I$ be a set such that $Card(I) = \aleph$. We show that the space $U = \mathbb{K}(I)$ discussed in section 3.1 has dimension $\aleph$ by finding a basis for $U$ which is in one-to-one correspondence with $I$. Let $S = \{e_\alpha\}_{\alpha \in I}$ be the set of canonical vectors in $\mathbb{K}(I)$; $S$ is clearly in one-to-one correspondence with $I$. We show that $S$ is a basis for $U$. Let $\{e_{\alpha_1}, \ldots, e_{\alpha_n}\}$ be a finite subset of $S$, and suppose that, for some scalars $a_1, \ldots, a_n$, $f = \sum_{k=1}^{n} a_k e_{\alpha_k} = 0$ (the zero function from $I \to \mathbb{K}$). For a fixed $1 \le j \le n, 0 = f(\alpha_j) = \sum_{k=1}^{n} a_k e_{\alpha_k}(\alpha_j) = a_j$. This shows the independence of $S$. Next we show that $S$ spans $U$. Let $f \in U$ and let $a_1 = f(\alpha_1), \ldots, a_n = f(\alpha_n)$ be all the nonzero values of $f$. Clearly, $f = \sum_{k=1}^{n} a_k e_{\alpha_k}$.* ∎

## Exercises

1. In this problem, the base filed is $\mathbb{R}$. Let $V_1$ be the set of real symmetric $n \times n$ matrices, and let $V_2$ be the set of **skew-symmetric matrices**. Show that $V_1$ and $V_2$ are subspaces of $\mathbb{R}_{n \times n}$, and find their dimensions. An $n \times n$ matrix $A$ is skew-symmetric if, for all $1 \le i, j \le n, a_{ij} = -a_{ji}$.
2. Let $V$ be a subspace of $U$. Show that $dim(V) \le dim(U)$.
3. Let $U$ be an $n$-dimensional vector space, and let $S$ be a subset of $U$ of exactly $n$ elements. Prove that the following are equivalent:
   (a) $S$ is a basis for $U$.
   (b) $S$ is independent.
   (c) $S$ spans $U$.
4. Let $V$ be a subspace of $U$. Show that if $V$ contains a basis for $U$, then $V = U$.
5. Show that a vector space $U$ is infinite dimensional if and only if it contains an infinite independent subset.
6. Let $U$ be an infinite-dimensional vector space. Show that there is a sequence $V_1 \supset V_2 \supset \ldots$ (proper containments) of subspaces of $U$ such that $dim(V_n) = dim(U)$ for all $n$.
7. Let $\{x_0, \ldots, x_n\}$ be a set of distinct real numbers. For $0 \le i \le n$, define the following set of polynomials in $\mathbb{P}_n$:

$$L_0(x) = \frac{(x - x_1)(x - x_2)\ldots(x - x_n)}{(x_0 - x_1)(x_0 - x_2)\ldots(x_0 - x_n)},$$

$$L_1(x) = \frac{(x - x_0)(x - x_2)\ldots(x - x_n)}{(x_1 - x_0)(x_1 - x_2)\ldots(x_1 - x_n)}, \ldots, \text{ and}$$

$$L_n(x) = \frac{(x - x_0)(x - x_1)\ldots(x - x_{n-1})}{(x_n - x_0)(x_n - x_1)\ldots(x_n - x_{n-1})}.$$

Show that the set $\{L_i\}_{i=0}^n$ is a basis for $\mathbb{P}_n$. Hint: For a polynomial $f \in \mathbb{P}_n$, show that $f = \sum_{i=0}^n f(x_i)L_i(x)$. Observe that $L_i(x_j) = \delta_{ij}$.

8. Let $I = [a, b]$ be a closed, bounded interval, and suppose $a = t_1 < t_2 < ... < t_n = b$ is a fixed set of points (also called nodes) in $I$. Define $V$ to be the set of continuous functions on $[a, b]$ whose restrictions to the subintervals $[t_i, t_{i+1}]$ are linear. Prove that $V$ is a vector space, and find a basis for it. A function in the space $V$ is known as a **continuous, piecewise linear function** with nodes $\{t_1, ..., t_n\}$.

9. **The space of continuous, piecewise linear functions.** Let $U$ be the collection of all continuous, piecewise linear functions on $[a, b]$. Prove that $U$ is an infinite-dimensional vector space.

10. Show that $\mathbb{R}$ is an infinite-dimensional vector space over $\mathbb{Q}$.

11. Let $M$ be a field, let $L$ be a subfield of $M$, and let $K$ be a subfield of $L$. We can consider $L$ as a vector space over $K$, and $M$ as a vector space over either $L$ or $K$. Prove that if $dim_L(M)$ and $dim_K(L)$ are finite, then $dim_K(M) = dim_L(M).dim_K(L)$.

## 3.4  Linear Mappings, Quotient Spaces, and Direct Sums

A proper understanding of this section is essential for a smooth transition to the rest of the book. While the early results in the section are elementary, a number of important concepts make their first debut later in the section. Specifically, this includes quotient spaces and quotient maps, direct sums, projections and algebraic complements, linear functionals and linear operators, maximal subspaces and the co-dimension of a subspace and, finally, the definition of an algebra over a field.

**Definition.** Let $U$ and $V$ be vector spaces over $\mathbb{K}$. A mapping $T : U \to V$ is said to be **linear** if, for all $u, v \in U$, and all $a \in \mathbb{K}$,

$$T(u + v) = T(u) + T(v), \text{ and } T(au) = aT(u).$$

The following are examples of linear mappings.

**Example 1.** Define $T : \mathbb{P} \to \mathbb{P}$ by $T(f) = f'$ (the derivative of $f$). ◆

**Example 2.** Let $A$ be an $m \times n$ matrix with entries in $\mathbb{K}$. The linear mapping $T : \mathbb{K}^n \to \mathbb{K}^m$, defined by $T(x) = Ax$, is known as the **mapping induced by the matrix** $A$. It is easy to check that every linear transformation $T : \mathbb{K}^n \to \mathbb{K}^m$ is induced by the $m \times n$ matrix $A$ whose columns are $T(e_1), ..., T(e_n)$. Here $\{e_1, ..., e_n\}$ is the canonical basis for $\mathbb{K}^n$. The matrix $A$ is called the **standard matrix of** $T$. ◆

**Theorem 3.4.1.** *If $T : U \rightarrow V$ is linear, then*

    *(a)* $T(0) = 0$;
    *(b)* $T(-u) = -T(u)$;
    *(c)* *if $a_1, \ldots, a_n \in \mathbb{K}$ and $u_1, \ldots, u_n \in U$, then $T(\sum_{i=1}^{n} a_i u_i) = \sum_{i=1}^{n} a_i T(u_i)$;*
    *(d)* *the image under $T$ of a subspace of $U$ is a subspace of $V$; and*
    *(e)* *the inverse image under $T$ of a subspace of $V$ is a subspace of $U$.* ∎

**Definition.** Let $T : U \rightarrow V$ be linear. The **kernel** (or **null-space**) of $T$, written $Ker(T)$ or $\mathcal{N}(T)$, is $T^{-1}(0)$. The **range** of $T$ is defined by $\mathcal{R}(T) = \{T(u) : u \in U\}$.

**Theorem 3.4.2.** *Let $T : U \rightarrow V$ be linear. Then*

    *(a)* $\mathcal{N}(T)$ *is a subspace of $U$,*
    *(b)* $\mathcal{R}(T)$ *is a subspace of $V$, and*
    *(c)* *$T$ is one-to-one if and only if $\mathcal{N}(T) = \{0\}$.* ∎

**Example 3.** Let $T : \mathbb{P} \rightarrow \mathbb{P}$ be defined by $T(f) = \int_0^x f(t)dt$. It is easy to verify directly that $\mathcal{N}(T) = \{0\}$ and that $T$ is one-to-one. ◆

**Definition.** Let $T : U \rightarrow V$ be linear. The **rank** of $T$ is the dimension of $\mathcal{R}(T)$ and the **nullity** of $T$ is the dimension of $\mathcal{N}(T)$. The rank and nullity of a linear mapping are particularly useful when they are finite.

**Theorem 3.4.3.** *Let $U$ be a vector space of dimension $n < \infty$, and let $T$ be a linear transformation from $U$ to a vector space $V$.*
    *Then $dim(Ker(T)) + dim(\mathcal{R}(T)) = n$. In other words,*

$$rank(T) + nullity(T) = n.$$

*Proof.* Let $S_1 = \{u_1, \ldots, u_r\}$ be a basis for $Ker(T)$. Augment $S_1$ to a basis $\{u_1, \ldots, u_n\}$ for $U$. We show that $rank(T) = n - r$ by showing that $\{T(u_{r+1}), \ldots, T(u_n)\}$ is a basis for $\mathcal{R}(T)$. Every element $y$ in $\mathcal{R}(T)$ has the form $T(x)$, where $x \in U$. Write $x = \sum_{i=1}^{n} a_i u_i$, then $y = T(\sum_{i=1}^{n} a_i u_i) = \sum_{i=r+1}^{n} a_i T(u_i)$. This shows that $T(u_{r+1}), \ldots, T(u_n)$ span $\mathcal{R}(T)$. Suppose, for some scalars $b_{r+1}, \ldots, b_n$, $\sum_{i=r+1}^{n} b_i T(u_i) = 0$. Then $T(\sum_{i=r+1}^{n} b_i u_i) = 0$, and $\sum_{i=r+1}^{n} b_i u_i \in Ker(T)$, so $\sum_{i=r+1}^{n} b_i u_i = \sum_{i=1}^{r} a_i u_i$ for some scalars $a_1, \ldots, a_r$. This would contradict the independence of $\{u_1, \ldots, u_n\}$, unless $a_1, \ldots, a_r$, and $b_{r+1}, \ldots, b_n$ are all zero. This shows the independence of $T(u_{r+1}), \ldots, T(u_n)$ and concludes the proof. ∎

**Example 4.** Let $T : \mathbb{P}_n \rightarrow \mathbb{P}_n$ be defined by $T(f) = f'$. Clearly, $\mathcal{N}(T)$ consists of all constant functions. Now $\mathcal{R}(T) = Span(\{1, x, \ldots, x^{n-1}\})$ because if

$g = \sum_{j=0}^{n-1} a_j x^j$, then $T(f) = g$, where $f = \sum_{j=0}^{n-1} \frac{a_j x^{j+1}}{j+1}$. Observe that $rank(T) = n$, and $nullity(T) = 1$, consistent with theorem 3.4.3. ◆

**Theorem 3.4.4.** *Let $S = \{u_\alpha\}_{\alpha \in I}$ be a basis for a vector space $U$, and let $\{v_\alpha\}_{\alpha \in I}$ be an arbitrary subset of a vector space $V$. Then there exists a unique linear mapping $T : U \to V$ such that, for every $\alpha \in I$, $T(u_\alpha) = v_\alpha$.*

*Proof.* *Every vector $x \in U$ has a unique representation as $x = \sum_{\alpha \in F} a_\alpha u_\alpha$ for some finite subset $F \subseteq I$. Define $T(x) = \sum_{\alpha \in F} a_\alpha v_\alpha$; $T$ is clearly linear (the interested reader is encouraged to formulate the notation needed to write out the details). To show that $T$ is unique, suppose $S : U \to V$ is another linear mapping such that $S(u_\alpha) = v_\alpha$. Let $x = \sum_{\alpha \in F} a_\alpha u_\alpha \in U$. Then*

$$S(x) = S\left(\sum_{\alpha \in F} a_\alpha u_\alpha\right) = \sum_{\alpha \in F} a_\alpha S(u_\alpha)$$

$$= \sum_{\alpha \in F} a_\alpha v_\alpha = \sum_{\alpha \in F} a_\alpha T(u_\alpha) = T\left(\sum_{\alpha \in F} a_\alpha u_\alpha\right) = T(x). \blacksquare$$

The above theorem says that a linear mapping is completely (and uniquely) determined by its values on a basis. Stated differently, an arbitrary function on a basis for $U$ can be uniquely extended to a linear function on $U$.

**Example 5.** Let $S = \{1, x, x^2, ...\}$ be the canonical basis for $\mathbb{P}$, and define $T : S \to \mathbb{P}$ by $T(1) = 0, T(x) = 0$, and, for $n \geq 2$, $T(x^n) = n(n-1)x^{n-2}$. It is clear that the unique linear mapping on $\mathbb{P}$ that extends $T$ is $T(f) = f''$ (the second derivative of $f$.) ◆

**Definition.** A linear mapping $T : U \to V$ is an **isomorphism** if it is a bijection. In this case, we say that $U$ and $V$ are **isomorphic**. Isomorphic spaces may have different underlying sets and different operations, but, from the algebraic point of view, they are essentially identical.

**Example 6.** $\mathbb{P}_n$ is isomorphic to $\mathbb{K}^{n+1}$ because the linear mapping $T(\sum_{i=0}^{n} a_i x^i) = (a_0, a_1, ..., a_n)$ is an isomorphism. ◆

**Example 7.** The space $\mathbb{P}$ of all polynomials is isomorphic to $\mathbb{K}(\mathbb{N})$.
Let $f = \sum_{i=0}^{n} a_i x^i \in \mathbb{P}$. The following linear mapping is an isomorphism: $T(f) = y$, where $y$ is the sequence $(y_0, y_1, y_2, ...)$ such that

$$y_i = \begin{cases} a_i & \text{if } 0 \leq i \leq n, \\ 0 & \text{if } i > n. \end{cases} ◆$$

In theorem 3.3.5, we established the existence of a vector space of any given dimension. We now show that such a space is unique up to an isomorphism.

**Theorem 3.4.5.** *An n-dimensional vector space U is isomorphic to $\mathbb{K}^n$.*

*Proof.* Let $\{u_1, \ldots, u_n\}$ be a basis for $U$, and define $T : U \to \mathbb{K}^n$ to be the unique linear mapping that extends $T(u_i) = e_i, 1 \leq i \leq n$; $T$ is clearly one-to-one, and it is onto because its range contains the canonical basis for $\mathbb{K}^n$. ∎

**Theorem 3.4.6.** *Let U be a vector space of infinite dimension $\aleph$, and let I be a set such that Card$(I) = \aleph$. Then U is isomorphic to $\mathbb{K}(I)$.*

*Proof.* Let $\{u_\alpha\}_{\alpha \in I}$ be a basis for $U$, and let $\{e_\alpha\}$ be the canonical basis for $\mathbb{K}(I)$. If $\sum_{\alpha \in F} a_\alpha u_\alpha$ is the unique representation of an element $x \in U$ as a finite linear combination of the basis elements, define $T : U \to \mathbb{K}(I)$ by $T(x) = \sum_{\alpha \in F} a_\alpha e_\alpha$. The proof that $T$ is an isomorphism is much like the proof of the previous theorem. ∎

## Quotient Spaces

Let $V$ be a subspace of a vector space $U$. Define a relation $R$ on $U$ by $xRy$ if $x - y \in V$. It is easy to verify that $R$ is an equivalence relation. The equivalence classes of $R$ are subsets of $U$ of the form $x + V = \{x + v : v \in V\}$. Such a set is called a **coset** of $V$.

For example, let $U = \mathbb{R}^2$ and let $V$ be a one-dimensional subspace of $U$. Then $V$ is a straight line containing the origin, and the cosets of $V$ are lines parallel to $V$. ◆

**Definition.** The **quotient space** $U/V$ (read $U$ modulo $V$) consists of the cosets of $V$, endowed with a vector space structure by the operations

$$(x + V) + (y + V) = (x + y) + V$$
$$\text{and}$$
$$a(x + V) = (ax) + V.$$

The above operations are well defined in the sense that they do not depend on the particular element $x$ chosen to represent the coset $x + V$. For example, if $x' + V = x + V$ and $y' + V = y + V$, then $x' - x \in V$, $y' - y \in V$, and $(x' + y') - (x + y) \in V$; hence $(x' + y') + V = (x + y) + V$. For brevity of notation, the coset $x + V$ will be denoted by $\bar{x}$.

**Definition.** Let $V$ be a subspace of a vector space $U$. The function $\pi : U \to U/V$, defined by $\pi(x) = \bar{x}$, is called the **quotient map**. It is easy to verify that $\pi$ is linear.

**Theorem 3.4.7.** *Let $T : U \to W$ be linear, and let $V = Ker(T)$. Then $U/V$ is isomorphic to $\mathfrak{R}(T)$ via the isomorphism $\overline{T}(\overline{x}) = T(x)$.*

*Proof.* We leave it to the reader to verify that $\overline{T}$ is well defined. Clearly, $\overline{T}$ is onto. We verify the linearity of $\overline{T}$:

$$\overline{T}(a\overline{x} + b\overline{y}) = \overline{T}(\overline{ax + by}) = T(ax + by) = aT(x) + bT(y) = a\overline{T}(\overline{x}) + b\overline{T}(\overline{y}).$$

To show that $\overline{T}$ is one-to-one, suppose $\overline{T}(\overline{x}) = 0$. Therefore $T(x) = 0$, and $x \in Ker(T) = V$; hence $\overline{x} = 0$. ∎

## Direct Sums

**Definition.** Let $U_1$ and $U_2$ be subspaces of a vector space $U$. The **sum** of $U_1$ and $U_2$ is the set $U_1 + U_2 = \{x + y : x \in U_1, y \in U_2\}$. It is clear that $U_1 + U_2$ is a subspace of $U$.

**Example 8.** Let $U = \mathbb{R}^3$, and let $U_1$ and $U_2$ be distinct lines containing the origin. Then the subspace $U_1 + U_2$ is the plane that contains $U_1$ and $U_2$. ◆

**Theorem 3.4.8.** $U_1 + U_2 = Span(U_1 \cup U_2)$. ∎

**Definition.** A vector space $U$ is the **direct sum** of two subspaces $U_1$ and $U_2$ if $U_1 + U_2 = U$, and $U_1 \cap U_2 = \{0\}$. In this case, we write $U = U_1 \oplus U_2$ and say that $U_2$ is an **algebraic complement** of $U_1$ in $U$.

**Example 9.** $\mathbb{R}^2 = \{(x_1, 0) : x_1 \in \mathbb{R}\} \oplus \{(0, x_2) : x_2 \in \mathbb{R}\}$. ◆

**Theorem 3.4.9.** *Let $U_1$ and $U_2$ be subspaces of a vector space $U$. Then $U = U_1 \oplus U_2$ if and only if every vector $u \in U$ can be written uniquely as $u = u_1 + u_2$, where $u_1 \in U_1, u_2 \in U_2$.*

*Proof.* Such a representation of $u$ is guaranteed by the definition of a direct sum. To prove the uniqueness, suppose $u_1 + u_2 = v_1 + v_2$, where $u_1, v_1 \in U_1$ and $u_2, v_2 \in U_2$. Then $u_1 - v_1 = v_2 - u_2 \in U_1 \cap U_2 = \{0\}$, so $u_1 - v_1 = v_2 - u_2 = 0$ and $u_1 = v_1$ and $u_2 = v_2$. The converse is straightforward. ∎

**Example 10.** Let $c$ be the space of all convergent sequences, and let $c_0$ be the space of all sequences that converge to 0. We show that $c = c_0 \oplus Span(\{e\})$, where $e = (1, 1, 1, ...)$. Let $x = (x_1, x_2, ...)$ be a convergent sequence, and let

$\xi = \lim_n x_n$. Define $y = (x_1 - \xi, x_2 - \xi, x_3 - \xi, \ldots)$, and let $z = \xi e = (\xi, \xi, \xi, \ldots)$. Clearly, $y$ converges to 0 (i.e., $y \in c_0$), and $x = y + z$. The representation of $x = y + z$ is unique because the only constant sequence that converges to 0 is the zero sequence. ◆

**Theorem 3.4.10.** *Every subspace $U_1$ of a vector space $U$ has a complement in $U$.*

*Proof.* We need to show that there is a subspace $U_2$ of $U$ such that $U = U_1 \oplus U_2$. Let $S_1$ be a basis for $U_1$. Augment $S_1$ to a basis $S$ of $U$, and let $S_2 = S - S_1$. If $U_2 = Span(S_2)$, then $U = U_1 \oplus U_2$. We leave it to the reader to write out the details. ∎

**Definition.** Let $U = U_1 \oplus U_2$. The **projection** $\pi_1 : U \to U_1$ is the linear mapping $\pi_1(u) = u_1$, where $u = u_1 + u_2$, and is the unique representation of $u$ provided by theorem 3.4.9. The projection $\pi_2$ onto $U_2$ is defined similarly. Some of the properties of projections are explored in the section exercises.

**Example 11.** $\mathbb{R}^2 = \{(x_1, 0) : x_1 \in \mathbb{R}\} \oplus \{(0, x_2) : x_2 \in \mathbb{R}\}$. The projection $\pi_1$ projects $\mathbb{R}^2$ onto the $x_1$-axis in the sense of elementary geometry. ◆

**Theorem 3.4.11.** *Let $U = U_1 \oplus U_2$. Then $Ker(\pi_1) = U_2$, and $U/U_2$ is isomorphic to $U_1$.*

*Proof.* To verify that $Ker(\pi_1) = U_2$, let $x = u_1 + u_2$, where $u_1 \in U_1, u_2 \in U_2$. $\pi_1(x) = 0$ if and only if $u_1 = 0$, if and only if $x = u_2 \in U_2$. The fact that $U/U_2$ is isomorphic to $U_1$ follows from theorem 3.4.7. ∎

## Linear Functionals and Operators

A particularly important set of linear transformations is that from a vector space $U$ to the base field $\mathbb{K}$.

**Definition.** A linear mapping from a vector space $U$ to the base field $\mathbb{K}$ is called a **linear functional** on $U$.

The following are examples of linear functionals.

**Example 12.** Define $\lambda : \mathbb{P} \to \mathbb{R}$ by $\lambda(f) = \int_0^1 f(x)dx$. (The base field is $\mathbb{R}$ and the polynomials have real coefficients.) ◆

**Example 13.** Define $\lambda : \mathbb{K}_{n \times n} \to \mathbb{K}$ by $\lambda(A) = \sum_{i=1}^{n} a_{ii}$. Here $A = (a_{ij})$ is an $n \times n$ matrix. The quantity $\sum_{i=1}^{n} a_{ii}$ is called the trace of $A$, often written $tr(A)$. ◆

**Theorem 3.4.12.** *Let M be a subspace of a vector space U. The following are equivalent:*

*(a) $M = Ker(\lambda)$ for some nonzero linear functional $\lambda$ on U.*
*(b) M has a one-dimensional complement.*

*Proof.* (a) implies (b). Let $x \in U$ be such that $\lambda(x) \neq 0$. By replacing $x$ with $x/\lambda(x)$, we may assume that $\lambda(x) = 1$. For $y \in U$, let $w = y - \lambda(y)x$. Then $\lambda(w) = \lambda(y) - \lambda(\lambda(y)x) = \lambda(y) - \lambda(y)\lambda(x) = 0$. This shows that $w \in Ker(\lambda) = M$; hence $y = w + \lambda(y)x \in M + Span(\{x\})$, and $U = M + Span(\{x\})$. Next we show that $M \cap Span(\{x\}) = \{0\}$. This will complete the proof. If $y \in M \cap Span(\{x\})$, then $y = ax$ for some $a \in \mathbb{K}$, and $\lambda(y) = 0$. But $\lambda(y) = a\lambda(x) = a$. Thus $a = 0$, and $y = 0$.

Conversely, suppose that $U = M \oplus Span(\{x\})$ for some nonzero $x \in U$. Let $S_1$ be a basis for $M$, and let $S = S_1 \cup \{x\}$. Then $S$ is a basis for $U$. Define $\lambda : S \to \mathbb{K}$ by $\lambda(x) = 1$, and $\lambda(u) = 0$ for all $u \in S_1$. Finally, extend $\lambda$ to a linear functional, which we also denote by $\lambda$, on $U$ according to theorem 3.4.4. The reader can easily verify that $Ker(\lambda) = M$. ∎

**Example 14.** Refer to example 12. Let $M = Ker(\lambda)$. The following facts are easy to verify: A basis for $M$ is $\{x^n - \frac{1}{n+1} : n \in \mathbb{N}\}$, and the one-dimensional subspace $N$ of constant polynomials is a complement of $M$. Every polynomial $f$ can be written as $f = g + c$, where $c = \int_0^1 f(t)dt$, and $g = f - c$. ◆

**Definition.** A proper subspace $M$ of a vector space $U$ is said to be a **maximal subspace** if it is not properly contained in any other proper subspace of $U$.

**Theorem 3.4.13.** *For a subspace M of a vector space U, each of the following is equivalent to each of the conditions (a) and (b) of the previous theorem:*

*(a) M is a maximal subspace of U.*
*(b) U/M has dimension 1.* ∎

**Definition.** If $dim(U/M) = 1$, $M$ is said to have co-dimension 1. More generally, if $V$ is a subspace of $U$, then $dim(U/V)$ is called the **co-dimension of** $V$ **in** $U$. The concept is particularly useful when $dim(U/V) < \infty$.

Another important vector space is the space of all linear transformations from one vector space $U$ to another space $V$.

**Notation.** Let $U$ and $V$ be vector spaces. The set of all linear transformations from $U$ to $V$ is denoted by $Hom(U, V)$. A linear mapping is also called a **homomorphism**, hence the notation $Hom(U, V)$.

It is easy to see that $Hom(U,V)$ is a vector space with the following operations: for $T_1, T_2 \in Hom(U,V)$ and $a \in \mathbb{K}, (T_1 + T_2)(u) = T_1(u) + T_2(u)$, and $(aT_1)(u) = aT_1(u)$.

An element of $Hom(U,U)$ is often called a **linear operator** on $U$. $Hom(U,U)$ has additional structure provided by the composition of linear operators; if $S, T \in Hom(U,U), (SoT)(u) = S(T(u))$. When there is no danger of ambiguity, we write $ST$ for $SoT$. The composition of linear operators satisfies a number of properties including, for example, $S(T_1 + T_2) = ST_1 + ST_2$.

**Definition.** A vector space $U$ over a field $\mathbb{K}$ is an **algebra** over $\mathbb{K}$, if $U$ possesses another binary operation, to be called multiplication, such that, for all $u, v, w \in U$, and all $a \in \mathbb{K}$, the following conditions are met:

(a) $u(vw) = (uv)w$
(b) $u(v + w) = uv + uw$, and $(u + v)w = uw + vw$
(c) $a(uv) = u(av) = (au)v$

The multiplication operation in an algebra is not necessarily commutative, and an algebra need not contain a multiplicative identity element, although many important algebras do. The simplest example of an algebra is the space of square matrices $\mathbb{K}_{n\times n}$, where the binary operations are addition and multiplication of matrices.

**Theorem 3.4.14.** *Suppose $U$ and $V$ are vector spaces over a field $\mathbb{K}$. Then $Hom(U,V)$ is a vector space, and $Hom(U,U)$ is an algebra over $\mathbb{K}$.* ∎

## Exercises

1. Prove theorem 3.4.1.
2. Prove theorem 3.4.2.
3. Let $T : U \to V$ be linear, and let $S_1$ be a basis for $Ker(T)$. Augment $S_1$ to a basis $S$ of $U$, and let $S_2 = S - S_1$. Prove that $T(S_2)$ is a basis for $\mathfrak{R}(T)$. This result is a generalization of theorem 3.4.3 when $dim(U) = \infty$.
4. Show that if there exists a linear mapping that maps $U$ onto $V$, then $dim(V) \leq dim(U)$.
5. Show that if there exists a one-to-one linear mapping from $U$ to $V$, then $dim(U) \leq dim(V)$.
6. Let $\{x_0, \ldots, x_n\}$ be a set of distinct real numbers. Show that the mapping $T : \mathbb{P}_n \to \mathbb{K}^{n+1}$ given by $T(f) = (f(x_0), \ldots, f(x_n))$ is an isomorphism.

7. Prove theorem 3.4.8.
8. Give an example to show that the algebraic complement of a subspace is not unique.
9. Prove that if $U = U_1 \oplus U_2$, then $dim(U) = dim(U_1) + dim(U_2)$.
10. In this problem, the base field is $\mathbb{R}$. Let $U_1$ be the subspace of $\mathbb{R}_{n\times n}$ of real symmetric $n \times n$ matrices, and let $U_2$ be the subspace of skew-symmetric matrices. Show that $\mathbb{R}_{n\times n} = U_1 \oplus U_2$.
11. Let $U = U_1 \oplus U_2$, and, for $i = 1,2$ let $T_i$ be a linear mapping from $U_i$ to a vector space $W$. Prove that there exists a unique linear mapping $T : U \to W$ such that $T|_{U_i} = T_i$.

**Definition.** Let $T : U \to U$ be linear. A subspace $V$ of U is said to be **T-invariant** if $T(V) \subseteq V$.

12. Prove that if $V$ is a $T$-invariant subspace of $U$, then the mapping $\overline{T} : U/V \to U/V$, defined by $\overline{T}(x + V) = T(x) + V$, is linear. Part of the exercise is to show that $\overline{T}$ is well defined. If $\pi : U \to U/V$ is the quotient map, show that $\pi o T = \overline{T} o \pi$.
13. Let $U_1$ and $U_2$ be subspaces of a vector space $U$ such that $U = U_1 \oplus U_2$, and let $\pi : U \to U_1$ be the projection of $U$ onto $U_1$. Prove that $\pi^2 = \pi$, that $U_1 = \{x \in U : \pi(x) = x\} = \mathfrak{R}(\pi)$, and that $U_2 = Ker(\pi)$. By definition, $\pi^2(x) = \pi(\pi(x))$.

**Definition.** Definition. A linear operator $T$ on a vector space $U$ is said to be **idempotent** if $T^2 = T$. The problem above says that the projection $\pi$ of $U$ onto $U_1$ is idempotent.

14. Let $\pi$ be a linear idempotent operator on a vector space $U$. Prove that $U = U_1 \oplus U_2$, where $U_1 = \{x \in U : \pi(x) = x\}$, and $U_2 = Ker(\pi)$.
15. Prove theorem 3.4.13.
16. Exhibit a basis for the null-space of the functional in example 13.
17. Prove theorem 3.4.14.
18. Let $U$ be an $n$-dimensional vector space, and let $U^* = Hom(U, \mathbb{K})$. Suppose $\{u_1, \ldots, u_n\}$ is a basis for $U$, and, for each $1 \leq i \leq n$, define a linear functional $\lambda_i \in U^*$ by $\lambda_i(u_j) = \delta_{ij}, (1 \leq j \leq n)$ (see theorem 3.4.4). Prove that $\{\lambda_i : 1 \leq i \leq n\}$ is a basis for $U^*$. The space $U^*$ is called the dual of $U$.
19. Let $\lambda_1, \ldots, \lambda_n$ be linear functionals on a vector space $U$, let $M_i = Ker(\lambda_i)$, and let $N = \cap_{i=1}^{n} M_i$. Prove that $dim(U/N) \leq n$. Hint: Define $T : U \to \mathbb{K}^n$ by $T(x) = (\lambda_1(x), \ldots, \lambda_n(x))$. Use theorem 3.4.7.

## 3.5  Matrix Representation and Diagonalization

A careful reading of example 2 in section 3.4 reveals that the set of linear mappings from $\mathbb{K}^n$ to $\mathbb{K}^m$ is in one-to-one correspondence with the set of $m \times n$ matrices. This section generalizes this result. Suppose $U$ and $V$ are finite-dimensional vector spaces and that $\{u_1, \ldots, n_n\}$ and $\{v_1, \ldots, v_m\}$ are bases for $U$ and $V$, respectively. Theorem 3.4.4 states that a linear mapping $T : U \to V$ is uniquely determined by the vectors $T(u_1), \ldots, T(u_n)$. Since each of the vectors $T(u_j)$ can be uniquely written as a linear combination of $\{v_1, \ldots, v_m\}$ with coefficients in $\mathbb{K}$, the set of coefficients determines $T$ uniquely. This observation is the basis for the opening definition of this section. The information in this section is standard, and we assume familiarity with its contents.

### Matrix Representations of Linear Mappings

Let $U$ and $V$ be finite-dimensional vector spaces, and let $n = dim(U), m = dim(V)$. Fix a pair of bases $B = \{u_1, \ldots, u_n\}$ and $C = \{v_1, \ldots, v_m\}$ for $U$ and $V$, respectively. If $T \in Hom(U, V)$, then, for every $1 \leq j \leq n, T(u_j)$ can be written as a linear combination of $C$, say, $T(u_j) = \sum_{i=1}^{m} a_{ij} v_i$.

**Definition.** Given the construction in the previous paragraph, the matrix $A = (a_{ij})$ is called the **matrix of $T$ relative to the base pair** $(B, C)$.

The matrix representing a linear mapping is totally dependent on the base pair $(B, C)$ and is even sensitive to the permutation of the elements in each basis. Thus the bases $B$ and $C$ are assumed to be ordered.

**Example 1.** Consider the linear transformation $T : \mathbb{K}^n \to \mathbb{K}^m$ induced by an $m \times n$ matrix $A$. It is clear that the matrix of $T$ relative to the base pair $(B_n, B_m)$ is $A$, where $B_n$ and $B_m$ are the canonical bases for $\mathbb{K}^n$ and $\mathbb{K}^m$, respectively. ◆

**Example 2.** Let $T : \mathbb{P}_n \to \mathbb{P}_n$ be the linear transformation $T(f) = \frac{df}{dx}$. If $B = \{1, x, \ldots, x^n\}$, then the matrix of $T$ relative to $(B, B)$ is

$$\begin{pmatrix} 0 & 1 & & & \\ & 0 & 2 & & \\ & & & \ddots & n \\ & & & & 0 \end{pmatrix}. \blacklozenge$$

**Theorem 3.5.1.** *The function $\Phi : Hom(U, V) \to \mathbb{K}_{m \times n}$ that assigns to an element $T \in Hom(U, V)$ its matrix relative to the base pair $(B, C)$ is an isomorphism.* ∎

Now we study how the matrix of the composition of two linear transformations relates to the matrices of the composed transformations.

**Theorem 3.5.2.** *Let U, V, B, C, and T be as in the above definition, let D = $\{w_1, \ldots, w_p\}$ be a basis for a third vector space W, and, finally, let $S \in Hom(V, W)$. If $A = (a_{ij})$ is the matrix of T relative to the base pair $(B, C)$, and $A' = (a'_{ij})$ is the matrix of S relative to the base pair $(C, D)$, then the matrix product $A'A$ is the matrix of SoT relative to the base pair $(B, D)$.*

*Proof.* Let $1 \leq j \leq n$. Then

$$(SoT)(u_j) = S(T(u_j)) = S\left(\sum_{i=1}^{m} a_{ij}v_i\right) = \sum_{i=1}^{m}\sum_{k=1}^{p} a_{ij}a'_{ki}w_k$$

$$= \sum_{k=1}^{p}\left(\sum_{i=1}^{m} a'_{ki}a_{ij}\right)w_k = \sum_{k=1}^{p} e_{kj}w_k, \text{ where } e_{kj} = \sum_{i=1}^{m} a'_{ki}a_{ij}.$$

*Thus the matrix of SoT relative to $(B, D)$ is $E = (e_{kj})$. By the definition of matrix multiplication, $e_{kj}$ is the $(k, j)$ entry of the product $A'A$.* ∎

The above theorem is the crucial piece of information needed to prove the following theorem, which is a special case of theorem 3.5.1 when $V = U$ and $C = B$.

**Theorem 3.5.3.** *The function $\Phi : Hom(U, U) \to \mathbb{K}_{n \times n}$ that assigns to an element $T \in Hom(U, U)$ its matrix relative to the base B (i.e., relative to the base pair $(B, B)$) is an algebra isomorphism. Thus, in addition to being linear, $\Phi$ satisfies $\Phi(SoT) = \Phi(S)\Phi(T)$.* ∎

**Definition.** Let $U$ be an $n$-dimensional vector space, and let $B = \{u_1, \ldots, u_n\}$ and $B' = \{u'_1, \ldots, u'_n\}$ be two bases for $U$. Every vector $u_j \in B$ is a linear combination of the base $B'$, $u_j = \sum_{i=1}^{n} p_{ij}u'_i$. The resulting matrix $P = (p_{ij})$ is called the matrix from $B$ to $B'$.

It is important to understand that the matrix $P$ from $B$ to $B'$ is the matrix of the identity transformation $I_U : U \to U$ relative to the base pair $(B, B')$.

**Example 3.** Notice that if $P$ is the matrix from $B$ to $B'$, then $P^{-1}$ is the matrix from $B'$ to $B$. Indeed, let $Q$ be the matrix from $B'$ to $B$, and consider the mapping $T = I_U$ with the base pair $(B, B')$ (its matrix is $P$), and the mapping $S = I_U$ with the base pair $(B', B)$ (its matrix is $Q$). Consider the matrix of the composition $SoT$. On the one hand, its matrix relative to $(B, B)$ is $QP$ by theorem 3.5.2. On the other hand, the matrix of $I_U$ relative to $(B, B)$ is the identity matrix $I_n$. Therefore $QP = I_n$, and $Q = P^{-1}$. ◆

**Example 4.** Given a basis $B$ for $U$ and an invertible $n \times n$ matrix $P$, there is a basis $B'$ for $U$ such that $P$ is the matrix from $B$ to $B'$. To see this, let $Q = (q_{ij})$ be the inverse of $P$, and define $u'_j = \sum_{i=1}^{n} q_{ij} u_i$. The set $B' = \{u'_1, \ldots, u'_n\}$ is a basis for $U$ (see problem 3 at the end of this section), and the matrix from $B'$ to $B$ is $Q$. By example 3, $P$ is the matrix from $B$ to $B'$. ◆

**Example 5.** As another application of problem 3, let $B' = \{P_0, \ldots, P_n\}$ be polynomials such that, for each $0 \le i \le n$, $P_i$ has exact degree $i$. Then $B'$ is a basis for $\mathbb{P}_n$. To see this, write $P_i = \sum_{j=0}^{i} q_{ij} x^j$. The lower triangular matrix $Q = (q_{ij})$ is invertible because its determinant is $\prod_{i=0}^{n} q_{ii} \neq 0$. Since $B = \{1, \ldots, x^n\}$ is a basis for $\mathbb{P}_n$, so is $B'$, by problem 3. ◆

The above discussion leads to the following.

**Theorem 3.5.4.** *Let $U$ be an $n$-dimensional vector space. Then the collection of bases for $U$ is in one-to-one correspondence with the collection of invertible $n \times n$ matrices.*

*Proof. Fix a basis $B$ for $U$. For another basis $B'$ for $U$, let $P$ be the matrix from $B$ to $B'$. The correspondence $\Psi : B' \mapsto P$ is the correspondence promised by the theorem. We leave the rest of the formalities to the reader. The examples preceding this theorem are relevant for verifying the details.* ∎

**Theorem 3.5.5 (change of base Formula).** *Let $U$ and $V$ be finite-dimensional vector spaces, and let $n = dim(U), m = dim(V)$. Let $B$ and $B'$ be bases for $U$, and let $C$ and $C'$ be bases for $V$. Let $T \in Hom(U, V)$ and suppose that $A$ is the matrix of $T$ relative to a base pair $(B, C)$ and that $A'$ is the matrix of $T$ relative to the base pair $(B', C')$. If $P$ is the matrix from $B'$ to $B$ and $Q$ is the matrix from $C'$ to $C$, then $A' = Q^{-1}AP$.*

*Proof. Consider diagram 1. Each corner contains a pair: a space and a basis. The top arrow prompts the reader to consider the mapping $T : U \to V$ and mind the bases indicated in the top corners of the diagram. Thus the matrix of the mapping $T$ represented by the top arrow is relative to the base pair $(B, C)$ and is therefore $A$. Likewise, the matrices representing the rest of mappings indicated on the diagram are $Q^{-1}$ for $I_V$, $P^{-1}$ for $I_U$, and $A'$ for the mapping depicted by the bottom arrow. Now $I_V \circ T = T \circ I_U$. Applying theorem 3.5.2 to each side of the above equation, we get $Q^{-1}A = A'P^{-1}$, or $A' = Q^{-1}AP$.* ∎

$$(U,B) \xrightarrow{\;T\;} (V,C)$$
$$I_U \downarrow \qquad\qquad \downarrow I_V$$
$$(U,B') \xrightarrow{\;T\;} (V,C')$$

Diagram 1

**Corollary 3.5.6.** *Let U be an n-dimensional vector space, let $T \in Hom(U,U)$, and let B and B' be bases for U. If A is the matrix of T relative to B, and A' is the matrix of T relative to B', then $A' = P^{-1}AP$, where P is the matrix from B' to B.*

*Proof. This is the special case of the above theorem when $V = U$, $C = B$, and $C' = B'$.* ∎

## Diagonalization

When the matrix representing a linear operator $T$ on a finite-dimensional vector space relative to a basis $B$ is diagonal, the action of $T$ on $B$ is quite simple: $T$ maps each element of $B$ to a multiple of itself. The following question is natural: given an operator $T \in Hom(U,U)$, can you find a basis for $U$ relative to which the matrix of $T$ is diagonal? By corollary 3.5.6, the matrix equivalent of the question is as follows: given an arbitrary square matrix $A$, can you find an invertible matrix $P$ such that $P^{-1}AP$ is diagonal? The answer to both questions is no. The following definitions formalize the discussion.

**Definition.** A linear operator $T$ on a finite-dimensional vector space $U$ is **diagonalizable** if $U$ contains a basis relative to which the matrix of $T$ is diagonal. Equivalently, $U$ possesses a basis $B$ consisting entirely of eigenvectors of $T$.

**Definition.** A square matrix $A$ is diagonalizable if there exists an invertible matrix $P$ such that $P^{-1}AP$ is diagonal.

The following theorem gives a necessary and sufficient condition for a square matrix (linear operator) to be diagonalizable.

**Theorem 3.5.7.** *A square matrix A is diagonalizable if and only if $\mathbb{K}^n$ has a basis consisting entirely of eigenvectors of A.*

*Proof. Suppose A is diagonalizable. Thus there exists an invertible matrix P such that $P^{-1}AP = D$, a diagonal matrix. Let $\lambda_1, \ldots, \lambda_n$ be the diagonal entries of D, and let $P = [u_1, \ldots, u_n]$ be a partitioning of A by its columns. The equation $P^{-1}AP = D$*

is equivalent to $A[u_1, \ldots, u_n] = PD$, or $[Au_1, \ldots, Au_n] = [\lambda_1 u_1, \ldots, \lambda_n u_n]$. Thus $Au_i = \lambda_i u_i$, for $1 \leq i \leq n$, and $\{u_1, \ldots, u_n\}$ is a basis of $\mathbb{K}^n$ consisting of eigenvectors of $A$. To prove the converse, we simply reverse the above argument. ∎

We will discuss in section 3.7 a class of matrices that can be diagonalized in a very spacial way. We will also extend the discussion to infinite-dimensional spaces in chapter 7. We conclude the section with two examples of linear operators on infinite-dimensional spaces. In the first example, the operator has uncountably many eigenvalues; in the second, it has none.

**Example 6.** Let $T$ be an operator on $\mathcal{C}^\infty(\mathbb{R})$, defined by $T(f) = \frac{d^2 f}{dx^2} + f$. It is easy to verify that, for every $\omega \in \mathbb{R}$, the function $f_\omega(x) = \sin(\omega x)$ is an eigenfunction of $T$ corresponding to the eigenvalue $\lambda_\omega = 1 - \omega^2$. ◆

**Example 7.** Let $T$ be an operator on $\mathcal{C}^\infty(\mathbb{R})$, defined by $(Tf)(x) = xf(x)$. We verify that $T$ has no eigenvalues. If $T(f) = \lambda f$, then $xf(x) = \lambda f(x)$ for all $x \in \mathbb{R}$. This implies that $f(x) = 0$ for all $x \neq \lambda$. The continuity of $f$ implies that $f(\lambda) = 0$; thus $f = 0$. ◆

## Exercises

1. Let $A$ and $B$ be $n \times n$ matrices such that $AB = I_n$. Prove that $A$ is invertible and hence $B = A^{-1}$.
2. Let $U$ be a finite-dimensional vector space, and let $T : U \to U$ be linear. Prove that if $V$ is an $r$-dimensional, $T$-invariant subspace of $U$, then there is a basis for $U$ relative to which the matrix of $T$ has the form

$$\left( \begin{array}{c|c} A_{11} & A_{12} \\ \hline 0 & A_{22} \end{array} \right),$$

   where $A_{11}$ is an $r \times r$ submatrix.
3. Given a basis $B = \{u_1, \ldots, u_n\}$ for $U$ and an invertible $n \times n$ matrix $Q = (q_{ij})$, define $u_j' = \sum_{i=1}^n q_{ij} u_i$. Show that the set $B' = \{u_1', \ldots, u_n'\}$ is a basis for $U$.
4. Let $A$ be an $n \times n$ matrix and let $P$ be an invertible $n \times n$ matrix. Show that $A$ and $P^{-1}AP$ have the same eigenvalues. It follows from this result that the eigenvalues of a linear operator $T$ on a finite-dimensional space are those of the matrix representing $T$ relative to any basis.
5. Let $U$ be a finite-dimensional vector space, and let $T \in Hom(U, U)$. Prove that $T$ is diagonalizable if and only if its matrix relative to any basis is diagonalizable.

6. Let $U$ be a finite-dimensional vector space, and let $T \in Hom(U, U)$. Define $det(T)$ to be the determinant of the matrix representing $T$ relative to some basis of $U$. Prove that $det(T)$ is independent of the choice of the basis.

7. Let $T$ be a linear mapping from a finite-dimensional vector space $U$ to a finite-dimensional vector space $V$. Prove that there exist a pair of bases for $U$ and $V$ relative to which the matrix of $T$ is diagonal. Hint: See theorem B.3 in appendix B.

8. Let $T : \mathbb{P}_n \to \mathbb{P}_n$ be the linear operator $T(f) = \frac{df}{dx} + f$. Show that $T$ is not diagonalizable.

## 3.6 Normed Linear Spaces

Let us examine the function $d : \mathbb{R}^2 \to \mathbb{R}$, which assigns to a point $(x_1, x_2) \in \mathbb{R}^2$ its distance from the origin. Thus $d(x) = \left(x_1^2 + x_2^2\right)^{1/2}$. The function $d$ has the following characteristics:

(1) $d(x) \geq 0$ and $d(x) = 0$ if and only if $x = 0$.
(2) For a real scalar $a$ and a point $x \in \mathbb{R}^2$, $d(ax) = |a| d(x)$.
(3) For $x, y \in \mathbb{R}^2$, $d(x + y) \leq d(x) + d(y)$.

The abstraction of the function $d$ to an arbitrary vector space yields the definition of a normed linear space. Instead of using the notation $d(x)$, we use the universally accepted notation $\|x\|$ for the length of a vector $x$, or its distance from the zero vector.

Normed linear spaces are the most common examples of metric spaces. What sets norms apart, still using the function $d$ on $\mathbb{R}^2$ as our prototype, is the fact that the distance function between two points in the plane is translation invariant in the sense that if $D : \mathbb{R}^2 \times \mathbb{R}^2 \to \mathbb{R}$ is the function $D(x, y) = \{(x_1 - y_1)^2 + (x_2 - y_2)^2\}^{1/2}$, then $D(x, y) = D(x - a, y - a)$ for all $x, y, a \in \mathbb{R}^2$. Equivalently, $D(x, y) = D(x - y, 0) = d(x - y)$. See the definition of a translation later on in this section. This property makes no sense for a general metric space because the underlying set of a metric space is not required to be a vector space.

**Definition.** A **normed linear space** is a vector space X over $\mathbb{K}$ together with a function $\|.\| : X \to \mathbb{R}$ such that, for all $x, y \in X$ and all $a \in \mathbb{K}$,

(a) $\|x\| \geq 0$ and $\|x\| = 0$ if and only if $x = 0$,
(b) $\|ax\| = |a| \|x\|$, and
(c) $\|x + y\| \leq \|x\| + \|y\|$.

The function $\|.\|$ is called a **norm** on $X$, and condition (c) in the above definition is known as the **triangle inequality**.

Motivated by the discussion about the translation invariance of the distance function in the plane, the following definition makes sense.

**Definition.** The distance between two points $x$ and $y$ in a normed linear space $X$ is the scalar $\|x - y\|$.

The reader can easily verify that the defining conditions of a norm are satisfied in each of the examples below.

**Example 1.** Let $X = \mathbb{K}^n$, and define the 1-norm of $x = (x_1, \ldots, x_n)$ by

$$\|x\|_1 = \sum_{i=1}^{n} |x_i|. \; \blacklozenge$$

**Example 2.** Let $X = \mathbb{K}^n$, and define the $\infty$-norm of $x$ by

$$\|x\|_\infty = max_{1 \leq i \leq n}|x_i|. \; \blacklozenge$$

**Example 3.** Let $l^\infty$ be the space of bounded sequences discussed in section 3.1. The norm of a bounded sequence $(x_n)$ is defined by

$$\|x\|_\infty = sup_{n \in \mathbb{N}}|x_n|. \; \blacklozenge$$

**Example 4.** In section 3.1, we defined the space $X = \mathcal{B}[a,b]$ to consist of all bounded functions on the interval $[a,b]$. The **supremum norm** (also the **uniform** or $\infty$-**norm**) of a function $f \in X$ is defined by

$$\|f\|_\infty = sup_{x \in [a,b]}|f(x)|.$$

We verify the triangle inequality here. If $f$ and $g$ are bounded functions on $[a,b]$ and $x \in [a,b]$, then $|(f+g)(x)| \leq |f(x)| + |g(x)| \leq \|f\|_\infty + \|g\|_\infty$. Thus $\|f+g\|_\infty = sup_{x \in [a,b]}|(f+g)(x)| \leq \|f\|_\infty + \|g\|_\infty. \; \blacklozenge$

**Example 5.** An important subspace of $\mathcal{B}[a,b]$ is the space $\mathcal{C}[a,b]$ of continuous functions on $[a,b]$. Both spaces are given the uniform norm. $\blacklozenge$

**Example 6.** Another useful norm on $\mathcal{C}[a,b]$ is the 1-norm defined by

$$\|f\|_1 = \int_a^b |f(x)|dx. \; \blacklozenge$$

**Example 7.** Consider the following sequence of functions in $\mathcal{C}[0,1]$:

$$f_n(x) = \begin{cases} 2n^3 x & \text{if } 0 \leq x \leq \frac{1}{2n^2}, \\ -2n^3(x - \frac{1}{n^2}) & \text{if } \frac{1}{2n^2} \leq x \leq \frac{1}{n^2}, \\ 0 & \text{if } \frac{1}{n^2} \leq x \leq 1. \end{cases}$$

It is clear that

$$\|f_n\|_\infty = f_n(1/2n^2) = n, \|f_n\|_1 = \frac{1}{2n}. \blacklozenge$$

**Example 8.** Define a function $\|.\|$ on the space $\mathbb{R}_{n \times n}$ of $n \times n$ matrices as follows: let $a_1, \ldots, a_n$ be the rows of a matrix $A$ and set

$$\|A\|_\infty = max_{1 \leq i \leq n} \|a_i\|_1.$$

The function defined above is a norm on $\mathbb{R}_{n \times n}$. We verify the triangle inequality: if $b_1, \ldots, b_n$ are the rows of another matrix $B$, then the rows of $A + B$ are $a_1 + b_1, \ldots, a_n + b_n$, and, for $1 \leq i \leq n$,

$$\|a_i + b_i\|_1 \leq \|a_i\|_1 + \|b_i\|_1 \leq max_{1 \leq i \leq n} \|a_i\|_1 + max_{1 \leq i \leq n} \|b_i\|_1 = \|A\|_\infty + \|B\|_\infty.$$

Thus

$$\|A + B\|_\infty \leq \|A\|_\infty + \|B\|_\infty. \blacklozenge$$

The matrix norm in the above example is compatible with the $\infty$-norm on $\mathbb{R}^n$ in the sense that, for $x \in \mathbb{R}^n$, $\|Ax\|_\infty \leq \|A\|_\infty \|x\|_\infty$, as the reader can easily verify.

## $l^p$ Spaces

We now define the rest of the $l^p$ spaces.

**Definition.** For every real number $1 \leq p < \infty$, define $l^p$ to be the set of all sequences $x = (x_1, x_2, \ldots)$ in $\mathbb{K}$ such that $\sum_{n=1}^{\infty} |x_n|^p < \infty$. For $x \in l^p$,

$$\|x\|_p = \left( \sum_{n=1}^{\infty} |x_n|^p \right)^{1/p}.$$

It is straightforward to verify that $l^1$ is a normed linear space. For example, the triangle inequality is obtained from the following inequality upon taking the limit as $n \to \infty$: $\sum_{i=1}^{n} |x_i + y_i| \leq \sum_{i=1}^{n} |x_i| + \sum_{i=1}^{n} |y_i|$.

Showing that $l^p$ (for $1 < p < \infty$) is a normed linear space is less straightforward and requires the development of two useful inequalities which are important in their own right.

**Definition.** Let $1 < p < \infty$. The **conjugate Hölder exponent** of $p$ is the number $q > 1$ such that $\frac{1}{p} + \frac{1}{q} = 1$. By definition, $p = 1$ and $q = \infty$ are conjugate Hölder exponents.

**Lemma 3.6.1.** *If $p > 1$, and $x, y \in \mathbb{C}$, then $|xy| \leq \frac{|x|^p}{p} + \frac{|y|^q}{q}$. Here $p$ and $q$ are conjugate Hölder exponents.*

*Proof. Consider the function $f(t) = t^{1/p} - \frac{t}{p} - \frac{1}{q}, t \geq 1; f'(t) = \frac{1}{p}t^{1/p-1} - \frac{1}{p} \leq 0$. Thus $f$ is decreasing for all $t \geq 1$, and since $f(1) = 0$, it follows that $f(t) \leq 0$ for all $t \geq 1$. Thus $t^{1/p} \leq \frac{t}{p} + \frac{1}{q}$ for $t \geq 1$. Now let $a, b > 0$ and, say, $\frac{a}{b} \geq 1$. By replacing $t$ with $a/b$, we obtain $\left(\frac{a}{b}\right)^{1/p} \leq \frac{1}{p}\left(\frac{a}{b}\right) + \frac{1}{q}$. Therefore, $\frac{ba^{1/p}}{b^{1/p}} \leq \frac{a}{p} + \frac{b}{q}$, or $a^{1/p}b^{1/q} \leq \frac{a}{p} + \frac{b}{q}$. Letting $a = |x|^p, b = |y|^q$, we obtain the inequality we seek.* ∎

**Theorem 3.6.2 (Hölder's inequality).** *If $x = (x_n) \in l^p$ and $y = (y_n) \in l^q$, then $z = (x_n y_n) \in l^1$ and $\|z\|_1 \leq \|x\|_p \|y\|_q$.*

*Proof. If $p = 1, q = \infty, x \in l^1$, and $y \in l^\infty$, then*

$$\sum_{n=1}^{\infty} |x_n y_n| \leq sup_n |y_n| \sum_{n=1}^{\infty} |x_n| = \|y\|_\infty \|x\|_1.$$

*Now let $1 < p, q < \infty$. Applying lemma 3.6.1, $\frac{|x_i|}{\|x\|_p}\frac{|y_i|}{\|y\|_q} \leq \frac{1}{p}\frac{|x_i|^p}{\|x\|_p^p} + \frac{1}{q}\frac{|y_i|^q}{\|y\|_q^q}$. Thus*

$$\frac{1}{\|x\|_p\|y\|_q}\sum_{i=1}^{n}|x_i y_i| \leq \frac{1}{p\|x\|_p^p}\sum_{i=1}^{n}|x_i|^p + \frac{1}{q\|y\|_q^q}\sum_{i=1}^{n}|y_i|^q$$

$$\leq \frac{1}{p\|x\|_p^p}\sum_{i=1}^{\infty}|x_i|^p + \frac{1}{q\|y\|_q^q}\sum_{i=1}^{\infty}|y_i|^q = \frac{1}{p} + \frac{1}{q} = 1.$$

*The summary of the above calculations is that $\frac{1}{\|x\|_p\|y\|_q}\sum_{i=1}^{n}|x_i y_i| \leq 1$. Taking the limit as $n \to \infty$ we obtain Hölder's inequality.* ∎

**Theorem 3.6.3 (Minkowsi's inequality).** *If* $x = (x_n), y = (y_n) \in l^p$, *then* $x + y \in l^p$, *and*

$$\|x + y\|_p \leq \|x\|_p + \|y\|_p.$$

*Proof. We already proved the theorem for $p = 1$ and $p = \infty$, so assume that $1 < p < \infty$, and let $q$ be the conjugate Hölder exponent of $p$. Then:*

$$\sum_{i=1}^{n} |x_i + y_i|^p \leq \sum_{i=1}^{n} |x_i + y_i|^{p-1}(|x_i| + |y_i|).$$

*Applying Hölder's inequality to the right side of the above inequality yields*

$$\sum_{i=1}^{n} |x_i + y_i|^{p-1}|x_i| + \sum_{i=1}^{n} |x_i + y_i|^{p-1}|y_i|$$

$$\leq \left[ \left( \sum_{i=1}^{n} |x_i|^p \right)^{1/p} + \left( \sum_{i=1}^{n} |y_i|^p \right)^{1/p} \right] \left( \sum_{i=1}^{n} |x_i + y_i|^{(p-1)q} \right)^{1/q}$$

$$\leq \left( \|x\|_p + \|y\|_p \right) \left( \sum_{i=1}^{n} |x_i + y_i|^p \right)^{1/q}.$$

*The summary of the above calculations is that*

$$\sum_{i=1}^{n} |x_i + y_i|^p \leq \left( \|x\|_p + \|y\|_p \right) \left( \sum_{i=1}^{n} |x_i + y_i|^p \right)^{1/q}.$$

*Thus*

$$\left( \sum_{i=1}^{n} |x_i + y_i|^p \right)^{1-1/q} \leq \|x\|_p + \|y\|_p.$$

*Taking the limit as $n \to \infty$, and recalling that $1 - 1/q = 1/p$, we have*

$$\left( \sum_{i=1}^{\infty} |x_i + y_i|^p \right)^{1/p} \leq \|x\|_p + \|y\|_p \text{ or, } \|x + y\|_p \leq \|x\|_p + \|y\|_p. \blacksquare$$

We have verified all the crucial details needed to prove the result below.

**Theorem 3.6.4.** *For $1 \leq p \leq \infty$, $l^p$ is a normed linear space.* $\blacksquare$

Observe that Hölder's inequality and the triangle inequality apply to finite sequences:

$$\sum_{i=1}^{n} |x_i y_i| \le \left( \sum_{i=1}^{n} |x_i|^p \right)^{1/p} \left( \sum_{i=1}^{n} |y_i|^q \right)^{1/q}, \text{ and}$$

$$\left( \sum_{i=1}^{n} |x_i + y_i|^p \right)^{1/p} \le \left( \sum_{i=1}^{n} |x_i|^p \right)^{1/p} + \left( \sum_{i=1}^{n} |y_i|^p \right)^{1/p}.$$

Thus, for $1 \le p \le \infty$, $(\mathbb{R}^n, \|.\|_p)$ is a normed linear space.

## Balls, Lines, and Convex Sets

**Definition.** Let $X$ be a normed linear space. The **open ball** of radius $r$ centered at $x \in X$ is the set

$$B(x, r) = \{y \in X : \|y - x\| < r\}.$$

**Example 9.** In $(\mathbb{R}^2, \|.\|_\infty)$, the open ball of radius $r$ centered at the point $(x_0, y_0)$ is the open square $\{(x, y) : |x - x_0| < r, |y - y_0| < r\}$. In $(\mathbb{R}^2, \|.\|_1)$, the open ball of radius $r$ centered at $(x_0, y_0)$ is the open square with vertices $(x_0 \pm r, y_0)$ and $(x_0, y_0 \pm r)$. ◆

**Definition.** Let $u$ be a unit vector (i.e., a vector of length 1) in a normed linear space $X$, and let $\xi$ be a fixed point in $X$. The line that contains $\xi$ and is parallel to $u$ is the set

$$\{\xi + tu : -\infty < t < \infty\}.$$

**Remark.** The vector $u$ in the above definition does not have to be a unit vector, but when it is, $t$ is the exact distance between $\xi + tu$ and $\xi$. An important special case is the equation of the line joining two points $\xi$ and $\eta$ in $X$. In this case, the line is the set of all points $x$ such that

$$x = \xi + t(\eta - \xi) = (1 - t)\xi + t\eta, \ -\infty < t < \infty.$$

The set $\{(1 - t)\xi + t\eta : 0 \le t \le 1\}$ is called the **line segment** joining $\xi$ and $\eta$.

**Example 10.** In $\mathbb{R}^n$, the above definition reduces to the familiar definition of a straight line, especially when $n = 2, 3$. Indeed, if $u = (u_1, \ldots, u_n)$, and $\xi = (\xi_1, \ldots, \xi_n)$, then the parametric equation of the line containing $\xi$ and parallel to $u$ is

$$x_1 = \xi_1 + tu_1, x_2 = \xi_2 + tu_2, \ldots, x_n = \xi_n + tu_n, -\infty < t < \infty. ◆$$

**Definition.** Let $E$ be a subset of a vector space $V$, and let $x \in V$. The set $E + x = \{x + y : y \in E\}$ is known as the **translation** of $E$ by $x$ (or in the direction of $x$).

The set $E + x$ can be visualized as rigidly moving $E$ in the direction of the vector $x$. The graph of the parabola $y = x^2 + 1$ is the translation of the graph of the parabola $y = x^2$ by the vector $(0,1)$. Figure 3.1 depicts the translation of the raindrop-like set $E$ by the vector $x = (0,-1)$.

Translating a set preserves most of its characteristics. Convexity is a good example.

**Definition.** A subset $C$ of a vector space $V$ is said to be **convex** if, for every $\xi, \eta \in C$, and all $0 \le t \le 1$, $(1-t)\xi + t\eta \in C$. Thus $C$ is convex if whenever it contains $\xi$ and $\eta$, it contains the line segment joining $\xi$ and $\eta$.

**Example 11.**

(a) An open ball in $\mathbb{R}^n$ is a convex set.
(b) Let $A$ be an $m \times n$ real matrix, and let $b \in \mathbb{R}^m$. The two sets
$\{x \in \mathbb{R}^n : Ax = b\}$ and $\{x \in \mathbb{R}^n : Ax > b\}$ are convex subsets of $\mathbb{R}^n$.[3]
(c) The union of the first and third quadrants in the plane is not convex.
(d) The raindrop region in figure 3.1 is not convex.
(e) The intersection of an arbitrary collection of convex sets is convex. ◆

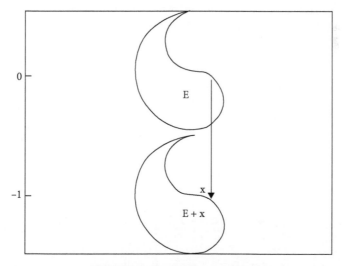

Figure 3.1 The falling raindrop

---

[3] The notation $Ax < b$ means that $\sum_{j=1}^{n} a_{ij}x_j < b_i$ for all $1 \le i \le m$.

## Excursion: Convex Hulls and Polytopes

We limit the discussion below to $\mathbb{R}^n$, although some of the statements are valid for an arbitrary vector space.

**Definition.** A **convex combination** of a finite set $\{x_1, \ldots, x_k\} \subseteq \mathbb{R}^n$ is a point of the form $x = \sum_{i=1}^{k} \lambda_i x_i$ such that $0 \le \lambda_i \le 1$ and $\sum_{i=1}^{k} \lambda_i = 1$.

**Theorem 3.6.5.** *A subset $C \subseteq \mathbb{R}^n$ is convex if and only if it contains all convex combinations of points of $C$.*

*Proof. It is enough to show that if $C$ is convex and $x = \sum_{i=1}^{k} \lambda_i x_i$ is a convex combination of points $x_1, \ldots, x_k \in C$, then $x \in C$. The converse is trivial. We use induction on $k$. The statement is true for $k = 2$ by the very definition of convexity. Without loss of generality, assume that $\lambda_1 < 1$, and write $x = \lambda_1 x_1 + (1 - \lambda_1) \sum_{i=2}^{k} \frac{\lambda_i x_i}{1 - \lambda_1}$. Now $\sum_{i=2}^{k} \frac{\lambda_i}{(1 - \lambda_1)} = 1$ and $y = \sum_{i=2}^{k} \frac{\lambda_i x_i}{1 - \lambda_1} \in C$ by the inductive hypothesis. By the convexity of $C$, $x = \lambda_1 x_1 + (1 - \lambda_1) y \in C$.* ∎

**Definition.** The **convex hull of a nonempty set** $A \subseteq \mathbb{R}^n$ is the smallest convex subset of $\mathbb{R}^n$ that contains $A$. The notation $conv(A)$ denotes the convex hull of $A$.

It is clear that $conv(A)$ is the intersection of all the convex subsets of $\mathbb{R}^n$ that contain $A$ and that $conv(A) \ne \emptyset$, since $A \subseteq \mathbb{R}^n$, and $\mathbb{R}^n$ is convex.

**Theorem 3.6.6.** *For a nonempty set $A \subseteq \mathbb{R}^n$, $conv(A)$ is the set of all convex combinations of points of $A$.*

*Proof. By the previous theorem, it is enough to show that the set of all convex combinations of points in $A$ is a convex set. Suppose that $x = \sum_{i=1}^{k} \lambda_i x_i$ and $y = \sum_{j=1}^{l} \mu_j y_j$ are, respectively, convex combinations of points $x_1, \ldots, x_k$ and $y_1, \ldots, y_l$ in $A$. If $\alpha \in [0, 1]$, then*

$$(1 - \alpha)x + \alpha y = \sum_{i=1}^{k} (1 - \alpha)\lambda_i x_i + \sum_{j=1}^{l} \alpha \mu_j y_j$$

*is a convex combination of $x_1, \ldots, x_k, y_1, \ldots, y_l$ because*

$$\sum_{i=1}^{k} (1 - \alpha)\lambda_i + \sum_{j=1}^{l} \alpha \mu_j = 1.$$ ∎

A natural question now is whether there is an upper bound on the length of the convex combinations of vectors in $A$ needed to generate all of $conv(A)$. The next theorem provides the answer.

**Theorem 3.6.7 (Carathéodory's theorem).** *Let $A \subseteq \mathbb{R}^n$, and let $C = conv(A)$. Then every point in $C$ is the convex combination of, at most, $n + 1$ points of $A$.*

*Proof.* Let $x \in C$. Then $x = \sum_{i=1}^{k} \lambda_i x_i$ for some $x_1, \ldots, x_k \in A$ and some $\lambda_1, \ldots, \lambda_k \in (0, 1]$ with $\sum_{i=1}^{k} \lambda_i = 1$. If $k > n + 1$, then the vectors $x_2 - x_1, \ldots, x_k - x_1$ are linearly dependent, so there are constants $\mu_2, \ldots, \mu_k$, not all zero, such that $\sum_{j=2}^{k} \mu_j(x_j - x_1) = 0$. If we set $\mu_1 = -\sum_{j=2}^{k} \mu_j$, then $\sum_{j=1}^{k} \mu_j x_j = 0$, and $\sum_{j=1}^{k} \mu_j = 0$. Observe that at least one of the numbers $\mu_1, \ldots, \mu_k$ is positive. Now, for all $\alpha \in \mathbb{R}$, $x = \sum_{j=1}^{k} (\lambda_j - \alpha \mu_j) x_j$. Let $i$ be such that $\frac{\lambda_i}{\mu_i} = min_{1 \le j \le k}\{\frac{\lambda_j}{\mu_j} : \mu_j > 0\}$, and choose $\alpha = \frac{\lambda_i}{\mu_i}$. Observe that $\alpha > 0$ and for $1 \le j \le k$, $\lambda_j - \alpha \mu_j \ge 0$. Now $x = \sum_{j=1}^{k} (\lambda_j - \alpha \mu_j) x_j$, $\lambda_j - \alpha \mu_j \ge 0$, and $\sum_{j=1}^{k} (\lambda_j - \alpha \mu_j) = 1$. Since $\lambda_i - \alpha \mu_i = 0$, $x$ is a convex combination of, at most, $k - 1$ points of $A$. We continue this process until $x$ is a convex combination of, at most, $n + 1$ vectors in $A$. ∎

**Example 12.** It is possible that $k < n + 1$. The closed unit disk $D$ in $\mathbb{R}^2$ is the convex hull of the unit circle $\mathcal{S}^1$, and every interior point in $D$ is a convex combination of two vectors in $\mathcal{S}^1$. However, $k = n + 1$ is the best possible bound. For example, if $x_0, x_1$, and $x_2$ are three noncollinear points in the plane, then an interior point in the triangle defined by the three points is not a convex combination of any two of the three points. ◆

**Definition.** A **polytope** in $\mathbb{R}^n$ is the convex hull of a finite subset of $\mathbb{R}^n$.

**Definition.** A point $x$ in a convex subset $C \subseteq \mathbb{R}^n$ is said to be an **extreme point** of $C$ if whenever $y, z \in C$, and $y \ne z$, then, for any $\lambda \in (0, 1)$, $x \ne \lambda y + (1 - \lambda)z$. The extreme points of a polytope are more specifically called its **vertices**.

A convex set may not have any extreme points. A simple example is the set $\{(x, y) \in \mathbb{R}^2 : 0 \le x \le 1, -\infty < y < \infty\}$.

**Example 13.** If $x_1$ and $x_2$ are distinct points in $\mathbb{R}^n$, then the polytope $Q = conv(x_1, x_2)$ is the line segment $\{(1 - t)x_1 + tx_2 : 0 \le t \le 1\}$. It is easy to verify that $x_1$ and $x_2$ are the vertices of $Q$. ◆

The number of vertices of a polytope in $\mathbb{R}^n$ is not related to the dimension $n$ of the space. For example, a regular polygon in $\mathbb{R}^2$ can have any number of vertices.

While it is intuitively obvious that a polytope is the convex hull of its vertices, this is not an entirely trivial fact. We prove this fact, together with the even more fundamental fact that polytopes do have vertices.

**Lemma 3.6.8.** *The vertices of a polytope $Q = conv(x_1, \ldots, x_k)$ in $\mathbb{R}^n$ are contained in $\{x_1, \ldots, x_k\}$.*

*Proof.* We show that if $x \in Q$ and $x \neq x_j$, then $x$ is not an extreme point of $Q$. By assumption, $x$ is a convex combination of $x_1, \ldots, x_k$, say, $x = \sum_{i=1}^{k} \lambda_j x_j$, and assume, without loss of generality, that $\lambda_1 > 0$. Since $x \neq x_1$, $\lambda_1 \neq 1$. Let $y = \sum_{i=2}^{k} \frac{\lambda_i}{1-\lambda_1} x_j$. Since $y$ is a convex combination of $x_2, \ldots, x_k$, $y \in Q$, and $x = \lambda_1 x_1 + (1 - \lambda_1)y$. Since $0 < \lambda_1 < 1$, $x$ is not an extreme point of $Q$. ■

**Lemma 3.6.9.** *Consider the polytope $Q = conv(x_1, \ldots, x_k)$. If $x_k$ is not a vertex of $Q$, then $x_k$ is a convex combination of $x_1, \ldots, x_{k-1}$. Consequently, $Q = conv(x_1, \ldots, x_{k-1})$.*

*Proof.* Since $x_k$ is not a vertex of $Q$, there exist convex combinations $y = \sum_{i=1}^{k} \beta_i x_i$ and $z = \sum_{i=1}^{k} \gamma_i x_i$ $(y \neq z)$, and a number $\lambda \in (0,1)$ such that $x_k = \lambda y + (1 - \lambda)z$. Now $x_k = \sum_{i=1}^{k} \alpha_i x_i$, where $\alpha_i = \lambda \beta_i + (1 - \lambda)\gamma_i$. If $\alpha_k = 0$, the proof is complete. It is easy to check that $\alpha_k = 1$ is possible only if $\beta_k = \gamma_k = 1$. But this would force $y = z = x_k$, which is a contradiction. Thus $0 < \alpha_k < 1$ and $x_k = \sum_{i=1}^{k-1} \frac{\alpha_i}{1-\alpha_k} x_i$, as desired. We leave it to the reader to check that $Q = conv(x_1, \ldots, x_{k-1})$. ■

**Theorem 3.6.10.** *The set of vertices of a polytope $Q = conv(x_1, \ldots, x_k)$ is not empty, and $Q$ is the convex hull of its vertices.*

*Proof.* We prove the result by induction on $k$. The result is true for $k = 2$ by example 13. Now consider the polytope $Q = conv(x_1, \ldots, x_k)$. If all the points $x_1, \ldots, x_k$ are vertices of $Q$, there is nothing to prove. Otherwise, one point, say, $x_k$ is not a vertex of $Q$. By the previous lemma, $Q = conv(x_1, \ldots, x_{k-1})$. By the inductive hypothesis, $Q$ is the convex hull of its vertices. ■

The fact that a polytope is the convex hull of its extreme points is the weakest version of the well-known **Krein-Millman theorem.**

An important special type of polytopes is the simplex.

**Definition.** The **standard $n$-simplex,** $T_n$, in $\mathbb{R}^n$ is the convex hull of the $n+1$ vectors $0, e_1, \ldots, e_n$. In general, if $\{x_0, \ldots, x_n\} \subseteq \mathbb{R}^n$ is such that $x_1 - x_0, \ldots, x_n - x_0$ are independent, the set $conv\{x_0, \ldots, x_n\}$ is called an $n$-simplex in $\mathbb{R}^n$.

The standard 2-simplex is a triangle with vertices $(0,0), (1,0)$, and $(0,1)$. The standard 3-simplex is a pyramid with vertices $(0,0,0)$, $(1,0,0)$, $(0,1,0)$, and $(0,0,1)$.

Every point $x$ in the standard $n$-simplex can be written uniquely as $x = \sum_{i=1}^{n} \lambda_i e_i$, where $\lambda_i \in [0,1]$, and $\sum_{i=1}^{n} \lambda_i \leq 1$. Set $\lambda_0 = 1 - \sum_{i=1}^{n} \lambda_i$. The numbers $\lambda_0, \ldots, \lambda_n$ are called the **barycentric coordinates** of $x$.

## Exercises

1. Show that, for elements $x$ and $Y$ of a normed linear space,

$$\left| \|x\| - \|y\| \right| \leq \|x \pm y\|.$$

2. Let $a_1, \ldots, a_n$ be the columns of a real $n \times n$ matrix $A$. Prove that the function

$$\|A\|_1 = max_{1 \leq i \leq n} \|a_i\|_1$$

defines a norm on $\mathbb{R}_{n \times n}$ and that, for $x \in \mathbb{R}^n$,

$$\|Ax\|_1 \leq \|A\|_1 \|x\|_1.$$

3. Show that if $1 \leq r < s \leq \infty$, then $l^r \subset l^s$.
4. Show that if $x \in l^1$, then $\lim_{p \to \infty} \|x\|_p = \|x\|_\infty$.
5. Show that in a normed linear space, $B(x,r) + B(y,s) = B(x+y, r+s)$.
6. Prove that the translation of a convex subset of $\mathbb{R}^n$ is convex.
7. Prove that $x$ is an extreme point of a convex set $C$ if and only if $C - \{x\}$ is convex.
8. Prove that $0, e_1, \ldots, e_n$ are the vertices of the standard $n$-simplex.

## 3.7 Inner Product Spaces

The concept of an inner product stems out of the need to have an instrument that determines the orthogonality of vectors in a normed linear space. Let us consider the Euclidean norm on $\mathbb{R}^n$. The orthogonality of two vectors $x = (x_1, \ldots, x_n)$ and $y = (y_1, \ldots, y_n)$ in $\mathbb{R}^n$ is equivalent to the condition that $\|x + y\|^2 = \|x\|^2 + \|y\|^2$. (the Pythagorean theorem). Now

$$\|x + y\|^2 = \sum_{i=1}^{n}(x_i + y_i)^2 = \sum_{i=1}^{n} x_i^2 + \sum_{i=1}^{n} y_i^2 + 2\sum_{i=1}^{n} x_i y_i = \|x\|^2 + \|y\|^2 + 2\sum_{i=1}^{n} x_i y_i.$$

Thus the orthogonality of $x$ and $y$ is equivalent to the condition $\sum_{i=1}^{n} x_i y_i = 0$.

This suggests that we examine the function $B : \mathbb{R}^n \times \mathbb{R}^n \to \mathbb{R}$ defined by $B(x,y) = \sum_{i=1}^{n} x_i y_i$. A little reflection reveals that $B$ is linear in each of its arguments, that $B(x,y) = 0$ if and only if $x$ and $y$ are orthogonal, and that $B$ defines the norm in the sense that $B(x,x) = \|x\|^2$. Therefore a function $B$ with the above properties may be a useful specialization of norms in that it provides a tool for defining orthogonality. Abstracting the above discussion leads directly to the definition of an inner product.

**Definition.** An **inner product** on a vector space $H$ is a function $\langle .,. \rangle : H \times H \to \mathbb{K}$ such that, for all $x, y, z \in H$ and all scalars $\alpha \in \mathbb{K}$,

(a) $\langle x, y \rangle = \overline{\langle y, x \rangle}$,
(b) $\langle x + y, z \rangle = \langle x, z \rangle + \langle y, z \rangle$,
(c) $\langle \alpha x, y \rangle = \alpha \langle x, y \rangle$, and
(d) $\langle x, x \rangle \geq 0$, and $\langle x, x \rangle = 0$ if and only if $x = 0$.

A vector space $H$ with an inner product is called an **inner product space**.

**Example 1. The standard inner product** on $\mathbb{C}^n$ is defined by

$$\langle x, y \rangle = \sum_{i=1}^{n} x_i \bar{y}_i = y^* x.$$

Consistent with matrix notation, we write vectors as columns and $y^*$ as the conjugate transpose of the vector $y$, while $y^* x$ is conveniently thought of as a matrix product. The standard inner product on $\mathbb{R}^n$ is defined by

$$\langle x, y \rangle = \sum_{i=1}^{n} x_i y_i = y^T x = x^T y. \;\blacklozenge$$

**Example 2.** The space $l^2$ is an inner product space with the inner product

$$\langle x, y \rangle = \sum_{n=1}^{\infty} x_n \bar{y}_n. \;\blacklozenge$$

**Example 3.** The space $\mathcal{C}[a, b]$ is an inner product space with the inner product

$$\langle f, g \rangle = \int_a^b f(x) \overline{g(x)} dx. \;\blacklozenge$$

Of particular interest to us is the special case when $a = -\pi$, $b = \pi$. In this case, we define

$$\langle f, g \rangle = \frac{1}{2\pi} \int_{-\pi}^{\pi} f(x) \overline{g(x)} dx.$$

The normalization constant $1/2\pi$ is included for convenience, as will become clear later in this section.

For an element $x$ in an inner product space $H$, we write $\|x\| = \sqrt{\langle x, x \rangle}$. We will see shortly that $\|.\|$ is indeed a norm on $H$.

**Theorem 3.7.1 (the Cauchy-Schwarz inequality).** *If H is an inner product space, then, for all $x, y \in H$,*

$$|\langle x, y \rangle| \leq \|x\| \|y\|.$$

*Equality holds if and only if x and y are linearly dependent.*

*Proof. Without loss of generality, assume that $y \neq 0$. For $\alpha \in \mathbb{K}$,*

$$
\begin{aligned}
0 \leq \|x + \alpha y\|^2 &= \langle x + \alpha y, x + \alpha y \rangle \\
&= \langle x, x \rangle + \alpha \langle y, x \rangle + \overline{\alpha} \langle x, y \rangle + |\alpha|^2 \langle y, y \rangle.
\end{aligned}
$$

*Substituting $\alpha = -\langle x, y \rangle / \|y\|^2$, we obtain $0 \leq \|x\|^2 - \frac{|\langle x, y \rangle|^2}{\|y\|^2}$, from which the Cauchy-Schwarz inequality follows.*

*It is easy to verify that if $y = \alpha x$, then $|\langle x, y \rangle| = \|x\| \|y\|$. Conversely, suppose that $|\langle x, y \rangle| = \|x\| \|y\|$. Now*

$$
\begin{aligned}
\|\|y\|^2 x - \langle x, y \rangle y\|^2 &= \langle \|y\|^2 x - \langle x, y \rangle y, \|y\|^2 x - \langle x, y \rangle y \rangle \\
&= \|y\|^4 \|x\|^2 - \|y\|^2 \langle x, y \rangle \langle y, x \rangle - \|y\|^2 \overline{\langle x, y \rangle} \langle x, y \rangle + |\langle x, y \rangle|^2 \|y\|^2 \\
&= \|y\|^2 \{ \|x\|^2 \|y\|^2 - |\langle x, y \rangle|^2 \} = 0.
\end{aligned}
$$

*Thus $\|y\|^2 x - \langle x, y \rangle y = 0$, and x and y are dependent.* ∎

**Corollary 3.7.2 (the triangle inequality).** *In an inner product space H,*

$$\|x + y\| \leq \|x\| + \|y\|.$$

*Proof. Using the Cauchy-Schwartz inequality,*

$$
\begin{aligned}
\|x + y\|^2 &= \langle x + y, x + y \rangle = \|x\|^2 + \langle x, y \rangle + \langle y, x \rangle + \|y\|^2 \\
&= \|x\|^2 + 2Re\langle x, y \rangle + \|y\|^2 \leq \|x\|^2 + 2\|x\| \|y\| + \|y\|^2 = (\|x\| + \|y\|)^2.
\end{aligned}
$$

*Taking the square roots of the extreme sides of the above string yields the triangle inequality.* ∎

It follows from the above corollary that the function $\|x\| = \langle x, x \rangle^{1/2}$ defines a norm on $H$. Therefore every inner product space is a normed linear space.

**Definition.** Two vectors $x$ and $y$ in an inner product space are said to be **orthogonal** if $\langle x, y \rangle = 0$. Symbolically, we write $x \perp y$ to indicate the orthogonality of $x$ and $y$.

**Theorem 3.7.3 (the Pythagorean theorem).** *If $x$ and $y$ are orthogonal vectors in an inner product space, then*

$$\|x+y\|^2 = \|x\|^2 + \|y\|^2.$$

*Proof.*

$$\|x+y\|^2 = \langle x+y, x+y \rangle = \langle x,x \rangle + \langle x,y \rangle + \langle y,x \rangle + \langle y,y \rangle$$
$$= \langle x,x \rangle + \langle y,y \rangle = \|x\|^2 + \|y\|^2. \blacksquare$$

The Pythagorean theorem can be easily generalized as follows: if $x_1, \ldots, x_n$ are mutually orthogonal elements of an inner product space, then

$$\|x_1 + \ldots + x_n\|^2 = \|x_1\|^2 + \ldots + \|x_n\|^2.$$

**Definition.** A subset $S$ of an inner product space $H$ is said to be **orthogonal** if the vectors in $S$ are pairwise orthogonal. If, in addition, each vector in $S$ is a unit vector, then $S$ is called an **orthonormal** subset of $H$. We always assume that an orthogonal subset excludes the zero vector.

**Example 4.** The canonical vectors in $\mathbb{K}^n$ form an orthonormal set. ◆

**Example 5.** The set of functions $u_n(t) = \{e^{int} : n \in \mathbb{Z}\}$ is orthogonal in $\mathcal{C}[-\pi, \pi]$ and the inner product $\langle f,g \rangle = \frac{1}{2\pi} \int_{-\pi}^{\pi} f(x)\overline{g(x)}dx$. (Here $e^{i\theta} = \cos\theta + i\sin\theta$.) Indeed, if $m$ and $n$ are distinct integers, then

$$\langle u_n, u_m \rangle = \frac{1}{2\pi} \int_{-\pi}^{\pi} e^{int} e^{-imt} dt = \frac{1}{2\pi i(n-m)} e^{i(n-m)t} \Big|_{-\pi}^{\pi} = 0,$$

while

$$\langle u_n, u_n \rangle = \frac{1}{2\pi} \int_{-\pi}^{\pi} e^{int} e^{-int} dt = 1.$$

Observe the convenience of including the factor $1/2\pi$ in the definition of the inner product. ◆

**Theorem 3.7.4.** *An orthogonal subset $S$ of an inner product space $H$ is independent.*

*Proof.* Let $\{u_1, \ldots, u_n\}$ be a finite subset of $S$, and suppose that, for scalars $a_1, \ldots, a_n$, $\sum_{i=1}^{n} a_i u_i = 0$. For a fixed $1 \leq j \leq n$,

$$\left\langle \sum_{i=1}^{n} a_i u_i, u_j \right\rangle = \sum_{i=1}^{n} a_i \langle u_i, u_j \rangle = a_j.$$

*But*

$$\left\langle \sum_{i=1}^{n} a_i u_i, u_j \right\rangle = \langle 0, u_j \rangle = 0.$$

*Therefore $a_j = 0$, and $\{u_1, \ldots, u_n\}$ is independent.* ■

We now pause briefly in order to get confirmation that the concepts we have developed so far are consistent with the geometry of $\mathbb{R}^n$. The Cauchy-Schwartz inequality, which we write as $\frac{|\langle x, y \rangle|}{\|x\|\|y\|} \leq 1$, implies that there exists a unique number $\theta \in [0, \pi]$ such that $\cos\theta = \frac{\langle x, y \rangle}{\|x\|\|y\|}$, or $\langle x, y \rangle = \|x\|\|y\|\cos\theta$. Now $\|y - x\|^2 = \|x\|^2 + \|y\|^2 - 2\langle x, y \rangle = \|x\|^2 + \|y\|^2 - 2\|x\|\|y\|\cos\theta$. When $n = 2$, the last identity is the well-known law of cosines in elementary trigonometry. The number $\theta$, of course, is the angle between $x$ and $y$.

We continue to exploit the geometry of vectors in $\mathbb{R}^2$ to get direction for the next step. An important concept in geometry (and in Hilbert space theory) is that of projecting a vector onto another. Let $x \in \mathbb{R}^2$ and let $u$ be a unit vector in the plane. The length of the projection of $x$ onto $u$ is given by $\|x\| \cos\theta = \langle x, u \rangle$; hence the vector projection of $x$ onto the line containing $u$ is the vector $y = \langle x, u \rangle u$. This is the closest vector in the line containing $u$ to the vector $x$. Since the projection of a vector $x \in \mathbb{R}^3$ onto the span $M$ of two orthonormal vectors $u_1$ and $u_2$ is the sum of the individual projections of $x$ onto $u_1$ and $u_2$, the projection of $x$ on $M$ is $\langle x, u_1 \rangle u_1 + \langle x, u_2 \rangle u_2$. The constructions involved in the next two theorems are now well motivated.

**Theorem 3.7.5.** *Let $S = \{u_1, \ldots, u_n\}$ be a finite orthonormal subset of an inner product space H, let $x \in Span(S)$, and write $\hat{x}_i = \langle x, u_i \rangle$. Then*

$$x = \sum_{i=1}^{n} \hat{x}_i u_i, \text{ and } \|x\|^2 = \sum_{i=1}^{n} |\hat{x}_i|^2.$$

*In particular, if $\langle x, u_i \rangle = 0$ for all $1 \leq i \leq n$, then $x = 0$.*

*Proof.* Since $x \in Span(S)$, there are scalars $a_1, \ldots, a_n$ such that $x = \sum_{i=1}^{n} a_i u_i$. For a fixed $1 \leq j \leq n$,

$$\hat{x}_j = \langle x, u_j \rangle = \sum_{i=1}^{n} a_i \langle u_i, u_j \rangle = a_j.$$

*By the Pythagorean theorem,*

$$\|x\|^2 = \sum_{i=1}^{n} \|\hat{x}_i u_i\|^2 = \sum_{i=1}^{n} |\hat{x}_i|^2. \ \blacksquare$$

**Definition.** Let $M$ be a subspace of an inner product space $H$. The **orthogonal complement** of $M$ is the set

$$M^\perp = \{z \in H : z \perp x \ \forall x \in M\}.$$

It is clear that $M^\perp$ is a subspace of $H$ and that $M \cap M^\perp = \{0\}$.

**Example 6.** Let $a = (a_1, \ldots, a_n)^T$ be a nonzero vector in $\mathbb{R}^n$. The orthogonal complement of $M = Span(\{a\})$ is the set of all vectors $x$ such that $a^T x = 0$. Thus $M^\perp$ can be viewed as the kernel of an $1 \times n$ matrix and is therefore a maximal subspace of $\mathbb{R}^n$. A translation of a maximal subspace of $\mathbb{R}^n$ is called a **hyperplane**. Thus the equation of a hyperplane is of the form $\sum_{i=1}^{n} a_i x_i = b$, $b \in \mathbb{R}$. Observe that the hyperplane $a^T x = b$ partitions $\mathbb{R}^n$ into the three sets $\{x \in \mathbb{R}^n : a^T x = b\}$, $\{x \in \mathbb{R}^n : a^T x < b\}$ and $\{x \in \mathbb{R}^n : a^T x > b\}$. The latter two sets are called the open **half-spaces** determined by the hyperplane $a^T x = b$. ◆

**Theorem 3.7.6.** *Let $S = \{u_1, \ldots, u_n\}$ be a finite orthonormal subset of $H$, and let $M = Span(S)$. Then every vector $x \in H$ can be written uniquely as*

$$x = y + z, \text{ where } y \in M \text{ and } z \in M^\perp.$$

*Additionally, $y$ is the closest vector in $M$ to $x$ in the sense that if $y' \in M$ and $y' \neq y$, then $\|x - y\| < \|x - y'\|$.*

*Proof.* *Define $y = \sum_{i=1}^{n} \langle x, u_i \rangle u_i$, and let $z = x - y$.*
*Clearly, $y \in M$. We show that $z \in M^\perp$. For $1 \leq j \leq n$,*

$$\langle z, u_j \rangle = \left\langle x - \sum_{i=1}^{n} \langle x, u_i \rangle u_i, u_j \right\rangle$$

$$= \langle x, u_j \rangle - \sum_{i=1}^{n} \langle x, u_i \rangle \langle u_i, u_j \rangle = \langle x, u_j \rangle - \langle x, u_j \rangle = 0.$$

*This shows that $H = M + M^\perp$. To show the uniqueness part, suppose that $x = y + z = y' + z'$, where $y, y' \in M$ and $z, z' \in M^\perp$. Thus $y - y' = z' - z$. Since $y - y' \in M$ and $z' - z \in M^\perp$, $y - y' \in M \cap M^\perp = \{0\}$.*

To prove the last assertion, suppose $y' \in M$, $y' \neq y$, and write $x - y' = (x - y) + (y - y')$. Now $x - y = z \in M^\perp$ and $y - y' \in M$. Using the Pythagorean theorem, we have $\|x - y'\|^2 = \|x - y\|^2 + \|y - y'\|^2 > \|x - y\|^2$. ■

The vector $y = \sum_{i=1}^{n} \langle x, u_i \rangle u_i$ is called the **orthogonal projection** of $x$ on $M$. We reiterate that $y$ is the closest vector of $M$ to $x$. We also say that $y$ is the **best approximation** of $x$ in $M$.

Sometimes the basis vectors $u_1, \ldots, u_n$ in the above theorem are merely orthogonal and not orthonormal. In this case, we find the orthogonal projection of $x$ on $M$ by using the formula

$$y = \sum_{i=1}^{n} \langle x, u_i \rangle \frac{u_i}{\|u_i\|^2},$$

which is the previously stated formula for $y$ when each $u_i$ is replaced with the normalized vector $\frac{u_i}{\|u_i\|}$.

The following question naturally imposes itself: does every finite-dimensional inner product space have an orthonormal basis? The following theorem delivers the answer.

**Theorem 3.7.7.** *Every finite-dimensional inner product space contains an orthonormal basis.*

*Proof.* *We use induction on* $\dim(H)$. *Let* $\{v_1, \ldots, v_n\}$ *be a basis for H. Use the inductive hypothesis to find an orthogonal basis* $\{u_1, \ldots, u_{n-1}\}$ *for the inner product space* $Span(\{v_1, \ldots, v_{n-1}\})$, *and define*

$$u_n = v_n - \sum_{j=1}^{n-1} \langle v_n, u_j \rangle \frac{u_j}{\|u_j\|^2}.$$

*Clearly,* $u_n \neq 0$ *because otherwise* $v_n \in Span(\{u_1, \ldots, u_{n-1}\}) = Span(\{v_1, \ldots, v_{n-1}\})$. *Observe that* $u_n$ *is nothing but the difference between* $v_n$ *and its orthogonal projection on* $Span(\{u_1, \ldots, u_{n-1}\})$ *(the vector z in the notation of theorem 3.7.6), and therefore* $u_n$ *is orthogonal to each of the vectors* $u_1, \ldots, u_{n-1}$. *By theorem 3.7.4, the orthogonal set* $\{u_1, u_2, \ldots, u_n\}$ *is independent and hence is a basis for H. To obtain the desired orthonormal basis, we simply normalize each of the vectors* $u_i$. ■

The above theorem and its proof deliver more than the mere existence of an orthonormal basis for an arbitrary finite-dimensional inner product space. The proof is inductive and constructive; hence it can be applied to an infinite independent sequence of vectors, recursively, as follows.

## The Gram-Schmidt Process

Given an infinite sequence $v_1, v_2, \ldots$ of independent vectors in an inner product space, the sequence defined below is orthogonal:

$$u_1 = v_1, \text{ and, for } n \geq 2, u_n = v_n - \sum_{j=1}^{n-1} \langle v_n, u_j \rangle \frac{u_j}{\|u_j\|^2}.$$

Additionally, for each $n \in \mathbb{N}$,

$$V_n = Span(\{v_1, \ldots, v_n\}) = Span(\{u_1, \ldots, u_n\}) = U_n.$$

**Example 7.** Consider the space $\mathcal{C}[-1,1]$ with the inner product $\langle f,g \rangle = \int_{-1}^{1} f(x)\overline{g(x)}dx$. Applying the Gram-Schmidt process to the infinite independent sequence of monomials $1, x, x^2 \ldots$, we obtain a sequence of orthogonal polynomials $P_0, P_1, \ldots$, that spans the space of polynomials such that

$$Span(\{1, x, \ldots, x^n\}) = Span(\{P_0, P_1, \ldots, P_n\}) \text{ for all } n \geq 0.$$

The polynomials $P_n$ are known as the **Legendre polynomials**. We will study some of the properties of these polynomials in section 4.10. ◆

The following observation is sometimes crucial for avoiding the often cumbersome calculations needed to compute the orthogonal sequence $u_1, u_2, \ldots$.

If $w_1, w_2, \ldots$ is an orthogonal sequence and, for each $n \in \mathbb{N}$, $Span(\{v_1, \ldots, v_n\}) = Span(\{w_1, \ldots, w_n\})$, then each $w_n$ is a multiple of the corresponding $u_n$. This is because the orthogonal complement $U_{n-1}^{\perp}$ of $U_{n-1}$ in $V_n$ is one-dimensional, hence any nonzero vector in $U_{n-1}^{\perp}$ is a multiple of any other nonzero vector in $U_{n-1}^{\perp}$. The following example exploits this idea to generate the Legendre polynomials.

**Example 8.** Consider the following set of polynomials:

$$Q_n(x) = D^n[(x^2 - 1)^n],$$

where $D^n f$ denotes the $n^{th}$ derivative of $f$. Each $Q_n$ is a polynomial of exact degree $n$; thus $Span(\{P_0, \ldots, P_n\}) = San(\{Q_0, \ldots, Q_n\}) = Span(\{1, x, \ldots, x^n\})$. If we show that $Q_n$ is orthogonal to each of the monomials $x^j$ for $0 \leq j \leq n-1$, then the polynomials $\{Q_n : n \in \mathbb{N}\}$ are orthogonal and, by the above observation, $P_n = c_n Q_n$.

Integration by parts yields

$$\int_{-1}^{1} x^j D^n(x^2 - 1)^n dx = -\int_{-1}^{1} jx^{j-1}D^{n-1}(x^2 - 1)^n dx + x^j D^{n-1}(x^2 - 1)^n \Big|_{-1}^{1}.$$

The second term is zero because if $k < n$, then $x^2 - 1$ is a factor of $D^k(x^2 - 1)^n$. The same reason coupled with integration by parts $j - 1$ times proves the desired result. ◆

**Example 9.** Let $\lambda$ be a linear functional on a finite-dimensional inner product $H$. Then there exists a unique vector $y \in H$ such that, for every $x \in H$, $\lambda(x) = \langle x, y \rangle$.

Let $\{u_1, \ldots, u_n\}$ be an orthonormal basis for $H$, and define $a_i = \lambda(u_i)$. We claim that $y = \sum_{i=1}^{n} \overline{a_i} u_i$ is the desired vector. For a vector $x \in H$, use theorem 3.7.5 to write $x = \sum_{i=1}^{n} \hat{x}_i u_i$. On the one hand,

$$\lambda(x) = \sum_{i=1}^{n} \hat{x}_i \lambda(u_i) = \sum_{i=1}^{n} \hat{x}_i a_i.$$

On the other hand,

$$\langle x, y \rangle = \left\langle \sum_{i=1}^{n} \hat{x}_i u_i, \sum_{j=1}^{n} \overline{a_j} u_j \right\rangle = \sum_{i,j=1}^{n} \hat{x}_i a_j \langle u_i, u_j \rangle = \sum_{i=1}^{n} \hat{x}_i a_i = \lambda(x). ◆$$

## The Spectral Decomposition of a Normal Matrix

The goal of this subsection is to derive the spectral decomposition of a normal matrix. An exact generalization of this decomposition is valid for compact self-adjoint (in fact, normal) operators on infinite-dimensional separable Hilbert spaces.

For a (complex) matrix $A$, we use the symbol $A^*$ to denote the **conjugate transpose** of $A$. Thus $(A^*)_{ij} = \overline{a_{ji}}$. The following theorem sums up the properties of conjugate transposition. We only verify part (c).

**Theorem 3.7.8.** *Let $A$ and $B$ be matrices of compatible sizes for matrix multiplication. Then*

(a) $A^{**} = A$,
(b) $(AB)^* = B^*A^*$, and
(c) *if $A$ is an $n \times n$ matrix, then, for all $x, y \in \mathbb{C}^n$, $\langle Ax, y \rangle = \langle x, A^*y \rangle$.*

*Proof.* $\langle x, A^*y \rangle = (A^*y)^*x = y^*A^{**}x = y^*Ax = \langle Ax, y \rangle$. ∎

**Definition.** An $n \times n$ matrix $P$ is said to be **unitary** if its columns form an orthonormal basis for $\mathbb{C}^n$. A real unitary matrix is specifically called an **orthogonal matrix**.

**Theorem 3.7.9.** *A unitary matrix $P$ is invertible, and $P^{-1} = P^*$.*

*Proof. Partition $P$ by its columns, and write $P = (u_1, \ldots, u_n)$. Then*

$$
P^*P = \begin{pmatrix} u_1^* \\ \cdot \\ \cdot \\ \cdot \\ u_n^* \end{pmatrix} (u_1, \ldots, u_n).
$$

*Thus the $(i,j)$ entry of the product $P^*P$ is $u_i^* u_j = \delta_{i,j}$. Hence $P^*P = I_n$.* ∎

The simplest example of an orthogonal (hence unitary) matrix is a rotation matrix.

**Example 10.** For $\theta \in [0, 2\pi)$, the $2 \times 2$ matrix

$$
P_\theta = \begin{pmatrix} \cos\theta & -\sin\theta \\ \sin\theta & \cos\theta \end{pmatrix}
$$

is an orthogonal matrix. We show that the linear mapping induced by $P_\theta$ is a rotation of the plane. Indeed, if we identify a point $(x, y) \in \mathbb{R}^2$ with the complex number $z = x + iy$ and write $z = re^{it}$, then multiplying $z$ by $e^{i\theta}$ produces the point $z_1 = x_1 + iy_1$, which is the rotation of $z$ through the origin by the angle $\theta$. Thus

$$
\begin{aligned}
z_1 = ze^{i\theta} = re^{i(t+\theta)} &= r[\cos(t+\theta) + i\sin(t+\theta)] \\
&= r[\cos t\cos\theta - \sin t\sin\theta + i(\sin t\cos\theta + \cos t\sin\theta)] \\
&= x\cos\theta - y\sin\theta + i(x\sin\theta + y\cos\theta).
\end{aligned}
$$

Equating the real and imaginary parts yields

$$
\begin{pmatrix} x_1 \\ y_1 \end{pmatrix} = \begin{pmatrix} x\cos\theta - y\sin\theta \\ x\sin\theta + y\cos\theta \end{pmatrix} = P_\theta \begin{pmatrix} x \\ y \end{pmatrix}. \; \blacklozenge
$$

Rotation matrices are important because their obvious geometrical properties typify most of the general properties of a unitary matrix. For example, it is clear that a rotation of the plane preserves distances between vectors as well as angles between them. More specifically, $\|P_\theta x\| = \|x\|$ and $\langle P_\theta x, P_\theta y \rangle = \langle x, y \rangle$. This is the reason many people, including this author, loosely think of orthogonal matrices as *rotations*, although this is inaccurate, even in two dimensions. For example, the following matrix is orthogonal, but it is not a rotation matrix:

$$
\begin{pmatrix} -1 & 0 \\ 0 & 1 \end{pmatrix}.
$$

The following innocent-sounding question leads to a whole set of interesting definitions and problems, in both finite and infinite-dimensional inner product spaces: which linear operators on $\mathbb{R}^2$ can be diagonalized by simply rotating the axes? The question is exactly equivalent to the question of which matrices can be diagonalized by a rotation matrix. The immediate generalization is the question of which (complex) matrices can be diagonalized by a unitary matrix. To answer the question, suppose that a matrix $A$ can be diagonalized by a unitary matrix $P = (u_1, \ldots, u_n)$. Thus $P^{-1}AP = P^*AP = D$, where $D$ is a diagonal matrix whose entries are the eigenvalues of $A$ (see the proof of theorem 3.5.7). Then

$$A = PDP^* = (u_1, \ldots, u_n) \begin{pmatrix} \lambda_1 & & \\ & \ddots & \\ & & \lambda_n \end{pmatrix} \begin{pmatrix} u_1^* \\ \vdots \\ u_n^* \end{pmatrix} = (u_1, \ldots, u_n) \begin{pmatrix} \lambda_1 u_1^* \\ \vdots \\ \lambda_n u_n^* \end{pmatrix}.$$

Thus $A = \sum_{i=1}^{n} \lambda_i u_i u_i^*$. Similarly, $A^* = \sum_{i=1}^{n} \overline{\lambda_i} u_i u_i^*$.

Now $AA^* = \sum_{i,j=1}^{n} \lambda_i \overline{\lambda_j} u_i (u_i^* u_j) u_j^* = \sum_{i=1}^{n} |\lambda_i|^2 u_i u_i^* = A^*A$.

The above calculation shows that a necessary condition for a matrix $A$ to be unitarily diagonalizable is that $A^*A = AA^*$. Such a matrix has a name.

**Definition.** A (complex) matrix $A$ is called **normal** if $A^*A = AA^*$. A matrix $A$ is called **Hermitian** if $A^* = A$. A Hermitian matrix is clearly normal. Observe that a real Hermitian matrix is simply a symmetric matrix.

Theorem 3.7.13 establishes the fact that normality is also a sufficient condition for the unitary diagonalization of a matrix.

**Lemma 3.7.10.** *If $A$ is normal, then, for all $x \in \mathbb{C}^n$, $\|Ax\| = \|A^*x\|$.*

*Proof.*
$$\begin{aligned}
\|Ax\|^2 - \|A^*x\|^2 &= \langle Ax, Ax \rangle - \langle A^*x, A^*x \rangle \\
&= \langle A^*Ax, x \rangle - \langle AA^*x, x \rangle \\
&= \langle (A^*A - AA^*)x, x \rangle = \langle 0, x \rangle = 0. \blacksquare
\end{aligned}$$

**Theorem 3.7.11.** *Let $A$ be a normal matrix. Then a vector $u$ is an eigenvector of $A$ with the corresponding eigenvalue $\lambda$ if and only if $u$ is an eigenvector of $A^*$ corresponding to the eigenvalue $\overline{\lambda}$.*

*Proof. It is easy to verify that $A - \lambda I$ is normal and that its conjugate transpose is $A^* - \overline{\lambda} I$. By the previous lemma, $\|(A - \lambda I)u\| = \|(A^* - \overline{\lambda} I)u\|$. Thus $(A - \lambda I)u = 0$ if and only if $(A^* - \overline{\lambda} I)u = 0$. $\blacksquare$*

**Theorem 3.7.12.** *If A is a normal matrix, then eigenvectors of A corresponding to distinct eigenvalues are orthogonal.*

*Proof.* Suppose $Au_1 = \lambda_1 u_1, Au_2 = \lambda_2 u_2$, where $u_1 \neq 0 \neq u_2$, and $\lambda_1 \neq \lambda_2$. Then

$$\lambda_1 \langle u_1, u_2 \rangle = \langle \lambda_1 u_1, u_2 \rangle = \langle Au_1, u_2 \rangle$$
$$= \langle u_1, A^* u_2 \rangle = \langle u_1, \overline{\lambda_2} u_2 \rangle = \lambda_2 \langle u_1, u_2 \rangle.$$

Thus $(\lambda_1 - \lambda_2)\langle u_1, u_2 \rangle = 0$, and $\langle u_1, u_2 \rangle = 0$. ∎

**Theorem 3.7.13 (diagonalization).** *Let A be a normal matrix. Then there exists a unitary matrix P and a diagonal matrix D such that*

$$P^* AP = D.$$

*Proof.* The proof is inductive. The base case (n=2) is left as an exercise. Let $\lambda_1$ be an eigenvalue of A, and let $v_1$ be a unit eigenvector corresponding to $\lambda_1$. Let $M = Span(\{v_1\})$, and let $\{v_2, \ldots, v_n\}$ be an orthonormal basis for $M^\perp$. By construction, the matrix $Q = (v_1, \ldots, v_n)$ is unitary.
    We claim that $Q^*AQ$ has the form

$$Q^* AQ = \begin{pmatrix} \lambda_1 & 0 & \ldots & 0 \\ \hline 0 & & & \\ \vdots & & A' & \\ 0 & & & \end{pmatrix}.$$

The $(i,j)$ entry of $Q^*AQ$ is $e_i^T Q^* AQe_j$.[4] But, for $1 \leq i \leq n$,

$$e_i^T Q^* AQe_1 = e_i^T Q^* Av_1 = \lambda_1 e_i^T Q^* v_1 = \lambda_1 (Qe_i)^* v_1 = \lambda_1 v_i^* v_1 = \lambda_1 \delta_{i,1}.$$

Thus the entries in the first column of $Q^*AQ$ are what we claim they are. The entries in the top row are computed similarly by examining the quantity $e_1^T Q^* AQe_j$, and using the fact that $A^* v_1 = \overline{\lambda_1} v_1$ (from theorem 3.7.11).
    Next we show that the matrix $Q^*AQ$ is normal:

$$(Q^*AQ)(Q^*AQ)^* = Q^*AQQ^*A^*Q = Q^*AA^*Q = Q^*A^*AQ$$
$$= Q^*A^*QQ^*AQ = (Q^*AQ)^*(Q^*AQ).$$

---

[4] Here $\{e_1, \ldots, e_n\}$ is the standard basis for $\mathbb{K}^n$.

*Now*

$$(Q^*AQ)(Q^*AQ)^* = \begin{pmatrix} |\lambda_1|^2 & 0 & \cdots & 0 \\ 0 & & & \\ \vdots & & A'(A')^* & \\ 0 & & & \end{pmatrix},$$

*while*

$$(Q^*AQ)^*(Q^*AQ) = \begin{pmatrix} |\lambda_1|^2 & 0 & \cdots & 0 \\ 0 & & & \\ \vdots & & (A')^*A' & \\ 0 & & & \end{pmatrix}.$$

*This shows that $A'$ is normal.*

*Invoking the inductive hypothesis, there is a unitary $(n-1)\times(n-1)$ matrix $Q_1$ such that*

$$Q_1^*A'Q_1 = \begin{pmatrix} \lambda_2 & & \\ & \ddots & \\ & & \lambda_n \end{pmatrix}.$$

*Now define*

$$P = Q \begin{pmatrix} 1 & 0 & \cdots & 0 \\ 0 & & & \\ \vdots & & Q_1 & \\ 0 & & & \end{pmatrix}.$$

*Being the product of two unitary matrices, $P$ is unitary. It is straightforward to verify that*

$$P^*AP = \begin{pmatrix} \lambda_1 & & \\ & \ddots & \\ & & \lambda_n \end{pmatrix} = D. \ \blacksquare$$

**Remarks.**  1. If we write $P = (u_1, \ldots, u_n)$, then, by the proof of theorem 3.5.7, the eigenvalues of $A$ are $\lambda_1, \ldots, \lambda_n$, and $u_1, \ldots, u_n$ are the corresponding eigenvectors.

2. Observe that $A = PDP^* = \sum_{i=1}^n \lambda_i u_i u_i^*$. Each of the matrices $P_i = u_i u_i^*$ is a rank 1 matrix and, in fact, $P_i$ is the projection of $\mathbb{C}^n$ onto the one-dimensional subspace generated by $u_i$, because $P_i x = (u_i u_i^*)x = u_i(u_i^* x) = \langle x, u_i \rangle u_i$. The representation

$$A = \sum_{i=1}^n \lambda_i P_i$$

is known as the **spectral decomposition of the normal matrix** $A$.

## Spectral Theory for Normal Operators

Let $H$ be a finite-dimensional inner product space, and let $T$ be a linear operator on $H$. For a fixed element $y \in H$, define a functional $\lambda_y$ on $H$ by $\lambda_y(x) = \langle Tx, y \rangle$. It is clear that $\lambda_y$ is linear. By example 9, there is a unique vector $T^*y \in H$ such that $\lambda_y(x) = \langle x, T^*y \rangle$. Therefore we have a function $T^* : H \to H$ defined by the requirement that

$$\langle Tx, y \rangle = \langle x, T^*y \rangle$$

for all $x, y \in H$.

It is straightforward to verify that $T^*$ itself is a linear operator on $H$. We call it the **adjoint operator** of $T$.

**Definition.** A linear operator $T$ on a finite-dimensional inner product $H$ is said to be **normal** if $T^*T = TT^*$. We say that $T$ is **self-adjoint** if $T = T^*$. A self-adjoint operator is clearly normal.

We will develop the analog of the spectral decomposition of a normal matrix for normal operators.

**Lemma 3.7.14.** *Let $T$ be a normal operator on a finite-dimensional inner product space $H$, and let $B = \{v_1, \ldots, v_n\}$ be an orthonormal basis for $H$. Then the matrix $A$ of $T$ relative to $B$ is a normal matrix.*

*Proof. By theorem 3.5.2, it is sufficient to prove that the matrix of $T^*$ relative to $B$ is $A^*$. By assumption, $T(v_k) = \sum_{i=1}^{n} a_{ik} v_i$ for $1 \leq k \leq n$. We need to show that, for $1 \leq j \leq n$, $T^*(v_j) - \sum_{i=1}^{n} \overline{a_{ji}} v_i = 0$. It is further sufficient to show that, for all $1 \leq j, k \leq n$, the quantity $q_{jk} = \langle T^*(v_j) - \sum_{i=1}^{n} \overline{a_{ji}} v_i, v_k \rangle$ is zero:*

$$q_{jk} = \langle T^*(v_j), v_k \rangle - \sum_{i=1}^{n} \overline{a_{ji}} \langle v_i, v_k \rangle = \langle v_j, Tv_k \rangle - \overline{a_{jk}}$$

$$= \left\langle v_j, \sum_{i=1}^{n} a_{ik} v_i \right\rangle - \overline{a_{jk}} = \sum_{i=1}^{n} \overline{a_{ik}} \langle v_j, v_i \rangle - \overline{a_{jk}} = \overline{a_{jk}} - \overline{a_{jk}} = 0. \blacksquare$$

**Theorem 3.7.15.** *A normal operator $T$ on a finite-dimensional inner product space $H$ is diagonalizable.*

*Proof. Fix an orthonormal basis $B = \{v_1, \ldots, v_n\}$ for $H$, and let $A$ be the matrix of $T$ relative to $B$. By the previous lemma, $A$ is normal and, by theorem 3.7.13, $A$ is* *diagonalizable by a unitary matrix P. Thus $P^*AP = D$. Let $B'$ be the basis of H such that P is the matrix from $B'$ to B. Such a basis exists by example 4 in section 3.5. By theorem 3.5.5, the matrix of T relative to $B'$ is $P^{-1}AP = P^*AP = D$, as desired. We leave it to the reader to verify that $B'$ is, in fact, an orthonormal basis for H.* ∎

The above theorem leads to the **spectral theorem for normal operators** on finite-dimensional inner product spaces.

**Theorem 3.7.16 (the spectral theorem).** *A normal operator T on a finite-dimensional inner product space can be written as*

$$T = \sum_{i=1}^{n} \lambda_i P_i,$$

*where $\lambda_1, \ldots, \lambda_n$ are the eigenvalues of T, and $P_1, \ldots, P_n$ are rank 1 operators.*

*Proof. In the notation of the previous theorem, let $u_1, \ldots, u_n$ be the columns of the matrix P, and let $\lambda_1, \ldots, \lambda_n$ be the diagonal entries of D. Then $T(u_i) = \lambda_i u_i$. Write an arbitrary element x of H as $x = \sum_{i=1}^{n} \hat{x}_i u_i$. Then $T(x) = \sum_{i=1}^{n} \hat{x}_i T(u_i) = \sum_{i=1}^{n} \lambda_i \hat{x}_i u_i$. Define $P_i(x) = \hat{x}_i u_i = \langle x, u_i \rangle u_i$. Each of the operators $P_i$ is the projection of H onto the one-dimensional subspace generated by $u_i$ and $T = \sum_{i=1}^{n} \lambda_i P_i$.* ∎

## Exercises

1. For functions $f, g \in \mathcal{C}^{\infty}[0,1]$, define $\langle f, g \rangle = f(0)\overline{g(0)} + \int_0^1 f'(x)\overline{g'(x)}dx$. Prove that $\langle ., . \rangle$ is an inner product on $\mathcal{C}^{\infty}[0,1]$.

2. Prove that the following generalization of the previous exercise also defines an inner product on $\mathcal{C}^{\infty}[0,1]$: for a fixed positive integer $n$,

$$\langle f, g \rangle = \sum_{i=0}^{n-1} f^{(i)}(0)\overline{g^{(i)}(0)} + \int_0^1 f^{(n)}(x)\overline{g^{(n)}(x)}dx.$$

3. Prove the following properties of inner products, which are often used without explicit mention:
   (a) If $x, y$ are vectors in an inner product space H such that $\langle x, w \rangle = \langle y, w \rangle$ for every $w \in H$, then $x = y$.

(b) For $u_1, \ldots, u_n, v_1, \ldots, v_m \in H$ and for scalars $\alpha_1, \ldots, \alpha_n, \beta_1, \ldots, \beta_m$,

$$\left\langle \sum_{i=1}^{n} \alpha_i u_i, \sum_{j=1}^{m} \beta_j v_j \right\rangle = \sum_{i=1}^{n} \sum_{j=1}^{m} \alpha_i \overline{\beta_j} \langle u_i, v_j \rangle.$$

(c) If $u_1, \ldots, u_n$ are mutually orthogonal, then

$$\left\| \sum_{i=1}^{n} \alpha_i u_i \right\|^2 = \sum_{i=1}^{n} |\alpha_i|^2 \|u\|_i^2.$$

4. Prove that, in an inner product space, $x \perp y$ if and only if $\|x + \alpha y\| = \|x - \alpha y\|$ for every $\alpha \in \mathbb{K}$.

5. Let $u_1, u_2, \ldots$ be an infinite orthonormal sequence in an inner product space $H$. Prove that, for $x \in H$, $\sum_{n=1}^{\infty} |\hat{x}_n|^2 \leq \|x\|^2$. Here $\hat{x}_n = \langle x, u_n \rangle$.

6. Prove that if $M$ is a subspace of an inner product space $H$, then $M^{\perp}$ is a subspace of $H$, and $M \cap M^{\perp} = \{0\}$.

7. Let $M$ be a finite-dimensional proper subspace of an inner product space $H$. Prove that $M^{\perp} \neq \{0\}$. In particular, there exists a unit vector $x$ orthogonal to $M$.

8. Consider the space $M = \mathbb{K}(\mathbb{N})$ of finite sequences. Clearly, $M$ is a subspace of $l^2$. Prove that $M^{\perp} = \{0\}$.

9. For real square matrices $A$ and $B$, define $\langle A, B \rangle = tr(AB^T)$. Prove that $\langle ., . \rangle$ is an inner product on $\mathbb{R}_{n \times n}$.

10. This is a continuation of the previous exercise. Prove the orthogonal complement of the space of symmetric matrices is the subspace of skew-symmetric matrices.

11. Let $M$ be a proper subspace of $\mathbb{R}^n$, and let $r = dim(M)$. Prove that there exists an $(n - r) \times n$ matrix $A$ whose null-space is $M$. What additional properties can $A$ have?

12. **QR factorization**. Prove that every real invertible matrix $A$ can be written as $A = QR$, where $Q$ is an orthogonal matrix, and $R$ is an upper triangular matrix. Hint: Apply the Gram-Schmidt process to the columns $a_1, \ldots, a_n$ of $A$ to find an orthonormal basis $q_1, \ldots, q_n$. For $1 \leq i \leq n$, $a_i$ is a linear combination of $q_1, \ldots, q_i$.

13. Let $P$ be an orthogonal matrix.
    (a) Prove that $\langle Px, Py \rangle = \langle x, y \rangle$ for every $x, y \in \mathbb{R}^n$.
    (b) Prove that $\|Px\| = \|x\|$ for every $x \in \mathbb{R}^n$.

14. Let $P$ be an orthogonal matrix.
    (a) Prove that $P^T$ is orthogonal.
    (b) Prove that $det(P) = \pm 1$.
    (c) Prove that the product of two orthogonal matrices is orthogonal.

15. Prove theorem 3.7.13 when $n = 2$.
16. Refer to theorem 3.7.16. Prove that the operators $P_i$ satisfy
    (a) $\sum_{i=1}^{n} P_i = I$ (the identity operator on $H$), and
    (b) $P_i P_j = \delta_{ij} P_i$.
17. Find an orthogonal matrix that diagonalizes the matrix $A$ below, and write down the spectral decomposition of $A$:

$$A = \begin{pmatrix} -1 & 0 & 0 \\ 0 & 0 & 2 \\ 0 & 2 & 3 \end{pmatrix}.$$

# 4
# The Metric Topology

*A mathematician who is not also something of a poet will never be a perfect mathematician.*

Karl Weierstrass

Felix Hausdorff. 1868–1942

Felix Hausdorff was born into a wealthy Jewish family and when he was still a young boy, the family moved to Leipzig. He studied at Leipzig University, graduating in 1891 with a doctorate in the applications of mathematics to astronomy. He published four papers on astronomy and optics over the next few years. Hausdorff remained in Leibzig, where he lectured until 1910. He then moved to Bonn, then to Greifswald in 1913, returning to Bonn in 1921, where he continued his work until 1935.

Hausdorff was the first to coin the definitions of metric and topological spaces. In 1914, building on work by Maurice Fréchet and others, he published his famous text *Grundzüge der Mengenlehre*. The book was the beginning point for studying metric and topological spaces, which are now core topics in modern mathematics. Among Hausdorff's numerous achievements, we count his introduction of the notion of the Hausdorff dimension, his study of the Gaussian law of errors, limit theorems and the problem of moments, and the strong law of large numbers. He introduced the concept of a partially ordered set and, from 1901 to 1909, he proved

*Fundamentals of Mathematical Analysis.* Adel N. Boules, Oxford University Press (2021). © Adel N. Boules.
DOI: 10.1093/oso/9780198868781.003.0004

a series of results on ordered sets. In 1907 he introduced special types of ordinals in an attempt to prove Cantor's continuum hypothesis, and he was also the first to pose the generalized continuum hypothesis.

Hausdorff sensed the oncoming calamity of Nazism but made no attempt to emigrate while it was still possible. Although he swore the necessary oath to Hitler in 1934, he was forced to give up his position in 1935. He continued to undertake research in topology and set theory but the results could not be published in Germany. As a Jew, Hausdorff's position grew increasingly more difficult. He lived under the constant threat of being deported to an internment camp. Bonn University requested that the Hausdorffs be allowed to remain in their home, and the request was granted. But, in 1941, they were forced to wear the "yellow star", and, in January 1942, the Hausdorffs were informed that they were to be interned in Endenich. Together with his wife and his wife's sister, Hausdorff committed suicide on 26 January.

Hausdorff was, according to a quote attributed to Weierstrass, a perfect mathematician. Indeed, he was something of a poet, according to the following excerpt:[1]

> Hausdorff pursued, especially during the early years in Leipzig, a kind of double identity: as Felix Hausdorff, the productive mathematician, and as Paul Mongré. Under this pseudonym, Hausdorff enjoyed remarkable recognition within the German intelligentsia at the end of the 19th century as a writer, philosopher and socially critical essayist. He fostered a circle of friends that consisted of well-known writers, artists and publishers including Hermann Conradi, Richard Dehmel, Otto Erich Hartleben, Gustav Kirstein, Max Klinger, Max Reger and Frank Wedekind. Between 1897 and 1904, Hausdorff reached the peak of his literary-philosophical accomplishment: during this period, 18 of a total of 22 works were published under his pseudonym. These included the volume of aphorisms Sant' Ilario: Thoughts from Zarathrustra's Country, his critique Das Chaos in kosmischer Auslese, a book of poems entitled Ekstases, the farce Der Arzt seiner Ehre, as well as numerous essays, most of which appeared in the leading journal of the day, "Neue Deutsche Rundschau (Freie Bühne)". The play was Hausdorff's greatest literary success, as it was performed over 300 times in 31 cities.

---

[1] Excerpted from Hausdorff Center for Mathematics, *Felix Hausdorff*, http://www.hcm.uni-bonn. de/about-hcm/felix-hausdorff/about-felix-hausdorff/, accessed Oct. 29, 2020.

## 4.1 Definitions and Basic Properties

Basic calculus concepts such as limits and continuity are heavily based on the concept of proximity. A metric is the most common tool for measuring proximity. The definition of a metric is a direct abstraction of the properties of the distance function in the plane. The most important characteristics of the Euclidean distance are:

(1)  the distance between two points in the plane is positive,
(2)  the distance is a symmetric function, and
(3)  the triangle inequality as it is understood in plane geometry.

These three characteristics are the ingredients of the definition of a metric. It is an amazing fact that so few axioms produce such a rich structure. The abstraction of a simple concept almost never produces a structure with properties identical to those of the concept. Indeed, there are fundamental differences between the properties of a general metric and those of the Euclidean distance. For example, you will see that there are metrics where a ball consists of a single point or the entire space. Of course, such metrics generally have much less importance than the most common metrics, those induced by a norm. Thus the fact that some metric properties violate our sense of geometry does not detract from the usefulness of metric spaces as one of the most powerful tools of mathematics.

**Definition.** A **metric space** is a nonempty set $X$ together with a function $d : X \times X \to \mathbb{R}$ such that, for all $x, y$ and $z \in X$,

(a)  $d(x,y) \geq 0$, and $d(x,y) = 0$ if and only if $x = y$,
(b)  $d(x,y) = d(y,x)$, and
(c)  $d(x,y) \leq d(x,z) + d(z,y)$

The function $d$ is called the distance function, or the metric. Property (c) is known as the triangle inequality.

**Example 1.** Let $X = R$, and let $d(x,y) = |x-y|$. The triangle inequality is indeed the inequality known by the same name in elementary mathematics. In general, the metric on $\mathbb{R}^n$ given by $d(x,y) = (\sum_{i=1}^{n} |x_i - y_i^2|)^{1/2} = \|x-y\|_2$ is called the **Euclidean** (or the **usual**) **metric** on $\mathbb{R}^n$. In this case, the triangle inequality follows from Minkowski's inequality with $p = 2$. ◆

**Example 2.** Normed linear spaces provide a rich source of metric spaces. If $(X, \|.\|)$ is a normed linear space, and the distance function is defined by $d(x,y) = \|x-y\|$, then $d(x,y) = \|x-y\| = \|(x-z)+(z-y)\| \leq \|x-z\| + \|z-y\| = d(x,z) +$

$d(z, y)$, and this proves the triangle inequality. The other properties are trivial to verify. Special cases include all $l^p$ spaces, $\mathbb{R}^n$ with any of the $l^p$ metrics, and the space $\mathcal{B}[0, 1]$. See section 3.6. ◆

**Example 3.** Let $X$ be a nonempty set and define the **discrete metric** on $X$ as follows:

$$d(x, y) = \begin{cases} 1 & \text{if } x \neq y, \\ 0 & \text{if } x = y. \end{cases} ◆$$

**Definition.** Let $(X, d)$ be a metric space. **The open ball** of radius $r$ centered at $x \in X$ is the set

$$B(x, r) = \{y \in X : d(x, y) < r\}.$$

The special case of this definition stated in section 3.6 when $X$ is a normned linear space is consistent with the above definition.

**Example 4.** In $\mathbb{R}$ with the usual metric, $B(x, r)$ is the open interval of radius $r$ centered at $x$. ◆

**Example 5.** In $(\mathbb{R}^2, \|.\|_2)$, the open ball of radius $r$ centered at $(x_0, y_0)$ is the open disk of radius $r$ centered at $(x_0, y_0)$. ◆

**Example 6.** In the space $\mathcal{B}[0, 1]$ of bounded real functions on $[0, 1]$, the ball $B(f, r)$ is the set of all bounded functions whose graphs on $[0, 1]$ are between the graphs of the functions $y = f(x) - r$ and $y = f(x) + r$. ◆

**Example 7.** In the discrete metric on a set $X$, $B(x, r) = \{x\}$ if $r \leq 1$, and $B(x, r) = X$ if $r > 1$. ◆

**Definition.** A subset of a metric space $X$ is said to be **open** if it is the union of open balls.

**Example 8.** An open ball in any metric space $X$ is an open subset of $X$. ◆

**Example 9.** Consider the discrete metric on a nonempty set $X$. Single-point subsets of $X$ are open because $B(x, 1) = \{x\}$. It follows that every subset of $X$ is an open set since a set is the union of its single-point subsets. ◆

**Example 10.** In $\mathbb{R}$ with the usual metric, the interval $(0, \infty)$ is open since $(0, \infty) = \cup_{i=0}^{\infty}(i, i + 2)$. ◆

**Theorem 4.1.1.** *A subset $U$ of a metric space $X$ is open if and only if, for every $x \in U$, there exists a positive number $\delta$ such that $B(x, \delta) \subseteq U$.*

*Proof.* Suppose $U$ is open, and let $x \in U$. Since $U$ is the union of open balls, there exists a ball $B(y, r)$ in $X$ such that $x \in B(y, r) \subseteq U$. Since $d(x, y) < r$, the number $\delta = r - d(x, y)$ is positive. We show that $B(x, \delta) \subseteq B(y, r)$. Let $z \in B(x, \delta)$. Then $d(z, y) \leq d(z, x) + d(x, y) < d(x, y) + \delta = r$. Conversely, if, for every $x \in U$, there is a positive number $\delta_x$ such that $B(x, \delta_x) \subseteq U$, then $U = \cup_{x \in U} B(x, \delta_x)$. ∎

**Example 11.** In $\mathbb{R}^2$, the first quadrant $U = \{(x, y) : x > 0, y > 0\}$ is open. If $(x_0, y_0) \in U$, then the disk centered at $(x_0, y_0)$ of radius $\delta < min(x_0, y_0)$ is contained in $U$. ◆

**Theorem 4.1.2.** *Let $X$ be a metric space. Then*

(a) *The union of an arbitrary collection of open subsets of $X$ is open.*
(b) *The intersection of two (hence any finite number of) open sets is open.*
(c) *$X$ is open.*

*Proof.* (a) Let $\{U_\alpha\}$ be a collection of open sets of $X$, and let $U = \cup_\alpha U_\alpha$. If $x \in U$, then $x \in U_\alpha$ for some $\alpha$. Therefore there exists $\delta > 0$ such that $B(x, \delta) \subseteq U_\alpha \subseteq U$. Thus $U$ is open by theorem 4.1.1.

(b) Let $U$ and $V$ be open subsets of $X$, and let $x \in U \cap V$. By theorem 4.1.1, there exist positive numbers $\delta_1$ and $\delta_2$ such that $B(x, \delta_1) \subseteq U$, and $B(x, \delta_2) \subseteq V$. Let $\delta = min\{\delta_1, \delta_2\}$. Clearly, $B(x, \delta) \subseteq U \cap V$. Again, by theorem 4.1.1, $U \cap V$ is open.

(c) Fix an element $x \in X$. Clearly, $X = \cup_{n=1}^{\infty} B(x, n)$. ∎

By definition, the empty set is also an open subset of any metric space. This is largely a useful convention. For example, the statement of theorem 4.1.2 (b) should read: "The intersection of two open sets is open or empty." If we declare the empty set to be open, the statement as it stands is correct.

**Definition.** A subset $F$ of a metric space $X$ is **closed** if its complement $X - F$ is open.

**Example 12.** In $\mathbb{R}$ with the usual metric, $[a, b], [a, \infty)$, and $\cup_{n \in \mathbb{Z}}[2n, 2n + 1]$ are all closed sets, as the complement of each set is open. ◆

The theorem below follows from theorem 4.1.2 and De Morgan's laws.

**Theorem 4.1.3.** *Let $X$ be a metric space. Then*

(a) *$X$ and $\emptyset$ are closed.*
(b) *A finite union of closed sets is closed.*
(c) *The intersection of an arbitrary collection of closed sets is closed.* ∎

**Theorem 4.1.4.** *Let $x$ and $y$ be distinct elements of a metric space $X$. Then there exist open sets $U$ and $V$ containing $x$ and $y$, respectively, such that $U \cap V = \emptyset$.*

*Proof.* Let $\delta = d(x,y)$, and set $U = B(x,\delta/2)$ and $V = B(y,\delta/2)$. We show that $U \cap V = \emptyset$. If $z \in U \cap V$, then $d(x,y) \leq d(x,z) + d(z,y) < \delta/2 + \delta/2 = \delta$, which is a contradiction. ∎

The property established by the above theorem, namely, that distinct points in a metric space are contained in disjoint open sets, is called the **Hausdorff property**. This is an important separation property of metric spaces. Common terminology used to describe the Hausdorff property is that distinct points in a metric space can be separated by disjoint open sets.

**Definition.** A **neighborhood** of a point $x$ of a metric space $X$ is a subset of $X$ that contains an open subset of $X$ that contains $x$. A neighborhood of a point need not be open. The concept is sometimes helpful in economizing on verbiage.

Now you see that distance functions are not created equal. An open neighborhood of a point in the discrete metric is either very small (a single point) or very large (the whole space), while the collection of neighborhoods of a point $x$ in a normed linear space includes all the open balls centered at $x$ and is therefore quite rich. There is another important distinction between a general metric and the metric generated by a norm. In the latter case, the collection of open neighborhoods of a point is exactly the translation of the collection of open neighborhoods of any other points. Thus the neighborhoods of a point are identical (up to a translation) to the neighborhoods of any other points. The open neighborhoods in a general metric space can be quite heterogeneous in the sense that knowledge of the neighborhoods of one point tells us nothing about the open neighborhoods of other points.

**Definition.** Let $(x_n)$ be a sequence in a metric space $X$, and let $x \in X$. We say that $(x_n)$ **converges** to $x$ if $\lim_n d(x_n,x) = 0$. In this case, we write $\lim_n x_n = x$. We also say that $x$ is the limit of $(x_n)$. Observe that if $X$ is a normed linear space, $\lim_n x_n = x$ is equivalent to the condition that $\lim_n \|x_n - x\| = 0$.

**Theorem 4.1.5.** *The limit of a convergent sequence $(x_n)$ is unique.*

THE METRIC TOPOLOGY 109

*Proof. Suppose that $\lim_n x_n = x$, $\lim_n x_n = y$, and $x \neq y$. By the Hausdorff property, there exists $\delta > 0$ such that $B(x,\delta) \cap B(y,\delta) = \emptyset$. There exist natural numbers $N_1$ and $N_2$ such that, for $n > N_1$, $d(x_n, x) < \delta$, and, for $n > N_2$, $d(x_n, y) < \delta$. If we choose an integer $n > \max\{N_1, N_2\}$, then $d(x_n, x) < \delta$ and $d(x_n, y) < \delta$, and $x_n \in B(x,\delta) \cap B(y,\delta) = \emptyset$, which is a contradiction.* ∎

Convergence in the spaces $(\mathcal{B}[0,1], \|.\|_\infty)$ and $(\mathcal{C}[0,1], \|.\|_\infty)$ is equivalent to uniform convergence. Explicitly stated, a sequence $(f_n)$ of bounded (respectively, continuous) functions converges in the uniform norm to a bounded (respectively, continuous) function $f$ if, for $\epsilon > 0$, there exists a positive integer $N$, dependent only on $\epsilon$, such that, for all $x \in [0,1]$ and all $n > N$, $|f_n(x) - f(x)| < \epsilon$. Clearly, the pointwise convergence of $(f_n)$ to $f$ is necessary for its uniform convergence to $f$. The following two examples illustrate that the converse is not true.

**Example 13.** Let

$$f_n(x) = \begin{cases} nx & \text{if } 0 \leq x \leq 1/n, \\ 1 & \text{if } 1/n \leq x \leq 1. \end{cases}$$

The pointwise limit of the sequence $(f_n)$ is clearly the bounded function

$$f(x) = \begin{cases} 0 & \text{if } x = 0, \\ 1 & \text{if } 0 < x \leq 1. \end{cases}$$

However, the sequence $(f_n)$ does not converge to $f$ in $\mathcal{B}[0,1]$ because, for every $n \in \mathbb{N}$, $\|f_n - f\|_\infty \geq |f_n(1/(2n)) - f(1/(2n))| = |1/2 - 1| = 1/2$. ♦

**Example 14.** Let $f_n(x) = \frac{1}{n^3(x-1/n)^2+1}$. Clearly, $0 \leq f_n(x) \leq 1$, and $\lim_n f_n(x) = 0$ for all $x \in [0,1]$. However, $f_n$ does not converge to the zero function in the uniform norm since $\|f_n\|_\infty = f_n(1/n) = 1$. ♦

The following theorem is occasionally useful. Its proof is left as an exercise.

**Theorem 4.1.6.** *Let $(x_n)$ be a sequence in a metric space $X$, and let $x \in X$. If every subsequence of $(x_n)$ contains a subsequence that converges to $x$, then $(x_n)$ converges to $x$.* ∎

## Exercises

1. Show that the intersection of an arbitrary collection of open sets need not be open.

2. Show that an arbitrary union of closed sets is not necessarily closed.
3. Show that a single-point subset of a metric space $X$ is closed. Hint: Use the Hausdorff property. Conclude that an arbitrary subset of $X$ is the union of closed sets.
4. Show that $|d(x,z) - d(y,z)| \leq d(x,y)$. Hence show that if $\lim_n x_n = x$, $\lim_n y_n = y$, then $\lim_n d(x_n, y_n) = d(x,y)$.
5. Prove that a convergent sequence in a metric space is bounded.
6. If, in a normed linear space $X$, $\lim_n x_n = x$, $\lim_n y_n = y$, and $(a_n)$ and $(b_n)$ are scalar sequences that converge to $a$ and $b$, respectively, then $\lim_n(a_n x_n + b_n y_n) = ax + by$.
7. Prove that if $\lim_n x_n = x$, then every subsequence of $(x_n)$ converges to $x$.
8. Prove theorem 4.1.6.
9. Show that $\lim_n x_n = x$ if and only if every neighborhood of $x$ contains all but finitely many terms of $(x_n)$.
10. Give an example of a normed linear space that contains an uncountable number of mutually disjoint balls of equal radii. Hint: Let $A$ be the subset of $l^\infty$ of all binary sequences. What is the distance between any pair of points in $A$?
11. Prove that the sphere $S^{n-1} = \{x \in \mathbb{R}^n : \|x\|_2 = 1\}$ is a closed subset of $\mathbb{R}^n$.

## 4.2  Interior, Closure, and Boundary

The notions of interior, closure, and boundary are quite familiar, and their meaning is rather obvious for simple sets. For example, the interior of the closed disk $D = \{x \in \mathbb{R}^2 : \|x\|_2 \leq 1\}$ is the open disk $U = \{x \in \mathbb{R}^2 : \|x\|_2 < 1\}$, and the boundary of $D$ is the unit circle. The fact that a concept is intuitively obvious is no substitute for a definition. It is often the case that the definition of a familiar concept deepens our realization that familiarity and simplicity are not synonymous. You will see in this section that the interior of $\mathbb{Q}$ is empty, that its boundary is the entire real line, and that important subsets of $\mathbb{R}$, such as the Cantor set, can come in infinitely many fragments. Intuitively speaking, one expects the definitions to capture the ideas that an interior point of a set $A$ must be *completely surrounded* by points of $A$ and that a boundary point of $A$ falls *on the edge* of $A$. Thus any ball centered at a boundary point of $A$ falls partially inside $A$ and partially outside it. This section formulates precise generalizations of those concepts. We will also see that disjoint closed sets can be separated in much the same way that the Hausdorff property separates points.

**Definition.** Let $A$ be a nonempty subset of a metric space $X$. **The interior of** $A$, denoted $int(A)$, is the union of all the open subsets of $X$ contained in $A$. A point of $int(A)$ is called **an interior point** of $A$.

**Example 1.** The interior of a nonempty subset may well be empty. The simplest example is the subset $\mathbb{Q}$ of the metric space $\mathbb{R}$; $int(\mathbb{Q}) = \emptyset$ because $\mathbb{Q}$ contains no open intervals and hence no open subsets of $\mathbb{R}$. ◆

The proofs of the following two theorems are straightforward.

**Theorem 4.2.1.** *The interior of a subset $A$ is the largest open subset of $X$ contained in $A$. A subset $A$ is open if and only if $int(A) = A$. Finally, if $A \subseteq B$, then $int(A) \subseteq int(B)$.* ∎

**Theorem 4.2.2.** *Let $A$ be a nonempty subset of $X$, and let $x \in X$. Then $x \in int(A)$ if and only if there exists $\delta > 0$ such that $B(x,\delta) \subseteq A$.* ∎

The above theorem captures what it means for *an interior point of $A$ to be totally surrounded by points of $A$*. In fact, the statement of theorem 4.2.2 can be taken as the definition of an interior point $x$ of $A$. One can then define the interior of $A$ to be the set of all interior points of $A$.

**Definition.** Let $A$ be a subset of a metric space $X$. The **closure**, $\overline{A}$, of $A$ is the intersection of all the closed subsets of $X$ containing $A$. Points of $\overline{A}$ are called **closure points** of $A$. Since $X$ is closed and it contains $A$, the closure of a nonempty set is nonempty. The following theorem is an immediate consequence of theorem 4.1.3.

**Theorem 4.2.3.** *The closure, $\overline{A}$, of $A$ is the smallest closed subset of $X$ containing $A$. A subset $A$ of $X$ is closed if and only if $\overline{A} = A$. Finally, if $A \subseteq B$, then $\overline{A} \subseteq \overline{B}$.* ∎

**Theorem 4.2.4.** *Let $A$ be a nonempty subset of $X$, and let $x \in X$. Then $x \in \overline{A}$ if and only if for every $\delta > 0, A \cap B(x,\delta) \neq \emptyset$.*

*Proof.* Suppose, $x \notin \overline{A}$. Then $x \in X - \overline{A}$, which is open. Thus there exists $\delta > 0$ such that $B(x,\delta) \subseteq X - \overline{A}$. In particular, $B(x,\delta) \cap A = \emptyset$. Conversely, if for some $\delta > 0$, $B(x,\delta) \cap A = \emptyset$, then $A \subseteq X - B(x,\delta)$, which is closed, so $\overline{A} \subseteq X - B(x,\delta) = X - B(x,\delta)$. In particular, $x \notin \overline{A}$. ∎

**Example 2.** In $\mathbb{R}$ with the usual metric, $\overline{\mathbb{Q}} = \mathbb{R}$. This is because every open interval in $\mathbb{R}$ contains rational points. ◆

**Example 3. (The Comb)** Consider the following subset of $\mathbb{R}^2$ :
$A = \bigcup_{n=1}^{\infty} \left(\{\frac{1}{n}\} \times [0,1]\right)$. The line segments $\{\frac{1}{n}\} \times [0,1]$ are called the teeth of $A$. We claim that $\overline{A} = A \cup (\{0\} \times [0,1])$. Since $A$ is contained in the closed unit

square $S = [0,1] \times [0,1]$, $\overline{A} \subseteq S$. Any point in $S$ that does not belong to $A$ or the line segment $\{0\} \times [0,1]$ must lie strictly between two consecutive teeth, and a small-enough disk centered at the point is strictly contained between the two teeth. Finally any disk centered at a point on $\{0\} \times [0,1]$ intersects all the teeth from some point $n$ on. ◆

**Theorem 4.2.5.** *Let $A$ be a nonempty subset of $X$, and let $x \in X$. Then $x \in \overline{A}$ if and only if there exists a sequence $(x_n)$ in $A$ such that $\lim_n x_n = x$.*

*Proof.* Suppose $\lim_n x_n = x$, where each $x_n \in A$, and let $\delta > 0$. There exists a natural number $N$ such that, for all $n \geq N, d(x_n, x) < \delta$. In particular, $x_N \in B(x,\delta) \cap A$; thus $x \in \overline{A}$, by theorem 4.2.4. Conversely, suppose $x \in \overline{A}$. By theorem 4.2.4, $B(x, 1/n) \cap A \neq \emptyset$ for all $n \in \mathbb{N}$. Choose a point $x_n \in B(x, 1/n) \cap A$. Clearly, $\lim_n x_n = x$. ∎

**Definition.** Let $A$ be a subset of a metric space $X$. A point $x \in X$ is called a **limit point** of $A$ if, for every $\delta > 0, B(x,\delta) \cap A$ contains a point other than $x$. A point of $A$ that is not a limit point of $A$ is called an **isolated point** of $A$.

Observe that a limit point of $A$ need not belong to $A$. A point $x$ of $A$ is isolated if and only if there is $\delta > 0$ such that $B(x,\delta) \cap A = \{x\}$.

**Example 4.** In $\mathbb{R}^2$ with the usual metric, points on the unit circle $\{(x,y) : x^2 + y^2 = 1\}$ are limit points of the open unit disk $\{(x,y) : x^2 + y^2 < 1\}$. ◆

**Example 5.** In $\mathbb{R}$, every point of $\mathbb{N}$ is an isolated point of $\mathbb{N}$. A little reflection shows that every point of the set $A = \{\frac{1}{n} : n \in \mathbb{N}\}$ is an isolated point of $A$. ◆

**Theorem 4.2.6.** *If $A$ is a nonempty subset of $X$, and $x \in X$, then $x$ is a limit point of $A$ if and only if there exists a sequence $(x_n)$ of distinct points of $A$ such that $\lim_n x_n = x$.*

*Proof.* Let $x$ be a limit point of $A$. There exists a point $x_1 \in A$ such that $0 < d(x_1, x) < 1$. Let $\delta_2 = \min\{d(x_1, x), 1/2\}$. There exists a point $x_2 \in A$ such that $0 < d(x_2, x) < \delta_2$. Note that $x_2 \neq x_1$ by construction. The rest of the construction is inductive. Having found points $x_1, \ldots, x_n$ such that $0 < d(x_n, x) < \ldots < d(x_1, x)$ such that $d(x_i, x) < 1/i$, let $\delta_n = \min\{\frac{1}{n+1}, d(x_n, x)\}$, then choose a point $x_{n+1} \in A$ such that $0 < d(x_{n+1}, x) < \delta_n$. Clearly, $\lim_n x_n = x$. The converse is straightforward. ∎

**Definition.** The **derived set** of a subset $A$ of a metric space $X$, denoted by $A'$, is the set of all limit points of $A$.

**Theorem 4.2.7.** $\overline{A} = A \cup A'$. *Thus A is closed if and only if it contains all its limit points.*

*Proof.* By theorems 4.2.5 and 4.2.6, $A' \subseteq \overline{A}$, and, by definition $A \subseteq \overline{A}$. Thus $A \cup A' \subseteq \overline{A}$. Now suppose $x \in \overline{A}$ and that $x \notin A$. Since $x \in \overline{A}$, $B(x,\delta) \cap A \neq \varnothing$ for every $\delta > 0$. Because $x \notin A$, $B(x,\delta) \cap A$ contains a point of $A$ other than $x$. This makes $x$ a limit point of $A$, by definition. ∎

**Definition.** The **boundary of a subset** $A$ of a metric space $X$ is the set $\partial A = \overline{A} \cap \overline{X - A}$. Points of $\partial A$ are called the **boundary points** of $A$. Observe that $x \in \partial A$ if and only if every neighborhood of $x$ intersects both $A$ and $X - A$.

**Example 6.** In $\mathbb{R}$ with the usual metric, $\partial \mathbb{Q} = \mathbb{R}$. This is because every open interval in $\mathbb{R}$ contains both rational and irrational points. ◆

**Theorem 4.2.8.** $\overline{A} = int(A) \cup \partial A$.

*Proof.* It is enough to show that $\overline{A} \subseteq int(A) \cup \partial A$. The reverse containment is obvious. Let $x \in \overline{A} - \partial A$. Since every open ball centered at $x$ intersects $A$, and since $x \notin \partial A$, there exists an open ball $B(x,\delta)$ that does not intersect $X - A$. This means that $B(x,\delta) \subseteq A$; hence $x \in int(A)$. ∎

**Theorem 4.2.9.** $\overline{A} = A \cup \partial A$.

*Proof.* By the previous theorem, $\overline{A} = int(A) \cup \partial A \subseteq A \cup \partial A \subseteq \overline{A}$. ∎

**Definition.** Let $A$ be a nonempty subset of a metric space $X$, and let $x \in X$. The distance from $x$ to $A$ is $dist(x,A) = inf\{d(x,a) : a \in A\}$.

**Definition.** Let $A$ and $B$ be nonempty subsets of a metric space $X$. The distance between $A$ and $B$ is $dist(A,B) = inf\{d(a,b) : a \in A, b \in B\}$.

Observe that $dist(x,A)$ and $dist(A,B)$ are always finite numbers.

**Example 7.** Let $A = \{n \in \mathbb{N} : n \geq 2\}$ and $B = \{n + \frac{1}{n} : n \geq 2\}$. Then $dist(A,B) = 0$. To see this, observe that $a_n = n \in A$, that $b_n = n + \frac{1}{n} \in B$, and that $|a_n - b_n| = 1/n \to 0$ as $n \to \infty$. ◆

**Theorem 4.2.10.** *Let $A$ be a nonempty subset of a metric space $X$. Then $\overline{A} = \{x \in X : dist(x,A) = 0\}$.*

*Proof.* Suppose $x \in \overline{A}$. By theorem 4.2.5, there exists a sequence $(x_n)$ in A such that $\lim_n x_n = x$. Thus $dist(x, A) = \inf\{d(x, a) : a \in A\} \leq d(x_n, x)$ for all $n \in \mathbb{N}$. Since $\lim_n d(x_n, x) = 0, dist(x, A) = 0$. Conversely, if $dist(x, A) = 0$, then there exists a sequence of points $x_n \in A$ such that $\lim_n d(x_n, x) = 0$. Thus $\lim_n x_n = x$, and $x \in \overline{A}$ by theorem 4.2.5. ∎

**Definition.** Let A be a nonempty subset of a metric space X. The **diameter** of A, $diam(A) = sup\{d(x, y) : x, y \in A\}$. If $diam(A)$ is finite, we say that A is a **bounded subset** of X. A sequence $(x_n)$ is said to be **bounded** if its range, $\{x_n\}$, is a bounded set. If $diam(X) < \infty$, we say that d is a **bounded metric**.

**Theorem 4.2.11.** $diam(A) = diam(\overline{A})$.

*Proof.* Clearly, $diam(A) \leq diam(\overline{A})$. To prove that $diam(\overline{A}) \leq diam(A)$, let $x, y \in \overline{A}$. We will show that $d(x, y) \leq diam(A)$. There exist sequences $(x_n)$ and $(y_n)$ in A such that $\lim_n x_n = x, \lim_n y_n = y$. Now $d(x, y) \leq d(x, x_n) + d(x_n, y_n) + d(y_n, y) \leq d(x_n, x) + diam(A) + d(y_n, y)$. The desired inequality follows from the above string of inequalities by taking the limit as $n \to \infty$. ∎

## Separation by Open Sets

Separation is a central idea in topology and analysis, and its importance cannot be exaggerated. The Hausdorff property is the simplest form of separation. We will see below that closed sets can be separated in much that same way points can be.

**Theorem 4.2.12.** Let F be a closed subset of X, and let $x \in X - F$. Then there exist open subsets U and V such that $x \in U$, $F \subseteq V$, and $U \cap V = \emptyset$.

*Proof.* Since $x \in X - F$, which is open, there exists $\delta > 0$ such that $B(x, \delta) \subseteq X - F$. For every $y \in F, d(x, y) \geq \delta$; hence $B(x, \delta/2) \cap B(y, \delta/2) = \emptyset$. The open sets $U = B(x, \delta/2)$ and $V = \cup\{B(y, \delta/2) : y \in F\}$ satisfy the conclusion of the theorem. ∎

An alternative (and commonly used) terminology to summarize theorem 4.2.12 is to say that there are disjoint open subsets that separate a closed subset of X and a point outside it.

**Theorem 4.2.13.** Let E and F be disjoint closed subsets of a metric space X. Then E and F can be separated by disjoint open subsets of X.

*Proof.* We need to find disjoint open subsets $U$ and $V$ such that $E \subseteq U$ and $F \subseteq V$. For $x \in E$, $dist(x,F) > 0$, by theorem 4.2.10. Let $\delta_x = dist(x,F)$. By the proof of theorem 4.2.12, for every $y \in F$, $B(x,\delta_x/2) \cap B(y,\delta_x/2) = \emptyset$. For $y \in F$, let $\delta_y = dist(y,E) > 0$. By the proof of theorem 4.2.12, for every $x \in E$, $B(x,\delta_y/2) \cap B(y,\delta_y/2) = \emptyset$. Let $U = \cup_{x \in E} B(x,\delta_x/2)$, and $V = \cup_{y \in F} B(y,\delta_y/2)$. Clearly, $U$ and $V$ are open, $E \subseteq U$, and $V \subseteq V$. It remains to show that $U$ and $V$ are disjoint. If $z \in U \cap V$, then $z \in B(x,\delta_x/2) \cap B(y,\delta_y/2)$ for some $x \in E$ and $y \in F$. Now, $d(x,y) \leq d(x,z) + d(z,y) < \delta_x/2 + \delta_y/2 \leq max\{\delta_x,\delta_y\}$. But $d(x,y) \geq dist(x,F) = \delta_x$ and $d(x,y) \geq dist(y,E) = \delta_y$. We have arrived at a contradiction that proves the theorem. ∎

**Example 8.** Let $E = \{(x,y) \in \mathbb{R}^2 : x > 0, y \geq 1/x^2\}$, and $F = \{(x,y) \in \mathbb{R}^2 : x < 0, y \geq 1/x^2\}$. Clearly, $E$ and $F$ are disjoint closed subsets of the plane. They are separated by the open right and left half planes. ◆

## Subspaces

Let $(X,d)$ be a metric space, and let $A$ be a subset of $X$. The defining conditions of the metric are clearly satisfied by the elements of $A$. Since the distance function is the only defining characteristic of a metric space, the pair $(A,d)$ is a metric space in its own right. We say that $(A,d)$ is a **subspace** of $(X,d)$, and the metric $d$ on $A$ is called the **restricted** (**induced**, or **subspace**) **metric**.

If $A$ is a subspace of a metric space $X$, we use the notation $B_A(x,\delta)$ to denote the ball in $A$ of radius $\delta$ centered at a point $x \in A$. Thus $B_A(x,\delta) = \{x \in A : d(x,a) < \delta\}$. We use the notation $\overline{B}_A$ to denote the closure of a subset $B$ of $A$ in the restricted metric.

**Theorem 4.2.14.** *Let $A$ be a subspace of a metric space $X$, and let $B \subseteq A$. Then*

(a) $B_A(x,\delta) = B(x,\delta) \cap A$.

(b) *$B$ is open in the restricted metric on $A$ if and only if there exists an open subset $U$ of $X$ such that $B = U \cap A$.*

(c) *$B$ is closed in the restricted metric on $A$ if and only if there exists a closed subset $E$ of $X$ such that $B = E \cap A$.*

*Proof.* We prove part (b) and leave the rest of the statements to the reader. If $B$ is an open subset of $A$, then $B$ is the union of open balls in $A$. Thus $B = \cup_{\alpha \in I} B_A(x_\alpha, \delta_\alpha)$. By part (a), $B_A(x_\alpha, \delta_\alpha) = B(x_\alpha, \delta_\alpha) \cap A$; thus, $B = \cup_{\alpha \in I}[B(x_\alpha, \delta_\alpha) \cap A] = [\cup_{\alpha \in I} B(x_\alpha, \delta_\alpha)] \cap A = U \cap A$, where $U = \cup_{\alpha \in I} B(x_\alpha, \delta_\alpha)$, which is open in $X$. We leave the proof of the converse as an exercise. ∎

## The Cantor Set

Consider the closed unit interval $I = [0, 1]$. Trisect $I$ and remove the open middle third $(1/3, 2/3)$. This leaves two closed intervals: $I_{1,1} = [0, 1/3]$ and $I_{1,2} = [2/3, 1]$. Let $C_1 = I_{1,1} \cup I_{1,2}$. Repeat the construction for each of the intervals $I_{1,1}$ and $I_{1,2}$, thus removing the middle open third of each of the two intervals. This leaves four closed intervals: $I_{2,1} = [0, 1/9]$, $I_{2,2} = [2/9, 1/3]$, $I_{2,3} = [2/3, 7/9]$, and $I_{2,4} = [8/9, 1]$. Define $C_2 = \cup_{j=1}^{4} I_{2,j}$. Repeating this construction yields, for every $n \in \mathbb{N}$, a sequence of closed intervals $I_{n,1}, \ldots, I_{n,2^n}$, each of length $3^{-n}$. Define $C_n = \cup_{j=1}^{2^n} I_{n,j}$.

The **Cantor set** is defined to be $C = \cap_{n=1}^{\infty} C_n$.

It is clear that $C$ is an infinite set because it contains the endpoints of each of the intervals $I_{n,j}$ for all $n \in \mathbb{N}$ and all $1 \leq j \leq 2^n$. What is less obvious is whether $C$ contains any additional points. The surprising answer is that $C$ is uncountable.

First we establish some topological properties of $C$.

**Definition.** A closed subset $A$ of a metric space $X$ is said to be a **perfect set** if every point of $A$ is a limit point of $A$. Thus $A$ is perfect if it is equal to its derived set.

**Example 9.** The closed unit interval $[0, 1]$ is a perfect set. ♦

**Definition.** A subset $A$ of a metric space $X$ is said to be **nowhere dense** if $int(\overline{A}) = \emptyset$.

**Example 10.** A hyperplane $M$ in $\mathbb{R}^n$ is nowhere dense in $\mathbb{R}^n$.

Without loss of generality, assume that the hyperplane contains the origin. Thus there is a nonzero vector $a$ such that $M = \{x \in \mathbb{R}^n : a^T x = 0\}$. For $x \in M$, and $\epsilon > 0$, the open ball $B = B(x, \epsilon)$ is not contained in $M$ because the point $x + \dfrac{\epsilon a}{2\|a\|} \in B - M$. ♦

**Lemma 4.2.15.** *The Cantor set is closed, perfect, and nowhere dense.*

*Proof.* Since each $C_n$ is closed, $C = \cap_{n=1}^{\infty} C_n$ is closed.

*We show that $C$ contains no open intervals. This proves that $int(C) = \emptyset$. Let $J$ be an open interval of length $\epsilon > 0$, and choose an integer $n$ such that $3^{-n} < \epsilon$. Since the length of each of the intervals $I_{n,j}$ is $3^{-n}$, none of the intervals $I_{n,j}$ contains $J$.*

*Since J is connected,[2] it cannot be contained in the (disconnected) union of two or more of the intervals $I_{n,j}$. Thus J is not contained in $\cup_{j=1}^{2^n} I_{n,j} = C_n$. Hence J is not contained in C.*

*Finally, let $x \in C$ and consider the interval $(x - \epsilon, x + \epsilon)$. Again choose an integer n such that $3^{-n} < \epsilon$. Since $x \in C_n$, $x \in I_{n,j}$ for some $1 \leq j \leq n$. Because the length of $I_{n,j}$ is $3^{-n} < \epsilon$, $I_{n,j} \subseteq (x - \epsilon, x + \epsilon)$. Thus $(x - \epsilon, x + \epsilon)$ intersects C at a point other than x (at least one of the endpoints of $I_{n,j}$ is not equal to x.) This shows that every point in C is a limit point of C; hence C is perfect. ∎*

In the rest of this subsection and in the section exercises, we need to consider the ternary (base 3) expansions of points in $[0,1]$. Every point $x \in [0,1]$ has a ternary representation $x = \sum_{i=1}^{\infty} \frac{a_i}{3^i}$ where each $a_i \in \{0,1,2\}$. The sum may well be finite, $x = \sum_{i=1}^{n} \frac{a_i}{3^i}$, and the ternary representation of a number may not be unique, but this point is of no immediate consequence. However, see the section exercises.

**Lemma 4.2.16.** *If $x = \sum_{i=1}^{n} \frac{a_i}{3^i}$, $a_i \in \{0,2\}$, then x is the left endpoint of an interval $I_{n,j}$ for some $1 \leq j \leq 2^n$.[3]*

*Proof. We prove the result by induction on n. When $n = 1$, $x = \frac{a_1}{3}$. If $a_1 = 0, x = 0$, which is the left endpoint of $I_{1,1}$. If $a_1 = 2$, $x = 2/3$, which is the left endpoint of $I_{1,2}$.*

*Now suppose the statement is true for a certain integer n. Consider a number $y = \sum_{i=1}^{n+1} \frac{a_i}{3^i}$, where $a_i \in \{0,2\}$. If $a_{n+1} = 0$, there is nothing to prove, so suppose $a_{n+1} = 2$. By the inductive hypothesis, the number $x = \sum_{i=1}^{n} \frac{a_i}{3^i}$ is the left endpoint of an interval $I_{n,j}$ for some $1 \leq j \leq 2^n$. Since $y = x + \frac{2}{3^{n+1}}$, y is the left endpoint of the right closed subinterval that results from the trisection of $I_{n,j}$. Thus y is the left endpoint of an interval $I_{n+1,k}$ for some $1 \leq k \leq 2^{n+1}$. ∎*

**Proposition 4.2.17.** *If $y = \sum_{i=1}^{\infty} \frac{a_i}{3^i}$, where $a_i \in \{0,2\}$, then $y \in C$.*

*Proof. It is enough to prove that $y \in C_n$ for every $n \in \mathbb{N}$. Let $x = \sum_{i=1}^{n} \frac{a_i}{3^i}$. By the previous lemma, x is the left endpoint of some interval $I_{n,j}$. Since the length of $I_{n,j}$ is $3^{-n}$ and $y - x \leq \sum_{i=n+1}^{\infty} \frac{2}{3^i} = 3^{-n}$, $y \in I_{n,j} \subseteq C_n$. ∎*

---

[2] To say that J is connected means that if $x, y \in J$, and $x < z < y$, then $z \in J$.
[3] Observe that the statement does not exclude the possibility that $a_n = 0$. This is because if a point x is the left endpoint of an interval $I_{m,k}$ for some m and some $1 \leq k \leq 2^m$, then x is the left endpoint of $I_{n,j}$ for every $n > m$ and some $1 \leq j \leq 2^n$. The reason is that the successive trisections of $I_{m,k}$ always result in an interval (the leftmost) whose left endpoint is x.

We now need the binary (base 2) representations of numbers in the interval $[0,1]$. In this system, every $x \in [0,1]$ can be written as a series $\sum_{i=1}^{\infty} \frac{a_i}{2^i}$, where $a_i \in \{0,1\}$. Again, such a representation may be finite; $x = \sum_{i=1}^{n} \frac{a_i}{2^i}$, and, in this case, $x$ does not have a unique representation. For example, the number $1/2$ can also be written as $1/4 + 1/8 + 1/16 + \dots$. In general, the number $x = \sum_{i=1}^{n} \frac{a_i}{2^i}$, where $a_n = 1$ can also be written as $x = \sum_{i=1}^{n-1} \frac{a_i}{2^i} + \sum_{i=n+1}^{\infty} \frac{1}{2^i}$. In order to avoid ambiguity, we use the latter representation of $x$ and not the finite sum representation.

**Theorem 4.2.18.** *The Cantor set has cardinality $\mathfrak{c}$.*

*Proof.* We define a function $f : [0,1] \to C$ as follows: $f(0) = 0$, and, for $x \in (0,1]$, write $x = \sum_{i=1}^{\infty} \frac{a_i}{2^i}$ and define $f(x) = \sum_{i=1}^{\infty} \frac{2a_i}{3^i}$. By the previous proposition, $f(x) \in C$. We leave it to the reader to verify that $f$ is one-to-one. Now lemma 2.2.3 implies that $C$ is equivalent to $[0,1]$; hence $Card(C) = Card([0,1]) = \mathfrak{c}$. ∎

## Exercises

1. Which of the following subsets of $\mathbb{R}^2$ are open?
   (a) $A = \{(x,y) \in \mathbb{R}^2 : x \neq 0, y < 1/x^2\}$
   (b) $B = \cup_{a>0}\{(x,y) \in \mathbb{R}^2 : x \in \mathbb{R}, y = a^x\}$
2. Find the closure of each of the sets $A$ and $B$ in the previous problem.
3. Let $X$ be a metric space, and let $x \in X$. Show that the set $B[x,\delta] = \{y \in X : d(x,y) \leq \delta\}$ is closed in $X$. The set $B[x,\delta]$ is called the closed ball of radius $\delta$ centered at $x$. Give an example to show that the closure of the open ball $B(x,\delta)$ is not necessarily the closed ball $B[x,\delta]$.
4. Show that if $X$ is a normed linear space, then the closure of the open ball $B(x,\delta)$ is the closed ball $B[x,\delta]$.
5. Let $\mathcal{H} = \{x \in l^2 : |x_n| \leq 1/n\}$. Prove that $\mathcal{H}$ is closed. This set is known as the **Hilbert cube**.
6. Prove that a subset $A$ of a metric space $X$ is bounded if and only if it is contained in a ball.
7. Let $(X,d)$ be a metric space, let $A \subseteq X$, and let $x,y \in X$. Prove that $dist(x,A) \leq d(x,y) + dist(y,A)$.
8. Let $X, A$, and $x$ be as in the previous exercise. Prove that $dist(x,A) = dist(x,\overline{A})$.
9. Let $(X,d)$ be a metric space, and let $A$ and $B$ be nonempty subsets of $X$. Prove that
   (a) $dist(A,B) = inf\{dist(a,B) : a \in A\} = inf\{dist(b,A) : b \in B\}$,
   (b) $dist(A,B) = dist(\overline{A},\overline{B})$, and
   (c) there are disjoint closed subsets $E$ and $F$ (of $\mathbb{R}$) such that $dist(E,F) = 0$.

10. Let $(X, d)$ be a metric space, and let $A$ and $B$ be subsets of $X$. Prove that
    (a) $int(A \cap B) = int(A) \cap int(B)$, and
    (b) $int(A \cup B) \supseteq int(A) \cup int(B)$, giving an example to show that the containment may be proper.

11. Let $A$ and $B$ be as in the previous exercise. Prove that
    (a) $\overline{A \cup B} = \overline{A} \cup \overline{B}$, and
    (b) $\overline{A \cap B} \subseteq \overline{A} \cap \overline{B}$, giving an example to show that the containment may be proper.

12. Show that if a sequence $(x_n)$ in a metric space $X$ converges to $x$, then $\{x_n : n \in \mathbb{N}\} \cup \{x\}$ is closed in $X$.

13. **Cantor-like sets.** Let $0 < \epsilon \leq 1$. From the unit interval $[0, 1]$, remove the open subinterval of length $\epsilon/3$ centered at $1/2$, leaving the two closed intervals $I_{1,1}$ and $I_{1,2}$. Then repeat the geometric construction of the Cantor set, except require that the removed open interval from $I_{n,j}$ be centered at the midpoint of $I_{n,j}$ and have length $\epsilon/3^{n+1}$. The resulting set, $C_\epsilon$, is known as a Cantor-like set. Prove that $C_\epsilon$ is closed, perfect, and nowhere dense.

14. Complete the proof of theorem 4.2.18. Hint: Modify the proof of theorem 2.1.15

15. Prove the converse of lemma 4.2.16.
    We now take a more careful look at the ternary representation of numbers in $[0, 1]$. Specifically, we address the issue of the nonuniqueness of the representation of a finite sum $x = \sum_{i=1}^{n} \frac{a_i}{3^i}$, where $a_n \neq 0$. If $a_n = 2$, we use the finite sum to represent $x$. If $a_n = 1$, we use the series $x = \sum_{i=1}^{n-1} \frac{a_i}{3^i} + \sum_{i=n+1}^{\infty} \frac{2}{3^i}$ and not the finite sum to represent $x$.

16. Prove the converse of proposition 4.2.17. Thus the Cantor set consists of exactly the points in $[0, 1]$ that have a ternary representation of the form $x = \sum_{i=1}^{\infty} \frac{a_i}{3^i}$, $a_i \in \{0, 2\}$. Hint: Prove that if $x = \sum_{i=1}^{\infty} \frac{a_i}{3^i}$, and any of the integers $a_i = 1$, then $x \notin C$.

17. Prove that a number $x \in C$ is a right endpoint of an interval $I_{n,j}$ if and only if the ternary representation of $x$ contains a finite number of zeros.

18. Prove that the interior of the standard $n$-simplex $T_n$ consists of all the points in $T_n$ with positive barycentric coordinates. Hence describe the boundary of $T_n$.

## 4.3 Continuity and Equivalent Metrics

Continuity, from the intuitive point of view, is about the gradual rather than the abrupt change of function values. In its simplest form, the graph of a continuous, real-valued function of a single real variable must be connected. Most functions in mathematics are too complicated for such a visual characterization of continuity, and a more rigorous and robust definition is needed. The ε-δ definition of continuity revolutionized calculus, and hence mathematics, in the early nineteenth century. It is based on the idea that the fluctuations of a continuous function *can be controlled in a sufficiently small neighborhood of a point of continuity.* Our definition of local continuity in the metric setting is an immediate generalization of the ε-δ definition. We then define the *global continuity of a function* on a metric space, an important concept seldom treated in undergraduate textbooks. You will see that continuity does not depend on the specific metric we use to measure proximity, but rather on the collection of open sets the metric induces. This leads us to the notion of equivalent metrics and, more generally, homeomorphisms.

**Definition.** Let $(X, d)$ and $(Y, \rho)$ be metric spaces. A function $f : X \to Y$ is said to be **continuous at a point** $x_0 \in X$ if, for every $\epsilon > 0$, there exists $\delta > 0$ such that $\rho(f(x), f(x_0)) < \epsilon$ whenever $d(x, x_0) < \delta$.

The following theorem is an obvious restatement of the definition.

**Theorem 4.3.1.** *Let $(X, d)$ and $(Y, \rho)$ be metric spaces. A function $f : X \to Y$ is continuous at $x_0$ if and only if the inverse image of an open ball in $Y$ centered at $f(x_0)$ contains an open ball in $X$ centered at $x_0$.* ∎

**Theorem 4.3.2.** *Let $f : (X, d) \to (Y, \rho)$. Then $f$ is continuous at $x_0$ if and only if, for a sequence $(x_n)$ in $X$ with $\lim_n x_n = x_0, \lim_n f(x_n) = f(x_0)$.*

*Proof. Suppose $f$ is continuous at $x_0$, and let $\lim_n x_n = x_0$. Given $\epsilon > 0$, there exists $\delta > 0$ such that $\rho(f(x), f(x_0)) < \epsilon$ whenever $d(x, x_0) < \delta$. Now there exists a natural number $N$ such that, for $n > N$, $d(x_n, x) < \delta$. Thus, for $n > N$, $\rho(f(x_n), f(x_0)) < \epsilon$, and $\lim_n f(x_n) = f(x_0)$. Conversely, if $f$ is not continuous at $x_0$, there exists $\epsilon > 0$ such that $f^{-1}(B(f(x_0), \epsilon))$ contains no open ball centered at $x_0$, and hence $B(x_0, 1/n) - f^{-1}(B(f(x_0), \epsilon)) \neq \emptyset$ for every $n \in \mathbb{N}$. Pick a point $x_n \in B(x_0, 1/n) - f^{-1}(B(f(x_0), \epsilon))$. Clearly, $\lim_n x_n = x_0$, but $\lim_n f(x_n) \neq f(x_0)$ because $\rho(f(x_n), f(x_0)) \geq \epsilon$ for all $n$.* ∎

Theorem 4.3.2 provides an extremely useful criterion for proving that a given function is continuous. It is called the **sequential characterization of continuity**. See examples 1 and 2 below.

**Definition.** A function $f$ from a metric space $(X, d)$ to a metric space $(Y, \rho)$ is **continuous** on $X$ if it is continuous at each point $x \in X$.

**Theorem 4.3.3.** *For a function f from a metric space $(X,d)$ to a metric space $(Y,\rho)$, the following are equivalent.*

*(a) f is continuous on X.*
*(b) The inverse image of an open subset of Y is an open subset of X.*
*(c) The inverse image of a closed subset of Y is a closed subset of X.*

*Proof.* (a) implies (b). Let $V$ be an open subset of $Y$, and let $x_0 \in f^{-1}(V)$. Since $V$ is open, there exists $\epsilon > 0$ such that $B(f(x_0),\epsilon) \subseteq V$. Since f is continuous at $x_0$, there exists $\delta > 0$ such that $f(B(x_0,\delta)) \subseteq B(f(x_0),\epsilon)$. Thus $f^{-1}(V) \supseteq f^{-1}(B(f(x_0),\epsilon)) \supseteq B(x_0,\delta)$. This proves that $f^{-1}(V)$ is open in $X$.

(b) implies (c). Let $F$ be a closed subset of $Y$. Then $Y - F$ is open in $Y$. By assumption, $f^{-1}(Y - F)$ is open in $X$. But $f^{-1}(Y - F) = X - f^{-1}(F)$; hence $f^{-1}(F)$ is closed in $X$.

(c) implies (a). Let $x_0 \in X$ and let $V = B(f(x_0),\epsilon)$; $Y - V$ is closed in $Y$, so, by assumption, $f^{-1}(Y - V) = X - f^{-1}(V)$ is closed in $X$, and hence $f^{-1}(V)$ is open. Because $x_0 \in f^{-1}(V)$, there exists $\delta$ such that $B(x_0,\delta) \subseteq f^{-1}(V)$. By theorem 4.3.1, f is continuous at $x_0$. ∎

**Example 1** (the **continuity of norms**). Let $(x_n)$ be a convergent sequence in a normed linear space, and suppose that $\lim_n x_n = x$. Then $\lim_n \|x_n\| = \|x\|$. This follows immediately from the fact that $|\|x_n\| - \|x\|| \leq \|x_n - x\|$. ◆

**Example 2** (the **continuity of inner products**). Let $(x_n)$ and $(y_n)$ be convergent sequences in an inner product space with limits $x$ and $y$, respectively. Then $\lim_n \langle x_n, y_n \rangle = \langle x, y \rangle$. First recall that convergent sequences are bounded. Thus there is a constant $M$ such that $\|y_n\| \leq M$. Now

$$
\begin{aligned}
|\langle x_n, y_n \rangle - \langle x, y \rangle| &= |\langle x_n - x, y_n \rangle + \langle x, y_n - y \rangle| \\
&\leq |\langle x_n - x, y_n \rangle| + |\langle x, y_n - y \rangle| \\
&\leq \|x_n - x\|\|y_n\| + \|x\|\|y_n - y\| \\
&\leq M\|x_n - x\| + \|x\|\|y_n - y\|] \to 0 \text{ as } n \to \infty. \blacklozenge
\end{aligned}
$$

**Definition.** Let $d_1$ and $d_2$ be metrics on the same underlying set $X$. We say that $d_1$ is **weaker** (or **coarser**) than $d_2$ if every $d_1$-open subset of $X$ is $d_2$-open. In this case, we also say that $d_2$ is **stronger** or **finer** than $d_1$.

**Example 3.** Let $(X, d_1)$ be any metric space, and let $d_2$ be the discrete metric on $X$. Clearly, $d_1$ is weaker than $d_2$. We will give more interesting examples later. ◆

**Theorem 4.3.4.** *A metric $d_1$ is weaker that another metric $d_2$ if and only if the identity function $I_X : (X, d_2) \to (X, d_1)$ is continuous.*

*Proof.* If $I_X : (X, d_2) \to (X, d_1)$ is continuous and $V$ is $d_1$-open, then $I_X^{-1}(V)$ is open in $d_2$. But $I_X^{-1}(V) = V$, so $d_1$ is weaker than $d_2$. The converse is proved by reversing the above reasoning. ∎

Now we discuss concrete criteria that guarantee that a metric $d_1$ is weaker than $d_2$. Since every $d_1$-open set is the union of $d_1$-open balls, it suffices to show that every $d_1$-open ball, $B_{d_1}(x, \delta)$ is $d_2$-open. Since every $y \in B_{d_1}(x, \delta)$ is the center of a ball $B_{d_1}(y, \delta') \subseteq B_{d_1}(x, \delta)$, it is further sufficient to show that every open ball $B_{d_1}(y, \delta')$ contains a $d_2$-open ball $B_{d_2}(y, \epsilon)$ for some $\epsilon > 0$. We apply the above strategy to prove the following theorem.

**Theorem 4.3.5.** *If there exists a real number $\alpha > 0$ such that $d_1(x, y) \leq \alpha d_2(x, y)$ for all $x, y \in X$, then $d_1$ is weaker than $d_2$.*

*Proof.* Consider a $d_1$-open ball $B_{d_1}(x, \delta)$, and let $\epsilon = \delta/\alpha$. It follows that $B_{d_2}(x, \epsilon) \subseteq B_{d_1}(x, \delta)$, because if $y \in B_{d_2}(x, \epsilon)$, then $d_1(x, y) \leq \alpha d_2(x, y) \leq \alpha \epsilon = \delta$. ∎

Now we look at more significant examples of the concepts we developed.

**Example 4.** It is clear that $l^1 \subseteq l^\infty$ since every absolutely convergent series is bounded. Thus the space $X = l^1$ has two metrics; the metric induced by the 1-norm $\|.\|_1$ and that induced by the infinity norm $\|.\|_\infty$. Since, for $x \in l^1$, $\|x\|_\infty \leq \|x\|_1$, and $d_\infty(x, y) = \|x - y\|_\infty \leq \|x - y\|_1 = d_1(x, y)$, the infinity metric on $X$ is weaker than the 1-metric on $X$. ◆

**Example 5.** Consider the space $X = \mathcal{C}[0, 1]$ under the uniform metric and the 1-metric. The identity function $I_X : (X, \|.\|_\infty) \to (X, \|.\|_1)$ is continuous since, for $f \in \mathcal{C}[0, 1]$, $\|f\|_1 \leq \|f\|_\infty$. By theorem 4.3.4, the 1-metric is weaker than the uniform metric. However, the identity function $I_X : (X, \|.\|_1) \to (X, \|.\|_\infty)$ is not continuous. To see this, consider the sequence (see section 3.6)

$$f_n(x) = \begin{cases} 2n^3 x & \text{if } 0 \leq x \leq \frac{1}{2n^2}, \\ -2n^3(x - \frac{1}{n^2}) & \text{if } \frac{1}{2n^2} \leq x \leq \frac{1}{n^2}, \\ 0 & \text{if } \frac{1}{n^2} \leq x \leq 1. \end{cases}$$

$\|f_n\|_1 = \frac{1}{2n}$; hence $f_n \to 0$ in the 1-norm, while $f_n$ does not converge in the uniform norm since $\|f_n\|_\infty = n$. ◆

**Definition.** Two metrics $d_1$ and $d_2$ on $X$ are **equivalent** if they generate the same collection of open sets. Thus $d_1$ and $d_2$ are equivalent if $d_1$ is weaker than $d_2$ and $d_2$ is weaker than $d_1$.

**Definition.** A bijection $f$ from a metric space $(X, d)$ to a metric space $(Y, \rho)$ is **bicontinuous** if $f$ and $f^{-1}$ are continuous.

By theorem 4.3.4, the following result is immediate.

**Theorem 4.3.6.** *Two metrics $d_1$ and $d_2$ on a set $X$ are equivalent if and only if the identity function $I_X : (X, d_1) \to (X, d_2)$ is bicontinuous.* ∎

Theorem 3.4.5 directly implies the following theorem.

**Theorem 4.3.7.** *If there exist positive constants $\alpha$ and $\beta$ such that $\beta d_2(x, y) \leq d_1(x, y) \leq \alpha d_2(x, y)$ for every $x, y \in X$, then $d_1$ and $d_2$ are equivalent.* ∎

The following theorem gives a sequential characterization of the equivalence of two metrics.

**Theorem 4.3.8.** *A necessary and sufficient condition for two metrics $d_1$ and $d_2$ on $X$ to be equivalent is that a sequence $(x_n)$ converges to $x$ in $d_1$ if and only if it converges to $x$ in $d_2$.*

*Proof.* Suppose $d_1$ and $d_2$ are equivalent, and let $\lim_n x_n = x$ in $d_1$. By theorem 4.3.6, $I_X : (X, d_1) \to (X, d_2)$ is continuous; hence, by the sequential characterization of continuity, $I_X(x_n) = x_n$ converges to $x$ in $d_2$. We leave the rest of the proof to the reader. ∎

**Example 6.** Let $X = \mathbb{R}^n$. The metrics induced by the 1-norm, the 2-norm, and the $\infty$-norm are all equivalent. To see this, we use theorem 4.3.7. The reader should work out the details. A partial list of the inequalities needed includes $\|x\|_1 \leq n\|x\|_\infty$ and $\|x\|_1 \leq \sqrt{n}\|x\|_2$. ♦

**Example 7.** Let $(X, d)$ be a metric space. Then the metric $\bar{d}(x, y) = min\{1, d(x, y)\}$ is equivalent to $d$. It is a simple exercise to show that $\bar{d}$ is a metric. The fact that the two metrics are equivalent follows from $B_d(x, \epsilon) \subseteq B_{\bar{d}}(x, \epsilon)$ and $B_{\bar{d}}(x, \delta) \subseteq B_d(x, \epsilon)$, where $\delta = min\{\epsilon, 1\}$. ♦

**Remarks.**

1. Important properties of metric spaces are often determined by the collection of open sets and not by the specific metric that generates the open sets. For

example, a function $f$ from a metric space $(X, d)$ to a metric space $(Y, \rho)$ is continuous if and only if the inverse image of an open subset of $Y$ is open in $X$. Clearly, if the metric $d$ is replaced with an equivalent metric $d_1$, then $f$ is continuous with respect to $d_1$. The same is true if $\rho$ is replaced with an equivalent metric. In many ways, the collection of open sets in a metric space is almost as intrinsic to the space as the specific metric that generates the open sets.

2. Not all metric properties are preserved under metric equivalence. Observe that the metric $\bar{d}$ in example 2 above is a bounded metric because $\bar{d}(X) = 1$, even though the metric $d$ may be unbounded. For example, the metric $\bar{d}$ on $\mathbb{R}$ is equivalent to the usual metric on $\mathbb{R}$. In particular, boundedness is not preserved under metric equivalence.

We include the following as another example of a bounded metric that is equivalent to an arbitrary metric.

**Example 8.** For an arbitrary metric space $(X, d)$, the metric $\bar{d}(x, y) = \frac{d(x,y)}{1+d(x,y)}$ is equivalent to $d$.

We show that $\bar{d}$ satisfies the triangle inequality and leave the rest of the details for the reader to verify. The function $f : [0, \infty) \to [0, 1)$ defined by $f(t) = \frac{t}{1+t}$ is increasing. Thus if $0 \le a \le b$, then $\frac{a}{1+a} \le \frac{b}{1+b}$. Replacing $a$ with $d(x, z)$, and $b$ with $d(x, y) + d(y, z)$, yields

$$
\begin{aligned}
\bar{d}(x, z) = \frac{d(x, z)}{1 + d(x, z)} &\le \frac{d(x, y) + d(y, z)}{1 + d(x, y) + d(y, z)} \\
&= \frac{d(x, y)}{1 + d(x, y) + d(y, z)} + \frac{d(y, z)}{1 + d(x, y) + d(y, z)} \\
&\le \frac{d(x, y)}{1 + d(x, y)} + \frac{d(y, z)}{1 + d(y, z)} = \bar{d}(x, y) + \bar{d}(y, z). \blacklozenge
\end{aligned}
$$

**Definition.** Let $(X, d)$ and $(Y, \rho)$ be metric spaces. A function $\varphi : X \to Y$ is said to be an **isometry** if, for every $x, y \in X$, $\rho(\varphi(x), \varphi(y)) = d(x, y)$. Notice that an isometry is always injective. We say that the metric spaces $(X, d)$ and $(Y, \rho)$ are **isometric** if there is a bijective isometry $\varphi : (X, d) \to (Y, \rho)$.

**Example 9 (linear isometries on $\mathbb{R}^n$).** Let $T$ be a linear isometry on $\mathbb{R}^n$. Then there exists an orthogonal matrix $P$ such that $T(x) = Px$ for every $x \in \mathbb{R}^n$. Observe that the converse was established in the exercises in section 3.7.

Let $P$ be the standard matrix of $T$. We prove that $P$ is orthogonal. The assumption is that, for $x, y \in \mathbb{R}^n$, $\|Px - Py\| = \|x - y\|$. In particular, taking

$y = 0$, we have $\|Px\| = \|x\|$. First we claim that, for $x, y \in \mathbb{R}^n$, $\langle Px, Py \rangle = \langle x, y \rangle$. This will conclude the proof because we then have

$$\langle P^T Px - x, y \rangle = \langle P^T Px, y \rangle - \langle x, y \rangle = \langle Px, Py \rangle - \langle x, y \rangle = 0.$$

Choosing $y = P^T Px - x$, we obtain $\langle P^T Px - x, P^T Px - x \rangle = 0$, or

$$(P^T P - I_n)x = 0.$$

Since $x$ is arbitrary, $P^T P - I_n = 0$.

We now prove the claim. The assumption that $\|Px - Py\|^2 = \|x - y\|^2$ yields $\langle Px - Py, Px - Py \rangle = \langle x - y, x - y \rangle$. Expanding the bilinear forms on the two sides of the last identity yields $\langle Px, Py \rangle = \langle x, y \rangle$, as claimed. ◆

Isometric spaces are virtually identical except for the nature of the elements of the spaces $X$ and $Y$ and the definition of the metrics $d$ and $\rho$. An isometry preserves all the metric properties of the space, including boundedness, which, as we saw, is not preserved under the equivalence of metrics. Another metric property that is preserved under isometries but not under metric equivalence is completeness. See section 4.6.

## Homeomorphisms

The concept of a homeomorphism is of central importance in topology. In the metric setting, isometry, although quite useful, is too stringent and does not preclude homeomorphisms from being useful. One can loosely think of a homeomorphism as a relaxation of the concept of isometry and an extension of the notion of metric equivalence.

**Definition.** Two metric spaces $(X, d)$ and $(Y, \rho)$ are **homeomorphic** if there exists a bicontinuous bijection $\varphi$ from $X$ to $Y$. The function $\varphi$ is called **a homeomorphism** from $X$ to $Y$.

**Example 10.** The open interval $(-1, 1)$ is homeomorphic to $\mathbb{R}$ (both sets have the usual metric). The function $f(t) = \frac{t}{1-t^2}$ maps $(-1, 1)$ bicontinuously onto $\mathbb{R}$. ◆

**Example 11.** The closed upper half plane $H = \mathbb{R} \times [0, \infty)$ is homeomorphic to the half-open strip $A = \mathbb{R} \times [0, 1)$. To see this, define $\varphi : H \to A$ by $\varphi(x, y) = (x, \frac{y}{1+y})$. It is a rather routine matter to verify that $\varphi$ is a bijection and that its inverse is $\varphi^{-1}(x, t) = (x, \frac{t}{1-t})$. ◆

**Example 12** (the **stereographic projection**).   Let $S^1 = \{(\xi_1, \xi_2) \in \mathbb{R}^2 : \xi_1^2 + \xi_2^2 - \xi_2 = 0\}$ be a circle of diameter 1 and centered at the point $(0, 1/2)$, and let $N = (0, 1)$ be the top point on the circle. Define the punctured circle to be the circle with the top point removed: $S^1_* = S^1 - \{N\}$. We give $S^1_*$ the Euclidean metric in the plane. Define a bijection $P : S^1_* \to \mathbb{R}$ as follows: for a point $\xi = (\xi_1, \xi_2) \in S^1_*$, $P(\xi)$ is the horizontal intercept of the line that contains the points $N$ and $\xi$, as shown in figure 4.1.

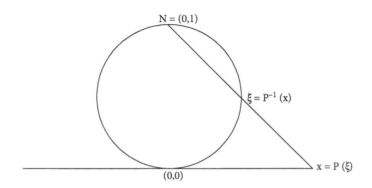

**Figure 4.1** The stereographic projection

The mapping $P$ is known as the stereographic projection of the punctured circle onto the real line. It is geometrically clear that $P$ is a bijection and that it is bicontinuous: the inverse image of a bounded open interval in $\mathbb{R}$ is an open arc on $S^1_*$, and conversely.

Explicit formulas exist for $P$ and $P^{-1}$. It is easier to derive the formula for $P^{-1}$ than to compute that for $P$. For a fixed $x \in \mathbb{R}$, the parametric equations of the line containing the points $N$ and $(x, 0)$ are $\xi_1 = xt, \xi_2 = 1 - t$, and $-\infty < t < \infty$. Finding the intersection point $\xi$ of the line and the circle yields the formula for $P^{-1}$ :

$$P^{-1}(x) = \xi = (\xi_1, \xi_2) = \left( \frac{x}{1 + x^2}, \frac{x^2}{1 + x^2} \right).$$

Inverting the above formulas, one obtains the following formula for the stereographic projection:

$$x = P(\xi) = \frac{\xi_1}{1 - \xi_2}.$$

We define the **chordal metric** $\chi(x, y)$ on $\mathbb{R}$ as follows: for two points $x, y \in \mathbb{R}$, $\chi(x, y)$ is the length of the chord of the circle that joins the points $P^{-1}(x)$ and $P^{-1}(y)$, hence the name *chordal metric*. Note that $\chi$ is the metric on $\mathbb{R}$ that makes the stereographic projection an isometry. Given the above formula for

$P^{-1}$, a direct calculation of the Euclidean distance between $P^{-1}(x)$ and $P^{-1}(y)$ yields

$$\chi(x,y) = \frac{|x-y|}{\sqrt{1+x^2}\sqrt{1+y^2}}.$$

Because the stereographic projection is a homeomorphism, the chordal metric $\chi$ is equivalent to the usual metric on $\mathbb{R}$. We will see in section 4.6 that the chordal metric is not complete. This illustrates the fact that completeness is not preserved under homeomorphisms. The reader will recall that boundedness is not preserved under metric equivalence. Saying that two metrics $d_1$ and $d_2$ on a space $X$ are equivalent is exactly the same as saying that the identity mapping $I_X : (X,d_1) \to (X,d_2)$ is a homeomorphism. The properties of a space that are preserved under homeomorphisms are called topological properties of the space. Compactness is the prime example of a topological property; see theorem 4.7.4. The fact that some metric properties, such as boundedness and completeness, fail to be hereditary under homeomorphisms is a rather inconvenient fact and does not diminish the usefulness of such properties. ◆

**Example 13.** Stereographic projections can be defined in all dimensions. Let $\mathcal{S}^n = \{\xi = (\xi_1, \xi_2, \ldots, \xi_{n+1}) \in \mathbb{R}^{n+1} : \sum_{i=1}^{n+1} \xi_i^2 - \xi_{n+1} = 0\}$ be the sphere in $\mathbb{R}^{n+1}$ of diameter 1 and center $(0,0,\ldots,0,1/2) \in \mathbb{R}^{n+1}$, and let $N = (0,0,\ldots,0,1) \in \mathcal{S}^n$. Define the **punctured sphere** $\mathcal{S}^n_* = \mathcal{S}^n - \{N\}$. The stereographic projection $P$ of $\mathcal{S}^n_*$ onto $\mathbb{R}^n$ maps a point $\xi = (\xi_1, \xi_2, \ldots, \xi_{n+1})$ on the punctured sphere to the intersection of the hyperplane $\xi_{n+1} = 0$ and the line that contains the points $\xi$ and $N$. As in the one-dimensional case, it is easier to compute the formula for $P^{-1}$ than that for $P$. The calculations needed for computing the formulas for $P$ and $P^{-1}$ are left as an exercise (see problem 18). The continuity of all the component functions shows that $P$ is a homeomorphism. See theorem 4.4.6. One can also define the chordal metric on $\mathbb{R}^n$ by $\chi(x,y) = \|P^{-1}(x) - P^{-1}(y)\|_2$. See problem 19 in the section exercises. ◆

One important special case is when $n = 2$. This is relevant to the one-point compactification of the plane; see section 5.10.

## Exercises

1. Let $\mathbb{K}$ denote the real or complex field with the usual metric. Prove that if $f$ and $g$ are continuous functions from a metric space $(X,d)$ to $\mathbb{K}$, then so are the functions $f \pm g$ and $fg$. If, in addition, $g(x) \neq 0$ for all $x \in X$, then $f/g$ is continuous.

2. Let $f$ be a continuous function from a metric space $X$ to a metric space $Y$, and let $g$ be a continuous function from $Y$ to a metric space $Z$. Prove that the composition $gof : X \to Z$ is continuous.

3. Let $f$ be a continuous function from a metric space $X$ to a metric space $Y$, and let $A \subseteq X$. Prove that the restriction of $f$ to $A$ is continuous when $A$ is given the restricted metric.

4. Fix an element $a$ of a metric space $X$, and define a function $f : X \to \mathbb{R}$ by $f(x) = d(x, a)$. Prove that $f$ is continuous.

5. Let $A$ be a fixed subset of a metric space $X$, and define a function $f : X \to \mathbb{R}$ by $f(x) = dist(x, A)$. Prove that $f$ is continuous.

6. Let $E$ be a closed subset of a metric space $X$, and let $a \in X - E$. Prove that there exists a continuous function $f : X \to \mathbb{R}$ such that $f(a) = 0, f(E) = 1$. Hint: Consider the function $f(x) = \dfrac{d(x,a)}{d(x,a) + dist(x,E)}$. Show how this result provides an alternative proof of theorem 4.2.12.

7. Let $E$ and $F$ be disjoint closed subsets of a metric space $X$. Prove that there exists a continuous function $f : X \to \mathbb{R}$ such that $f(E) = 0, f(F) = 1$. Show how this result provides an alternative proof of theorem 4.2.13.

8. Let $(X, d)$ be a metric space, and let $E$ and $F$ be disjoint closed subsets of $X$. Set $U = \{x \in X : dist(x, E) < dist(x, F)\}$, and $V = \{x \in X : dist(x, F) < dist(x, E)\}$. Show that $U$ and $V$ are open sets that separate $E$ and $F$.

9. Let $f$ and $g$ be continuous functions from a metric space $X$ to a metric space $Y$. Prove that $\{x \in X : f(x) \neq g(x)\}$ is an open subset of $X$.

10. Let $f$ and $g$ be continuous functions from a metric space $X$ to a metric space $Y$, and let $A$ be a subset of $X$ such that $f(x) = g(x)$ for every $x \in A$. Prove that $f(x) = g(x)$ for all $x \in \overline{A}$.

11. Prove that the metric $\overline{d}$ in example 8 is equivalent to $d$.

12. Show that the converse of theorem 4.3.5 is false by finding a metric $d_1$ that is weaker that another metric $d_2$ but where exists no constant $\alpha > 0$ such that $d_1(x, y) \leq \alpha d_2(x, y)$ for all $x, y \in X$.

13. Fix an element $x_0$ of a normed linear space $X$, and define a function $\varphi : X \to X$ by $\varphi(x) = x + x_0$. Show that $\varphi$ is an isometry.

14. Define a function $\varphi : X \to X$ on a normed linear space $X$ by $\varphi(x) = -x$. Show that $\varphi$ is an isometry.

15. Show that the open unit disk $U = \{(x, y) \in \mathbb{R}^2 : x^2 + y^2 < 1\}$ is homeomorphic to the plane $\mathbb{R}^2$. Hence show that the punched disk $U - \{(0, 0)\}$ is homeomorphic to the punctured plane $\mathbb{R}^2 - \{(0, 0)\}$. The same results extend to $\mathbb{R}^n$.

16. Let $f, g : \mathbb{R} \to \mathbb{R}$ be continuous functions such that $f(x) < g(x)$ for all $x \in \mathbb{R}$. Show that the region between the graphs, $\{(x, y) : x \in \mathbb{R}, f(x) < y < g(x)\}$, is homeomorphic to the open strip $\mathbb{R} \times (0, 1)$.

17. Let $X$ be a normed linear space. Prove that any two open balls in $X$ are homeomorphic. The same is true of closed balls.

18. The parametric equations of the line containing the points $(0, 0, \ldots, 1)$ and $(x_1, x_2, \ldots, x_n, 0)$ in $\mathbb{R}^{n+1}$ are

$$\xi = (\xi_1, \ldots, \xi_{n+1}) = (tx_1, tx_2, \ldots, tx_n, 1 - t).$$

Find the point of intersection of the line and the sphere,

$$\{\xi = (\xi_1, \xi_2, \ldots, \xi_{n+1}) : \sum_{i=1}^{n+1} \xi_i^2 - \xi_{n+1} = 0\},$$

to derive the formula for the inverse of the stereographic projection, $P^{-1}(x_1, \ldots, x_n) = (\xi_1, \ldots, \xi_{n+1})$, where

$$\xi_i = \frac{x_i}{1 + \|x\|_2^2}, 1 \leq i \leq n, \xi_{n+1} = \frac{\|x\|_2^2}{1 + \|x\|_2^2}.$$

Hence, by inverting the above formulas, derive the formula for the stereographic projection,

$$P(\xi_1, \ldots, \xi_{n+1}) = (x_1, \ldots, x_n), \text{ where } x_i = \frac{\xi_i}{1 - \xi_{n+1}}.$$

19. Derive the formula for the chordal metric on $\mathbb{R}^n$,

$$\chi(x, y) = \frac{\|x - y\|_2}{\sqrt{1 + \|x\|_2^2}\sqrt{1 + \|y\|_2^2}}.$$

## 4.4  Product Spaces

The Euclidean plane $\mathbb{R}^2$, as the product of two copies of $\mathbb{R}$, is the simplest example of a product space. We saw in section 4.3 that the Euclidean metric in the plane, although the most natural, is equivalent to several other metrics, including the $\infty$-metric, which, according to the definition below, is the product metric on $\mathbb{R}^2$. It is only natural to expect that the product of two open intervals should be an open subset of $\mathbb{R}^2$, and the definition we adopt for the product metric smoothly guarantees that. When we identify the complex field with $\mathbb{R}^2$, the convergence of a complex sequence $z_n = x_n + iy_n$ is equivalent to the convergence of its real and imaginary parts in $\mathbb{R}$, and one expects that product metrics in general should extend this property. Not only does the product metric preserve the componentwise convergence in the factor spaces, it is *characterized by it*. You will see that the product metric is the weakest metric that guarantees componentwise convergence in the factor spaces. Additionally, we will show that the product metric admits the continuity of the projections on the factor spaces and, once again, is characterized by it. We therefore think of the product metric as the most economical metric that generalizes the properties of Euclidean space in relation to its factor spaces.

Let $\{(X_i, d_i)\}_{i=1}^n$ be a finite set of metric spaces, and let $X = \prod_{i=1}^n X_i = \{(x_1, \ldots, x_n) : x_i \in X_i\}$ be the Cartesian product of the underlying sets $X_i$.

**Definition.** The **product metric** $D$ on $X$ is defined by

$$D(x, y) = max_{1 \leq i \leq n} d_i(x_i, y_i).$$

Here $x = (x_1, \ldots, x_n)$ and $(y_1, \ldots, y_n)$ are points in $X$. The verification that $D$ is a metric is straightforward.

**Example 1.** For $1 \leq i \leq n$, take $X_i = \mathbb{R}$, and let $d_i$ be the usual metric on $\mathbb{R}$. The product metric $D$ on $\prod_{i=1}^n X_i = \mathbb{R}^n$ is exactly the $\infty$-metric on $\mathbb{R}^n$. ♦

For $x \in X$ and $\delta > 0$, we denote the D-ball in $X$ of radius $\delta$ centered at $x$ by $B_D(x, \delta)$.

**Theorem 4.4.1.** *If $x \in X$ and $\delta > 0$, then*

$$B_D(x, \delta) = B_{d_1}(x_1, \delta) \times \ldots \times B_{d_n}(x, \delta).$$

*Proof. A point $y = (y_1, \ldots, y_n)$ is in $B_D(x, \delta)$ if and only if $max_{1 \leq i \leq n} d_i(x_i, y_i) < \delta$, if and only if $d_i(x_i, y_i) < \delta$ for each $1 \leq i \leq n$, if and only if $y_i \in D_{d_i}(x_i, \delta)$ for each $1 \leq i \leq n$, if and only if $y \in \prod_{i=1}^n B_{d_i}(x_i, \delta)$.* ∎

**Theorem 4.4.2.** *If $U_i$ is open in $X_i$ for each $1 \leq i \leq n$, then the set $U = \prod_{i=1}^n U_i$ is open in $(X, D)$.*

*Proof. Let $x \in \prod_{i=1}^n U_i$. Then $x_i \in U_i$, and hence there exists $\delta_i > 0$ such that $B_{d_i}(x_i, \delta_i) \subseteq U_i$. Let $\delta = min_{1 \leq i \leq n} \delta_i$. Clearly, $\prod_{i=1}^n B_{d_i}(x_i, \delta) \subseteq \prod_{i=1}^n B_{d_i}(x_i, \delta_i) \subseteq \prod_{i=1}^n U_i$. By theorem 4.4.1, $\prod_{i=1}^n B_{d_i}(x_i, \delta) = B_D(x, \delta)$. Thus $x \in B_D(x, \delta) \subseteq U$, which proves that $U$ is open.* ∎

**Remark.** For a fixed $1 \leq i \leq n$, let $U_i$ be an open subset of $X_i$. As an immediate consequence of theorem 4.4.2, the set $X_1 \times \ldots \times X_{i-1} \times U_i \times X_{i+1} \times \ldots \times X_n$ is open in $X$. It follows that $X - [X_1 \times \ldots \times X_{i-1} \times U_i \times X_{i+1} \times \ldots \times X_n] = X_1 \times \ldots \times X_{i-1} \times (X_i - U_i) \times X_{i+1} \times \ldots \times X_n$ is closed in $X$.

**Theorem 4.4.3.** *If* $F_1, \ldots, F_n$ *are closed subsets of* $X_1, \ldots, X_n$, *respectively, then* $F_1 \times \ldots \times F_n$ *is closed in* $X$.

*Proof.* Let $U_i = X_i - F_i$. Then $F = \prod_{i=1}^{n} F_i = \prod_{i=1}^{n}(X_i - U_i) = \cap_{i=1}^{n} X_1 \times \ldots \times X_{i-1} \times (X_i - U_i) \times X_{i+1} \times \ldots \times X_n]$. *By the above remark, each of the sets* $X_1 \times \ldots \times X_{i-1} \times (X_i - U_i) \times X_{i+1} \times \ldots \times X_n$ *is closed, and hence F is closed.* ∎

**Theorem 4.4.4.** *Suppose* $(X, D) = \prod_{i=1}^{n}(X_i, d_i)$. *Let* $(x^{(k)})_{k=1}^{\infty}$ *be a sequence in X, and let* $x = (x_1, \ldots, x_n) \in X$. *Write* $x^{(k)} = (x_1^{(k)}, \ldots, x_n^{(k)})$. *Then* $\lim_k x^{(k)} = x$ *in D if and only if* $\lim_k x_i^{(k)} = x_i$ *in* $d_i$ *for each* $1 \leq i \leq n$.

*Proof.* Because $d_i(x_i^{(k)}, x_i) \leq D(x^{(k)}, x)$, $\lim_k D(x^{(k)}, x) = 0$ *implies that* $\lim_k d_i(x_i^{(k)}, x_i) = 0$. *Conversely, if* $\lim_k x_i^{(k)} = x_i$ *in* $d_i$ *for each* $1 \leq i \leq n$, *then* $\lim_k max_{1 \leq i \leq n} d_i(x_i^{(k)}, x_i) = 0$, *and hence* $\lim_k x^{(k)} = x$. ∎

Theorem 4.4.4 says that the convergence of a sequence in the product metric $D$ is equivalent to the convergence of each of the component sequences (**componentwise convergence**). In fact, componentwise convergence characterizes all the metrics on $X$ that are equivalent to the product metric $D$, as the following theorem shows.

**Theorem 4.4.5.** *Suppose* $D^*$ *is a metric on the product space X where convergence in* $D^*$ *is equivalent to componentwise convergence. Then* $D^*$ *is equivalent to D.*

*Proof.* We use theorem 4.3.8. The metrics $D$ and $D^*$ are equivalent if and only if convergence of a sequence in one metric occurs if and only if it occurs in the other metric. Clearly, this is the case for $D$ and $D^*$, since convergence in either metric is equivalent to componentwise convergence. ∎

**Example 2.** To illustrate the importance of the above theorem, note that each of the following metrics are equivalent to the product metric $D$ on $X$:

$$D_1(x, y) = \sum_{i=1}^{n} d_i(x_i, y_i),$$

$$D_2(x, y) = \left( \sum_{i=1}^{n} d_i^2(x_i, y_i) \right)^{1/2}.$$

It is clear that convergence of a sequence in $(X, D_1)$ or $(X, D_2)$ occurs exactly when the component sequences converge.

Therefore we can use either of the metrics $D_1$ or $D_2$ or any other metric equivalent to $D$ as a definition of the product space, since they all generate the same collection of open sets. It is common to use whatever metric happens to be convenient in any particular situation. Theorem 4.4.5 also yields an equivalent metric for the product space if the metrics $d_1, \ldots, d_n$ are replaced with equivalent metrics. ◆

The following theorem is very useful in characterizing continuity of vector functions (functions into a product space). It says that a vector function is continuous exactly when its component functions are.

**Theorem 4.4.6.** *Let $(Y, \rho)$ be a metric space, let $f : Y \to \prod_{i=1}^{n} X_i$, and write $f(y) = (f_1(y), \ldots, f_n(y))$. Then $f$ is continuous if and only if each of the component functions $f_i : Y \to X_i$ is continuous.*

*Proof. Let $(y^{(k)})$ be a sequence in $Y$, and suppose $\lim_k y^{(k)} = y$. By theorem 4.4.4, $\lim_k f(y^{(k)}) = f(y)$ if and only if $\lim_k f_i(y^{(k)}) = f_i(y)$ for all $1 \le i \le n$.* ■

**Example 3.** We used the previous theorem to prove the continuity of the stereographic projections. See example 12 on section 4.3, and problem 18 on the same section.

## Exercises

1. Let $\{(X_i, d_i)\}_{i=1}^{n}$ be a finite set of metric spaces. Prove that $X_1 \times \ldots \times X_n$ is isometric to $X_1 \times (X_2 \times \ldots \times X_n)$.
2. Let $\{(X_i, d_i)\}_{i=1}^{n}$ be a finite set of metric spaces, and let $(X, D)$ be their product.
   (a) Prove directly, using the definition of $D$, that the projections $\pi_i : X \to X_i$ are continuous. It follows that $\pi_i^{-1}(U_i)$ is open in $X$ for every open subset $U_i$ of $X_i$.
   (b) It then follows from part (a) that, for open subsets $U_i \subseteq X_i$, $1 \le i \le n$, $\cap_{i=1}^{n} \pi_i^{-1}(U_i)$ is open. What is $\cap_{i=1}^{n} \pi_i^{-1}(U_i)$?
3. When you have solved exercise 2 above, you will have an alternative proof of theorem 4.4.2. Do you see it?
4. Consider the metric $d$ on a metric space $X$ as a function on the product space $X \times X$, endowed with the product metric. Prove that $d : X \times X \to \mathbb{R}$ is continuous.
5. Let $\{(X_i, d_i)\}_{i=1}^{n}$ be a finite set of metric spaces, and let $X = \prod_{i=1}^{n} X_i$ be the Cartesian product of the underlying sets. Prove that the product metric is the

weakest metric relative to which all the projections $\pi_i$ are continuous. More explicitly stated, show that if $D^*$ is a metric on $X$ and each $\pi_i : (X, D^*) \rightarrow (X_i, d_i)$ is continuous, then the product metric is weaker that $D^*$.

## 4.5  Separable Spaces

Although the rigorous definition of the real line was a giant leap in the development of mathematics, it would not be nearly as useful an invention had it not been for the fact that it contains the rational numbers as a dense subset. Indeed, all practical computations, including machine calculations, are done exclusively using rational numbers. The simplicity of rational numbers is enhanced by their countability. Thus $\mathbb{Q}$ *is numerous enough, simple enough, but not too enormous* to be a useful approximation of $\mathbb{R}$. It is a reasonable quest to study metric spaces that contain a countable dense subset (*of simpler elements*). Such spaces are, by definition, separable. You will see that many (but not all) metric spaces are separable. The classical example is the space $\mathcal{C}[0,1]$. It is well known that (see section 4.8) the set of polynomials with rational coefficients, which is countable, is dense in $\mathcal{C}[0,1]$. What can be a nicer approximation of a continuous function than a rational polynomial! Separability of a metric space turns out to be equivalent to the existence of a countable collection of open sets that generate all open sets, which is an added benefit and an important characterization of separability.

**Definition.** A subset $A$ of a metric space $X$ is **dense** in $X$ if $\overline{A} = X$. By theorem 4.2.5, $A$ is dense in $X$ if and only if every point in $X$ is the limit of a sequence in $A$. Equivalently, $A$ is dense in $X$ if and only if for every $x \in X$ and every $\epsilon > 0$, there is an element $a \in A$ such that $d(x, a) < \epsilon$.

**Example 1.** Given a function $f \in \mathcal{C}[0,1]$ and a number $\epsilon > 0$, there exists a continuous, piecewise linear function $g$ such that $\|f - g\|_\infty < \epsilon$.

We use the uniform continuity of $f$ (see example 8 on section 1.2). Let $\delta > 0$ be such that $|f(x) - f(y)| < \epsilon$ whenever $|x - y| < \delta$. Choose a natural number $n$ such that $1/n < \delta$, and, for $0 \leq j \leq n$, let $x_j = j/n$. Define the function $g$ to be the continuous, piecewise linear function such that $g(x_j) = f(x_j)$ for $0 \leq j \leq n$. By construction, $\|f - g\|_\infty < \epsilon$. Observe that this example says that the space of continuous, piecewise linear functions is dense in $\mathcal{C}[0,1]$. ◆

**Definition.** A metric space is **separable** if it contains a countable dense subset.

**Example 2.** Since $\mathbb{Q}$ is dense in $\mathbb{R}$, $\mathbb{R}$ is separable. More generally, $\mathbb{Q}^n$ is countable and dense in $\mathbb{R}^n$; hence $\mathbb{R}^n$ is separable. ◆

**Example 3.** The metric space $l^1$ is separable. Let $A = \{(a_1, a_2, \ldots, a_n, 0, 0, \ldots) : n \in \mathbb{N}, a_i \in \mathbb{Q}\}$ be the subset of $l^1$ that contains all of the sequences with finitely many nonzero rational terms. As $A$ is countable, we need to show that it is dense in $l^1$. Let $x = (x_i) \in l^1$, and let $\epsilon > 0$. Since $\sum_{i=1}^{\infty} |x_i|$ is convergent, there exists $N \in \mathbb{N}$ such that $\sum_{i=N+1}^{\infty} |x_i| < \epsilon/2$. For $1 \leq i \leq N$, choose $a_i \in \mathbb{Q}$ such that $|x_i - a_i| < \epsilon/(2N)$, and let $a = (a_1, \ldots, a_n, 0, 0, \ldots)$. Now $\|x - a\|_1 = \sum_{i=1}^{N} |x_i - a_i| + \sum_{i=N+1}^{\infty} |x_i| < N\frac{\epsilon}{2N} + \epsilon/2 = \epsilon.$ ◆

**Example 4.** The space $c$ of convergent sequences is separable.

Let $A = \{(a_1, a_2, \ldots, a_n, a, a, a, \ldots) : n \in \mathbb{N}, a_i \in \mathbb{Q}, a \in \mathbb{Q}\}$ be the set of rational sequences that are eventually constant. We show that $A$ is dense in $c$. Let $x = (x_n)$ be a convergent sequence, and let $\xi = \lim_n x_n$. For $\epsilon > 0$, there is an integer $N$ such that, for $n > N$, $|x_n - \xi| < \epsilon/2$. For $1 \leq i \leq N$, choose $a_i \in \mathbb{Q}$ such that $|x_i - a_i| < \epsilon$, and then choose a rational number $a$ such that $|\xi - a| < \epsilon/2$. Finally, set $y = (a_1, \ldots, a_n, a, a, \ldots)$. By construction, $\|x - y\|_\infty < \epsilon.$ ◆

**Definition.** A collection $\mathfrak{B}$ of open subsets of a metric space $X$ is said to be an **open base** for $X$ if every open subset of $X$ is the union of members of $\mathfrak{B}$.

**Example 5.** The collection of open sets $\mathfrak{B} = \{B(x, r) : x \in \mathbb{Q}^n, r \in \mathbb{Q}\}$ of open balls in $\mathbb{R}^n$ that have rational centers and rational radii is an open base for the Euclidean metric on $\mathbb{R}^n$. This takes some verification, and we urge the reader to work out the details. ◆

**Definition.** A metric space $X$ is **second countable** if it has a countable open base.

The above example shows that $\mathbb{R}^n$ is second countable because the collection $\mathfrak{B}$ is countable.

**Definition.** A collection of open subsets $\mathcal{U} = \{U_\alpha\}_{\alpha \in I}$ of a metric space $X$ covers $X$ if $\cup_{\alpha \in I} U_\alpha = X$. We also say that $\{U_\alpha\}$ is an **open cover** of $X$. A subset of $\mathcal{U}$ that also covers $X$ is said to be a **subcover** of $\mathcal{U}$.

**Example 6.** The collection $\mathcal{U} = \{(-n, n) : k \in \mathbb{N}\}$ is an open cover of $\mathbb{R}$. The subset $\{(-2n, 2n) : n \in \mathbb{N}\}$ is a subcover of $\mathcal{U}$. ◆

**Definition.** A metric space $X$ is said to be a **Lindelöf space** if every open cover of $X$ contains a countable subcover of $X$.

**Theorem 4.5.1.** *The following are equivalent for a metric space $X$*

(a) X is separable.

(b) X is second countable.

(c) X is Lindelöf.

*Proof.* (a) implies (b). Let $A = \{a_1, a_2, \ldots\}$ be a countable dense subset of X. We claim that the countable collection $\mathfrak{B} = \{B(a_n, r) : n \in \mathbb{N}, r \in \mathbb{Q}\}$ is an open base for X. To prove that every open subset of X is the union of members of $\mathfrak{B}$, it is sufficient to show that if $x \in X$ and $\delta > 0$, there exist an element $a_n \in A$ and a rational number r such that $x \in B(a_n, r) \subseteq B(x, \delta)$. Pick an element $a_n \in A$ such that $d(x, a_n) < \delta/4$, and choose a rational number r such that $\delta/4 < r < \delta/2$. Then $x \in B(a_n, r)$, and if $y \in B(a_n, r)$, then $d(x, y) \leq d(x, a_n) + d(a_n, y) < \delta/4 + r < \delta$, so $B(a_n, r) \subseteq B(x, \delta)$.

(b) implies (c). Let $\mathfrak{B} = \{B_n : n \in \mathbb{N}\}$ be a countable open base for X. Suppose, for some collection $\mathcal{U} = \{U_\alpha : \alpha \in I\}$ of open subsets of X, $X = \cup_{\alpha \in I} U_\alpha$. For each natural number n, pick an element $V_n$ in $\mathcal{U}$ that contains $B_n$. If no element of $\mathcal{U}$ contains $B_n$, define $V_n = \varnothing$. We claim that $\{V_n\}_{n \in \mathbb{N}}$ covers X. If $x \in X$, then $x \in U_\alpha$ for some $\alpha \in I$. There exists $B_n$ such that $x \in B_n \subseteq U_\alpha$; thus, $V_n \neq \varnothing$ and $x \in V_n$.

(c) implies (a). For a fixed $n \in \mathbb{N}, X = \cup_{x \in X} B(x, 1/n)$. By assumption, there exists a set $\{x_{n,1}, x_{n,2} \ldots\}$ such that $X = \cup_{j=1}^{\infty} B(x_{n,j}, 1/n)$. We claim that $\{x_{n,j} : n, j \in \mathbb{N}\}$ is dense in X. Let $x \in X$ and let $\delta > 0$. Choose $n \in \mathbb{N}$ such that $1/n < \delta$. Because $x \in \cup_{j=1}^{\infty} B(x_{n,j}, 1/n), x \in B(x_{n,j}, 1/n)$ for some $j \in \mathbb{N}$. Now $d(x_{n,j}, x) < 1/n < \delta$, and the proof is complete. ∎

The following example shows that a separable metric space is, in a way, *not too large*.

**Example 7.** The cardinality of a separable metric space is, at most, $\mathfrak{c}$.

Let $A = \{a_n : n \in \mathbb{N}\}$ be a countable dense subset of a separable metric space X. For each $x \in X$, define a real sequence $S_x = (d(x, a_1), d(x, a_2), d(x, a_3), \ldots)$ We prove that the function $x \mapsto S_x$ is an injection from X to the space $\mathbb{R}^{\mathbb{N}}$ of real sequences. Let x and y be distinct elements of X, and let $r = d(x, y)$. Since A is dense in X, there exists an element $a_n$ of A such that $d(x, a_n) < r/3$. It is easy to see that $d(y, a_n) \geq 2r/3$. In particular, $d(x, a_n) \neq d(y, a_n)$; thus S is an injection. It follows that $Card(X) \leq Card(\mathbb{R}^{\mathbb{N}}) = \mathfrak{c}^{\aleph_0} = \mathfrak{c}$. ◆

## Exercise

1. Show that $l^p$ is separable for all $1 \leq p < \infty$.
2. Show that the space $c_0$ of null sequences is separable.

3. Prove that a subset $A$ of a metric space $X$ is dense if and only if it intersects every nonempty open subset of $X$.

4. Prove that if $X$ is separable, then any collection of pairwise disjoint open subsets of $X$ is countable. Conclude that $l^\infty$ is not separable. See problem 10 on section 4.1.

5. Show that a normed linear space $X$ is separable if and only if the unit ball $\{x \in X : \|x\| \le 1\}$ is separable.

6. Show that a subspace of a second countable space is second countable.

7. Show that a subspace of a separable space is separable and that a subspace of a Lindelöf space is Lindelöf.

8. Show that the product of finitely many separable metric spaces is separable.

9. Let $f$ be a function from a metric space $X$ to a metric space $Y$, and let $\mathfrak{B}$ be an open base for $Y$. Prove that $f$ is continuous if and only if $f^{-1}(V)$ is open for every $V \in \mathfrak{B}$.

10. Prove that every open subset $V$ of $\mathbb{R}$ is the countable union of disjoint open intervals. Hint: For $x \in V$, let $a_x = inf\{t \in V : (t,x] \subseteq V\}$, and $b_x = sup\{t \in V : [x,t) \subseteq V\}$. Show that $I_x = (a_x, b_x) \subseteq V$, then prove that, for $x, y \in V$, either $I_x = I_y$ or $I_x \cap I_y = \emptyset$.

## 4.6 Completeness

The mathematicians of antiquity had a clear understanding of the existence of irrational numbers, and mathematicians through the ages understood that irrational numbers are gaps inside the rational number field. Thus it was quite well understood that the rational field is not complete. It took some twenty-four centuries for a rigorous definition of the real number field as a complete ordered field to materialize. The definitions and some of the results in this section parallel those in section 1.2. For example, the proof of the Bolzano-Weierstrass property of bounded sets (theorem 1.2.10) includes a proof of the nested interval theorem, which is a very special case of the Cantor intersection theorem. Another highlight of this section is Baire's theorem, which is one of the cornerstones upon which functional analysis is built. We will establish the completeness of the $l^p$ spaces as well as the function space $\mathcal{C}[a, b]$, which will pave the way for a number of interesting applications begun in the section and continued in the section exercises.

**Definition.** A sequence $(x_n)$ in a metric space $X$ is said to be a **Cauchy sequence** if, for every $\epsilon > 0$, there is a natural number $N$ such that,

$$\text{for all } m, n > N, d(x_n, x_m) < \epsilon.$$

**Theorem 4.6.1.** *A convergent sequence is a Cauchy sequence.*

*Proof.* Let $\lim x_n = x$, and let $\epsilon > 0$. There exists a natural number $N$ such that, for $n > N$, $d(x_n, x) < \epsilon/2$. Now, for $m, n > N$, $d(x_n, x_m) \leq d(x_n, x) + d(x, x_m) < \epsilon$. ∎

**Theorem 4.6.2.** *A Cauchy sequence is bounded.*

*Proof.* Let $\epsilon = 1$. There exists a positive integer $N$ such that, for $m, n \geq N$, $d(x_n, x_m) < 1$. In particular, $d(x_n, x_N) < 1$ for all $n \geq N$. Therefore, for all $n \in \mathbb{N}$, $d(x_n, x_N) \leq \max\{1, d(x_1, x_N), \ldots, d(x_{N-1}, x_N)\}$. ∎

**Theorem 4.6.3.** *If a Cauchy sequence $(x_n)$ contains a subsequence $x_{n_k}$ that converges to $x$, then $(x_n)$ converges to $x$.*

*Proof.* Let $\epsilon > 0$. There exists a natural number $N$ such that, for $m, n \geq N$, $d(x_n, x_m) < \epsilon/2$. Since $\lim_k x_{n_k} = x$, there exists an integer $K$ such that, for $k \geq K$, $d(x_{n_k}, x) < \epsilon/2$. Without loss of generality, we may assume that $K > N$ and thus $n_K \geq K > N$. Taking $m = n_K$ and using the triangle inequality, for $n > N$, $d(x_n, x) \leq d(x_n, x_{n_K}) + d(x_{n_K}, x) < \epsilon$. ∎

**Definition.** A metric space $X$ is **complete** if every Cauchy sequence in $X$ converges to a point in $X$.

Before we look at major examples of complete spaces, we look at an example of an incomplete one.

**Example 1.** Consider the chordal metric $\chi$ on $\mathbb{R}$. We will show that although the sequence $x_n = n$ is a Cauchy sequence in $(\mathbb{R}, \chi)$, it does not converge.

$$\chi(n, m) = \frac{|n - m|}{\sqrt{1 + n^2}\sqrt{1 + m^2}} = \frac{|\frac{1}{n} - \frac{1}{m}|}{\sqrt{1 + \frac{1}{n^2}}\sqrt{1 + \frac{1}{m^2}}} \leq |\frac{1}{n} - \frac{1}{m}| \to 0 \text{ as } m, n \to \infty.$$

To prove that the sequence does not converge to any $x \in \mathbb{R}$, we observe that

$$\lim_n \chi(n, x) = \lim_n \frac{|n - x|}{\sqrt{1 + n^2}\sqrt{1 + x^2}} = \lim_n \frac{|1 - \frac{x}{n}|}{\sqrt{1 + \frac{1}{n^2}}\sqrt{1 + x^2}} = \frac{1}{\sqrt{1 + x^2}} \neq 0. \blacklozenge$$

**Theorem 4.6.4.**
(a) *A closed subspace $A$ of a complete metric space is complete.*
(b) *A complete subspace $A$ of a metric space is closed.*

*Proof.* (a) Let $(x_n)$ be a Cauchy sequence in A. Since X is complete, there exists $x \in X$ such that $\lim_n x_n = x$. Since A is closed, theorem 4.2.5 guarantees that $x \in A$.

(b) Let $x \in \overline{A}$. By theorem 4.2.5, there exists a sequence $(x_n)$ in A such that $\lim_n x_n = x$. Now $(x_n)$ is Cauchy (theorem 4.6.1), and A is complete, so $(x_n)$ converges to a point y in A. By the uniqueness of limits, $x = y$. ∎

**Example 2.** The space $c$ of convergent sequences is complete. We show that $c$ is complete by showing that it is closed in $l^\infty$. See problem 2 in the section exercises. Let $x = (x_n) \in l^\infty$ be a closure point of $c$. We show that $x$ is convergent by showing that it is Cauchy. Let $\epsilon > 0$, and choose a convergent sequence $y = (y_n)$ such that $\|x - y\|_\infty < \epsilon$. Because $(y_n)$ is Cauchy, there exists an integer $N$ such that, for $n, m > N$, $|y_n - y_m| < \epsilon$. Now if $n, m > N$, then

$$|x_n - x_m| \leq |x_n - y_n| + |y_n - y_m| + |y_m - x_m| < 3\epsilon. \blacklozenge$$

**Theorem 4.6.5.** *The sequence spaces $l^p$ are complete for $1 \leq p \leq \infty$.*

*Proof.* We leave the proof of the case $p = \infty$ to the reader (also see theorem 4.8.1). Fix $1 \leq p < \infty$, let $(x_n)$ be a Cauchy sequence in $l^p$, and write $x_n = (x_{n,1}, x_{n,2}, \ldots, x_{n,k}, \ldots)$. Given $\epsilon > 0$, there exists $N \in \mathbb{N}$ such that, for $n, m > N$, $\|x_n - x_m\|_p^p = \sum_{k=1}^\infty |x_{n,k} - x_{m,k}|^p < \epsilon^p$. In particular, if $k$ is a fixed positive integer, then, for every $n, m > N$, $|x_{n,k} - x_{m,k}| < \epsilon$. Thus $(x_{n,k})_{n=1}^\infty$ is a Cauchy sequence in $\mathbb{K}$. By the completeness of $\mathbb{K}$, $x_k = \lim_n x_{n,k}$ exists for every $k \in \mathbb{N}$. Set $x = (x_k)_{k=1}^\infty$. We will show that $x \in l^p$ and that $\lim_n \|x_n - x\|_p = 0$. For an arbitrary positive integer K,

$$\sum_{k=1}^K |x_k|^p = \lim_n \sum_{k=1}^K |x_{n,k}|^p \leq \limsup_n \sum_{k=1}^\infty |x_{n,k}|^p = \limsup_n \|x_n\|_p^p.$$

Because $(x_n)$ is Cauchy, $\|x_n\|_p$ is bounded by theorem 4.6.2; hence $\limsup_n \|x_n\|_p^p < \infty$. This shows $\sum_{k=1}^\infty |x_k|^p < \infty$, and hence $x \in l^p$.

Finally we show that $\|x_n - x\|_p \to 0$, as $n \to \infty$. For arbitrary positive integers n and K,

$$\sum_{k=1}^K |x_{n,k} - x_k|^p = \lim_{m \to \infty} \sum_{k=1}^K |x_{n,k} - x_{m,k}|^p \leq \limsup_{m \to \infty} \|x_n - x_m\|_p^p.$$

Taking the limit as $K \to \infty$ of the extreme left side of the above string of inequalities, we have

$$\|x_n - x\|_p^p = \sum_{k=1}^\infty |x_{n,k} - x_k|^p \leq \limsup_{m \to \infty} \|x_n - x_m\|_p^p.$$

*Observe that the above inequalities hold for an arbitrary positive integer n. Now let $\epsilon > 0$. There exists $N \in \mathbb{N}$ such that, for $m, n > N, \|x_n - x_m\|_p < \epsilon$. Thus, for $n > N$, $\limsup_{m \to \infty} \|x_n - x_m\|_p \leq \epsilon$. We have shown that, for $n > N, \|x_n - x\|_p \leq \epsilon$. This completes the proof.* ∎

**Theorem 4.6.6 (the Cantor intersection theorem).** *Suppose $X$ is a complete metric space. If $\{F_n\}_{n=1}^{\infty}$ is a descending sequence of closed nonempty subsets of $X$ such that $\lim_n diam(F_n) = 0$, then $\cap_{n=1}^{\infty} F_n$ is a one-point set.*

*Proof. For every $n \in \mathbb{N}$, choose a point $x_n \in F_n$. Let $\epsilon > 0$. There exists a natural number $N$ such that, for $n > N, diam(F_n) < \epsilon$. Now if $m \geq n > N$, then $F_n \supseteq F_m$ and $x_n, x_m \in F_n$, and hence $d(x_n, x_m) < \epsilon$. This makes $(x_n)$ a Cauchy sequence and hence convergent to, say, $x$. Each of the sets $F_n$ contains all but a finite number of terms of $(x_k)$. Since each $F_n$ is closed, $x \in F_n$ for all $n$, and $x \in \cap_{n=1}^{\infty} F_n$. Now $diam(\cap_{n=1}^{\infty} F_n) \leq diam(F_n) \to 0$. Hence $\cap_{n=1}^{\infty} F_n = \{x\}$.* ∎

**Definition.** A subset $A$ of a metric space $X$ is **nowhere dense** if $int(\overline{A}) = \emptyset$.

**Example 3.** $\mathbb{N}$ is nowhere dense in $\mathbb{R}$. The reader is cautioned that $\overline{int(A)} \neq int(\overline{A})$; for example, $\overline{int(\mathbb{Q})} = \emptyset$, while $int(\overline{\mathbb{Q}}) = \mathbb{R}$. ♦

**Theorem 4.6.7 (Baire's theorem).** *A complete metric space cannot be expressed as a countable union of nowhere dense subsets.*

*Proof. Let $\{A_n\}$ be a countable family of nowhere dense subsets of $X$. Without loss of generality, assume that each $A_n$ is closed. Since $X - A_1$ is open and nonempty, there exists a ball $B_1 = B(x_1, \delta_1)$ such that $B_1 \cap A_1 = \emptyset$. By reducing the radius $\delta_1$, if necessary, we may assume that $\delta_1 < 1$ and that $\overline{B_1} \cap A_1 = \emptyset$. Since $B_1 - A_2$ is open and nonempty, we can find a ball $B_2 = B(x_2, \delta_2)$ such that $B_2 \cap A_2 = \emptyset$. As before, we may assume that $\delta_2 < 1/2$ and $\overline{B_2} \cap A_2 = \emptyset$. We can continue this process and construct a sequence of balls $\{B_n\}$ such that $\overline{B_n} \cap A_n = \emptyset$, $\overline{B_1} \supseteq \overline{B_2} \supseteq \ldots$, and $diam(\overline{B_n}) \leq 2/n$. By the Cantor intersection theorem, $\cap_{n=1}^{\infty} \overline{B_n} = \{x\}$. Since $\overline{B_n} \cap A_n = \emptyset, x \notin A_n$ for all $n \in \mathbb{N}$, and $\cup_{n=1}^{\infty} A_n \neq X$.* ∎

The following two results are powerful consequences of Baire's theorem.

**Theorem 4.6.8.** *Let $\{A_n\}$ be a countable family of closed nowhere dense subsets of a complete metric space $X$, and let $U_0$ be a nonempty open subset of $X$. Then $U_0 - \cup_{n=1}^{\infty} A_n \neq \emptyset$.*

*Proof. Modify the proof of Baire's theorem by requiring that $B_1 \subseteq U_0 - A_1$.* ∎

This result generalizes Baire's theorem: not only is the set $A = \cup_{n=1}^{\infty} A_n$ not equal to $X$, but $A$, in fact, has an empty interior.

**Theorem 4.6.9.** *If $\{U_n\}$ is a countable collection of open dense subsets of a complete metric space, then $\cap_{n=1}^{\infty} U_n$ is dense.*

*Proof.* *If there is a nonempty open subset $U_0$ such that $U_0 \cap \cap_{n=1}^{\infty} U_n = \emptyset$, then $U_0 \subseteq X - \cap_{n=1}^{\infty} U_n = \cup_{n=1}^{\infty}(X - U_n)$. This contradicts the previous theorem because each $X - U_n$ is closed and nowhere dense. See problem 10 at the end of the section.* ∎

**Theorem 4.6.10.** *The product of a finite number $\{(X_i, d_i)\}_{i=1}^{n}$ of complete metric spaces is complete.*

*Proof.* *Since $X_1 \times \dots \times X_n$ is isometric to $X_1 \times (X_2 \times \dots \times X_n)$, it is enough to show that the product of two complete metric spaces, $(X_1, d_1)$ and $(X_2, d_2)$, is complete. Let $(x^{(k)})$ be a Cauchy sequence in $X_1 \times X_2$, and write $x^{(k)} = (x_1^{(k)}, x_2^{(k)})$. Since $d_i(x_i^{(n)}, x_i^{(m)}) \leq D(x^{(n)}, x^{(m)})$, each of the sequences $(x_i^{(k)})$ is Cauchy. Therefore, for $i = 1, 2$, $\lim_k x_i^{(k)} = x_i$ exists. Clearly, $\lim_k x^{(k)} = (x_1, x_2)$.* ∎

Before we embark on the application subsection, we prove the following result.

**Theorem 4.6.11.** *The space $(\mathcal{C}[a,b], \|.\|_{\infty})$ is complete.*

*Proof.* *Let $(f_n)$ be a Cauchy sequence in $\mathcal{C}[a,b]$. For $\epsilon > 0$, there is a positive integer $N$ such that $\|f_n - f_m\|_{\infty} < \epsilon$ for every $m, n \geq N$. Thus, for every $x \in [a,b]$ and every $m, n \geq N$, $|f_n(x) - f_m(x)| < \epsilon$; hence $(f_n(x))$ is a Cauchy sequence for every $x \in [a,b]$. By the completeness of $\mathbb{K}$, $f(x) = \lim_n f_n(x)$ exists for every $x$.*

*We claim that $\lim_n \|f_n - f\|_{\infty} = 0$. Let $\epsilon$ and $N$ be as in the previous paragraph. Then $|f_n(x) - f_m(x)| < \epsilon$ for every $x \in [a,b]$ and every $n, m \geq N$. Taking the limit as $m \to \infty$, we obtain $|f_n(x) - f(x)| < \epsilon$ for every $x \in [a,b]$ and every $n \geq N$. This means that $\|f_n - f\|_{\infty} < \epsilon$, as claimed.*

*Finally, we need to show that $f$ is continuous. Suppose that $x_k \in [a,b]$ and that $\lim_k x_k = x$. Let $\epsilon > 0$. By the previous paragraph, there is an integer $N$ such that $\|f_N - f\|_{\infty} < \epsilon$. By the continuity of $f_N$ at $x$, there exists an integer $K$ such that, for $k > K$, $|f_N(x_k) - f_N(x)| < \epsilon$. Now, for $k > K$,*

$$|f(x) - f(x_k)| \leq |f(x) - f_N(x)| + |f_N(x) - f_N(x_k)| + |f_N(x_k) - f(x_k)| < 3\epsilon. \quad \blacksquare$$

**Example 4** (the **Weierstrass M-test**). Let $f_n$ be a sequence in $\mathcal{C}[a,b]$, and suppose that there exists a real sequence $(M_n)$ such that, for every $n \in \mathbb{N}$, $\|f_n\|_\infty \leq M_n$ and $\sum_{n=1}^{\infty} M_n < \infty$. Then the series of functions $\sum_{n=1}^{\infty} f_n(x)$ converges in $\mathcal{C}[a,b]$.

We prove that the sequence of partial sums $S_n(x) = \sum_{i=1}^{n} f_i(x)$ is a Cauchy sequence in $\mathcal{C}[a,b]$. Let $\epsilon > 0$. By the convergence of the positive series $\sum_{n=1}^{\infty} M_n$, there is an integer $N$ such that, for $m > n > N$, $\sum_{i=n+1}^{m} M_i < \epsilon$.[4] Thus, for $m > n > N$, and for every $x \in [a,b]$, $|S_m(x) - S_n(x)| \leq \sum_{i=n+1}^{m} |f_i(x)| \leq \sum_{i=n+1}^{m} M_i < \epsilon$, or $\|S_m - S_n\|_\infty < \epsilon$. This shows that $S_n$ is a Cauchy sequence and hence, by the completeness of $\mathcal{C}[a,b]$, is convergent to a function $f \in \mathcal{C}[a,b]$. ♦

In fact, the series $\sum_{=1}^{\infty} f_n(x)$ converges to $f$ absolutely as well as uniformly to $f$ on $[a,b]$.

**Example 5.** If a a sequence $f_n$ converges to $f$ in $\mathcal{C}[a,b]$, then

$$\lim_n \int_a^b f_n(x)dx = \int_a^b f(x)dx.$$

In particular, if the series $\sum_{n=1}^{\infty} g_n(x)$ converges in $\mathcal{C}[a,b]$, then

$$\int_a^b \sum_{n=1}^{\infty} g_n(x)dx = \sum_{n=1}^{\infty} \int_a^b g_n(x)dx.$$

Let $\epsilon > 0$. There exists an integer $n$ such that for $n > N$ and all $x \in [a,b]$, $|f_n(x) - f(x)| < \epsilon$. Now if $n > N$, then

$$\left| \int_a^b f_n(x)dx - \int_a^b f(x)dx \right| \leq \int_a^b |f_n(x) - f(x)|dx \leq \epsilon(b-a). ♦$$

Applications of Completeness, Part 1: **Contraction Mappings and Applications**

In this application we prove the contraction mapping theorem,[5] which is one of the simplest fixed point theorems. Then we apply it to derive the existence and uniqueness of solutions of certain types of differential and integral equations.

---

[4] The sequence of partial sums $\sum_{i=1}^{n} M_i$ is Cauchy because its limit, $\sum_{n=1}^{\infty} M_n$, is convergent.
[5] Also called Banach's fixed point theorem.

**Definition.** Let $X$ be a metric space. A function $T : X \to X$ is called a **contraction** if there exists a constant $0 < k < 1$ such that,

$$\text{for all } x, y \in X, \; d(T(x), T(y)) \leq k d(x, y).$$

**Theorem 4.6.12 (contraction mapping theorem).** *Let $T : X \to X$ be a contraction on a complete metric space $X$. Then $T$ has a unique **fixed point**. Thus there is a unique point $z$ in $X$ such that $T(z) = z$.*

*Proof.* Let $x_0$ be an arbitrary point in $X$, and define a sequence $(x_n)$ in $X$ by $x_{n+1} = T(x_n)$. First we show that $(x_n)$ is a Cauchy sequence.

For $n \geq 1, d(x_{n+1}, x_n) = d(T(x_n), T(x_{n-1})) \leq k d(x_n, x_{n-1})$, and, by induction, $d(x_{n+1}, x_n) \leq k^n d(x_1, x_0)$. Now, for $m, n \in \mathbb{N}$ with $m < n$,

$$d(x_n, x_m) \leq d(x_n, x_{n-1}) + d(x_{n-1}, x_{n-2}) + \dots + d(x_{m+1}, x_m)$$
$$\leq (k^{n-1} + k^{n-2} + \dots + k^m) d(x_1, x_0) = k^m (1 + k + \dots + k^{n-m-1}) d(x_1, x_0)$$
$$\leq k^m \sum_{j=0}^{\infty} k^j d(x_1, x_0) = \frac{k^m}{1-k} d(x_1, x_0).$$

Since $\lim_m k^m = 0, (x_n)$ is a Cauchy sequence. By the completeness of $X$, $z = \lim_n x_n$ exists. We claim that $z$ is the unique fixed point of $T$.

Now $z = \lim_n x_n = \lim_n T(x_{n-1}) = T(\lim_n x_{n-1}) = T(z)$. To show that $z$ is unique, suppose $w$ is a fixed point of $T$. Then $d(z, w) = d(T(z), T(w)) \leq k d(z, w)$. This would be a contradiction unless $d(z, w) = 0$, that is, $z = w$. ∎

**Definition.** A function $f : [a, b] \times \mathbb{R} \to \mathbb{R}$ is said to be a **Lipschitz function** in its second variable if there is a constant $L > 0$ such that

$$|f(x, y) - f(x, z)| \leq L|y - z|,$$
$$\text{for all } x \in [a, b] \text{ and all } y, z \in \mathbb{R}.$$

**Theorem 4.6.13.** *Consider the initial value problem*

$$\frac{dy}{dx} = f(x, y(x)), y(a) = y_0.$$

*Suppose that $f : [a, b] \times \mathbb{R} \to \mathbb{R}$ is continuous and that it satisfies the Lipschitz condition in its second argument: $|f(x, y) - f(x, z)| \leq L|y - z|$. Then the initial value problem has a unique solution $y(x)$ on the interval $[a, b]$.*

*Proof.* Choose a constant $K > L$, and define a metric on $X = \mathcal{C}[a,b]$ by

$$d(y,z) = \sup_{x \in [a,b]} \exp\{-K(x-a)\}|y(x) - z(x)|.$$

It is relatively easy to check that $d$ is a complete metric on $X$. Note that, for all $x \in [a,b]$, and all $y, z \in X$, $|y(x) - z(x)| \leq e^{K(x-a)}d(y,z)$. The initial value problem in question is equivalent to the integral equation

$$y(x) = y_0 + \int_a^x f(s, y(s))ds.$$

Define a function $T : X \to X$ by $y \mapsto T_y$, where $T_y(x) = y_0 + \int_a^x f(s, y(s))ds$. The proof will be complete if we show that $T$ has a unique fixed point $y \in X$. For two functions $y, z \in X$,

$$e^{-K(x-a)}|T_y(x) - T_z(x)| = e^{-K(x-a)}\left|\int_a^x f(s, y(s)) - f(s, z(s))ds\right|$$

$$\leq e^{-K(x-a)}\int_a^x |f(s, y(s)) - f(s, z(s))|ds \leq Le^{-K(x-a)}\int_a^x |y(s) - z(s)|ds$$

$$\leq Le^{-K(x-a)}d(y,z)\int_a^x e^{K(s-a)}ds = \frac{L}{K}e^{-K(x-a)}d(y,z)[e^{K(x-a)} - 1]$$

$$< \frac{L}{K}e^{-K(x-a)}d(y,z)e^{K(x-a)} = \frac{L}{K}d(y,z) = kd(y,z), \text{ where } k = \frac{L}{K} < 1.$$

The above inequalities show that $d(T_y, T_z) \leq kd(y,z)$; hence $T$ is a contraction. We now invoke the contraction mapping theorem to conclude that the initial value problem has a unique solution in $\mathcal{C}[a,b]$. ∎

**Theorem 4.6.14.** *If $X$ is a complete metric space and $T : X \to X$ is such that $T^n$ is a contraction for some positive integer $n$, then the unique fixed point of $T^n$ is the unique fixed point of $T$.*

*Proof.* Let $x$ be the unique fixed point of $T^n$. Thus $T^n(x) = x$. Now $T(x) = T^{n+1}(x) = T^n(T(x))$. Thus $T(x)$ is a fixed point of $T^n$. But the fixed point ot $T^n$ is unique, so $T(x) = x$. We leave it to the reader to show that $x$ is the only fixed point of $T$. ∎

**Theorem 4.6.15.** *Consider the **nonlinear Volterra equation***

$$u(x) = \int_a^x K(x, y, u(y))dy + f(x).$$

*Suppose $f \in \mathcal{C}[a,b]$ and that $K$ is continuous on $[a,b] \times [a,b] \times (-\infty, \infty)$ and satisfies the Lipschitz condition $|K(x,y,z_1) - K(x,y,z_2)| \leq L|z_1 - z_2|$ for all $x, y \in [a,b]$ and all $z_1, z_2 \in \mathbb{R}$. Then the above integral equation has a unique solution $u \in \mathcal{C}[a,b]$.*

*Let $X = \mathcal{C}[a,b]$, equipped with the uniform metric. Define a function $T : X \to X$ by*

$$u \mapsto T_u, \text{ where } T_u(x) = f(x) + \int_a^x K(x,y,u(y))dy.$$

*We leave it to the reader to verify that $T_u \in \mathcal{C}[a,b]$. For $u, v \in X$,*

$$|T_u(x) - T_v(x)| \leq L\|u - v\|_\infty(x - a).$$

*One more application of $T$ yields*

$$|T_u^2(x) - T_v^2(x)| \leq L^2\|u - v\|_\infty(x - a)^2/2,$$

*and, by induction,*

$$|T_u^n(x) - T_v^n(x)| \leq \frac{L^n}{n!}\|u - v\|_\infty(x - a)^n \leq \frac{L^n}{n!}\|u - v\|_\infty(b - a)^n.$$

*For sufficiently large $n$, $k = \frac{L^n(b-a)^n}{n!} < 1$, and, for such an $n$, $T^n$ is a contraction. By theorem 4.6.14, $T$ has a unique fixed point, and the Volterra equation has a unique solution.* ■

Applications of Completeness Part 2: **Continuous, Nowhere Differentiable Functions**

The first example of a continuous, nowhere differentiable function was produced by Weierstrass in 1872. Until that time, it was generally believed that continuous functions could fail to be differentiable at an isolated set of points. The main result in this application establishes an extreme contrast to the Weierstrass polynomial approximation theorem. Like polynomials, which are simple, well-behaved functions, the very erratic continuous, nowhere differentiable functions are also dense in $\mathcal{C}[0,1]$.

**Definition.** For a fixed integer $n \geq 1$, let $\mathfrak{F}_n$ be the set of functions $f \in \mathcal{C}[0,1]$ for which there is a point $x_0 \in [0,1]$ such that, for all $x \in [0,1]$, $|f(x) - f(x_0)| \leq n|x - x_0|$.

The geometric meaning of the condition in the above definition is that the slope of the line joining the point $(x_0, f(x_0))$ and an arbitrary point $(x, f(x))$ on the graph of $f$ cannot exceed $n$ in absolute value. There is no shortage of functions in $\mathfrak{F}_n$. For example, the functions $f_\alpha(x) = \alpha x$ are in $\mathfrak{F}_n$ if $|\alpha| \leq n$.

**Example 6.** If $f \in \mathcal{C}[0,1]$ has a continuous derivative and $\|f'\|_\infty \leq n$, then $f \in \mathfrak{F}_n$.
  Fix a point $x_0 \in [0,1]$. For any $x \in [0,1]$, $|f(x) - f(x_0)| = |f'(\xi)(x - x_0)| \leq n|x - x_0|$. Here $\xi$ is a point between $x$ and $x_0$. ◆

The assumptions of the previous example are much stronger than they need to be. The differentiablility of $f$ at a single point in $(0,1)$ is enough to guarantee that $f \in \mathfrak{F}_n$ for some $n$, as the following example illustrates.

**Example 7.** If $f \in \mathcal{C}[0,1]$ is differentiable at $x_0 \in (0,1)$, then $f \in \mathfrak{F}_n$ for some $n \geq 1$.
  By assumption, there exists a number $\delta > 0$ such that, for $|x - x_0| < \delta$, $\left|\frac{f(x) - f(x_0)}{x - x_0} - f'(x_0)\right| < 1$. Thus, for $|x - x_0| < \delta$, $|f(x) - f(x_0)| \leq (|f'(x_0)| + 1)|x - x_0|$. If $|x - x_0| \geq \delta$, then $|f(x) - f(x_0)| \leq 2\|f\|_\infty = \frac{2\|f\|_\infty}{\delta}\delta \leq \frac{2\|f\|_\infty}{\delta}|x - x_0|$. Now, for any integer $n > max\{|f'(x_0)| + 1, 2\|f\|_\infty/\delta\}$, and all $x \in [a, b]$, $|f(x) - f(x_0)| \leq n|x - x_0|$. ◆

**Remark 1.** A direct consequence of example 7 is that if $f \in \mathcal{C}[0,1]$ is not in any $\mathfrak{F}_n$, then $f$ is nowhere differentiable in $(0,1)$. To prove the existence of a single nowhere differentiable function, we need to show that $\mathcal{C}[0,1] - \cup_{n=1}^\infty \mathfrak{F}_n \neq \varnothing$. Theorem 4.6.9 provides the plan of attack if we wish to prove that there is an abundance of continuous nowhere differentiable functions: Prove that each $\mathfrak{F}_n$ is closed and nowhere dense. Then $U_n = \mathcal{C}[0,1] - \mathfrak{F}_n$ is open and dense in $\mathcal{C}[0,1]$, and hence $\cap_{n=1}^\infty U_n$ is also dense. The set $\cap_{n=1}^\infty U_n$ consists entirely of nowhere differentiable functions.

**Lemma 4.6.16.** *The set $\mathfrak{F}_n$ is closed.*

*Proof. Let $(f_k)$ be a sequence in $\mathfrak{F}_n$, and suppose $f_k \to f$ in the uniform norm. For each $k \in \mathbb{N}$, let $x_k$ be such that $|f_k(x) - f_k(x_k)| \leq n|x - x_k|$. By the Bolzano-Weierstrass theorem (theorem 1.2.8), $(x_k)$ contains a convergent subsequence $x_{k_p}$. For simplicity of notation, write $x_p$ for $x_{k_p}$ and $f_p$ for $f_{k_p}$, and let $x_0 = \lim_{p\to\infty} x_p$. For $x \in [0,1]$, $|f(x) - f(x_0)| = \lim_p |f_p(x) - f_p(x_p)| \leq n\lim_p |x - x_p| = n|x - x_0|.$ ∎*

We need to construct continuous functions that change direction steeply and frequently. Continuous, piecewise linear functions of this type exist. A continuous, piecewise linear function $f$ has one-sided derivatives at each $x$ in $[0,1]$. We denote the right and left derivatives of $f$ by $D^+f(x)$ and $D^-f(x)$, respectively. We use the notation $|Df|$ to denote the minimum (absolute) value of the (one-sided) derivatives of $f$. Simply put, $|Df|$ is the minimum absolute value of the slope of any straight line segment of the graph of $f$. Figure 4.2 shows the graph of the type of functions of interest to us. In that graph, the slope of any straight line segment of the graph is $\pm 4$, and $|D\psi| = 4$.

**Example 8.** Given an arbitrary interval $[a, b]$ and an integer $k \geq 1$, there exists a continuous, piecewise linear function $\psi$ on $[a, b]$ such that $\psi(a) = 0 = \psi(b)$, $\|\psi\|_\infty = 2^{-k}$, and $|D\psi| > 2^k$.

Choose an integer $m$ such that $2 \cdot 4^m > 4^k(b - a)$, and divide the interval $[a, b]$ into $4^m$ subintervals of equal length. For $0 \leq j \leq 4^m$, let $x_j = a + \frac{j(b-a)}{4^m}$. Define $\psi$ to be the continuous, piecewise linear function such that, for $0 \leq j \leq 4^m$, $\psi(x_j) = 0$, and, for $0 \leq j \leq 4^m - 1$, $\psi(\frac{x_j + x_{j+1}}{2}) = 2^{-k}$. The magnitude of the slope of any straight line segment of the graph of $\psi$ is equal to $\frac{2^{-k}}{\frac{1}{2}\frac{(b-a)}{4^m}} > 2^k$. ♦

The idea behind the construction in example 8 is geometrically simple. If we want the short function $\psi$ (its height is $2^{-k}$) to have very steep slopes, we must make the base of each triangle very small, about $4^{-k}$. Then the slopes would have magnitudes $2^{-k}/4^{-k} = 2^k$. Figure 4.2 depicts the function $\psi$ on $[0, 1]$ with $k = 1 = m$.

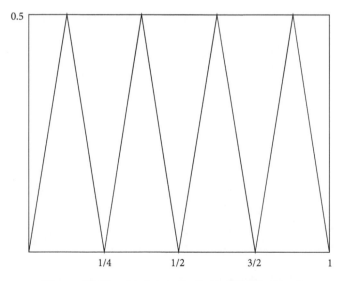

**Figure 4.2** The short, narrow, but spiky function $\psi$

**Remark 2.** The function $\psi$ constructed in example 8 is not in $\mathfrak{F}_n$ for any $n \leq 2^k$. For any point $(x_0, \psi(x_0))$ on the graph of $\psi$, there is another point $(x, \psi(x))$ on the same straight line segment of the graph. Thus the slope of that line segment is $\pm|D\psi|$, which is greater than $2^k$ in absolute value.

**Example 9.** Let $[a, b]$ be an arbitrary interval, and let $h$ be the linear function $h(x) = mx + c$. For every $\epsilon > 0$ and for every $n \geq 1$, there exists a continuous, piecewise linear function $\varphi$ on $[a, b]$ such that $\varphi(a) = h(a), \varphi(b) = h(b)$, $\|h - \varphi\|_\infty < \epsilon$, and $|D\varphi| > n$.

Choose an integer $k$ such that $2^{-k} < \epsilon$ and $2^k - |m| > n$. Using example 8, we find a continuous, piecewise linear function $\psi$ such that $\psi(a) = 0 = \psi(b)$, $\|\psi\|_\infty = 2^{-k}, |D\psi| > 2^k$. Define $\varphi = h + \psi$. Clearly, $\|h - \varphi\|_\infty = \|\psi\|_\infty = 2^{-k} < \epsilon$ and, for $x \in [a, b]$, $|D^\pm \varphi(x)| = |D^\pm \psi(x) + m| \geq |D^\pm \psi(x)| - |m| = |D\psi| - |m| > 2^k - |m| > n$. ◆

**Lemma 4.6.17.** *For each $n \geq 1$, $\mathfrak{F}_n$ is nowhere dense in $\mathcal{C}[0, 1]$.*

*Proof. Let $f \in \mathcal{C}[0, 1]$ and let $\epsilon > 0$. We will show that there is a continuous, piecewise linear function $g$ such that $\|f - g\|_\infty < \epsilon$ and $g \notin \mathfrak{F}_n$. Since continuous, piecewise linear functions are dense in $\mathcal{C}[0, 1]$ (see example 1 in section 4.5), let $h$ be a continuous, piecewise linear function such that $h(x_j) = f(x_j)$ for $0 \leq j \leq M$ and $\|f - h\|_\infty < \epsilon/2$. For $0 \leq j \leq M - 1$, let $h_j$ be the restriction of $h$ to $[x_j, x_{j+1}]$. By example 9, for each $j$ we construct a piecewise linear function $\varphi_j$ such that $\|h_j - \varphi_j\|_\infty < \epsilon/2$ and $|D\varphi_j| > n$. We define the required function $g$ by pasting together the functions $\varphi_j$. The function $g$ is continuous because $\varphi_j(x_j) = f(x_j) = \varphi_{j+1}(x_j)$. Now $\|f - g\|_\infty \leq \|f - h\|_\infty + \|h - g\|_\infty < \epsilon$.* ∎

The following result follows from remark 1, lemma 4.6.16, lemma 4.6.17, and theorem 4.6.9.

**Theorem 4.6.18.** *Continuous, nowhere differentiable functions are dense in $\mathcal{C}[0, 1]$.* ∎

## Exercises

1. Prove that $\mathbb{R}^n$ and $\mathbb{C}^n$ are complete.
2. Prove that $l^\infty$ is complete.
3. Prove that the space $c_0$ of null sequences is complete.

4. Define a metric on $\mathbb{N}$ as follows: $d(m,n) = |\frac{1}{n} - \frac{1}{m}|$. Prove that $d$ is an incomplete metric.

5. For $x, y \in \mathbb{R}$, define $d(x,y) = |tan^{-1}x - tan^{-1}y|$. Prove that $d$ is an incomplete metric on $\mathbb{R}$. Use the identity $tan^{-1}x - tan^{-1}y = tan^{-1}(\frac{x-y}{1+xy})$.

6. Let $A$ be a dense subset of a metric space $X$ such that every Cauchy sequence in $A$ is convergent to a point in $X$. Prove that $X$ is complete.

7. Prove that if $(x_n)$ and $(y_n)$ are Cauchy sequences in a metric space, then $d(x_n, y_n)$ converges.

8. Prove the converse of the Cantor intersection theorem. Hint: Let $(x_n)$ be a Cauchy sequence. For each $n \in \mathbb{N}$, let $A_n = \{x_n, x_{n+1}, \ldots\}$, and let $F_n = \overline{A_n}$. Show that $\lim_n diam(F_n) = 0$.

9. Prove that a subset $A$ of a metric space is nowhere dense if and only if every nonempty open set $U$ contains a nonempty open subset $V$ such that $V \cap A = \emptyset$.

10. Show that a closed subset $F$ of a metric space $X$ is nowhere dense, if and only if $X - F$ is dense.

11. Show that the boundary of a closed subset $F$ of a metric space $X$ is nowhere dense and give an example to show that the assumption that $F$ is closed cannot be omitted.

12. Let $X$ be a complete metric space, and let $\{F_n\}$ be a countable collection of closed, nowhere dense subsets of $X$. Is $\cup_{n=1}^{\infty} F_n$ necessarily nowhere dense?

13. Prove that a contraction on a metric space is continuous. Notice that this fact was used in the proof of theorem 4.6.12

14. Prove that the metric $d$ in the proof of theorem 4.6.13 is complete.

15. Prove that the function $T_y$ in the proof of theorem 4.6.13 is continuous.

16. Let $g : [a,b] \to \mathbb{R}$ and $K : [a,b] \times [a,b] \to \mathbb{R}$ be continuous functions. Show that when $|\alpha|$ is small enough, the integral equation

$$y(x) = \alpha \int_a^b K(x,t)y(t)dt + g(x)$$

has a unique solution in $\mathcal{C}[a,b]$.

17. Show that the fixed point of $T$ found in the proof of theorem 4.6.14 is unique.

18. Show that the function $T_u$ in the proof of theorem 4.6.15 is continuous.

**Definition.** An $n \times n$ matrix $A = (a_{ij})$ is said to be **diagonally dominant** if, for each $1 \leq i \leq n$, $|a_{ii}| > \sum_{j \neq i} |a_{ij}|$.

19. Prove that a diagonally dominant matrix is invertible. Hint: If $0 \neq x \in \mathbb{R}^n$ is such that $Ax = 0$, let $i$ be such that $|x_i| = max_{1 \leq j \leq n} |x_j|$. Now write down the $i^{th}$ equation of the system $Ax = 0$.

Numerical approximations of linear elliptic partial differential equations often lead to matrix equations with a very large, sparse, diagonally dominant matrix. Iterative solutions are practical in this situation, and the method described in the problem below, the **Jacobi iteration**, is one of the simplest (and the slowest).

20. Let $A$ be a diagonally dominant matrix, and consider the system $Ax = b$. Write $A$ as follows: $A = D + L + U$, where

$$U = \begin{pmatrix} 0 & a_{1,2} & \cdots & & a_{1,n} \\ \vdots & 0 & \ddots & & \\ \vdots & & \ddots & & a_{n-1,n} \\ 0 & \cdots & & & 0 \end{pmatrix}, L = \begin{pmatrix} 0 & \cdots & & & 0 \\ a_{2,1} & 0 & & & \\ \vdots & & \ddots & & \\ a_{n,1} & \cdots & a_{n,n-1} & & 0 \end{pmatrix},$$

$$D = \begin{pmatrix} a_{11} & & \\ & \ddots & \\ & & a_{nn} \end{pmatrix}.$$

Define $J = -D^{-1}(L + U)$. Show that the function $T : \mathbb{R}^n \to \mathbb{R}^n$ defined by $Tx = Jx + D^{-1}b$ is a contraction. Conclude that the iteration $x_k = Tx_{k-1}, k \geq 1$, converges to the solution of the system $Ax = b$. Hint: Examine the matrix norm $\|J\|_\infty$ defined in section 3.6.

## 4.7 Compactness

A clear manifestation of sequential compactness can be seen in examples 7 and 8 in section 1.2, where we proved the boundedness of continuous functions and their uniform continuity on a compact interval. We urge the reader to re-examine these two examples. This section opens with the topological (non-sequential) definition of compactness and the establishment of the general characteristics of compact spaces. This is done in order to avoid the duplication of definitions and results in the corresponding section in chapter 5. The various equivalent characterizations of compact metric spaces are discussed, and then we prove two famous theorems: Tychonoff's theorem and the Heine-Borel theorem. The section concludes with an illuminating application on closed convex sunsets of $\mathbb{R}^n$.

**Definition.** A metric space $X$ is said to be **compact** if every open cover of $X$ contains a finite subcover of $X$. The definitions of open covers and subcovers have been stated in section 4.5.

**Example 1.** The collection $\mathcal{U} = \{(-n, n) : n \in \mathbb{N}\}$ of open subsets of $\mathbb{R}$ is an open cover of $\mathbb{R}$ that contains no finite subcover. Therefore $\mathbb{R}$ is not compact. ♦

**Example 2.** The sequence of open intervals $\mathcal{U} = \{I_n = (1/n, 1 - 1/n) : n \geq 3\}$ covers the open interval $(0, 1)$, but no finite subset of $\mathcal{U}$ covers $(0, 1)$. Thus $(0, 1)$ is not compact. ◆

**Definition.** Let $K$ be a subset of a metric space $X$. We say that $K$ is a compact subset (or a compact subspace) of $X$ if it is compact in the restricted metric.

**Theorem 4.7.1.** *A subset $K$ of a metric space $X$ is compact if and only if it satisfies the following condition: if $\mathcal{U}$ is a collection of open subsets of $X$ such that $K \subseteq \cup \{U : U \in \mathcal{U}\}$, then there exists a finite subcollection $\{U_1, U_2, \ldots, U_n\}$ of $\mathcal{U}$ such that $K \subseteq \cup_{i=1}^{n} U_n$.*

*Proof. Suppose $K$ is compact and that $K \subseteq \cup \{U : U \in \mathcal{U}\}$. Then $K = \cup \{K \cap U : U \in \mathcal{U}\}$, and each of the sets $K \cap U$ is open in $K$. Therefore, for a finite subcollection $\{U_1, U_2, \ldots, U_n\}$ of $\mathcal{U}$, $K = \cup \{K \cap U_i : 1 \leq i \leq n\}$. Thus $K \subseteq \cup_{i=1}^{n} U_i$. The proof of the converse is left as an exercise.* ∎

**Example 3.** The set $K = \{1/n : n \in \mathbb{N}\} \cup \{0\}$ is compact. Suppose $\{U_\alpha : \alpha \in I\}$ is an open cover of $K$, and let $0 \in U_{\alpha_0}$. Since $\lim_n 1/n = 0$, there is an integer $N$ such that $a_n \in U_{\alpha_0}$ for all $n > N$. For $i = 1, \ldots, N$, choose members $U_{\alpha_1}, \ldots, U_{\alpha_N}$ of $\mathcal{U}$ such that $x_i \in U_{\alpha_i}$. Clearly, $K \subseteq \cup_{i=0}^{N} U_{\alpha_i}$. ◆

**Theorem 4.7.2.** *A closed subspace $K$ of a compact space $X$ is compact.*

*Proof. Let $\mathcal{U}$ be a collection of open subsets of $X$ whose union contains $K$. Then $\mathcal{U}^* = \mathcal{U} \cup \{X - K\}$ is an open cover of $X$. Therefore there exists a finite subcollection $\mathcal{U}'$ of $\mathcal{U}^*$ that covers $X$. There is no loss of generality in assuming that $X - K \in \mathcal{U}'$. Thus $X = (X - K) \cup \cup_{i=1}^{n} U_i$, where each $U_i \in \mathcal{U}$. Since $K \subseteq X$, and $K$ does not intersect $X - K$, $K \subseteq \cup_{i=1}^{n} U_i$. This proves that $K$ is compact.* ∎

**Theorem 4.7.3.** *A compact subspace $K$ of a metric space $X$ is closed and bounded.*

*Proof. We prove that $X - K$ is open. Let $x \in X - K$. For every point $y \in K$, there exist disjoint open subsets $U_y$ and $V_y$ of $X$ such that $x \in U_y$ and $y \in V_y$. Now $K \subseteq \cup_{y \in K} V_y$, so, by the compactness of $K$, there are finitely many points $y_1, y_2, \ldots, y_n \in K$ such that $K \subseteq \cup_{i=1}^{n} V_{y_i}$. Now let $U = \cap_{i=1}^{n} U_{y_i}$. Clearly, $K \cap U = \emptyset$, thus $x \in U \subseteq X - K$, and hence $X - K$ is open. We leave the proof that $K$ is bounded as an exercise.* ∎

**Theorem 4.7.4.** *The continuous image of a compact space is compact.*

*Proof.* Let $(X,d)$ be a compact space, and let $(Y,\rho)$ be a metric space. We show that if $f: X \to Y$ is a continuous surjection, then $(Y,\rho)$ is compact. Let $\{V_\alpha\}$ be an open cover of $Y$. Since $f$ is continuous, $f^{-1}(V_\alpha)$ is open in $X$ for each $\alpha$, and hence $\{f^{-1}(V_\alpha)\}$ is an open cover of $X$. The compactness of $X$ yields a finite subcover $\{f^{-1}(V_{\alpha_i})\}_{i=1}^n$ of $X$. Clearly, $\{V_{\alpha_i}\}_{i=1}^n$ covers $Y$. ∎

**Definition.** A metric space $X$ is **sequentially compact** if every sequence in $X$ contains a subsequence that converges in $X$.

**Definition.** A metric space $X$ has the **Bolzano-Weierstrass property** if every infinite subset of $X$ has a limit point.

**Theorem 4.7.5.** *A metric space $X$ is sequentially compact if and only if it has the Bolzano-Weierstrass property.*

*Proof.* Let $A$ be an infinite subset of a sequentially compact space $X$. Then $A$ contains a sequence $(x_n)$ of distinct points. By assumption, $(x_n)$ contains a subsequence that converges to a point, $x$. By theorem 4.2.6, $x$ is a limit point of $A$.

Conversely, suppose $X$ has the Bolzano-Weierstrass property, and let $(x_n)$ be a sequence in $X$. If the range $A = \{x_1, x_2, \ldots\}$ of $(x_n)$ is finite, then $(x_n)$ contains a constant subsequence, which is clearly convergent. So, suppose $A$ is infinite. By assumption, $A$ has a limit point, $x$. Let $n_1 \in \mathbb{N}$ be such that $d(x_{n_1}, x) < 1$. Having found positive integers $n_1 < n_2 < \ldots < n_k$ such that $d(x_{n_i}, x) < 1/i$, for $1 \leq i \leq k$, we pick an integer $n_{k+1} > n_k$ such that $d(x_{n_{k+1}}, x) < 1/(k+1)$. Such an integer exists because otherwise, for every $n > n_k$, we would have $d(x, x_n) \geq \frac{1}{k+1}$, which is impossible since $x$ is a limit point of $A$. By construction, $\lim_k x_{n_k} = x$. ∎

**Definition.** A metric space $X$ is **totally bounded** if, for every $\epsilon > 0$, there exists a finite subset $\{x_1, \ldots, x_n\}$ of $X$ such that $X = \cup_{i=1}^n B(x_i, \epsilon)$. The set $\{x_1, \ldots, x_n\}$ is called an $\epsilon$-dense subset of $X$.

**Theorem 4.7.6.** *A metric space $X$ is sequentially compact if and only if it is complete and totally bounded.*

*Proof.* Suppose $X$ is sequentially compact. Let $(x_n)$ be a Cauchy sequence in $X$. By assumption, $(x_n)$ contains a subsequence $(x_{n_k})$ that converges to a point $x$. By theorem 4.6.3, $\lim_n x_n = x$, and $X$ is complete. If $X$ is not totally bounded, then there exists $\epsilon > 0$ such that if $F$ is a finite subset of $X$, then $X \neq \cup_{x \in F} B(x, \epsilon)$. Pick a point $x_1 \in X$, and a point $x_2 \notin B(x_1, \epsilon)$. Since $B(x_1, \epsilon) \cup B(x_2, \epsilon) \neq X$, there exists a point $x_3 \in X$ such that $d(x_3, x_i) \geq \epsilon, i = 1, 2$. Continuing this construction yields a sequence $(x_1, x_2, \dots)$ of $X$ such that $d(x_n, x_m) \geq \epsilon$ for all $n, m \in \mathbb{N}, n \neq m$. Clearly, $(x_n)$ contains no convergent subsequence, which contradicts the sequential compactness of $X$.

Suppose $X$ is complete and totally bounded. We claim that $X$ has the Bolzano-Weierstrass property. The proof will be complete by theorem 4.7.5. Let $A$ be an infinite subset of $X$. The total boundedness of $X$ allows us to cover $X$ by a finite collection closed balls of radius 1. One of the balls, $B_1$, contains infinitely many points of $A$. Define $F_1 = B_1$, and $A_1 = A \cap B_1$. Now cover $X$ by a finite collection of closed balls of radius $1/2$. One of those balls, $B_2$, contains infinitely many points of $A_1$ and hence infinitely many points of $A$. Define $F_2 = B_2 \cap F_1$, and $A_2 = A_1 \cap B_2$. Continue by induction to construct a sequence of closed subsets $F_1 \supseteq F_2 \supseteq F_3 \supseteq \dots$ such that $\mathrm{diam}(F_n) \leq 2/n$ and each $F_n$ contains infinitely many points of $A$. By the Cantor intersection theorem, let $\{x\} = \cap_{n=1}^{\infty} F_n$. Since $\lim_n \mathrm{diam}(F_n) = 0$, any ball centered at $x$ contains $F_n$ for sufficiently large $n$. Since $F_n$ contains infinitely many points of $A$, $x$ is a limit point of $A$. ∎

**Definition.** Let $\mathcal{U} = \{U_\alpha\}$ be an open cover of a metric space $X$. A **Lebesgue number** for $\mathcal{U}$ is a positive number $a$ such that every subset $A$ of $X$ of diameter less that $a$ is contained in one member of $\mathcal{U}$.

**Theorem 4.7.7.** *In a sequentially compact metric space $X$, every open cover of $X$ has a Lebesgue number.*

*Proof.* Suppose that there is an open cover $\mathcal{U} = \{U_\alpha\}$ of $X$ that does not have a Lebesgue number. We show that $X$ is not sequentially compact. By assumption, for every $n \in \mathbb{N}$, there exists a subset $A_n$ of $X$ such that $\mathrm{diam}(A_n) < 1/n$, and $A_n$ is not contained in any member of $\mathcal{U}$. For each $n \in \mathbb{N}$, pick a point $x_n \in A_n$. We claim that $(x_n)$ has no convergent subsequence. Suppose, contrary to our claim, that some subsequence $(x_{n_k})$ of $(x_n)$ converges to $x$. Since $\cup_\alpha U_\alpha = X$, there exists a member $U_\alpha$ of $\mathcal{U}$ that contains $x$, and since $U_\alpha$ is open, there is a number $\delta > 0$ such that $B(x, \delta) \subseteq U_\alpha$. Now choose a positive integer $K$ such that $d(x_{n_K}, x) < \delta/2$, and $1/n_K < \delta/2$. If $y \in A_{n_K}$, then $d(x, y) \leq d(x, x_{n_K}) + d(x_{n_K}, y) < \delta/2 + \mathrm{diam}(A_{n_K}) < \delta/2 + \delta/2 = \delta$. This implies that $A_{n_K} \subseteq B(x, \delta) \subseteq U_\alpha$, which is a contradiction. ∎

**Theorem 4.7.8.** *Every sequentially compact metric space is compact.*

*Proof.* Let $X$ be sequentially compact, and let $\mathcal{U} = \{U_\alpha\}$ be an open cover of $X$. By theorem 4.7.7, $\mathcal{U}$ has a Lebesgue number, $a$. Let $\epsilon = a/3$. By theorem 4.7.6, there exists a finite subset $\{x_1, \dots, x_n\}$ of $X$ such that $\cup_{i=1}^{n} B(x_i, \epsilon) = X$. For each

$1 \leq i \leq n, diam(B(x_i, \epsilon)) \leq 2\epsilon < a$. Therefore each ball $B(x_i, \epsilon)$ is contained in a member $U_{\alpha_i}$ of $\mathcal{U}$. Clearly, $X = \cup_{i=1}^{n} U_{\alpha_i}$. ∎

**Theorem 4.7.9.** *For a metric space $X$, the following are equivalent:*

(a) *$X$ is compact.*
(b) *$X$ is sequentially compact.*
(c) *$X$ has the Bolzano-Weierstrass property.*
(d) *$X$ is complete and totally bounded.*

*Proof. In light of theorems 4.7.5, 4.7.6, and 4.7.8, we only need to show that (a) implies (b). Let $(x_n)$ be a sequence in $X$. Define $A_n = \{x_n, x_{n+1}, \ldots\}$, and let $F_n = \overline{A_n}$. Clearly, $\{F_n\}$ is a descending sequence of closed nonempty sets. If $\cap_{n \in \mathbb{N}} F_n = \emptyset$, then $\cup_{n \in \mathbb{N}} (X - F_n) = X$. Thus $(X - F_n)$ is an ascending sequence of open subsets that covers $X$. Therefore $X = X - F_n$, for some positive integer $n$, and hence $F_n = \emptyset$. This contradiction shows that $\cap_{n \in \mathbb{N}} F_n \neq \emptyset$. Let $x \in \cap_{n \in \mathbb{N}} F_n$. Observe that $x$ is a closure point of each of the sets $A_n$. Since $x \in \overline{A_1}$, there exists an integer $n_1 \geq 1$ such that $d(x_{n_1}, x) < 1$. Now $x \in \overline{A_{n_1 + 1}}$; thus there is an integer $n_2 \geq n_1 + 1$ such that $d(x_{n_2}, x) < 1/2$. Having found a sequence of positive integers $n_1 < n_2 < \ldots < n_k$ such that, for $1 \leq i \leq k$, $d(x_{n_i}, x) < 1/i$, choose an integer $n_{k+1} \geq n_k + 1$ such that $d(x_{n_{k+1}}, x) < \frac{1}{k+1}$. This is possible because $x \in \overline{A_{n_k + 1}}$. By construction, $\lim_k x_{n_k} = x$. ∎*

**Theorem 4.7.10 (Tychonoff's theorem).** *The product of finitely many compact metric spaces is compact.*

*Proof. It is enough to show that the product of two compact metric spaces $X$ and $Y$ is compact. Let $(x_n, y_n)$ be a sequence in $X \times Y$. Since $X$ is compact, there is a subsequence $(x_{n_k})$ of $(x_n)$ that converges to $x \in X$. Since $Y$ is compact, there exists a subsequence $(y_{n_{k_p}})$ of $(y_{n_k})$ that converges to a point $y \in Y$. Now $(x_{n_{k_p}}, y_{n_{k_p}})$ is a subsequence of $(x_n, y_n)$ that converges to $(x, y)$ as $p \to \infty$. ∎*

**Example 4.** The convex hull, $C$, of a compact subset $K \subseteq \mathbb{R}^n$ is compact.

Let $T_n$ be the standard $n$-simplex, which is compact by problem 21 at the end of this section. Consider the function $F : T_n \times K^{n+1} \to C$ defined by

$$F(\lambda_0, \ldots, \lambda_n, x_0, \ldots, x_n) = \sum_{i=0}^{n} \lambda_i x_i, \text{ where } (\lambda_0, \ldots, \lambda_n) \in T_n, \text{ and } x_0, \ldots, x_n \in K.$$

The continuity of $F$ is straightforward, and $F$ is surjective since, by Carathéodory's theorem, every point in $C$ is a convex combination of at most $n+1$ points in $K$. By Tychonoff's theorem, the set $T_n \times K^{n+1}$ is compact. The result now follows from theorem 4.7.4. ◆

**Theorem 4.7.11 (the Heine-Borel theorem).** *A subset $K$ of $\mathbb{R}^n$ is compact if and only if it is closed and bounded.*

*Proof. A compact subset of any metric space is closed and bounded by theorem 4.7.3. Conversely, suppose $K \subseteq \mathbb{R}^n$ is closed and bounded; $K$ is contained in some rectangle $I_1 \times \ldots \times I_n$, where each $I_i$ is a closed bounded interval in $\mathbb{R}$. By theorem 1.2.10, each $I_i$ has the Bolzano-Weierstrass property and hence is compact by theorem 4.7.9. By Tychonoff's theorem, $I_1 \times \ldots \times I_n$ is compact and, by theorem 4.7.2, $K$ is compact.* ∎

Remark 1. In the above proof of the Henie-Borel theorem, it is tacitly assumed that the metric involved is the product metric as defined in section 4.4. This is largely a matter of convenience. We may show that $K$ is closed and bounded in the 1-norm or the Euclidean norm.[6] See example 6 following theorem 4.3.8.

**Theorem 4.7.12.** *A continuous real-valued function $f$ on a compact space $X$ is bounded and attains its maximum and minimum values.*

*Proof. By theorem 4.7.4, $f(X)$ is a compact subset of $\mathbb{R}$. By the Heine-Borel theorem, $f(X)$ is closed and bounded. Therefore $f$ is a bounded function. Since $f(X)$ is closed, it contains its least upper and greatest lower bounds. Therefore $\max_{x \in X} f(x) = \sup_{x \in X} f(x)$ is in $f(X)$, and hence the maximum value of $f$ is attained. The same reasoning shows that the minimum value of $f$ is attained in $X$.* ∎

**Definition.** A metric space $X$ is **locally compact** if every point in $X$ belongs to the interior of a compact subset of $X$. Thus, for every $x \in X$, there is an open subset $V$ of $X$ such that $x \in V$ and $\overline{V}$ is compact.

**Theorem 4.7.13.** $\mathbb{R}^n$ *is locally compact.*

*Proof. Any point $x = (x_1, \ldots, x_n) \in \mathbb{R}^n$ is contained in the open rectangle*

$$V = (x_1 - 1, x_1 + 1) \times \ldots \times (x_n - 1, x_n + 1)$$

---

[6] We will see in section 6.1 that all norms on $\mathbb{R}^n$ are equivalent. Thus if a set $K$ is closed and bounded in one norm on $\mathbb{R}^n$, then it is closed and bounded in any norm on $\mathbb{R}^n$.

*and*

$$\overline{V} = [x_1 - 1, x_1 + 1] \times \ldots \times [x_n - 1, x_n + 1]$$

*is compact.* ■

**Example 5.** $\mathbb{Q}$ is not locally compact.

Since every open subset of $\mathbb{Q}$ is the union of sets of the form $(a, b) \cap \mathbb{Q}$, where $a, b \in \mathbb{R}$, it is enough to show that a set of the form $I = [a, b] \cap \mathbb{Q}$ is not sequentially compact. Choose an irrational number $r \in (a, b)$, then choose a sequence $x_n \in I$ such that $\lim_n x_n = r$. Clearly, no subsequence of $(x_n)$ is convergent in $I$. ♦

**Example 6.** The metric space $l^\infty$ is not locally compact. It is enough to show that the closed unit ball $B = \{x \in l^\infty : \|x\|_\infty \leq 1\}$ is not compact (see problem 8 in the section exercises). As $B$ contains the canonical vectors $e_n$ of $\mathbb{K}(\mathbb{N})$, since $d(e_n, e_m) = 1$ if $n \neq m$, the sequence $(e_n)$ in $l^\infty$ does not contain a convergent subsequence. ♦

The proof of the following theorem is left as an exercise.

**Theorem 4.7.14.** *The product of finitely many locally compact spaces is locally compact.* ■

### Excursion: Closed Convex Subsets of $\mathbb{R}^n$

**Example 7.** Let $K$ be a compact subset of $\mathbb{R}^n$, and let $a \in \mathbb{R}^n - K$. Then there exists a point $z \in K$ such that $\|z - a\|_2 = dist(a, K)$. The point $z$ is the closest point in $K$ to $a$.

Define a function $f : K \to \mathbb{R}$ by $f(x) = \|x - a\|_2$. Clearly, $f$ is continuous. By theorem 4.7.12, $f$ is bounded and attains its minimum value in $K$. Thus there is a point $z \in E$ such that $f(z) = \|z - a\|_2 = min\{f(x) : x \in K\} = dist(a, K)$. ♦

**Example 8.** Let $C$ be a closed subset of $\mathbb{R}^n$, and let $a \in \mathbb{R}^n - C$. Then there exists a point $z \in C$ such that $\|z - a\|_2 = dist(a, C)$. If, in addition, $C$ is convex, then $z$ is unique.

Let $B$ be a closed ball of radius $r$ centered at $a$, and assume $r$ is large enough so that $B \cap C \neq \emptyset$. The set $K = B \cap C$ is a closed and bounded subset of $\mathbb{R}^n$. By the Heine-Borel theorem, $K$ is compact. By the previous example, there is a point $z \in K$ such that $d = \|z - a\|_2 = dist(a, K)$. Since $d \leq r$ and $\|x - a\|_2 > r$ for every vector $x \in C - B$, $\|a - z\|_2 = dist(a, C)$. We leave it as an exercise to show that $z$ is unique when $C$ is convex. ◆

**Example 9** (the **obtuse angle criterion**). Let $C$ be a closed convex subset of $\mathbb{R}^n$, let $a \in \mathbb{R}^n - C$, and let $z$ be the closest element of $C$ to $a$. Then, for every $y \in C$, $\langle a - z, y - z \rangle \leq 0$. Here $\langle .,. \rangle$ is the Euclidean inner product on $\mathbb{R}^n$.

Without loss of generality, assume that $y \neq z$. Consider the quadratic function

$$\varphi(t) = \|(1 - t)z + ty - a\|_2^2, 0 \leq t \leq 1.$$

Observe that

$$\varphi(t) = \|(z - a) + t(y - z)\|_2^2 = \|z - a\|_2^2 + 2t\langle z - a, y - z \rangle + t^2\|y - z\|_2^2.$$

Because $C$ is convex, $(1 - t)z + ty \in C$ for every $0 \leq t \leq 1$. Since $z$ is the closest point in $C$ to $a$, $\varphi(0) \leq \varphi(t)$ for every $0 \leq t \leq 1$, and $\varphi$ is increasing on $[0, 1]$. This can happen only if $\varphi'(0) \geq 0$. Thus $2\langle z - a, y - z \rangle \geq 0$, and hence $\langle a - z, y - z \rangle \leq 0$. ◆

Observe that if $\theta$ is the angle between $a - z$ and $y - z$, then $cos \theta = \frac{\langle a - z, y - z \rangle}{\|a - z\|_2\|x - z\|_2}$. The condition $\langle a - z, y - z \rangle \leq 0$ is equivalent to saying that $\theta$ is at least 90°, hence the name *obtuse angle criterion*. Figure 4.3 illustrates the geometry. The wedge-shaped region depicts the convex set C, and the rest of the diagram is self-explanatory.

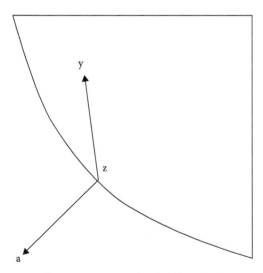

**Figure 4.3** The obtuse angle criterion

We know (see theorem 4.2.12) that a closed subset in a metric space can be separated from a point outside it by disjoint open sets. In $\mathbb{R}^n$, a closed convex subset can be separated from a point outside it in a much stronger and more specific way. They can be separated by a hyperplane, as the next example illustrates.

**Example 10.** Let $C$ be a closed convex subset of $\mathbb{R}^n$, and let $a \in \mathbb{R}^n - C$. Then there exists a hyperplane $n^T x = b$ such that $n^T y < b$ for every $y \in C$, and $n^T a > b$. Thus $C$ is contained in one of the open half-spaces determined by the hyperplane, and $a$ is contained in the other open half-space.

Let $z \in C$ be the closest point in $C$ to $a$, and let $m = (a + z)/2$. We show that the hyperplane we seek is the hyperplane that contains $m$ and is normal to the vector $n = a - z$. Without loss of generality, assume that $m = 0$ or, equivalently, $z = -a$ and $n = 2a$.[7]

By the previous example, for every $y \in C$, $\langle y - z, a - z \rangle \leq 0$, and hence $\langle y - z, n \rangle \leq 0$. Thus $\langle y, n \rangle \leq \langle z, n \rangle = \langle -a, 2a \rangle = -2\|a\|_2^2 < 0$, and $n^T y < 0$. On the other hand, $n^T a = \langle a, n \rangle = \langle a, 2a \rangle = 2\|a\|_2^2 > 0$. ♦

Remark 2. A direct consequence of the above example is the following. Under the assumptions of example 10, there exists a unit vector $u$ and a real number $b$ such that, for all $y \in C$, $u^T y < b < u^T a$.

**Theorem 4.7.15.** *A closed convex subset $C$ of $\mathbb{R}^n$ is the intersection of the closed half-spaces containing $C$.*

*Proof. We only need to show that $C$ contains the intersection of the closed half-spaces containing $C$. The reverse containment is obvious. If $a \notin C$, then, by the previous example, there is a hyperplane $n^T x = b$ such that $n^T y < b$ for all $y \in C$ and $n^T a > b$. Thus $C$ is contained in the closed half-space $H = \{x \in \mathbb{R}^n : n^T x \leq b\}$, but $a \notin H$.* ■

**Definition.** A hyperplane $M$ is said to be a **supporting hyperplane** of the closed convex set $C \subseteq \mathbb{R}^n$ if $C$ is contained in one of the closed half-spaces determined by $M$, and $C \cap M \neq \emptyset$. The closed half-space determined by $M$ that contains $C$ is called a **supporting half-space** of $C$. Observe that every point in $C \cap M$ is necessarily a boundary point of $C$.

---

[7] We can translate $C$ by $-m$. Specifically, we look at the set $C' = \{x - m : x \in C\}$ and the point $m' = 0$. This translation preserves all the properties of $C$ but has the advantage that the hyperplane we seek has a homogeneous equation. This simplifies the algebra.

**Example 11.** Every tangent line to the unit circle is a supporting line of the closed unit disk. The line $y = x + 1$ is a supporting line of the closed unit square $S = [0,1] \times [0,1]$. Slight rotations of the line about the point $(0,1)$ are also supporting lines of $S$. Thus there are infinitely many supporting lines of $S$ at the point $(0,1)$. The line $x = 1$ is also a supporting line of the square.

**Example 12.** In the notation of example 10, the hyperplane $n^T x = n^T z$ is a supporting hyperplane of $C$. ◆

We conclude this section with the following fine application of compactness.

**Theorem 4.7.16 (the supporting hyperplane theorem).** *Suppose $z$ is a boundary point of a closed convex set $C \subseteq \mathbb{R}^n$. Then there exists a supporting hyperplane $M$ of $C$ that contains $z$.*

*Proof.* Let $a_n$ be a sequence in $\mathbb{R}^n - C$ such that $\lim_n a_n = z$. By remark 2, there is a sequence $u_n$ of unit vectors such that,

$$\text{for every } y \in C, u_n^T y < u_n^T a_n.$$

By the compactness of the unit sphere in $\mathbb{R}^n$, $(u_n)$ contains a convergent subsequence, which we continue to call $(u_n)$ for simplicity. Let $u = \lim_n u_n$. Taking the limit of the two sides of the above inequality, we have $u^T y \leq u^T z$ for all $y \in C$. The hyperplane $M$ orthogonal to $u$ and containing $z$ is the one we seek. ∎

## Exercises

1. Prove directly that a compact metric space $X$ is bounded.
2. Let $(x_n)$ be a convergent sequence in a metric space $X$, and let $x = \lim_n x_n$. Prove that the set $\{x_n\}_{n=1}^{\infty} \cup \{x\}$ is compact.
3. Let $f : [0,1] \to \mathbb{R}$ be continuous. Prove that the graph of $f$, $\{(x, f(x)) : x \in [0,1]\}$, is compact in $\mathbb{R}^2$.
4. Prove that if $X$ is a compact metric space and $F_1 \supseteq F_2 \supseteq \ldots$ is a descending sequence of nonempty closed subsets of $X$, then $\cap_{n=1}^{\infty} F_n \neq \varnothing$.
5. Consider the space $c_0$ of null sequences, endowed with the supremum norm. Prove that a bounded subset $A$ of $c_0$ is totally bounded if and only if, for every $\epsilon > 0$, there exists a natural number $N$ such that $|x_n| < \epsilon$ for every $n > N$ and every $x \in A$.
6. Prove that if a subset $A$ of a metric space is totally bounded, then $\overline{A}$ is totally bounded.
7. Prove that a totally bounded metric space is separable.

8. Prove that, in a normed linear space $X$, the closed unit ball

$$B = \{x \in X : \|x\| \leq 1\}$$

is compact if and only if any closed ball in $X$ is compact.

9. Prove that a normed linear space $X$ is locally compact if and only if the closed unit ball is compact. In this case, show that the unit sphere $\{x \in X : \|x\| = 1\}$ is compact.

10. Prove that the product of finitely many locally compact spaces is locally compact.

11. In connection with example 8, prove the point $z$ is unique when $C$ is convex.

12. Let $F$ be a compact subset of a metric space $X$, and let $a \in X - F$. Prove that there exists a point $z \in F$ such that $d(z, a) = dist(a, F)$. Also give an example to show that $z$ is not necessarily unique.

13. Let $F$ be a closed subset of a locally compact normed linear space $X$, and let $a \in X - F$. Prove that there exists a point $z \in F$ such that $d(z, a) = dist(a, F)$.

14. Let $K$ be a compact subset of a metric space $X$. Prove that there exist points $x, y \in K$ such that $d(x, y) = diam(K)$.

15. Show that if $E$ is a compact subset of a metric space $X$ and $F$ is closed in $X$ and disjoint from $E$, then $dist(E, F) > 0$.

16. Let $A$ be a subset of a metric space $(X, d)$. For $\epsilon > 0$, define

$$A_\epsilon = \cup_{x \in A} B(x, \epsilon).$$

Prove that

$$A_\epsilon = \{x \in X : dist(x, A) < \epsilon\}.$$

Also show that if $E$ is a compact subset of $X$ and $F$ is closed in $X$ and disjoint from $E$, then $E_\epsilon \cap F_\epsilon = \varnothing$ for some $\epsilon > 0$.

17. Show that if $E$ and $F$ are disjoint compact subsets of $\mathbb{R}^n$, then there are points $x \in E$ and $y \in F$ such that $d(x, y) = dist(E, F)$.

18. Let $E$ and $F$ be disjoint compact convex subsets of $\mathbb{R}^n$. Show that there exists a hyperplane $u^T x = b$ such that $u^T x > b$ for every $x \in E$, and $u^T x < b$ for every $x \in F$.

19. Prove that a closed convex subset $C$ of $\mathbb{R}^n$ is the intersection of the closed supporting half-spaces that contain $C$.

20. Find a countable set of closed supporting half-planes whose intersection is the closed unit disk.

21. Prove that the standard $n$-simplex in $\mathbb{R}^n$ is compact.

## 4.8  Function Spaces

We already encountered several examples of function spaces. We mention two examples here before we embark on a more general study of function spaces. An early example is the space $l^\infty$, which is nothing but the space of bounded functions from $\mathbb{N}$ to the base field $\mathbb{K}$. A little reflection reveals that the same definition makes sense when $\mathbb{N}$ is replaced with an arbitrary set $X$. Another space we studied in some detail is the space $\mathcal{C}[0, 1]$. We also studied several of the properties of $l^\infty$ and $\mathcal{C}[0, 1]$, such as completeness and the lack of local compactness. We start the section with the definition of a number of important function spaces that generalize $l^\infty$ and $\mathcal{C}[0, 1]$ in particular.

**Definition.** Let $X$ be a nonempty set, and define $\mathcal{B}(X)$ to be the **set of all bounded, real or complex, functions** on $X$. Define vector addition and scalar multiplication in $\mathcal{B}(X)$ by $(f + g)(x) = f(x) + g(x), (af)(x) = af(x)$. Here $f$ and $g$ are bounded functions, $x \in X$, and $a \in \mathbb{K}$. **The supremum norm** (also the **uniform** or $\infty$-**norm**) of a function $f \in \mathcal{B}(X)$ is defined by

$$\|f\|_\infty = sup_{x \in X} |f(x)|.$$

It is a straightforward exercise to verify that $\mathcal{B}(X)$ is a vector space and that the function $\|.\|_\infty$ is a norm. Observe that it is not assumed that $X$ is necessarily a metric space. It is sometimes necessary to specify the scalar field. In this case, we use the notations $\mathcal{B}(X, \mathbb{R})$, and $\mathcal{B}(X, \mathbb{C})$ to indicate whether we wish to consider real or complex valued functions.

**Definition.** Let $X$ be a metric space, and define $\mathcal{C}(X)$ to be the set of **continuous real or complex functions on** $X$. The operations on $\mathcal{C}(X)$ are defined pointwise as in the above definition. This clearly makes $\mathcal{C}(X)$ into a vector space; see problem 1 on section 4.3. However, since a continuous function is not necessarily bounded, the supremum norm is not necessarily defined on $\mathcal{C}(X)$.

**Definition.** Let $X$ be a metric space. The **space of continuous bounded functions**, denoted by $\mathcal{BC}(X)$, is the intersection of $\mathcal{B}(X)$ and $\mathcal{C}(X)$. It is a normed subspace of $\mathcal{B}(X)$. In the special case when $X$ is a compact metric space, $\mathcal{C}(X) \subseteq \mathcal{B}(X)$, and $\mathcal{C}(X) = \mathcal{BC}(X)$.

**Theorem 4.8.1.** *The space $\mathcal{B}(X)$ of bounded functions on a set $X$ is a complete normed linear space.*

*Proof.* We only prove the completeness of $\mathcal{B}(X)$. Let $(f_n)$ be a Cauchy sequence in $\mathcal{B}(X)$, and let $\epsilon > 0$. There exists a natural number $N$ such that, for $m, n > N$,

$\|f_n - f_m\|_\infty = \sup_{x \in X}|f_n(x) - f_m(x)| < \epsilon$. In particular, $(f_n(x))$ is a Cauchy sequence in $\mathbb{K}$ for each $x \in X$. Therefore $\lim_n f_n(x)$ exists. Define $f(x) = \lim_n f_n(x)$.

We claim that $f \in \mathcal{B}(X)$. Since $f_n$ is Cauchy, there exists $N \in \mathbb{N}$ such that, for $n > N, \|f_n - f_N\|_\infty < 1$. Consequently, for all $x \in X$, and all $n > N$, $|f_n(x)| \leq |f_n(x) - f_N(x)| + |f_N(x)| \leq \|f_n - f_N\|_\infty + \|f_N\|_\infty < 1 + \|f_N\|_\infty$. Taking the limit of the quantity on the extreme left of the above string of inequalities, we obtain $|f(x)| \leq 1 + \|f_N\|_\infty$. Thus $f$ is a bounded function.

Finally, we show that $\lim_n f_n = f$ in $\mathcal{B}(X)$. Let $\epsilon > 0$. There exists $N \in \mathbb{N}$ such that, for $n, m > N$ and for all $x \in X, |f_n(x) - f_m(x)| < \epsilon$. Taking the limit as $m \to \infty$, we obtain $|f_n(x) - f(x)| \leq \epsilon$ for all $x \in X$, and all $n > N$. This means that $\|f_n - f\|_\infty < \epsilon$ for all $n > N$, and the proof is now complete. ■

**Theorem 4.8.2.** *If $X$ is a metric space, then the space $\mathcal{BC}(X)$ of continuous bounded functions on $X$ is a complete normed linear space.*

*Proof.* Since $\mathcal{BC}(X)$ is a subspace of $\mathcal{B}(X)$, it suffices, by theorems 4.8.1 and 4.6.4, to show that $\mathcal{BC}(X)$ is closed in $\mathcal{B}(X)$. Let $f \in \mathcal{B}(X)$ be a closure point of $\mathcal{BC}(X)$. We need to show that $f$ is continuous. For $\epsilon > 0$, there exists a function $g \in \mathcal{BC}(X)$ such that $\|f - g\|_\infty < \epsilon/3$. Fix $x_0 \in X$, and let $\delta > 0$ be such that $d(x, x_0) < \delta$ implies that $|g(x) - g(x_0)| < \epsilon/3$. Now if $d(x, x_0) < \delta$, then

$$|f(x) - f(x_0)| \leq |f(x) - g(x)| + |g(x) - g(x_0)| + |g(x_0) - f(x_0)| < \epsilon.$$

This proves that $f$ is continuous at $x_0$. ■

**Definition.** A function $f : X \to \mathbb{K}$ from a metric space $X$ to the base field $\mathbb{K}$ is **uniformly continuous** if, for every $\epsilon > 0$, there exists a number $\delta > 0$ such that, for all $x, y \in X$ with $d(x, y) < \delta$, $|f(x) - f(y)| < \epsilon$.

What distinguishes uniform continuity from continuity is that $\delta$ in the above definition does not depend on $x$.

**Theorem 4.8.3.** *A continuous (real or complex) function $f$ on a compact metric space $X$ is uniformly continuous.*

*Proof.* Let $\epsilon > 0$. For every $x \in X$, there exists $\delta_x > 0$ such that whenever $d(x, \xi) < \delta_x, |f(x) - f(\xi)| < \epsilon/2$. Now $X = \cup_{x \in X} B(x, \delta_x)$. Let $3\delta$ be a Lebesgue number for the open cover $\{B(x, \delta_x) : x \in X\}$. For each $\xi, \eta \in X$ with $d(\xi, \eta) < \delta$, $B(\xi, \delta)$ contains $\eta$ and has diameter $< 3\delta$. By the definition of a Lebesgue number, *there exists $x \in X$ such that $B(\xi, \delta) \subseteq B(x, \delta_x)$. In particular, $d(\xi, x) < \delta_x$, and $d(\eta, x) < \delta_x$. Consequently, $|f(\xi) - f(\eta)| \le |f(\xi) - f(x)| + |f(x) - f(\eta)| < \epsilon$.* ∎

The next result uses function spaces to provide an elegant and succinct proof of the existence of the completion of an arbitrary (incomplete) metric space.

**Theorem 4.8.4.** *Let $X$ be a metric space. Then there exists a complete metric space $\overline{X}$ and an isometry $\varphi : X \to \overline{X}$ such that $\varphi(X)$ is dense in $\overline{X}$.*

*Proof. We know that $\mathcal{B}(X, \mathbb{R})$ is a complete metric space. We will find an isometry $\varphi : X \to \mathcal{B}(X, \mathbb{R})$. The theorem follows by taking $\overline{X} = \overline{\varphi(X)}$. To this end, fix an element $a \in X$. For $\xi \in X$, define a function $\varphi_\xi : X \to \mathbb{R}$ by*

$$\varphi_\xi(x) = d(x, \xi) - d(x, a).$$

*By the triangle inequality, $|\varphi_\xi(x)| = |d(x, \xi) - d(x, a)| \le d(a, \xi)$ for all $x \in X$. Therefore $\varphi_\xi$ is bounded. We now show that the map $\xi \mapsto \varphi_\xi$ from $X$ to $\mathcal{B}(X, \mathbb{R})$ is an isometry. Specifically, we need to show that, for $\xi, \eta \in X$, $\|\varphi_\xi - \varphi_\eta\|_\infty = d(\xi, \eta)$.*
*Now $\|\varphi_\xi - \varphi_\eta\|_\infty = sup_{x \in X}|\varphi_\xi(x) - \varphi_\eta(x)| = sup_{x \in X}|d(x, \xi) - d(x, \eta)| \le d(\xi, \eta)$. Therefore $\|\varphi_\xi - \varphi_\eta\|_\infty \le d(\xi, \eta)$.*
*Since $|\varphi_\xi(\xi) - \varphi_\eta(\xi)| = d(\xi, \eta)$, $\|\varphi_\xi - \varphi_\eta\|_\infty = d(\xi, \eta)$, as desired.* ∎

Commonly used language to describe the conclusion of the previous theorem is that $X$ is isometrically embedded in $\overline{X}$. By identifying a point $\xi \in X$ with $\varphi_\xi \in \overline{X}$, we often think of $X$ as a subset of $\overline{X}$. We employ this convenience in the next theorem.

**Definition.** The space $\overline{X}$ that we just constructed is called the **completion** of $X$. It is the (unique) smallest complete metric space that contains $X$ as a dense subspace. The following theorem frames that concept.

**Theorem 4.8.5.** *Let $(Y, \rho)$ be a complete metric space space, and let $\varphi : X \to Y$ be an isometry from a metric space $X$ into $Y$ such that $\varphi(X)$ is dense in $Y$. Then $\varphi$ can be uniquely extended to an isometry $\overline{\varphi} : \overline{X} \to Y$.*

*Proof. Let $x \in \overline{X}$, and choose a sequence $(x_n)$ in $X$ such that $\lim_n x_n = x$. In particular, $(x_n)$ is Cauchy. Because $\varphi$ is an isometry, $\varphi(x_n)$ is a Cauchy sequence of $Y$. By the completeness of $Y$, $\varphi(x_n)$ converges to a point dependent on $x$. Define $\overline{\varphi}(x) = \lim_n \varphi(x_n)$. The reader should verify that the function $\overline{\varphi} : \overline{X} \to Y$ is well defined in the sense that it depends only on $x$ and not on the particular choice of the sequence $(x_n)$. See problem 4 on section 4.1*

*To show that $\overline{\varphi}$ is onto, let $y \in Y$. Since $\varphi(X)$ is dense in $Y$, there exists a sequence $x_n$ in $X$ such that $\lim \varphi(x_n) = y$. Again, because $\varphi$ is an isometry, $(x_n)$ is Cauchy in $X$. Since $\overline{X}$ is complete, $(x_n)$ converges to a point $x \in \overline{X}$. By the very definition of $\overline{\varphi}$, $\overline{\varphi}(x) = y$.*

*Finally, we verify that $\overline{\varphi}$ an isometry. Let $x, y \in \overline{X}$ and choose sequences $(x_n)$ and $(y_n)$ in $X$ such that $\lim_n x_n = x$, and $\lim_n y_n = y$. Since $\varphi$ is an isometry, $\rho(\varphi(x_n), \varphi(y_n)) = d(x_n, y_n)$. Taking the limit of the two sides of the last identity gives*

$$\rho(\overline{\varphi}(x), \overline{\varphi}(y)) = \rho(\lim_n \varphi(x_n), \lim_n \varphi(y_n)) = \lim_n \rho(\varphi(x_n), \varphi(y_n))$$

$$= \lim_n d(x_n, y_n) = d(x, y). \ \blacksquare$$

**Example 1.** We know that the chordal metric $\chi$ on $\mathbb{R}$ is not a complete metric. It make sense to ask if the completion of $(\mathbb{R}, \chi)$ can be described in concrete terms. The answer is rather obvious now. Since $(\mathbb{R}, \chi)$ is isometric to $\mathcal{S}^1_*$, which is a dense subset of $\mathcal{S}^1$, the completion of $(\mathbb{R}, \chi)$ is (isometric to) the circle $\mathcal{S}^1$. We did use here the fact that the completion of an incomplete metric space is unique. See theorems 4.8.4 and 4.8.5. More generally, the completion of $(\mathbb{R}^n, \chi)$ is the sphere $\mathcal{S}^n$. ◆

We now prove two theorems of great utility: Ascoli's theorem, which gives necessary and sufficient conditions for the compactness of a subset of continuous functions on a compact space in the uniform metric, and the Weierstrass polynomial approximation theorem. Later in the book, we will encounter several applications of the two theorems.

**Definition.** Let $X$ be a metric space. A subset $\mathfrak{F}$ of $\mathcal{C}(X)$ is said to be **equicontinuous** at $x \in X$ if, for every $\epsilon > 0$, there exists $\delta > 0$ such that, for every $y \in X$ with $d(x, y) < \delta$, and every $f \in \mathfrak{F}, |f(x) - f(y)| < \epsilon$. We say that $\mathfrak{F}$ is equicontinuous if it is equicontinuous at every $x \in X$.

**Definition.** A subset $\mathfrak{F}$ of $\mathcal{C}(X)$ is said to be **uniformly equicontinuous** if, for every $\epsilon > 0$, there exists $\delta > 0$ such that, for every $x, y \in X$ with $d(x, y) < \delta$, and every $f \in \mathfrak{F}, |f(x) - f(y)| < \epsilon$.

**Theorem 4.8.6.** *If $X$ is a compact metric space and $\mathfrak{F}$ is an equicontinuous subset of $\mathcal{C}(X)$, then $\mathfrak{F}$ is uniformly equicontinuous.*

*Proof. The proof mimics that of theorem 4.8.3 and is left as an exercise.* ■

**Theorem 4.8.7 (Ascoli's theorem).** [8] *Let $X$ be a compact metric space. A subset $\mathfrak{F}$ of $\mathcal{C}(X)$ is compact in the uniform metric if and only if $\mathfrak{F}$ is closed, bounded, and equicontinuous.*

*Proof.* Suppose $\mathfrak{F}$ is compact. By theorem 4.7.3, $\mathfrak{F}$ is closed and bounded. We show that $\mathfrak{F}$ is equicontinuous. Let $\epsilon > 0$. The total boundedness of $\mathfrak{F}$ (theorem 4.7.9) guarantees a finite set of functions $\{f_1, \ldots, f_n\}$ in $\mathfrak{F}$ such that $\mathfrak{F} \subseteq \cup_{i=1}^n B(f_i, \epsilon/3)$. By theorem 4.8.3, each $f_i$ is uniformly continuous, so there exists $\delta_i > 0$ such that if $d(x,y) < \delta_i$, then $|f_i(x) - f_i(y)| < \epsilon/3$. Let $\delta = \min_{1 \le i \le n} \delta_i$. If $f \in \mathfrak{F}$, there exists $f_i$ such that $\|f_i - f\|_\infty < \epsilon/3$. Now if $x, y \in X$ are such that $d(x,y) < \delta$, then $|f(x) - f(y)| \le |f(x) - f_i(x)| + |f_i(x) - f_i(y)| + |f_i(y) - f(y)| < \epsilon$. This proves that $\mathfrak{F}$ is equicontinuous. We now prove the converse.

Because of theorems 4.6.4 and 4.7.9, it is sufficient to show that $\mathfrak{F}$ is totally bounded. Let $\epsilon > 0$. Since $\mathfrak{F}$ is equicontinuous, for every $x \in X$, there exists $\delta_x > 0$ such that, for all $y \in B(x, \delta_x)$ and all $f \in \mathfrak{F}$, $|f(x) - f(y)| < \epsilon/4$. By the compactness of $X$, there exists a subset $A = \{x_1, \ldots, x_n\}$ of $X$ such that $X = \cup_{i=1}^n B(x_i, \delta_{x_i})$. By the boundedness of $\mathfrak{F}$, the set $R = \cup_{i=1}^n \{f(x_i) : f \in \mathfrak{F}\}$ is a bounded subset of the complex plane, and therefore $\overline{R}$ is compact and hence totally bounded. Thus there is a finite set $B = \{z_1, \ldots, z_m\}$ of complex numbers such that $R \subseteq \cup_{j=1}^m B(z_j, \epsilon/4)$. The following observation is crucial. Consider an arbitrary function $f$ in $\mathfrak{F}$. For every $x_i \in A$, there is a point $z_j \in B$ such that $|f(x_i) - z_j| < \epsilon/4$. The assignment $x_i \mapsto z_j$ clearly defines a function from $A$ to $B$.[9] This suggests that we look at the finite set $B^A$ of all functions from $A$ to $B$. For each $\varphi \in B^A$, we define a set $\mathfrak{F}_\varphi = \cap_{i=1}^n \{f \in \mathfrak{F} : |f(x_i) - \varphi(x_i)| < \epsilon/4\}$.[10] By the above observation, $\mathfrak{F} = \cup\{\mathfrak{F}_\varphi : \varphi \in A^B\}$. We claim that each of the sets $\mathfrak{F}_\varphi$ has a diameter less than $\epsilon$. This will complete the proof because we can choose a function $f_\varphi$ from each nonempty $\mathfrak{F}_\varphi$, and then we will have an $\epsilon$-dense subset of $\mathfrak{F}$.

To prove the claim, let $f, g \in \mathfrak{F}_\varphi$. Since $|f(x_i) - \varphi(x_i)| < \epsilon/4$ and $|g(x_i) - \varphi(x_i)| < \epsilon/4$ for every $x_i \in A$, $|f(x_i) - g(x_i)| < \epsilon/2$ for every $x_i \in A$. Now let $x \in X$. Then $x \in B(x_i, \delta_{x_i})$ for some $x_i \in A$, and

$$|f(x) - g(x)| \le |f(x) - f(x_i)| + |f(x_i) - g(x_i)| + |g(x_i) - g(x)| < \epsilon. \ \blacksquare$$

**Remark.** Observe that we did not use the full force of the assumption that $\mathfrak{F}$ is bounded, just that it is pointwise bounded. Problem 8 at the end of this section is relevant here.

---

[8] Also widely known as the Arzela-Ascoli theorem.
[9] The point $z_j$ may not be unique, so we pick one such.
[10] It is possible that $\mathfrak{F}_\varphi = \emptyset$.

**Theorem 4.8.8 (the Weierstrass polynomial approximation theorem).** *Let $g \in \mathcal{C}[0,1]$ and let $\epsilon > 0$. Then there exists a polynomial $P$ such that*

$$\|g - P\|_\infty < \epsilon.$$

*Proof. Observe that the theorem says that the space of polynomials is dense in $\mathcal{C}[0,1]$. Without loss of generality, we may replace $g$ with the function $f(x) = g(x) - [g(0) + x(g(1) - g(0))]$. This is because $g(0) + x(g(1) - g(0))$ is a polynomial. Replacing $g$ with $f$ has the advantage that $f(0) = f(1) = 0$. Extend $f$ to $\mathbb{R}$ by defining $f(x) = 0$ when $x \notin [0,1]$.*

*Define $L_n(x) = c_n \int_{-1}^{1} f(x+t)(1-t^2)^n dt$, where $c_n^{-1} = \int_{-1}^{1}(1-t^2)^n dt$. Since $f(x) = 0$ for $x \notin [0,1]$,*

$$L_n(x) = c_n \int_{-x}^{1-x} f(x+t)(1-t^2)^n dt = c_n \int_{0}^{1} f(\xi)[1 - (\xi - x)^2]^n d\xi. (\xi = x + t.)$$

*The last expression makes it clear that $L_n(x)$ is a polynomial of degree $\leq 2n$.*

*It is a simple induction exercise to show that, for all $n \in \mathbb{N}, (1 - t^2)^n \geq 1 - nt^2$, so $\int_{-1}^{1}(1-t^2)^n dt \geq \int_{-1/\sqrt{n}}^{1/\sqrt{n}}(1 - nt^2)dt = \frac{4}{3\sqrt{n}} > \frac{1}{\sqrt{n}}$. In particular, $c_n < \sqrt{n}$.*
*Now let $\epsilon > 0$. The uniform continuity of $f$ yields a number $0 < \delta < 1$ such that, for all $x, y$ with $|x - y| < \delta, |f(x) - f(y)| < \epsilon$. Now $c_n \int_{\delta}^{1}(1 - t^2)^n dt < \sqrt{n}(1 - \delta^2)^n$. Since $\lim_n \sqrt{n}(1 - \delta^2)^n = 0$, we can pick an integer $n$ such that $c_n \int_{\delta}^{1}(1 - t^2)^n dt < \epsilon$.*

*Since $\int_{-1}^{1} c_n(1-t^2)^n dt = 1$ and, for $|t| < 1, c_n(1-t^2)^n \geq 0$,*

$$|L_n(x) - f(x)| = |c_n \int_{-1}^{1} [f(x+t) - f(x)](1-t^2)^n dt|$$

$$\leq c_n \int_{-1}^{1} |f(x+t) - f(x)|(1-t^2)^n dt$$

$$= c_n \int_{-\delta}^{\delta} |f(x+t) - f(x)|(1-t^2)^n dt + c_n \int_{|t|>\delta} |f(x+t) - f(x)|(1-t^2)^n dt$$

$$< \epsilon c_n \int_{-\delta}^{\delta} (1-t^2)^n dt + 2\|f\|_\infty c_n \int_{|t|>\delta} (1-t^2)^n dt$$

$$< \epsilon + 4\|f\|_\infty \int_{\delta}^{1} (1-t^2)^n dt < \epsilon + 4\epsilon\|f\|_\infty. \blacksquare$$

**Example 2.** $\mathcal{C}[0, 1]$ is separable.

Let $A$ be the set of polynomials with rational coefficients. Clearly, $A$ is countable. We show it is dense in $\mathcal{C}[0, 1]$. Let $f \in \mathcal{C}[0, 1]$, and let $\epsilon > 0$. By the Weierstrass theorem, there is a polynomial $q = \sum_{i=0}^{n} a_i x^i$ such that $\|f - q\|_\infty < \epsilon/2$. For each $0 \leq i \leq n$, choose a rational number $r_i$ such that $|a_i - r_i| < \epsilon/2(n+1)$, and define $p = \sum_{i=0}^{n} r_i x^i$. Now $\|f - p\|_\infty \leq \|f - q\|_\infty + \|q - p\|_\infty \leq \epsilon/2 + \sum_{i=0}^{n} |a_i - r_i| < \epsilon$. ◆

The last example and the Weierstrass polynomial approximation theorem have far-reaching generalizations. Their proofs require the full power of the Stone-Weierstrass theorem. The proof of all three theorems can be found in section 4.9.

**Example 3.** Let $f \in \mathcal{C}[0, 1]$ be such that $\int_0^1 x^n f(x)dx = 0$, for every nonnegative integer $n$. Then $f = 0$.

Without loss of generality, assume that $f$ is a real function. The assumption implies that $\int_0^1 f(x)p(x)dx = 0$ for any polynomial $p$. By the Weierstrass theorem, there is a polynomial $p$ such that $\|f - p\|_\infty < \epsilon$. Now $|\int_0^1 f^2 dx| = |\int_0^1 f(f - p)dx| \leq \int_0^1 |f(x)||f(x) - p(x)|dx \leq \epsilon \int_0^1 |f|dx \leq \epsilon \|f\|_\infty$. Since $\epsilon$ is arbitrary, $\int_0^1 f^2 dx = 0$. The continuity of $f$ forces $f = 0$. ◆

The discussion so far has been focused on scalar-valued functions, and all the function spaces we have studied are normed linear spaces. We now expand the discussion and consider functions that take values in a general metric space. The next two examples are extensions of theorems 4.8.1 and 4.8.2.

**Example 4.** Let $(Y, \rho)$ be a bounded metric space, and let $X$ be an arbitrary nonempty set. For functions $f, g : X \to Y$, define

$$D(f, g) = \sup_{x \in X} \rho(f(x), g(x)).$$

Then $D$ is a metric on the set $Y^X$ of all functions from $X$ to $Y$. If, in addition, $\rho$ is a complete metric, so is $D$.

Observe that the definition of $D$ makes sense because $\rho$ is a bounded metric. The verification that $D$ is a metric on $Y^X$ is straightforward. Now assume that $\rho$ is complete, and let $f_n$ be a Cauchy sequence in $Y^X$. For $\epsilon > 0$, there is a natural number $N$ such that, for $m, n > N$, $\sup_{x \in X} \rho(f_n(x), f_m(x)) < \epsilon$. In particular, for an arbitrary $x \in X$, the sequence $(f_n(x))$ is a Cauchy sequence in $Y$. By the completeness of $\rho$, $f(x) = \lim_n f_n(x)$ exists. We now show that $f_n$ converges to $f$ in the metric $D$.

Let $\epsilon, N, n$, and $m$ be as in the last paragraph. Taking the limit as $m \to \infty$, we obtain $\rho(f_n(x), f(x)) < \epsilon$. Since the last inequality holds for all $x \in X$, $D(f_n, f) < \epsilon$ for all $n > N$. This shows that $\lim_n D(f_n, f) = 0$. ◆

**Example 5.** Let $(Y, \rho)$ be as in the previous example, and assume that $\rho$ is complete. If $(X, d)$ is a metric space, then the space $(\mathcal{C}(X, Y), D)$ of continuous functions from $(X, d)$ to $(Y, \rho)$ is complete.

We show that $\mathcal{C}(X, Y)$ is closed in $(Y^X, D)$. Let $f \in Y^X$ be a closure point of $\mathcal{C}(X, Y)$. We need to show that $f$ is continuous. For $\epsilon > 0$, there exists a function $g \in \mathcal{C}(X, Y)$ such that $D(f, g) < \epsilon/3$. Fix $x_0 \in X$, and let $\delta > 0$ be such that $d(x, x_0) < \delta$ implies that $\rho(g(x), g(x_0)) < \epsilon/3$. Now if $d(x, x_0) < \delta$, then $\rho(f(x), f(x_0)) \leq \rho(f(x), g(x)) + \rho(g(x), g(x_0)) + \rho(g(x_0), f(x_0)) < \epsilon$. This proves that $f$ is continuous at $x_0$. ◆

Let $I = [0, 1]$ be the closed unit interval, and let $I^2 = [0, 1] \times [0, 1]$ be the closed unit square; $I$ is given the usual metric on $\mathbb{R}$, and we give $I^2$ the product metric $\rho((r, s), (u, v)) = max\{|r - u|, |s - v|\}$. In theorem 4.8.9, we will make use of the space $\mathcal{C}(I, I^2)$ defined in example 5 with the complete metric $D$ defined in example 4.

## Application: A Space-Filling Curve

Let $J = [a, b]$ be an arbitrary closed interval, and let $S$ be an arbitrary closed square. We will refer to a function in $\mathcal{C}(J, S)$ as a path. We are particularly interested in the four types of triangular paths $g$ shown in figure 4.4. The triangles differ only in orientation. Specifically, the intervals $[a, (a + b)/2]$ and $[(a + b)/2, b]$ are mapped linearly onto the straight line segments of the triangle such that $g(a)$ and $g(b)$ are adjacent corners of $S$ and $g((a + b)/2)$ is the center of $S$. See the formula defining the path $f_0$ in the proof of theorem 4.8.9.

Before we embark on the task of finding the space-filling curve, we describe a special type of operation we need in the proof of the next theorem. Observe that the paths $g$ intersect only two of the four sub-squares that result from bisecting the sides of $S$. We define the modified paths $g'$ as follows. Divide $[a, b]$ into four congruent subintervals $J_j = [a + j(b - a)/4, a + (j + 1)(b - a)/4]$, $0 \leq j \leq 3$, and map the subinterval $J_j$ linearly onto the four triangular paths that make up the path $g'$, as shown in figure 4.5. Observe that the paths $g'$ intersect all the sub-squares of $S$.

We are now ready to find the space-filling curve. The statement of the theorem below justifies the term *space filling*.

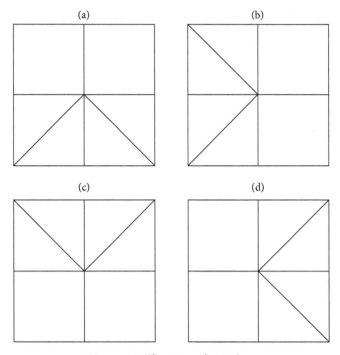

**Figure 4.4** The triangular paths g

**Theorem 4.8.9.** *There exists a continuous surjection f from I to $I^2$.*

*Proof.* We apply the operation discussed above the theorem to construct a sequence $(f_n)$ that converges to the desired function f.

Define the path $f_0 : I \to I^2$ by

$$f_0(t) = \begin{cases} (t,t) & \text{if } 0 \le t \le 1/2, \\ (t, 1-t) & \text{if } 1/2 \le t \le 1. \end{cases}$$

*Figure 4.4 (a) depicts the path $f_0$.*

*Applying the operation described above the theorem, we can find a path $f_1$ consisting of the four triangular paths shown in figure 4.5(a). Next we apply the operation to each of the triangular pieces of $f_1$ to produce the path $f_2$. Observe that this requires dividing each of the subintervals $I_j = [j/4, (j+1)/4]$ into four congruent sub-intervals and modifying the restriction of $f_1$ to each $I_j$, depending on its orientation, according to figure 4.5. The repeated application of the operation produces a sequence of paths $(f_n)$, which we show converges to the space-filling curve. The path $f_n$ consists of $4^n$ triangular paths, and each triangle is contained in a square of length $2^{-n}$. The triangular pieces correspond to partitioning I into the*

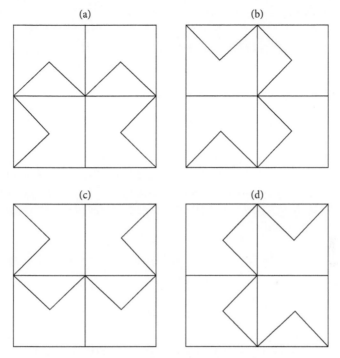

**Figure 4.5** The modified paths $g'$

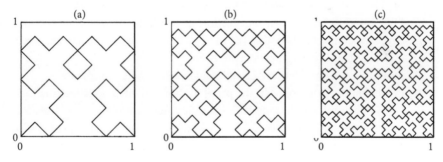

**Figure 4.6** The paths $f_2, f_3$, and $f_4$

subintervals $[j/4^n, (j+1)/4^n]$, $0 \leq j \leq 4^n - 1$. The paths $f_2, f_3$, and $f_4$ are shown in figure 4.6.

A crucial feature of the sequence $(f_n)$ is that if a triangular piece $T$ of the path $f_n$ is contained in a square $S$ of length $2^{-n}$, then the four triangular pieces of $f_{n+1}$ obtained by modifying $T$ are contained in the same square $S$. Thus, for every $t \in I$, $\rho(f_{n+1}(t), f_n(t)) < 2^{-n}$. Consequently, $D(f_{n+1}, f_n) < 2^{-n}$. This is the crux of the proof.

*Now, for positive integers $m, n$, if $m > n$, then*

$$D(f_m, f_n) \leq D(f_m, f_{m-1}) + D(f_{m-1}, f_{m-2}) + \cdots + D(f_{n+1}, f_n)$$
$$< 2^{-(m-1)} + \cdots + 2^{-n} < 2^{-n+1} \to 0 \text{ as } n \to \infty.$$

*Thus the sequence $(f_n)$ is Cauchy, and the completeness of $(\mathcal{C}(I, I^2), D)$ guarantees that $f_n$ converges to a function $f \in \mathcal{C}(I, I^2)$.*

*Since $I$ is compact, the range of $f$ is compact and hence closed in $I^2$. The proof will be complete if we show that the range of $f$ is dense in $I^2$. Let $x \in I^2$, and let $\epsilon > 0$. Choose an integer $n$ such that $2^{-n} < \epsilon/2$ and $D(f_n, f) < \epsilon/2$. The point $x$ belongs to one of the $4^n$ squares that contain the triangular pieces of $f_n$. Let $S$ be such a square. If $t \in [0, 1]$ is such that $f_n(t)$ is on the triangular piece contained in $S$, then $\rho(f_n(t), x) < 2^{-n}$ and*

$$\rho(f(t), x) \leq \rho(f(t), f_n(t)) + \rho(f_n(t), x) < D(f_n, f) + \epsilon/2 < \epsilon. \ \blacksquare$$

## Exercise

1. Let $Y$ be a complete metric space, let $A$ be a dense subset of a metric space $X$, and suppose that $f : A \to Y$ is a uniformly continuous function.
   (a) Show that $f$ maps Cauchy sequences into Cauchy sequences.
   (b) Show that $f$ admits a unique uniformly continuous extension $\bar{f} : X \to Y$.
2. Give an example to show that the mere continuity of $f$ in the above exercise is not enough to guarantee an extension.
3. Prove that the completion of a separable metric space is separable.
4. Prove that the function $\bar{\varphi}$ in the proof of theorem 4.8.5 is well defined.
5. Let $\mathfrak{F}$ be a pointwise bounded family of continuous, scalar-valued functions on a complete metric space $X$. Thus, for each $x \in X$, $sup\{|f(x)| : f \in \mathfrak{F}\} < \infty$. Prove that there exists an open subset $V \subseteq X$ such that $sup\{|f(x)| : x \in V, f \in \mathfrak{F}\} < \infty$. Hint: Let $A_n = \{x \in X : |f(x)| \leq n \text{ for every } f \in \mathfrak{F}\}$.
6. Show that if $X$ is compact and $\mathfrak{F} \subseteq \mathcal{C}(X)$ is equicontinuous, then $\overline{\mathfrak{F}}$ is equicontinuous.
7. **Ascoli's theorem** is often applied in the following form. Let $X$ be a compact metric space. If a sequence $(f_n)$ of functions in $\mathcal{C}(X)$ is bounded and equicontinuous, then $(f_n)$ contains a subsequence that converges in $\mathcal{C}(X)$. Prove this version of Ascoli's theorem.
8. Let $X$ be a compact metric space, and let $\mathfrak{F}$ be an equicontinuous family of functions in $\mathcal{C}(X)$. Prove that if $\mathfrak{F}$ is pointwise bounded, then $\mathfrak{F}$ is uniformly bounded. Hint: Let $g(x) = sup\{|f(x)| : f \in \mathfrak{F}\}$. For $n \in \mathbb{N}$, let

$U_n = \{x \in X : g(x) < n\}$. Prove that $U_n$ is open and that $\{U_n : n \in \mathbb{N}\}$ is an open cover of $X$.

9. Let $X$ be a compact metric space, and let $(f_n)$ be a sequence of equicontinuous functions that converges pointwise to a function $f$. Prove that $f$ is continuous and that $(f_n)$ converges uniformly to $f$.

10. **Dini's theorem.** Let $X$ be a compact metric space, and let $f_n : X \to \mathbb{R}$ be a sequence of continuous functions such that, for $x \in X, f_1(x) \geq f_2(x) \geq \ldots$, and $\lim_n f_n(x) = 0$. Prove that $f$ converges uniformly to the zero function.

11. Let $X$ be a compact metric space, and let $f_n : X \to \mathbb{R}$ be a sequence of continuous functions such that, for $x \in X, f_1(x) \leq f_2(x) \leq \ldots$, and $\lim_n f_n(x) = f(x)$, where $f$ is continuous. Prove that $f_n$ converges uniformly to $f$.

12. Prove that $\mathcal{C}[a, b]$ is not locally compact.

13. Prove that $\mathcal{C}[a, b]$ is separable and that polynomials are dense in $\mathcal{C}[a, b]$.

14. Let $X = \mathcal{C}(\mathbb{R}^n)$, and let $K_i$ be the closed ball of radius $i$ and centered at the origin. Clearly, $K_1 \subseteq K_2 \subseteq \ldots$, and $\cup_{i=1}^{\infty} K_i = \mathbb{R}^n$. Let $\|.\|_i$ denote the uniform norm on $\mathcal{C}(K_i)$. For a continuous function $f : \mathbb{R}^n \to \mathbb{C}, \|f\|_i = sup\{|f(x)| : x \in K_i\}$ denotes the norm of the restriction of $f$ to $K_i$. Define a metric $d$ on $X$ as follows:

$$d(f,g) = \sum_{i=1}^{\infty} 2^{-i} min\{1, \|f - g\|_i\}.$$

(a) Show that $d$ is a metric on $X$.

(b) Show that, for each $i \in \mathbb{N}$, $d(f,g) \leq \|f - g\|_i + 2^{-i}$.

(c) Show that a sequence of functions $(f_m)$ in $X$ converges in the metric $d$ to $f \in X$ if and only if $(f_m)$ converges uniformly to $f$ on compact subsets of $\mathbb{R}^n$.

(d) Show that $d$ is a complete metric.

15. Prove that, for every natural number $n$, there is a continuous surjection from $I$ to $I^n$.

16. Prove that a countable metric space can be isometrically injected in $l^{\infty}$. Hint: Examine the proof of theorem 4.8.4.

17. Prove that every separable metric space can be isometrically injected in $l^{\infty}$.

## 4.9 The Stone-Weierstrass Theorem

Like the Weierstrass theorem, the Stone-Weierstrass theorem is an approximation theorem. However, the Stone-Weierstrass theorem allows us to prove far-reaching generalizations of the results we obtained in section 4.8. Powerful theorems often require the development of elaborate machinery and this section demonstrates that.

Throughout this section, $X$ is a compact metric space, $\mathcal{C}(X,\mathbb{R})$ is the space of continuous, real-valued functions on $X$, and $\mathcal{C}(X,\mathbb{C})$ is the space of continuous, complex-valued functions on $X$. We use the notation $\mathcal{C}(X)$ for either of the two spaces when the distinction is immaterial; $\mathcal{C}(X)$ is a endowed with the uniform norm and, as such, is a Banach space. Additionally, $\mathcal{C}(X)$ is an algebra with the pointwise multiplication of functions: $(fg)(x) = f(x)g(x)$. See the definition of an algebra in section 3.4

Let $\mathcal{A}$ be a subalgebra of $\mathcal{C}(X)$ satisfying the following standing assumptions:

(SA1) $\mathcal{A}$ contains all constant functions.
(SA2) $\mathcal{A}$ separates points in $X$ in the sense that if $x,y$ are distinct points in $X$, then there exists a function $h \in \mathcal{A}$ such that $h(x) \neq h(y)$.

**Lemma 4.9.1.** *Let $\mathcal{A}$ be a subalgebra of $\mathcal{C}(X,\mathbb{R})$ satisfying SA1 and SA2. Then, for $f,g \in \mathcal{A}$, the functions $\max\{f,g\}$ and $\min\{f,g\}$ are in $\overline{A}$ (the closure of $\mathcal{A}$).*

*Proof.* Since $\max\{f,g\} = \frac{1}{2}(f+g) + \frac{1}{2}|f-g|$, and $\min\{f,g\} = \frac{1}{2}(f+g) - \frac{1}{2}|f-g|$, and since $\overline{A}$ is a subspace of $\mathcal{C}(X,\mathbb{R})$, it is sufficient to prove that $|f| \in \overline{A}$ whenever $f \in \mathcal{A}$. Let $M = \|f\|_\infty$, and let $\epsilon > 0$. By the Weierstrass approximation theorem applied to the function $g(t) = |t|$, there exists a polynomial $p(t) = \sum_{j=0}^{n} a_j t^j$ ($a_j \in \mathbb{R}$) such that, for all $t \in [-M,M]$, $| |t| - p(t)| < \epsilon$. Consider the function $p \circ f = \sum_{j=0}^{n} a_j f^j$. Since $\mathcal{A}$ is an algebra, $p \circ f \in \mathcal{A}$, and since $| |f(x)| - p(f(x))| < \epsilon$ for all $x \in X$, $\||f| - p \circ f\|_\infty < \epsilon$, and $|f| \in \overline{A}$. ∎

**Lemma 4.9.2.** *Let $\mathcal{A}$ be a subalgebra of $\mathcal{C}(X,\mathbb{R})$ satisfying SA1 and SA2, and let $f \in \mathcal{C}(X,\mathbb{R})$. For every $y,z \in X$, there exists a function $g_{yz} \in \mathcal{A}$ such that $g_{yz}(y) = f(y)$ and $g_{yz}(z) = f(z)$.*

*Proof.* If $y = z$, define $g_{yz}(x) = f(y)$ (a constant function). Otherwise, by SA2, there exists a function $h \in \mathcal{A}$ such that $h(y) \neq h(z)$. The following function is in $\mathcal{A}$ and satisfies the requirements:

$$g_{yz}(x) = f(z) + (f(y) - f(z))\frac{h(x) - h(z)}{h(y) - h(z)}. \blacksquare$$

**Theorem 4.9.3 (the Stone-Weierstrass theorem).** *Let $\mathcal{A}$ be a subalgebra of $\mathcal{C}(X,\mathbb{R})$ satisfying SA1 and SA2. Then $\mathcal{A}$ is dense in $\mathcal{C}(X,\mathbb{R})$.*

*Proof.* It is sufficient to show that $\overline{A}$ is dense in $\mathcal{C}(X,\mathbb{R})$. Observe that $\overline{A}$ is a subalgebra of $\mathcal{C}(X,\mathbb{R})$ that satisfies SA1 and SA2. Let $f \in \mathcal{C}(X,\mathbb{R})$, and let $\epsilon > 0$. We will show that there is a function $g \in \overline{A}$ such that $\|f - g\|_\infty < \epsilon$. For $y,z \in X$, let $g_{y,z}$ be as in lemma 4.9.2, and let $U_{y,z} = (f - g_{y,z})^{-1}(-\epsilon,\epsilon)$. By the continuity of $f - g_{y,z}$, $U_{y,z}$ is open and, clearly, $y, z \in U_{y,z}$. In particular, for every $x \in U_{yz}$, $f(x) < g_{y,z}(x) + \epsilon$ and $f(x) > g_{y,z}(x) - \epsilon$. The collection $\{U_{y,z} : y \in X\}$ covers X. Thus there exists a finite subset $\{y_1, \ldots, y_{n_z}\}$ of X such that $\cup_{i=1}^{n_z} U_{y_i,z} = X$.

Define $g_z = max\{g_{y_i,z} : 1 \leq i \leq n_z\}$, and let $V_z = \cap_{i=1}^{n_z} U_{y_i,z}$. The function $g_z$ is in $\overline{\mathcal{A}}$ by lemma 4.9.1. Observe that

$$f(x) < g_z(x) + \epsilon \text{ for all } x, z \in X,$$

and

$$f(x) > g_z(x) - \epsilon \text{ for all } z \in X \text{ and all } x \in V_z.$$

Now each $V_z$ is an open neighborhood of z; hence the collection $\{V_z : z \in X\}$ covers X. Thus there exists a finite subset $\{z_1, \ldots, z_m\}$ of X such that $\cup_{j=1}^{m} V_{z_j} = X$. Finally, let

$$g = min\{g_{z_j} : 1 \leq j \leq m\}.$$

A little reflection reveals that $g(x) - \epsilon < f(x) < g(x) + \epsilon$ for all $x \in X$. ∎

The following corollary is a far-reaching generalization of the Weierstrass polynomial approximation theorem.

**Corollary 4.9.4.** *If X is a compact subset of $\mathbb{R}^n$, then the set $\mathcal{A}$ of polynomials with real coefficients in n variables is dense in $\mathcal{C}(X, \mathbb{R})$.*

*Proof.* Clearly, $\mathcal{A}$ is an algebra, and it contains all constant functions. To show that $\mathcal{A}$ separates points in X, let $x = (x_1, \ldots, x_n)$ and $y = (y_1, \ldots, y_n)$ be distinct points in X. The polynomial $p(t_1, \ldots, t_n) = \sum_{i=1}^{n} (t_i - x_i)^2$ satisfies $p(x) = 0$, and $p(y) > 0$. By the Stone-Weierstrass theorem, $\mathcal{A}$ is dense in $\mathcal{C}(X, \mathbb{R})$. ∎

The following result is the promised generalization of example 2 in section 4.8.

**Corollary 4.9.5.** *If X is a compact metric space, then $\mathcal{C}(X, \mathbb{C})$ is separable.*

*Proof.* Since compact metric spaces are separable, let $\{\xi_n : n \in \mathbb{N}\}$ be a countable dense subset of X. For $n \in \mathbb{N}$, define $f_n(x) = d(x, \xi_n)$, and define $f_0(x) = 1$. Let $\mathcal{M}$ be the set of all finite products of the functions $f_0, f_1, \ldots$, and let $\mathcal{A}$ be the set of all linear combinations with real coefficients of elements in $\mathcal{M}$. Clearly, $\mathcal{A}$ is a subalgebra of $\mathcal{C}(X, \mathbb{R})$.[11] We show that $\mathcal{A}$ separates points in X. If x and y are

[11] In fact, $\mathcal{A}$ is the subalgebra generated by the set $\{f_n\}_{n=0}^{\infty}$, that is, the smallest subalgebra of $\mathcal{C}(X, \mathbb{R})$ that contains $\{f_n\}_{n=0}^{\infty}$.

*distinct, let $\delta = d(x,y)/4$. There exists a natural number $n$ such that $d(x,\xi_n) < \delta$. The function $f_n$ separates $x$ and $y$ since $f_n(x) < \delta$, and $f_n(y) > 3\delta$.*

*By theorem 4.9.3, $\mathcal{A}$ is dense in $\mathcal{C}(X,\mathbb{R})$. We now show that the countable set $\mathcal{A}_1 = \{\sum_{i=1}^n q_i g_i : n \in \mathbb{N}, q_i \in \mathbb{Q}, g_i \in \mathcal{M}\}$ is dense in $\mathcal{C}(X,\mathbb{R})$. By the first part of the proof, it is enough to show that if $f = \sum_{i=1}^n a_i g_i \in \mathcal{A}$ and $\epsilon > 0$, then there exists an element $h \in \mathcal{A}_1$ such that $\|f - h\|_\infty < \epsilon$. Let $M = \max\{\|g_i\|_\infty : 1 \le i \le n\}$ and choose rational numbers $q_i$ such that $|a_i - q_i| < \epsilon/(nM)$. Set $h = \sum_{i=1}^n q_i g_i$. Clearly, $\|f - h\|_\infty < \epsilon$.*

*To show that $\mathcal{C}(X,\mathbb{C})$ is separable, let $f = f_1 + if_2 \in \mathcal{C}(X,\mathbb{C})$, and choose functions $h_1$ and $h_2$ in $\mathcal{A}_1$ such that $\|f_1 - h_1\|_\infty < \epsilon/2$, and $\|f_2 - h_2\|_\infty < \epsilon/2$. The function $h = h_1 + ih_2$ is in $\mathcal{A}_1 + i\mathcal{A}_1$ and satisfies $\|f - h\|_\infty < \epsilon$. Since $\mathcal{A}_1 + i\mathcal{A}_1$ is countable, the proof is complete.* ∎

Theorem 4.9.3 does not extend to $\mathcal{C}(X,\mathbb{C})$, as we show in example 1 below. First we need a definition.

**Definition.** Let $\mathcal{C}(\mathcal{S}^1,\mathbb{C})^{12}$ be the space of all continuous complex functions on $[-\pi,\pi]$ such that $f(-\pi) = f(\pi)$. It is clear that $\mathcal{C}(\mathcal{S}^1,\mathbb{C})$ is a closed subspace of $\mathcal{C}[-\pi,\pi]$ when both spaces are given the uniform norm.

Another way to view the space $\mathcal{C}(\mathcal{S}^1,\mathbb{C})$ is as follows. The restriction of any continuous, $2\pi$-periodic function $g : \mathbb{R} \to \mathbb{C}$ to the interval $[-\pi,\pi]$ is in the space $\mathcal{C}(\mathcal{S}^1,\mathbb{C})$. Conversely, any function $f \in \mathcal{C}(\mathcal{S}^1,\mathbb{C})$ can be extended by periodicity to a continuous, $2\pi$-periodic function. Thus the space $\mathcal{C}(\mathcal{S}^1,\mathbb{C})$ is also the space of continuous, $2\pi$-periodic functions. Every point $\theta \in [-\pi,\pi)$ corresponds to a unique point $e^{i\theta}$ on the unit circle $\mathcal{S}^1$ in the complex place, and, for every function $f \in \mathcal{C}(\mathcal{S}^1,\mathbb{C})$, there corresponds a function $\tilde{f} : \mathcal{S}^1 \to \mathbb{C}$, where $\tilde{f}(e^{i\theta}) = f(\theta)$ (here $\theta \in [-\pi,\pi)$). The correspondence $f \leftrightarrow \tilde{f}$ is unambiguous because of the condition $f(-\pi) = f(\pi)$. Therefore the space of $2\pi$-periodic functions can also be thought of as the space of continuous functions on the unit circle $\mathcal{S}^1$. We adopt any of the three equivalent characterizations of $\mathcal{C}(\mathcal{S}^1,\mathbb{C})$, as convenience dictates.

**Example 1.** For $n = 0, 1, \ldots$ let $u_n(t) = e^{int}$. The set $\{u_n\}_{n=0}^\infty$ separates point in $[-\pi,\pi]$. Thus the set $\mathcal{A} = \{\sum_{i=0}^n a_i u_i : a_i \in \mathbb{C}\}$ is a subalgebra of $\mathcal{C}(\mathcal{S}^1,\mathbb{C})$ that separates points in $[-\pi,\pi]$ and contains all constant functions. However, $\mathcal{A}$ is not dense in $\mathcal{C}(\mathcal{S}^1,\mathbb{C})$. We show that for the function $f(t) = e^{-it}$, $\|f - p\|_\infty \ge 1$ for all $p \in \mathcal{A}$. First observe that for any $p = \sum_{j=0}^n a_j u_j \in \mathcal{A}$,

---

[12] The reason for the notation will be justified in the next paragraph.

$\int_0^{2\pi} \overline{f} p \, dt = \sum_{j=0}^{n} a_j \int_0^{2\pi} e^{i(j+1)t} dt = 0$. Because $|f| = \overline{f}f = 1$,

$2\pi = \left| \int_0^{2\pi} \overline{f}f dt \right| = \left| \int_0^{2\pi} \overline{f}(f - p) dt \right| \leq \int_0^{2\pi} |f - p| dt \leq 2\pi \|f - p\|_\infty$.   Thus   $\|f - p\|_\infty \geq 1$, as claimed. ◆

The following is the generalization of theorem 4.9.3 to the complex case.

**Theorem 4.9.6 (the Stone-Weierstrass theorem).** *Let $X$ be a compact metric space and let $\mathcal{A}$ be a subalgebra of $\mathcal{C}(X, \mathbb{C})$ that satisfies SA1 and SA2. If $\mathcal{A}$ is closed under complex conjugation, then $\mathcal{A}$ is dense in $\mathcal{C}(X, \mathbb{C})$.*

*Proof. Let $\mathcal{R} = \{Re(f) : f \in \mathcal{A}\}$, and let $\mathcal{I} = \{Im(f) : f \in \mathcal{A}\}$. If $f = f_1 + if_2 \in \mathcal{A}$, then $if = -f_2 + if_1 \in \mathcal{A}$. Thus $f_2 \in \mathcal{R}$ and $f_1 \in \mathcal{I}$. It follows that $\mathcal{I} = \mathcal{R}$. First we show that $\mathcal{R}$ satisfies SA1 and SA2. It is clear that $\mathcal{R}$ contains all constant functions. If $x$ and $y$ are distinct points of $X$, then there exists $f \in \mathcal{A}$ such that $f(x) \neq f(y)$. Thus $f_1(x) \neq f_1(y)$ or $f_2(x) \neq f_2(y)$. Because $f_1$ and $f_2$ are in $\mathcal{R}$, $\mathcal{R}$ separates points in $X$. Theorem B.3 implies that $\mathcal{R}$ is dense in $\mathcal{C}(X, \mathbb{R})$. Because $\mathcal{A}$ is closed under complex conjugation, $f_1 = (f + \overline{f})/2 \in \mathcal{A}$; thus $\mathcal{R} \subseteq \mathcal{A}$, and hence $\mathcal{R} + i\mathcal{R} \subseteq \mathcal{A}$. We show that $\mathcal{R} + i\mathcal{R}$ is dense in $\mathcal{C}(X, \mathbb{C})$. By the density of $\mathcal{R}$ in $\mathcal{C}(X, \mathbb{R})$, there are functions $h_1, h_2 \in \mathcal{R}$ such that $\|f_1 - h_1\|_\infty < \epsilon/2$ and $\|f_2 - h_2\|_\infty < \epsilon/2$. The function $h = h_1 + ih_2$ is in $\mathcal{R} + i\mathcal{R}$ and $\|f - h\|_\infty < \epsilon$.* ∎

**Example 2.** For $n \in \mathbb{Z}$, let $u_n(t) = e^{int}$, and consider the set $\mathcal{T} = Span(\{u_n : n \in \mathbb{Z}\})$. $\mathcal{T}$ is clearly a subalgebra of $\mathcal{C}(\mathcal{S}^1, \mathbb{C})$ that satisfies the assumptions of theorem 4.9.6. Therefore $\mathcal{T}$ is dense in $\mathcal{C}(\mathcal{S}^1, \mathbb{C})$.

The last example is really a well-known theorem. We will expand this discussion in a more focused manner in the next section.

## 4.10  Fourier Series and Orthogonal Polynomials

In section 3.7 we studied the geometry of inner product spaces more than their metric properties. We now have a bigger toolbox with which we can tackle inner product spaces. Before we pose the central questions of this section, let us summarize the highlights of section 3.7, upon which this section rests heavily. Let $\{u_1, u_2, \ldots\}$ be an infinite orthonormal sequence of vectors in an inner product space $H$. The orthogonal projection of an element $x \in H$ on the finite-dimensional space $M_n = Span(\{u_1, \ldots, u_n\})$ is, by definition, the vector $S_n x = \sum_{i=1}^{n} \langle x, u_i \rangle u_i$. We know from theorem 3.7.6 that the vector $S_n x$ is the closest vector in $M_n$ to $x$, and we also say that $S_n x$ is the best approximation of $x$ in $M_n$. Now that we have studied convergence in metric spaces, it is natural to ask whether $\lim_n S_n x = x$. Unfortunately, we are still not in a position to state an exact set of conditions under which a general answer can be provided because the answer depends on the space $H$ and the sequence $\{u_1, u_2, \dots\}$. The reader should suspect that completeness is relevant here, and it is. The spaces we study in this section *are not complete*, and this is precisely the reason we cannot decisively settle the question posed above about the convergence of the sequence $S_n x$. In two of the major examples we consider in this section, we will answer this question satisfactorily but not completely. The full picture will materialize in sections 7.2 and 8.9.

## Fourier series

In section 3.7, we defined the inner product $\langle f, g \rangle = \frac{1}{2\pi} \int_{-\pi}^{\pi} f(x)\overline{g(x)}dx$ on the space $\mathcal{C}[-\pi, \pi]$. The sequence

$$\{u_n(t) = e^{int} : n \in \mathbb{Z}\}$$

is an orthonormal sequence with respect to the above inner product. The norm of a function $f$ induced by the inner product will be denoted by $\|f\|_2$ in order to distinguish it from the uniform norm on $\mathcal{C}[-\pi, \pi]$, which will also play a prominent role in this section. Thus the uniform norm of a function $f \in \mathcal{C}[a, b]$ will be denoted by the usual notation $\|f\|_\infty$, while

$$\|f\|_2 = \left( \frac{1}{2\pi} \int_{-\pi}^{\pi} |f(x)|^2 dx \right)^{1/2}.$$

It is clear that $\|f\|_2 \le \|f\|_\infty$.

In section 4.9, we introduced the space $\mathcal{C}(\mathcal{S}^1, \mathbb{C})$ (which we now abbreviate $\mathcal{C}(\mathcal{S}^1)$) of $2\pi$-periodic functions on $[-\pi, \pi]$. It is clear that $\mathcal{C}(\mathcal{S}^1)$ is a closed subspace of $(\mathcal{C}[-\pi, \pi], \|.\|_\infty)$.

For a function $f \in \mathcal{C}[-\pi, \pi]$, we define the **Fourier series** of $f$ to be the formal series

$$\sum_{n=-\infty}^{\infty} \hat{f}(n)e^{inx} \text{ where } \hat{f}(n) = \frac{1}{2\pi} \int_{-\pi}^{\pi} f(t)e^{-int} dt.$$

The numbers $\hat{f}(n), n \in \mathbb{Z}$ are called the **Fourier coefficients** of $f$. It is clear that the partial sum of the Fourier series,

$$S_n f(x) = \sum_{j=-n}^{n} \hat{f}(j)e^{ijx}$$

is the orthogonal projection of $f$ on $M_n = Span(\{u_i : -n \le i \le n\})$.

The question now is whether the sequence $S_n f$ converges to $f$ in the 2-norm. We answer the question affirmatively after we establish a few facts. Convergence in the 2-norm is sometimes called the **mean square convergence** in order to distinguish it from the uniform convergence of $S_n f$ to $f$, which is also a valid question.

**Definition.** A **trigonometric polynomial** is a linear combination of the functions $\{u_n = e^{int} : n \in \mathbb{Z}\}$. Thus a trigonometric polynomial is a function of the form

$$p(t) = \sum_{j=-n}^{n} c_j e^{ijt}, \text{ where } c_j \in \mathbb{C}, n \in \mathbb{N}.$$

The collection $\mathcal{T}$ of all trigonometric polynomials is clearly the span of the sequence $\{u_n : n \in \mathbb{Z}\}$ and is a subspace of $\mathcal{C}(\mathcal{S}^1)$.

Using the terminology we just established, example 2 in section 4.9 can be stated as follows.

**Theorem 4.10.1.** *The space of trigonometric polynomials is dense in the space $(\mathcal{C}(\mathcal{S}^1), \|.\|_\infty)$. Explicitly, for every $f \in \mathcal{C}(\mathcal{S}^1)$ and every $\epsilon > 0$, there exists a trigonometric polynomial $p$ such that $\|f - p\|_\infty < \epsilon$.* ∎

**Lemma 4.10.2.** *For a function $f \in \mathcal{C}[-\pi, \pi]$ (not necessarily periodic), and for every $\epsilon > 0$, there exists a $2\pi$-periodic function $g$ such that $\|f - g\|_2 < \epsilon$.*

*Proof.* Let $M = \|f\|_\infty$ and define $\delta = \dfrac{\epsilon^2}{8M^2}$. Define $g \in \mathcal{C}(\mathcal{S}^1)$ as follows:

$$g(x) = \begin{cases} \frac{f(-\pi+\delta)}{\delta}(x+\pi) & \text{if } -\pi \le x \le -\pi+\delta, \\ f(x) & \text{if } -\pi+\delta \le x \le \pi-\delta, \\ \frac{f(\pi-\delta)}{-\delta}(x-\pi) & \text{if } \pi-\delta \le x \le \pi. \end{cases}$$

*Figure 4.7 below shows how $f$ is modified on the subintervals $[-\pi, -\pi+\delta]$ and $[\pi-\delta, \pi]$ to produce $g$. We replace the graph of $f$ on the subinterval $[-\pi, -\pi+\delta]$ with the straight line that interpolates the points $(-\pi+\delta, f(-\pi+\delta))$ and $(-\pi, 0)$, and similarly on the subinterval $[\pi-\delta, \pi]$. The dotted lines in figure 4.7 indicate the modification of $f$ to produce $g$. By construction, $g$ is continuous and periodic. Also, for $x \in [-\pi, \pi]$, $|f(x) - g(x)| < 2M$. Now*

$$\|f-g\|_2^2 = \frac{1}{2\pi}\int_{-\pi}^{-\pi+\delta} |f(x)-g(x)|^2 dx + \frac{1}{2\pi}\int_{\pi-\delta}^{\pi} |f(x)-g(x)|^2 dx$$

$$\le \frac{4M^2}{2\pi}\int_{-\pi}^{-\pi+\delta} dx + \frac{4M^2}{2\pi}\int_{\pi-\delta}^{\pi} dx = \frac{8M^2\delta}{2\pi} = \frac{\epsilon^2}{2\pi} < \epsilon^2. \blacksquare$$

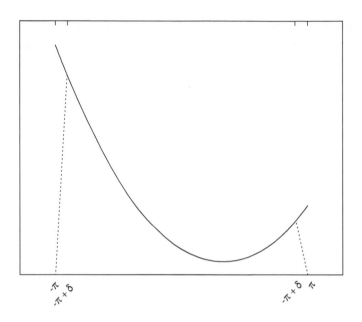

**Figure 4.7** The modified function $g$

**Corollary 4.10.3.** *Trigonometric polynomials are dense in $(\mathcal{C}[-\pi,\pi],\|.\|_2)$*

*Proof. Let $f\in \mathcal{C}[-\pi,\pi]$, and let $\epsilon > 0$. By lemma 4.10.2, there is a function $g\in \mathcal{C}(\mathcal{S}^1)$ such that $\|f-g\|_2 < \epsilon$. By theorem 4.10.1, there exists a trigonometric polynomial $p$ such that $\|g-p\|_\infty < \epsilon$. Now*

$$\|f-p\|_2 \leq \|f-g\|_2 + \|g-p\|_2 \leq \|f-g\|_2 + \|g-p\|_\infty < 2\epsilon. \blacksquare$$

Observe that the set of trigonometric polynomials with rational coefficients is dense in $(\mathcal{C}[-\pi,\pi],\|.\|_2)$; hence $(\mathcal{C}[-\pi,\pi],\|.\|_2)$ is separable.

We are now able to settle a question posed in the preamble to this section.

**Theorem 4.10.4.** *For every function $f\in \mathcal{C}[-\pi,\pi]$, the sequence of partial sums $S_n f$ converges in the mean square to $f$.*

*Proof. We need to show that $\lim_n \|f-S_n f\|_2 = 0$. Let $\epsilon > 0$. By corollary 4.10.3, there exists a trigonometric polynomial $p = \sum_{j=-N}^{N} c_j u_j$ such that $\|f-p\|_2 < \epsilon$. For every $n \geq N$, $p \in M_n = Span(\{u_j : -n \leq j \leq n\})$. Because $S_n f$ is the best approximation of $f$ in $M_n$, it follows that, for every $n \geq N$, $\|f-S_n f\|_2 \leq \|f-p\|_2 < \epsilon$. $\blacksquare$*

We take a short detour to discuss the sum of a two-sided sequence. The concept framed in the following, more general, definition is sometimes useful. See the excursion in section 7.2.

**Definition.** Let $\{a_\alpha : \alpha \in I\}$ be an indexed set of nonnegative numbers, where $I$ is an infinite set, possibly uncountable. The sum $\sum\{a_\alpha : \alpha \in I\}$ is, by definition

$$\sum\{a_\alpha : \alpha \in I\} = \sup\{\sum_{\alpha \in F} a_\alpha\}.$$

where the supremum is taken over all finite subsets $F \subseteq I$.

**Example 1.** If $\sum\{a_\alpha : \alpha \in I\} < \infty$, then the set $J = \{\alpha \in I : a_\alpha > 0\}$ is countable.

Fix a positive integer $n$. If the set $J_n = \{\alpha \in I : a_\alpha > 1/n\}$ is infinite, we can choose a sequence $\alpha_1, \alpha_2, \ldots$ of distinct elements of $J_n$. For the finite subset $F_j = \{\alpha_1, \ldots, \alpha_j\}$, $\sum_{\alpha \in F_j} a_\alpha > j/n$. Since $j$ is arbitrary, $\sum\{a_\alpha : \alpha \in I\}$ would be infinite. This proves that $J_n$ is a finite set for every $n \in \mathbb{N}$.
    Since $(0, \infty) = \cup_{n=1}^{\infty}(1/n, \infty)$, $J = \cup_{n=1}^{\infty} J_n$. Thus $J$ is countable. ◆

**Example 2.** Let $J$ be a countable set and suppose that $\sum\{a_\alpha : \alpha \in J\} < \infty$, where $a_\alpha \geq 0$. If $\alpha_1, \alpha_2, \ldots$ is any enumeration of $J$, then $\sum\{a_\alpha : \alpha \in J\} = \sum_{i=1}^{\infty} a_{\alpha_i}$.

For an integer $N$, $\sum_{i=1}^{N} a_{\alpha_i} \leq \sum\{a_\alpha : \alpha \in J\}$. Thus the partial sums of the series $\sum_{i=1}^{\infty} a_{\alpha_i}$ are bounded by $\sum\{a_\alpha : \alpha \in J\}$ and hence $\sum_{n=1}^{\infty} a_{\alpha_n} \leq \sum\{a_\alpha\}$. Conversely, if $F$ is an arbitrary finite subset of $J$, then there is an integer $N$ such that $F \subseteq \{\alpha_1, \ldots, \alpha_N\}$. Therefore $\sum\{a_\alpha : \alpha \in F\} \leq \sum_{n=1}^{N} a_{\alpha_n} \leq \sum_{n=1}^{\infty} a_{\alpha_n}$. Thus $\sum\{a_\alpha : \alpha \in J\} \leq \sum_{n=1}^{\infty} a_{\alpha_n}$. ◆

A special case of the above examples is when $(a_n)_{n \in \mathbb{Z}}$ is a two-sided sequence of nonnegative numbers.[13] The series $\sum_{-\infty}^{\infty} a_n$ can be defined (for example) as $\lim_{n \to \infty} \sum_{i=-n}^{n} a_i$, which corresponds to the following enumeration of $\mathbb{Z}$: $0, -1, 1, -2, 2, -3, 3, \ldots$.

We can now define, for $1 \leq p < \infty$, the space $l^p(\mathbb{Z})$ to be the set of all two-sided sequences $x = (x_n)_{n \in \mathbb{Z}}$ such that $\sum_{n=-\infty}^{\infty} |x_n|^p < \infty$. It is easy to check that $l^p(\mathbb{Z})$ is a complete normed linear space with the norm $\left(\sum_{n=-\infty}^{\infty} |x_n|^p\right)^{1/p}$.
    Similarly, we define $l^\infty(\mathbb{Z})$ to be the space of all bounded scalar functions $x : \mathbb{Z} \to \mathbb{K}$, which is a complete normed linear space with the norm $\|x\|_\infty = \sup\{|x(n)| : n \in \mathbb{Z}\}$.

---

[13] More accurately, functions from $\mathbb{Z}$ to the base field $\mathbb{K}$.

We also define the space $c_0(\mathbb{Z})$ as the subspace of $l^\infty(\mathbb{Z})$ of all two-sided sequences $x \in l^\infty(\mathbb{Z})$ such that $\lim_{|n|\to\infty} x_n = 0$.

**Theorem 4.10.5.** *For every $f \in \mathcal{C}[-\pi,\pi]$,*

$$\|f\|_2^2 = \frac{1}{2\pi}\int_{-\pi}^{\pi}|f(x)|^2 dx = \sum_{n=-\infty}^{\infty}|\hat{f}(n)|^2.$$

*Proof. By the continuity of norms (see section 4.3), $\lim_n \|S_n f\|_2^2 = \|f\|_2^2$. By the Pythagorean theorem, $\|S_n f\|_2^2 = \|\sum_{j=-n}^{n}\hat{f}(j)u_j\|_2^2 = \sum_{j=-n}^{n}|\hat{f}(j)|^2$. We obtain the required result by taking the limit of both sides as $n \to \infty$.* ∎

**Example 3.** Let $f(x) = x^2$. We compute the Fourier coefficients of $f$. Since $x^2 \sin(nx)$ is an odd function and $x^2 \cos(nx)$ is an even function, $\frac{1}{\pi}\int_{-\pi}^{\pi} x^2 \sin(nx)dx = 0$ and thus $\hat{f}(n) = \hat{f}(-n) = \frac{1}{\pi}\int_0^{\pi} x^2 \cos(nx)dx$.

Integration by parts now yields

$$\frac{1}{\pi}\int_0^{\pi} x^2 \cos(nx)dx = \frac{-2}{\pi n}\int_0^{\pi} x \sin(nx)dx = \frac{2x}{\pi n^2}\cos(nx)\Big|_0^{\pi}$$

$$= \frac{2\cos n\pi}{n^2} = \frac{2(-1)^n}{n^2}.$$

Thus

$$\hat{f}(n) = \hat{f}(-n) = \frac{2(-1)^n}{n^2}.$$

We also have

$$\hat{f}(0) = \frac{1}{2\pi}\int_{-\pi}^{\pi} x^2 dx = \frac{\pi^2}{3}.$$

Theorem 10.4.5 now yields

$$\frac{\pi^4}{5} = \frac{1}{2\pi}\int_{-\pi}^{\pi} x^4 dx = |\hat{f}(0)|^2 + \sum_{|n|=1}^{\infty}|\hat{f}(n)|^2$$

$$= \frac{\pi^4}{9} + 2\sum_{n=1}^{\infty}|\hat{f}(n)|^2 = \frac{\pi^4}{9} + \sum_{n=1}^{\infty}\frac{8}{n^4}.$$

Rearranging the extreme sides of the above string we obtain

$$\sum_{n=1}^{\infty}\frac{1}{n^4} = \frac{\pi^4}{90}. \blacklozenge$$

The next result says that a function in $\mathcal{C}[-\pi,\pi]$ is determined by its Fourier coefficients.

**Corollary 4.10.6 (the uniqueness theorem).** *If $f,g \in \mathcal{C}[-\pi,\pi]$ and $\hat{f}(n) = \hat{g}(n)$ for every $n \in \mathbb{Z}$, then $f = g$.*

*Proof.* Let $h = f - g$. By assumption, $\hat{h}(n) = 0$ for every $n \in \mathbb{Z}$. By theorem 4.10.5, $\frac{1}{2\pi}\int_{-\pi}^{\pi}|h(x)|^2 dx = \sum_{n=-\infty}^{\infty}|\hat{h}(n)|^2 = 0$. Since $h$ is continuous, $h = 0$. ∎

We now turn to the question of uniform convergence of Fourier series, which is a more complicated problem than the mean square convergence. Since $\mathcal{C}(\mathcal{S}^1)$ is closed in $(\mathcal{C}[-\pi,\pi], \|.\|_\infty)$, the Fourier series of a non-periodic function $f$ on $[-\pi,\pi]$ cannot converge uniformly to $f$. There are, however, simple criteria that guarantee the uniform convergence of the Fourier series of $2\pi$-periodic functions. The next theorem is a sample. See also example 5 below.

**Theorem 4.10.7.** *If $f \in \mathcal{C}(\mathcal{S}^1)$ is such that $\sum_{-\infty}^{\infty}|\hat{f}(n)| < \infty$, then the Fourier series of $f$ converges uniformly to $f$. In particular,*

$$f(x) = \sum_{n=-\infty}^{\infty} \hat{f}(n)e^{inx} \text{ for every } x \in [-\pi,\pi].$$

*Proof.* For $n \in \mathbb{N}$, define $F_n(x) = \sum_{j=-n}^{n}\hat{f}(j)e^{ijx}$. For positive integers $m > n$,

$$\|F_m - F_n\|_\infty = \sup_{x\in[-\pi,\pi]}\left|\sum_{|j|=n+1}^{m}\hat{f}(j)e^{ijx}\right| \leq \sum_{|j|=n+1}^{m}|\hat{f}(j)| \to 0 \text{ as } n \to \infty.$$

*Thus the sequence of functions $F_n(x)$ is Cauchy in $\mathcal{C}(\mathcal{S}^1)$ and hence converges uniformly on $[-\pi,\pi]$ to some function $F \in \mathcal{C}(\mathcal{S}^1)$. Thus $\lim_n \|F_n - F\|_\infty = 0$. But $\|F_n - F\|_2 \leq \|F_n - F\|_\infty$, and hence $F_n$ converges to $F$ in $\|.\|_2$. Since $F_n$ also converges to $f$ in $\|.\|_2$ (theorem 4.10.4), $F = f$ by the uniqueness of limits.* ∎

**Example 4.** This is a continuation of example 3. For the function $f(x) = x^2$, $\sum_{|n|=1}^{\infty}|\hat{f}(n)| = \sum_{|n|=1}^{\infty}\frac{2}{n^2} < \infty$, The Fourier series of $x^2$ converges uniformly (and absolutely) to $x^2$ on $[-\pi,\pi]$. Since $\hat{f}(n) = \hat{f}(-n)$,

$$\sum_{|n|=1}^{\infty}\hat{f}(n)e^{inx} = \sum_{n=1}^{\infty}\hat{f}(n)(e^{inx} + e^{-inx}) = 2\sum_{n=1}^{\infty}\hat{f}(n)\cos(nx). \text{ Therefore}$$

$$x^2 = \frac{\pi^2}{3} + 4\sum_{n=1}^{\infty}\frac{(-1)^n\cos(nx)}{n^2}, -\pi \leq x \leq \pi.$$

Substituting $x = 0$ and then $x = \pi$ in this identity, we obtain, respectively,

$$\sum_{n=1}^{\infty}\frac{(-1)^{n+1}}{n^2} = \frac{\pi^2}{12}, \text{ and } \sum_{n=1}^{\infty}\frac{1}{n^2} = \frac{\pi^2}{6}. \blacklozenge$$

**Example 5.** If a function $F \in \mathcal{C}(\mathcal{S}^1)$ has a continuous derivative, then the Fourier series of $F$ converges uniformly to $F$ on $[-\pi,\pi]$.

Let $F' = f$. If $n \neq 0$, integration by parts yields

$$\hat{F}(n) = \frac{-1}{2\pi i n} F(t)e^{-int}\Big|_{-\pi}^{\pi} + \frac{1}{2\pi i n}\int_{-\pi}^{\pi} f(t)e^{-int}\,dt = \frac{1}{in}\hat{f}(n).$$

Using the inequality $|ab| \leq \frac{1}{2}[|a|^2 + |b|^2]$, we have

$$|\hat{F}(n)| \leq \frac{1}{2}\left[\frac{1}{n^2} + |\hat{f}(n)|^2\right]$$

and

$$\sum_{|n|=1}^{\infty} |\hat{F}(n)| \leq \frac{1}{2}\left[2\sum_{n=1}^{\infty}\frac{1}{n^2} + \sum_{|n|=1}^{\infty} |\hat{f}(n)|^2\right] < \infty.$$

The result now follows from theorem 4.10.7. ◆

## Orthogonal Polynomials: The General Construction

Let $(a, b)$ be an interval. A function $\omega : (a, b) \to \mathbb{R}$ is said to be a **weight function** if

(a) $\omega$ is continuous and strictly positive on $(a, b)$,
(b) $\int_a^b \omega(x)dx < \infty$, and
(c) for every integer $n \geq 0$, $\int_a^b x^n\omega(x)dx < \infty$.

Consequently, $\int_a^b p(x)\omega(x)dx < \infty$ for every polynomial $p$. Neither the function $\omega$ nor the interval $(a, b)$ is assumed to be bounded.

When either $\omega$ or $(a, b)$ is unbounded, we interpret the integrals involved as improper Riemann integrals according to the standard definitions. Observe that if $(a, b)$ is a bounded interval, then $\omega$ can be unbounded if and only if $\lim_{x\downarrow a}\omega(x) = \infty$ or $\lim_{x\uparrow b}\omega(x) = \infty$. See the weight function for the Tchebychev polynomials later on in this section.

Let $H$ be the collection of all continuous functions on $(a, b)$ such that

$$\int_a^b |f(x)|^2\omega(x)dx < \infty.$$

It is obvious that $H$ is closed under complex conjugation and scalar multiplication. The following estimates show that $H$ is a vector space. If $f, g \in H$, then

$$\int_a^b |f(x)g(x)|\omega(x)dx \leq \frac{1}{2}\int_a^b \left[|f(x)|^2 + |g(x)|^2\right]\omega(x)dx < \infty$$

and

$$\int_a^b |f+g|^2 \omega(x)dx \le \int_a^b \left(|f| + |g|\right)^2 \omega(x)dx$$

$$= \int_a^b |f|^2 \omega(x) + 2|fg|\omega(x) + |g|^2\omega(x)dx < \infty.$$

We call $H$ the space of continuous, **square integrable functions** with respect to the weight function $\omega$. It now makes sense to define the following inner product on $H$:

$$\langle f,g \rangle = \int_a^b f(x)\overline{g(x)}\omega(x)dx.$$

The Gram-Schmidt process can be applied to the sequence of independent functions $\{1,x,x^2,\ldots\}$ to yield a sequence of **orthogonal polynomials** $\phi_0,\phi_1,\ldots$. The orthogonal polynomials in this general construction (regardless of the weight function or the interval $(a,b)$) share broad characteristics, which we will not discuss further. See the section exercises for some of the general features of orthogonal polynomials. In the remainder of this section, we give three major examples of orthogonal polynomials.

## The Legendre Polynomials

In this special case, we take

$$(a,b) = (-1,1)$$

and

$$\omega(x) = 1.$$

Observe that the space $H$ of continuous, square integrable functions on $(-1,1)$ contains the entire space $\mathcal{C}[-1,1]$. The resulting orthogonal polynomials are the well-known Legendre polynomials. In section 3.7, we derived the following formula for the Legendre polynomials (up to a multiplicative constant):

$$Q_n(x) = D^n(x^2 - 1)^n.$$

The first two Legendre polynomials are obvious: $Q_0(x) = 1$, and $Q_1(x) = 2x$. We establish the properties of the Legendre polynomials in a number of steps:

1. The parity of the Legendre polynomials. Since the binomial expansion of $(x^2 - 1)^n$ contains only even powers of $x$, $Q_n$ is an even polynomial if $n$ is even, and conversely.

2. The normalized Legendre polynomials $P_n$. It is customary to normalize the Legendre polynomials so that they take the value 1 at $x = 1$. Thus $P_n = c_n Q_n$, and we find $c_n$ by imposing the condition $P_n(1) = 1$. Using the Leibnitz rule,

$$Q_n(x) = D^n(x^2 - 1)^n = D^{n-1}\left[2nx(x^2 - 1)^{n-1}\right]$$
$$= 2nxD^{n-1}(x^2 - 1)^{n-1} + (n-1)2nD^{n-2}(x^2 - 1)^{n-1}.$$

Evaluating the last identity at $x = 1$, we obtain $Q_n(1) = 2nQ_{n-1}(1)$ and, by induction, $Q_n(1) = 2^n n!$. Therefore the polynomials

$$P_n(x) = \frac{1}{2^n n!} D^n(x^2 - 1)^n$$

are orthogonal and satisfy the normalization condition $P_n(1) = 1$.

3. The leading coefficient of $P_n$. We use the symbol $\alpha_n$ to indicate the leading coefficient of $P_n$. The leading coefficient in $D^n(x^2 - 1)^n$ is the result of differentiating $x^{2n}$ exactly $n$ times. Therefore

$$\alpha_n = \frac{1}{2^n n!}(2n)(2n-1)\ldots(n+1) = \frac{(2n)!}{2^n (n!)^2}.$$

4. **The three-term recurrence relation.** The recurrence relation we derive below facilitates the generation of the sequence $P_n$. We will use the brief notation $\beta_n = \frac{\alpha_{n+1}}{\alpha_n} = \frac{2n+1}{n+1}$. The polynomial $P_{n+1} - \frac{\alpha_{n+1}}{\alpha_n}xP_n$ is of degree at most $n$. Therefore,

$$P_{n+1} - \beta_n xP_n = \sum_{i=1}^{n} c_i P_i,$$

where

$$c_j\|P_j\|_2^2 = \langle P_{n+1} - \beta_n xP_n, P_j\rangle = -\beta_n\langle xP_n, P_j\rangle$$

If $j < n - 1$, $c_j\|P_j\|_2^2 = -\beta_n\langle xP_n, P_j\rangle = -\beta_n\langle P_n, xP_j\rangle = 0$, since $xP_j$ has degree less than $n$. Thus $P_{n+1} - \beta_n xP_n = c_n P_n + c_{n-1}P_{n-1}$. Now

$$c_n\|P_n\|_2^2 = -\beta_n\langle xP_n, P_n\rangle = -\beta_n\int_{-1}^{1} xP_n^2(x)dx = 0,$$

since $xP_n^2(x)$ is an odd function. Therefore

$$P_{n+1} - \beta_n xP_n = c_n P_{n-1}.$$

Evaluating the last identity at $x = 1$, we obtain $c_n = 1 - \beta_n = \frac{-n}{n+1}$, and we have the recurrence relation:

$$P_{n+1} = \frac{2n+1}{n+1}xP_n - \frac{n}{n+1}P_{n-1}.$$

Here is a list of the first Legendre polynomials:

$$P_0(x) = 1,$$
$$P_1(x) = x,$$
$$P_2(x) = \frac{1}{2}(3x^2 - 1),$$
$$P_3(x) = \frac{1}{2}(5x^3 - 3x),$$
$$P_4(x) = \frac{1}{8}(35x^4 - 30x^2 + 3).$$

5. The norm of $P_n$. Next we show that

$$\int_{-1}^{1} \left[P_n(x)\right]^2 dx = \frac{2}{2n+1}.$$

Let $a_n = \int_{-1}^{1} \left[P_n(x)\right]^2 dx$. Taking the inner product of $P_n$ with both sides of the identity $P_n = \frac{2n-1}{n}xP_{n-1} - \frac{n-1}{n}P_{n-2}$, we obtain $a_n = \frac{2n-1}{n}\int_{-1}^{1}(xP_n)P_{n-1}dx$. Using the recurrence relation again, $xP_n = \left[(n+1)P_{n+1} + nP_{n-1}\right]/(2n+1)$, and hence

$$a_n = \frac{2n-1}{n}\frac{1}{2n+1}\int_{-1}^{1}P_{n-1}\left[(n+1)P_{n+1} + nP_{n-1}\right] = \frac{2n-1}{2n+1}a_{n-1}.$$

Now $a_0 = \int_{-1}^{1} dx = 2$. By induction, one obtains

$$a_n = \frac{2}{2n+1}.$$

It follows that the polynomials below are orthonormal in $(\mathcal{C}[-1,1], \|.\|_2)$:

$$\tilde{P}_n = \sqrt{\frac{2n+1}{2}}P_n.$$

**Theorem 4.10.8 (mean square convergence).** *For every $f \in \mathcal{C}[-1,1]$, the sequence $S_nf = \sum_{i=0}^{n}\langle f, \tilde{P}_i\rangle\tilde{P}_i$ converges to $f$ in the sense that $\lim_n \|S_nf - f\|_2 = 0$.*

*Proof.* Let $\epsilon > 0$. By the Weierstrass approximation theorem, there exists a polynomial $q$ such that $\|f - q\|_\infty < \epsilon/\sqrt{2}$. Now $\|f - q\|_2 \leq \sqrt{2}\|f - q\|_\infty < \epsilon$. Let $N$ be the *degree of q. For every $n \geq N$, $q \in \mathbb{P}_n$, and since $S_n f$ is the best approximation of $f$ in $\mathbb{P}_n$, $\|f - S_n f\|_2 \leq \|f - q\|_2 < \epsilon$, as required.* ∎

Observe the resemblance between the proof of the last theorem and that of theorem 4.10.4. See also the examples in section 6.1.

## The Tchebychev Polynomials

In this special case, we take

$$(a, b) = (-1, 1)$$

and

$$\omega(x) = \frac{1}{\sqrt{1 - x^2}}.$$

Observe that the space $H$ of square integrable functions with respect to $\omega$ contains the entire space $\mathcal{C}[-1, 1]$.

A simple and direct derivation of the orthogonal polynomials is possible because of the observation that, for an integer $n \geq 0$, $cos(nx)$ can be expressed as a polynomial of $cos x$. For example, $cos(2x) = 2cos^2 x - 1$. The next lemma proves the existence of such polynomials and establishes the three-term recurrence relation among them.

**Lemma 4.10.9.** *For $n \geq 0$, there exists a polynomial $T_n$ of exact degree $n$ such that, for all $x \in \mathbb{R}$, $cos(nx) = T_n(cos x)$.*

*Proof.* For $n = 0, 1$, the polynomials $T_0(x) = 1$ and $T_1(x) = x$ trivially satisfy the requirements. The rest of the construction is inductive. Suppose that there are polynomials $T_0, \ldots, T_n$ that satisfy the statement we wish to prove. For $n \geq 1$, we have $cos(n + 1)x + cos(n - 1)x = 2cos(nx)cos x$. Therefore

$$cos(n + 1)x = 2cos(nx)cos x - cos(n - 1)x = 2cos x T_n(cos nx) - T_{n-1}(cos x).$$

*The last identity dictates the definition of $T_{n+1}$ and concludes the proof:*

$$T_{n+1}(x) = 2xT_n(x) - T_{n-1}(x). ■$$

**Definition.** The polynomials $T_n$ in the previous lemma are called the **Tchebychev polynomials**. A list of the next three Tchebychev polynomials appears below:

$$T_2(x) = 2x^2 - 1,$$
$$T_3(x) = 4x^3 - 3x,$$
$$T_4(x) = 8x^4 - 8x^2 + 1.$$

**Theorem 4.10.10.** *The Tchebychev polynomials are orthogonal with respect to the weight function $\omega$. Additionally,*

$$\|T_0\|_2^2 = \pi$$

*and, for $n \geq 1$,*

$$\|T_n\|_2^2 = \frac{\pi}{2}.$$

*Proof. We use the change of variable $x = \cos\theta$. If $m \neq n$, then*

$$\langle T_m, T_n \rangle = \int_{-1}^{1} \frac{T_n(x)T_m(x)}{\sqrt{1-x^2}} dx = \int_0^{\pi} \cos(n\theta)\cos(m\theta)d\theta$$

$$= \frac{1}{2}\int_0^{\pi} \cos(m+n)\theta + \cos(m-n)\theta \, d\theta = 0.$$

*Finally,* $\|T_n\|_2^2 = \int_{-1}^1 \frac{[T_n(x)]^2}{\sqrt{1-x^2}} dx = \int_0^{\pi} \cos^2(n\theta)d\theta = \frac{1}{2}\int_0^{\pi} 1 + \cos(2n\theta)d\theta = \pi/2.$ ■

The basic properties of the Tchebychev polynomials appear below. The first three follow from the three-term recurrence relation and induction:

1. $T_n$ is even if and only if $n$ is even.
2. The leading term of $T_n$ is $2^{n-1}$.
3. $T_n(1) = 1$, and $T_n(-1) = (-1)^n$.
4. For all $n \geq 0$, $\|T_n\|_\infty = max\{|T_n(x)| : -1 \leq x \leq 1\} = 1$. For every $x \in [-1,1]$, there is a number $\theta$ such that $x = \cos\theta$. Thus $|T_n(x)| = |\cos(n\theta)| \leq 1$. Since $T_n(1) = 1$, it follows that $\|T_n\|_\infty = 1$.
5. The roots of $T_n$ are $x_k = \cos\frac{(2k-1)\pi}{2n}$, $1 \leq k \leq n$. This can be verified directly, or one can write $x = \cos\theta$. If $T_n(x) = 0$, then $\cos(n\theta) = 0$. So $n\theta$ is an odd multiple of $\pi/2$; hence the stated values of $x_k$ are all the roots of $T_n$.

6. The extreme values of $T_n$ in $[-1,1]$ are attained at the points $y_k = \cos\frac{\pi k}{n}$, $0 \leq k \leq n$. Additionally, $T_n(y_k) = (-1)^k$. Again a direct verification is the simplest or, as before, we write $x = \cos\theta$, then $T_n(x) = \cos(n\theta)$, and $\frac{dT_n(x)}{dx} = \frac{d\cos(n\theta)}{d\theta}\frac{d\theta}{dx} = \frac{n\sin(n\theta)}{\sqrt{1-x^2}}$. The interested reader can work out the calculus and arrive at the points $y_k$.

For $n \geq 1$, let $\tilde{T}_n = \frac{1}{2^{n-1}}T_n$. From the above properties of $T_n$, $\tilde{T}_n$ is a monic polynomial,[14] and

$$\tilde{T}_n(x_k) = 0 \text{ for } 1 \leq k \leq n,$$

$$\tilde{T}(y_k) = \frac{(-1)^k}{2^{n-1}} \text{ for } 0 \leq k \leq n,$$

$$\text{and } \|\tilde{T}_n\|_\infty = \frac{1}{2^{n-1}}.$$

The following theorem establishes the curious fact that, among all monic polynomials of degree $n$, $\tilde{T}_n$ has the least uniform norm on $[-1,1]$. This result is important for understanding the error when a sufficiently differentiable function is interpolated by a polynomial.

**Theorem 4.10.11.** *Suppose $p$ is a monic polynomial of degree $n$. Then*

$$\|p\|_\infty = max\{|p(x)| : -1 \leq x \leq 1\} \geq \frac{1}{2^{n-1}}.$$

*Proof. Suppose, for a contradiction, that $\|p\|_\infty < \frac{1}{2^{n-1}}$. Consider the integers $0 \leq k \leq n$.*

*If $k$ is odd, then $p(y_k) > \frac{-1}{2^{n-1}} = \tilde{T}_n(y_k)$. If $k$ is even, then $p(y_k) < \frac{1}{2^{n-1}} = \tilde{T}_n(y_k)$.*

*Thus the polynomial $q = p - \tilde{T}_n$ alternates sign at the points $y_0, \ldots, y_n$; hence $q$ has a root in each of the $n$ open intervals $(y_0, y_1), \ldots, (y_{n-1}, y_n)$. This is a contradiction because $q$ has degree at most $n-1$.* ∎

## The Hermite Polynomials

For our last example of orthogonal polynomials, we take

$$(a, b) = (-\infty, \infty)$$

[14] A monic polynomial is one whose leading coefficient is 1.

and

$$\omega(x) = e^{-x^2}.$$

We will show that the polynomials defined below are orthogonal with respect to $\omega$:

$$H_n(x) = (-1)^n e^{x^2} D^n[e^{-x^2}].$$

Since $H_0 = 1$, and $H_1(x) = 2x$, $\langle H_0, H_1 \rangle = \int_{-\infty}^{\infty} 2x e^{-x^2} dx = 0$. We now use induction on $n$. If $0 \le j < n$, then integration by parts yields

$$\langle x^j, H_n \rangle = \int_{-\infty}^{\infty} x^j (-1)^n e^{x^2} D^n[e^{-x^2}] e^{-x^2} dx = (-1)^n \int_{-\infty}^{\infty} x^j D^n[e^{-x^2}] dx$$

$$= (-1)^n x^j D^{n-1} e^{-x^2} \Big|_{-\infty}^{\infty} - (-1)^n \int_{-\infty}^{\infty} j x^{j-1} D^{n-1} e^{-x^2} dx.$$

The first term of the last expression is 0 because $x^j D^{n-1} e^{-x^2}$ is the product of a polynomial and $e^{-x^2}$, and the second term is 0 by the inductive hypothesis.

We leave some of the properties of the Hermite polynomials as exercises for the interested reader.

## Exercises

1. Let $f, g \in C[-\pi, \pi]$. Prove that $\frac{1}{2\pi} \int_{-\pi}^{\pi} f(x)\overline{g(x)}dx = \sum_{n=-\infty}^{\infty} \hat{f}(n)\overline{\hat{g}(n)}$. Hint: Use theorem 4.10.4 and the continuity of inner products. See section 4.3.

2. Show that $|x| = \frac{\pi}{2} - \frac{4}{\pi} \sum_{n=1}^{\infty} \frac{\cos(2n-1)x}{(2n-1)^2}, -\pi \le x \le \pi$. Conclude that $\sum_{n=1}^{\infty} \frac{(-1)^{n+1}}{(2n-1)^2} = \frac{\pi^2}{8}$.

3. Show that $\frac{\pi^2 x - x^3}{12} = \sum_{n=1}^{\infty} \frac{(-1)^{n+1} \sin(nx)}{n^3}$.

4. Use the previous problem to show that $\sum_{n=1}^{\infty} \frac{1}{n^6} = \frac{\pi^6}{945}$.

5. This exercise furnishes the three-term recurrence relation for general orthogonal polynomials with respect to a weight function $\omega$ on an interval $(a, b)$, and the inner product $\langle f, g \rangle = \int_a^b f(x)\overline{g(x)}\omega(x)dx$. Let $\phi_0, \phi_1, \dots$ be the orthogonal monic polynomials with respect to the weight function $\omega$, where $\phi_0 = 1$, and $\phi_n$ has degree $n$.[15] Prove the three term recurrence relation below:

[15] Observe that these are precisely the orthogonal polynomials generated by applying the Gram-Schmidt process to the monomials $1, x, x^2, \dots$.

$$\phi_{n+1}(x) = (x - a_{n+1})\phi_n(x) - b_{n+1}^2\phi_{n-1},$$

where

$$a_{n+1} = \frac{\langle x\phi_n, \phi_n \rangle}{\|\phi_n\|^2}$$

and

$$b_{n+1} = \frac{\|\phi_n\|}{\|\phi_{n-1}\|}.$$

Here $n \geq 0$, and, for notational convenience, define $\phi_{-1} = 0$, $b_1 = 0$.

6. In the notation of the previous exercise, prove that the roots of $\phi_n$ are the eigenvalues of the tri-diagonal matrix

$$J_n = \begin{pmatrix} a_1 & b_2 & & \\ b_2 & a_2 & \ddots & \\ & \ddots & \ddots & b_n \\ & & b_n & a_n \end{pmatrix}.$$

7. Prove that all the roots of $\phi_n$ are real and simple and lie in the interval $(a, b)$. Outline: Since $\langle \phi_0, \phi_n \rangle = 0$, $\int_a^b \phi_n \omega \, dx = 0$. Thus $\phi_n$ changes sign in $(a, b)$, and hence it has at least one root of odd multiplicity. Let $x_1, \ldots, x_r$ be the roots of $\phi_n$ of odd multiplicity in $(a, b)$, and let $q = (x - x_1)\ldots(x - x_r)$. If $r < n$, examine $\langle q, \phi_n \rangle$.

8. Prove that the Legendre polynomial $P_n$ satisfies the differential equation $(x^2 - 1)P_n'' + 2xP_n' - n(n + 1)P_n = 0$.

9. Prove that the sum of the coefficients of any Legendre polynomial is 1. The same is true for the Tchebychev polynomials.

10. Prove that $\mathcal{C}[-1, 1]$ is contained in the space of continuous square integrable functions on $(-1, 1)$ with respect to $\omega(x) = \frac{1}{\sqrt{1-x^2}}$. The integrals involved are improper Riemann integrals.

11. Define the normalized Tchebychev polynomials $\overline{T}_0 = \frac{1}{\sqrt{\pi}}, \overline{T}_n = \sqrt{\frac{2}{\pi}} T_n$. For a function $f \in \mathcal{C}[-1, 1]$ let $S_n f = \sum_{j=0}^n \langle f, \overline{T}_j \rangle \overline{T}_j$. Prove that $\lim_n \|S_n f - f\|_2 = 0$.

12. Prove that $H_{n+1} = 2xH_n - 2nH_{n-1}$. Conclude that $H_n$ is even if and only if $n$ is even.

13. Prove that $H_n' = 2xH_n - H_{n+1}$. Conclude that $H_n' = 2nH_{n-1}$.

14. Compute $H_2$ and $H_3$.

15. Show that $\|H_n\|_2^2 = \int_{-\infty}^\infty [H_n(x)]^2 e^{-x^2} dx = n! 2^n \sqrt{\pi}$.

16. Prove that the Hermite polynomial $H_n$ satisfies the differential equation $H_n''(x) - 2xH_n'(x) + 2nH_n(x) = 0$.

# 5

# Essentials of General Topology

*Considering that he only had three years to devote to topology, he made his mark in his chosen field with brilliance and passion. He transformed the subject into a rich domain of modern mathematics. How much more might there have been, had he not died so young?*[1]

Crilly and Johnson wrote of Pavel Urysohn

Pavel Urysohn. 1898–1924

In 1915 Urysohn entered the University of Moscow to study physics. However, his interest in physics soon took second place, for, after attending lectures by Luzin and Egoroff, he began to concentrate on mathematics. Urysohn graduated in 1919 and continued working toward his doctorate. In June 1921, he became an assistant professor at the University of Moscow.

Urysohn soon turned to topology. Egoroff gave him two problems in 1921. These were difficult problems that had been around for some time. Egoroff was not to be disappointed. Near the end of August, even before working out the details, Urysohn had the correct ideas for solving the problems. During the following year, Urysohn worked through the details, building a whole new area of dimension theory in topology. It was an exciting time for topologists in Moscow, for Urysohn lectured on the topology of continua, and often his latest results were presented in the course shortly after he had proved them. He published a series of short notes on this topic during 1922. The complete theory was presented in an article

---

[1] T. Crilly and D. Johnson, "The emergence of topological dimension theory," in I. M. James (ed.), *History of Topology* (New York: Elsevier, 1999), 1–24.

*Fundamentals of Mathematical Analysis.* Adel N. Boules, Oxford University Press (2021). © Adel N. Boules.
DOI: 10.1093/oso/9780198868781.003.0005

that Lebesgue accepted for publication in the *Comptes rendus* of the Academy of Sciences in Paris. This gave Urysohn an international platform for his ideas, which immediately attracted the interest of mathematicians such as Hilbert, Hausdorff, and Brouwer. In addition to advancing dimension theory, Urysohn is credited for an important metrization theorem. He is particularly remembered for "Urysohn's lemma," which establishes the existence of a continuous function taking the values 0 and 1 on disjoint closed subsets of a normal space.

Urysohn published a full version of his dimension theory in *Fundamenta Mathematicae*. He wrote a major paper in two parts in 1923, but they did not appear in print until 1925 and 1926. Sadly, Urysohn died in a drowning accident before even the first part was published. His untimely death generated much sadness in the mathematical community.

In the summer of 1924, Urysohn set off with Alexandroff on a European trip through Germany, Holland, and France. The two mathematicians visited Hilbert. After they left, Hilbert wrote to Urysohn, informing him that his paper with Alexandroff had been accepted for publication in *Mathematische Annalen*, and expressing the hope that Urysohn would visit again the following summer. They then met Hausdorff, who was impressed with Urysohn's results. He also wrote a letter to Urysohn, which was dated August 11, 1924. The letter discusses Urysohn's metrization theorem and his construction of a universal separable metric space (one into which any separable metric space can be injected), which was one of Urysohn's last results. Like Hilbert, Hausdorff expressed the hope that Urysohn would visit again the following summer. Van Dalen writes about their final mathematical visit, which was to Brouwer:[2] *"This time [Urysohn and Alexandroff] visited Brouwer, who was most favourably impressed by the two Russians. He was particularly taken with Urysohn, for whom he developed something like the attachment to a lost son."*

## 5.1 Definitions and Basic Properties

While the metric topology is often sufficient for most introductory courses in analysis, a good understanding of the elements of general topology is essential for any advanced study of analysis. An attempt to define topology in a paragraph is quite difficult and not likely to be successful, but we offer the following narrative

---

[2] J. J. O'Connor and E. F. Robertson, "Pavel Samuilovich Urysohn," in *MacTutor History of Mathematics*, (St Andrews: University of St Andrews, 1998), http://mathshistory.st-andrews.ac.uk/Biographies/Urysohn/, accessed Oct. 31, 2020.

for the satisfaction of the the the reader who insists on an overview of the subject. We saw in chapter 4 that the collection of open sets generated by a metric has many intrinsic properties independent of the defining metric. In this section, we study the arrangement of the collection of open sets, or the topology, in a metric-free context. Every metric space is a topological space; hence all results for topological spaces (which are meaningful in the metric setting) are also valid for metric spaces, but not conversely. We often fall back on the metric case to gain insight into both subjects. We will encounter in this section many of the definitions that appeared in chapter 4, such as closure, interior, and boundary. We include those definitions again in this chapter for ease of reference. However, the proofs that duplicate those in chapter 4 are omitted. The amount of duplication is small and does not rise to the level of redundancy. We encourage the reader to compare results in this section to their counterparts in the previous chapter. The exercise is insightful.

Let $X$ be a nonempty set, and let $\mathcal{T}$ be a collection of subsets of $X$; $\mathcal{T}$ is called **a topology** on $X$ if

   (a) $\emptyset$ and $X$ are in $\mathcal{T}$,
   (b) the union of an arbitrary family of members of $\mathcal{T}$ is a member of $\mathcal{T}$, and
   (c) the intersection of two members of $\mathcal{T}$ is a member of $\mathcal{T}$.

Thus $\mathcal{T}$ is closed under the formation of arbitrary unions and finite intersections. The members of $\mathcal{T}$ are called **the open subsets of** $X$, and the pair $(X, \mathcal{T})$ is called a **topological space**.

**Example 1.** Let $X$ be a nonempty set, and let $\mathcal{T} = \mathcal{P}(X)$. Clearly, $(X, \mathcal{T})$ is a topological space. In fact, $\mathcal{P}(X)$ is the largest topology one can define on $X$. In this topology, every subset of $X$ is open. This topology is known as the **discrete topology** on $X$. It is clear that the discrete topology is too large to be useful. ◆

**Example 2.** Let $X$ be a nonempty set, and let $\mathcal{T} = \{\emptyset, X\}$. This topology is called the **trivial** or **indiscrete topology** on $X$. ◆

**Example 3.** Let $X$ be an infinite set, and define a subset $U$ of $X$ to be open if $U = \emptyset$ or if $X - U$ is finite. We verify that the collection of open sets we just defined is a topology. If $\{U_\alpha\}_\alpha$ is a collection of open sets, then, for each $\alpha$, $F_\alpha = X - U_\alpha$ is finite. Now $\cup_\alpha U_\alpha$ is open because $X - \cup_\alpha U_\alpha = \cap_\alpha F_\alpha$, which is finite. It is easy to verify that the intersection of two open subsets is open. This topology is called the **co-finite topology** on $X$ (or the finite complement topology.) ◆

**Example 4.** Let $X = (0, \infty)$, and let $\mathcal{T}$ consist of $\emptyset$ and all intervals of the form $(a, \infty)$, for all $a \geq 0$. It is easy to verify that $\mathcal{T}$ is a topology. ◆

**Example 5.** The most common topologies are the metric topologies. Thus every metric space is a topological space in accordance with the following definition: a subset $U$ of a metric space $(X, d)$ is open if it is the union of open balls. Theorem 4.1.2 says precisely that the collection of open sets thus defined is a topology. The reader can look up sections 3.6, 3.7, 4.1, and 4.8 for a variety of examples of metric spaces and hence topological spaces. ◆

**Example 6.** The most important topological space is $\mathbb{R}^n$, where the topology is the metric topology generated by the Euclidean metric (or any equivalent metric.) We will call this the **usual topology** on $\mathbb{R}^n$. ◆

If the topology $\mathcal{T}$ on a topological space $X$ is understood, we simply say that $X$ is a topological space and omit the reference to $\mathcal{T}$. If more than one topology on $X$ is being considered or if there is a danger of ambiguity, we will specifically state which topology applies to the situation in hand.

**Definition.** Let $(X, \mathcal{T})$ be a topological space. A subset $F$ of $X$ is said to be **closed** if its complement is open.

**Theorem 5.1.1.** *Let $X$ be a topological space. Then*

*(a) $X$ and $\emptyset$ are closed,*
*(b) the union of finitely many closed sets is closed, and*
*(c) the intersection of an arbitrary collection of closed sets is closed.* ■

**Definition.** Let $A$ be a subset of a topological space $X$. The **interior** of $A$, denoted $int(A)$, is the union of all the open sets contained in $A$. A point of $int(A)$ is called an **interior point** of $A$. The **closure** of $A$, denoted $\overline{A}$, is the intersection of all the closed sets containing $A$. A point of $\overline{A}$ is called a **closure point** of $A$.

The following properties of interiors and closures are straightforward. See the corresponding results in chapter 4.

**Theorem 5.1.2.** *Let $A$ and $B$ be subsets of a topological space $X$. Then*

*(a) $int(A)$ is the largest open subset of $A$;*
*(b) $A$ is open if and only if $int(A) = A$;*
*(c) if $A \subseteq B$, then $int(A) \subseteq int(B)$;*
*(d) $\overline{A}$ is the smalled closed subset of $X$ containing $A$;*
*(e) $A$ is closed if and only if $\overline{A} = A$; and*
*(f) if $A \subseteq B$, then $\overline{A} \subseteq \overline{B}$.* ■

**Definition.** Let $X$ be a topological space, and let $x \in X$. A **neighborhood** of $x$ is a subset $A$ of $X$ that contains an open set that contains $x$. A neighborhood of a set $E \subseteq X$ is a subset $A$ of $X$ that contains an open set that contains $E$.

Observe that the definition does not require a neighborhood of a point or a set to be open. If $A$ is open, we specifically refer to it as an **open neighborhood** of $x$ (respectively, $E$). For example, an open set is a neighborhood of each of its points.

The following theorem provides a useful criterion for characterizing the closure of a set. Compare its statement and proof to those of theorem 4.2.4.

**Theorem 5.1.3.** *Let $A$ be a subset of a topological space $X$. Then $x \in \overline{A}$ if and only if every open neighborhood of $x$ intersects $A$.*

*Proof.* Suppose $x \notin \overline{A}$. Then $x$ is in the open set $U = X - \overline{A}$, and, clearly, $U \cap A = \varnothing$. Conversely, if $U$ is an open neighborhood of $x$ that does not intersect $A$, then $A$ is contained in the closed set $F = X - U$. Therefore $\overline{A} \subseteq F$. In particular, $x \notin \overline{A}$. ∎

**Theorem 5.1.4.** *Let $A$ be a subspace of a topological space $(X, \mathcal{T})$. Then*

(a) $int(A) = X - \overline{(X - A)}$, and
(b) $\overline{A} = X - int(X - A)$.

*Proof.* The proof is as follows:

$$int(A) = \cup\{U : U \in \mathcal{T}, U \subseteq A\} = X - [X - \cup\{U : U \in \mathcal{T}, U \subseteq A\}]$$
$$= X - \cap\{X - U : U \in \mathcal{T}, U \subseteq A\} = X - \cap\{F : (X - A) \subseteq F, F \text{ closed}\}$$
$$= X - \overline{X - A}.$$

*To prove part (b), let $B = X - A$. Then, by part (a),*

$$X - int(B) = X - [X - \overline{(X - B)}] = \overline{X - B} = \overline{A}. \quad \blacksquare$$

**Definition.** Let $A$ be a subset of a topological space $X$. The **boundary** of $A$, denoted $\partial A$, is the set $\overline{A} \cap \overline{X - A}$. A point of $\partial A$ is called a **boundary point** of $A$.

By theorem 5.1.3, a point $x$ is a boundary point of $A$ if and only if every open neighborhood of $x$ intersects $A$ and its complement.

The proofs of the the following statements strongly resemble their metric counterparts. See, for example, theorems 4.2.8 and 4.2.9.

**Theorem 5.1.5.** *Let A be a subset of a topological space X. Then*

(a) $int(A) \cap \partial A = \emptyset$,
(b) $\overline{A} = int(A) \cup \partial A$,
(c) $\overline{A} = A \cup \partial A$, and
(d) *A is closed if and only if $\partial A \subseteq A$.* ∎

## Subspace Topology

Let $(X, \mathcal{T})$ be a topological space, and let $Y$ be a subset of $X$. Define $\mathcal{T}_Y = \{Y \cap U : U \in \mathcal{T}\}$. It is easy to verify that $\mathcal{T}_Y$ is a topology on $Y$. For example, if $\{Y \cap U_\alpha\}_\alpha$ is a collection of members of $\mathcal{T}_Y$, then $\cup_\alpha(Y \cap U_\alpha) = Y \cap (\cup_\alpha U_\alpha)$, which is in $\mathcal{T}_Y$ because $\cup_\alpha U_\alpha \in \mathcal{T}$. Verifying that $\mathcal{T}_Y$ is closed under the formation of finite intersections is straightforward.

**Definition.** The topology $\mathcal{T}_Y$ is known as the **relative, subspace, or restricted** topology on $Y$ induced by the topology $\mathcal{T}$.

**Theorem 5.1.6.** *Let $A \subseteq Y$, and let $\overline{A}_Y$ denote the closure of A in $(Y, \mathcal{T}_Y)$. Then $\overline{A}_Y = \overline{A} \cap Y$.*

*Proof.* Since $\overline{A}$ is closed in $X$, $\overline{A} \cap Y$ is closed in $Y$. Since $A \subseteq \overline{A} \cap Y$, $\overline{A}_Y \subseteq \overline{A} \cap Y$. We prove the reverse containment. Since $\overline{A}_Y$ is closed in $Y$, there exists a closed subset $F$ of $X$ such that $\overline{A}_Y = F \cap Y$. Thus $F$ is a closed subset of $X$, and $A \subseteq F$. Hence $\overline{A} \subseteq F$, and $\overline{A} \cap Y \subseteq F \cap Y = \overline{A}_Y$. ∎

## Exercises

1. Let $\mathcal{T} = \{[n, \infty) : n \in \mathbb{Z}\}$. Is $\mathcal{T}$ a topology on $\mathbb{R}$?
2. Let $X$ be an infinite set, and let $\mathcal{T}$ be the collection of subsets $U$ of $X$ such that $U = \emptyset$ or $X - U$ is countable. Is $\mathcal{T}$ a topology?
3. Define $\mathcal{T}$ to be the following collection of subsets of $\mathbb{R}$: $U \in \mathcal{T}$ if $U$ is empty or if $0 \in U$. Verify that $\mathcal{T}$ is a topology. Prove that $\overline{\{0\}} = X$ and that the restriction of $\mathcal{T}$ to $\mathbb{R} - \{0\}$ is the discrete topology.
4. Let $(X, \mathcal{T})$ be a topological space, and let $\omega$ be an object not in $X$. Define a collection $\mathcal{F}$ of subsets of $Y = X \cup \{\omega\}$ as follows: a subset $A$ of $Y$ is in $\mathcal{F}$ if $A = \emptyset$ or if $A = \{\omega\} \cup U$, where $U \in \mathcal{T}$. Is $\mathcal{F}$ a topology on $Y$?
5. Prove that the intersection of an arbitrary collection of topologies on a set $X$ is a topology on $X$.

6. (a) Prove that if $A$ and $B$ are subsets of a topological space $X$, then $\overline{A \cup B} = \overline{A} \cup \overline{B}$.

   (b) Let $\{A_\alpha\}_\alpha$ be an arbitrary collection of subsets of $X$. Prove that $\cup_\alpha \overline{A_\alpha} \subseteq \overline{\cup_\alpha A_\alpha}$ and give an example to show that strict inclusion is possible.

7. Let $Y$ be a subspace of a topological space $X$. Show that a subset $A$ of $Y$ is closed in $Y$ if and only if there exists a closed subset $F$ of $X$ such that $A = F \cap Y$.

8. Let $U$ be an open subset of $X$, and let $A \subseteq X$. Show that if $U \cap \overline{A} \neq \emptyset$, then $U \cap A \neq \emptyset$.

**Definition.** A point $x$ is said to be a **limit point** of a subset $A$ of a topological space $X$ if every open neighborhood of $x$ intersects $A$ at a point other than $x$. The set of limit points of $A$ is denoted by $A'$.

9. (a) Prove that $x$ is a limit point of $A$ if and only if $x \in \overline{A - \{x\}}$.

   (b) Prove that theorem 4.2.7 is valid for a general topological space.

10. Let $A$ and $B$ be subsets of a topological space $X$. Which of the following is true?

    (a) $(A \cup B)' = A' \cup B'$
    (b) $(A \cap B)' = A' \cap B'$

**Definition.** A subset $A$ of a topological space $X$ is said to be **nowhere dense** in $X$ if $int(\overline{A}) = \emptyset$.

11. Show that the results of problems 9 and 10 on section 4.6 are valid for a general topological space.

## 5.2  Bases and Subbases

Some topologies are quite difficult to define directly, and it is frequently the case that we want to define a topology on a set $X$ that includes a certain collection $\mathfrak{S}$ of subsets of $X$. The existence of such a topology is obvious because $\mathcal{P}(X)$ is such a topology. However, $\mathcal{P}(X)$ is useless because it is too large. This immediately suggests the question of finding the smallest topology $\mathcal{T}$ on $X$ that contains $\mathfrak{S}$. Fortunately, such a unique smallest topology $\mathcal{T}$ exists.

The reader may wonder what situations would compel us to *"want"* the members of $\mathfrak{S}$ to be open. The prime such situation is when we need a certain class of functions from $X$ to another topological space $Y$ to be continuous, which is the overarching idea behind the definition of product and weak topologies. See sections 5.4 and 6.7.

The set $\mathfrak{S}$ in the above discussion is called a subbase for $\mathcal{T}$, and a closely connected concept is that of a *base* for the topology $\mathcal{T}$, which is our first definition. Bases and subbases have a wide range of applications. In addition to providing the means to define useful topologies, bases and subbases give us easy ways to prove the continuity of functions and to characterize closures. See theorems 5.2.2 and 5.3.1.

**Definition.** An **open base** for a topology $\mathcal{T}$ on a set $X$ is a collection $\mathfrak{B}$ of open subsets of $X$ such that every nonempty open subset in $X$ is the union of members of $\mathfrak{B}$. If $\mathfrak{B}$ is an open base for $\mathcal{T}$, we say that $\mathfrak{B}$ **generates** $\mathcal{T}$.

See problem 2 at the end of this section for an equivalent, more explicit formulation of the definition of an open base.

**Example 1.** The collection $\mathfrak{B} = \{(r,s) : r,s \in \mathbb{Q}, r < s\}$ is an open base for the usual topology on $\mathbb{R}$. This is because every open subset of $\mathbb{R}$ is the union of open bounded intervals, and any such interval is the union of members of $\mathfrak{B}$: $(a,b) = \cup\{(r,s) : r \in \mathbb{Q}, s \in \mathbb{Q}, a < r < s < b\}$. See section 4.5 for a more general version of this example. ♦

The collection of open balls in a metric space is an open base for the metric topology. This follows immediately from the definition of open sets in a metric space.

Caution: Not every collection $\mathfrak{C}$ of subsets of $X$ such that $\cup\{U : U \in \mathfrak{C}\} = X$ is the open base for some topology on $X$, as the next example illustrates.

**Example 2.** Let $X = \{a,b,c\}$, and let $\mathfrak{C} = \{\varnothing, X, \{a,b\}, \{b,c\}\}$. The collection $\mathfrak{C}$ is not the base for any topology on $X$ because if it were, that topology would be $\mathfrak{C}$ because the union of two members of $\mathfrak{C}$ is in $\mathfrak{C}$. However, $\mathfrak{C}$ is not a topology, because $\{a,b\} \cap \{b,c\} \notin \mathfrak{C}$. ♦

**Theorem 5.2.1.** *Let $X$ be a nonempty set, and let $\mathfrak{B}$ be a collection of subsets of $X$ such that $\cup\{U : U \in \mathfrak{B}\} = X$. Then $\mathfrak{B}$ is an open base for some topology on $X$ if and only if, for every $U, V \in \mathfrak{B}$, and every $x \in U \cap V$, there exists a member $W \in \mathfrak{B}$ such that $x \in W \subseteq U \cap V$.*

*Proof.* If $\mathfrak{B}$ is an open base for some topology $\mathcal{T}$, and $x, U$, and $V$ are as in the statement of the theorem, then $U \cap V$ is a nonempty open set. By the definition of an open base, there is member $W$ of $\mathfrak{B}$ such that $x \in W \subseteq U \cap V$.

*Conversely, suppose $\mathfrak{B}$ satisfies the assumptions of the theorem. Define a family of subsets $\mathcal{T}$ of $X$ as follows: $U \in \mathcal{T}$ if and only if $U$ is the union of members of*

$\mathfrak{B}$. We claim that $\mathcal{T}$ is a topology. Suppose $\{U_\alpha\}_\alpha$ is a collection of members of $\mathcal{T}$, and let $x \in U = \cup_\alpha U_\alpha$. Then $x \in U_\alpha$ for some $\alpha$. By the very definition of $\mathcal{T}$, there is a member $B$ of $\mathfrak{B}$ such that $x \in B \subseteq U_\alpha \subseteq U$. This makes $U$ the union of members of $\mathfrak{B}$, that is, $U \in \mathcal{T}$. Now consider two members $U_1$ and $U_2$ of $\mathcal{T}$, and let $x \in U_1 \cap U_2$. By the definition of $\mathcal{T}$, there are members $B_1$ and $B_2$ of $\mathfrak{B}$ such that $x \in B_1 \subseteq U_1$ and $x \in B_2 \subseteq U_2$. By assumption, there is a member $W$ of $\mathfrak{B}$ such that $x \in W \subseteq B_1 \cap B_2$. Thus $x \in W \subseteq U_1 \cap U_2$, and $U_1 \cap U_2 \in \mathcal{T}$. We have proved that $\mathcal{T}$ is a topology. It is clear that $\mathfrak{B}$ is an open base for $\mathcal{T}$. ∎

**Example 3.** Let $X = \mathbb{R}$, and let $\mathfrak{B}$ be the collection of intervals in $\mathbb{R}$ of the form $[a, b)$, where $a, b \in \mathbb{R}$ and $a < b$. The nonempty intersection of two members of $\mathfrak{B}$ is a member of $\mathfrak{B}$. By theorem 5.2.1, $\mathfrak{B}$ is the base for a topology on $\mathbb{R}$ called the **lower limit topology**. The real line with the lower limit topology is sometimes referred to as the **Sorgenfrey line** and is denoted by $\mathbb{R}_l$. The lower limit topology is a rich and complicated topology that provides a number of interesting counterexamples. See problem 3 at the end of this section, and the exercises on section 5.7. ♦

The following theorem serves as an early indicator of the importance and typical uses of open bases.

**Theorem 5.2.2.** Let $X$ be a topological space, and let $\mathfrak{B}$ be an open base for the topology on $X$. If $A$ be a subset of $X$, and $x \in X$, then $x \in \overline{A}$ if and only if every basis element containing $x$ intersects $A$.

*Proof.* Use theorem 5.1.3 and problem 2 at the end of this section. ∎

**Definition.** A topology $\mathcal{T}$ on a set $X$ is said to be **weaker** (or **smaller**, or **coarser**) than a topology $\mathcal{F}$ on $X$ if $\mathcal{T} \subseteq \mathcal{F}$. We also say that $\mathcal{F}$ is **stronger** (or **finer**) than $\mathcal{T}$.

**Example 4.** If $X$ is an infinite set, then the indiscrete topology on $X$ is weaker than the co-finite topology, which, in turn, is weaker than the topology $\mathcal{P}(X)$. The lower limit topology is strictly stronger than the usual topology on $\mathbb{R}$. See problem 3 at the end of this section. ♦

**Definition.** An **open subbase** for the topology $\mathcal{T}$ on $X$ is a collection $\mathfrak{S}$ of open sets such that the collection of finite intersections of members of $\mathfrak{S}$ is an open base for $\mathcal{T}$. If $\mathfrak{S}$ is a subbase for $\mathcal{T}$, we say that $\mathfrak{S}$ generates $\mathcal{T}$.

**Example 5.** The collection of intervals $\{(-\infty, b) : b \in \mathbb{Q}\} \cup \{(a, \infty) : a \in \mathbb{Q}\}$ is an open subbase for the usual topology on $\mathbb{R}$. ♦

The following theorem provides an important mechanism for constructing a topology that contains a predetermined collection of subsets, as described in the preamble to the section.

**Theorem 5.2.3.** *Let $\mathfrak{S}$ be a collection of subsets of a nonempty set $X$ such that $\cup\{S : S \in \mathfrak{S}\} = X$. Then there exists a unique smallest topology on $X$ that contains $\mathfrak{S}$ as a subbase.*

*Proof. Let $\mathfrak{B}$ be the collection of finite intersections of members of $\mathfrak{S}$. If $U$ and $V$ are in $\mathfrak{B}$, then clearly $U \cap V$ is in $\mathfrak{B}$. By theorem 5.2.1, $\mathfrak{B}$ is the base of a topology, $\mathcal{T}$. Notice that the members of $\mathcal{T}$ are unions of finite intersections of members of $\mathfrak{S}$. If $\mathcal{F}$ is another topology that contains $\mathfrak{S}$, then $\mathcal{F}$, being a topology, contains all finite intersections of $\mathfrak{S}$ and hence all unions of such intersections. Thus $\mathcal{F}$ contains $\mathcal{T}$. This makes $\mathcal{T}$ the weakest topology that contains $\mathfrak{S}$.* ■

## Exercises

1. Show that the collection of open boxes $\{\prod_{i=1}^{n}(a_i, b_i) : a_i, b_i \in \mathbb{Q}\}$ is an open base for the usual topology on $\mathbb{R}^n$.
2. Prove that a collection $\mathfrak{B}$ of open subsets of a topological space $X$ is an open base if and only if, for every open set $U$ and every $x \in U$, there exists a member $B$ of $\mathfrak{B}$ such that $x \in B \subseteq U$.
3. (a) Prove that the usual topology on $\mathbb{R}$ is weaker that the lower limit topology.
   (b) Prove that each of the following intervals is both open and closed in the lower limit topology: $[a, b)$, $(-\infty, a)$, and $[a, \infty)$. Conclude that the usual topology is strictly weaker than the lower limit topology.
4. Let $\mathfrak{B}_1$ and $\mathfrak{B}_2$ be bases for the topologies $\mathcal{T}_1$ and $\mathcal{T}_2$ on the same set $X$. Show that if, for every $B \in \mathfrak{B}_1$ and every $x \in B$, there exists an element $B' \in \mathfrak{B}_2$ such that $x \in B' \subseteq B$, then $\mathcal{T}_1 \subseteq \mathcal{T}_2$.
5. Let $\mathfrak{B}$ be an open base for a topology $\mathcal{T}$ on a set $X$. Prove that $\mathcal{T}$ is the intersection of all the topologies on $X$ that contain $\mathfrak{B}$.
6. What topology on $\mathbb{R}$ is generated by the open subbase $\{(-\infty, a) : a \in \mathbb{R}\}$?
7. Let $\{\mathcal{T}_\alpha\}_\alpha$ be a collection of topologies on a set $X$. Prove that there is a unique smallest topology $\mathcal{T}$ that contains $\cup_\alpha \mathcal{T}_\alpha$.

## 5.3 Continuity

In section 4.3, we studied the definition of local continuity of functions on metric spaces. It is clear that the $\epsilon$-$\delta$ definition provides no clues to generalizing the definition to the topological case. However, theorem 4.3.1 provides a metric-free characterization of local continuity which, with very slight changes, produces the following definition.

**Definition.** Let $X$ and $Y$ be topological spaces. A function $f : X \to Y$ is said to be **continuous at a point** $x_0 \in X$ if, for every open subset $V$ of $Y$ containing $f(x_0)$, $f^{-1}(V)$ contains an open neighborhood of $x_0$.

We point out here an important distinction between metric and general topologies. Theorem 4.3.2 established the fact that, in the metric case, continuity is equivalent to sequential continuity. This is not the case for a general topological space. See problem 11 at the end of this section.

As in the metric case, we can define a function from a topological space $X$ to another space $Y$ to be continuous if it is continuous at each point of $X$. However, theorem 4.3.3 suggests a more convenient, and widely used, definition of global continuity.

**Definition.** Let $(X, \mathcal{T}_X)$ and $(Y, \mathcal{T}_Y)$ be topological spaces. A function $f : X \to Y$ is said to be **continuous** if the inverse image of every open subset of $Y$ is an open subset of $X$. Symbolically, $V \in \mathcal{T}_Y$ implies $f^{-1}(V) \in \mathcal{T}_X$.

Continuity depends entirely on the topologies on $X$ and $Y$. Let $X = \mathbb{R}$, $\mathcal{T}_1$ be the discrete topology on $X$, and let $\mathcal{T}_2$ be the usual topology on $\mathbb{R}$. The identity function $I_X : (X, \mathcal{T}_1) \to (X, \mathcal{T}_2)$ is continuous, but the very same function $I_X : (X, \mathcal{T}_2) \to (X, \mathcal{T}_1)$ is not continuous because not every subset of $\mathbb{R}$ is open in the usual topology of $\mathbb{R}$.

**Theorem 5.3.1.** *Using the notation of the above definition, the following are equivalent:*

*(a) $f$ is continuous.*
*(b) The inverse image of a closed subset of $Y$ is a closed subset of $X$.*
*(c) If $\mathfrak{B}$ is an open base for $\mathcal{T}_Y$, then $f^{-1}(B)$ is open in $X$ for every $B \in \mathfrak{B}$.*
*(d) If $\mathfrak{S}$ is an open subbase for $\mathcal{T}_Y$, then, for every $S \in \mathfrak{S}$, $f^{-1}(S)$ is open in $X$.*

*Proof.* *Parts (a) and (b) are equivalent because of the identity $f^{-1}(F) = X - f^{-1}$ $(Y - F)$ and the fact that a subset $F$ of $Y$ is closed if and only if $Y - F$ is open.*
    *Clearly, (a) implies (c), and (c) implies (d). Now (d) implies (c) by virtue of the identity $f^{-1}(S_1 \cap \dots \cap S_n) = f^{-1}(S_1) \cap \dots \cap f^{-1}(S_n)$, and (c) implies (a) because of the identity $f^{-1}(\cup_\alpha B_\alpha) = \cup_\alpha f^{-1}(B_\alpha)$.* ∎

**Example 1.** Suppose $f$ is a real-valued function on a topological space $X$. If $f$ is continuous at $x_0$, then there exists an open neighborhood $U$ of $x_0$ such that $f|_U$ is a bounded subset of $\mathbb{R}$.

Consider the open interval $V = (f(x_0) - 1, f(x_0) + 1)$. By the definition of continuity, there exists an open neighborhood $U$ of $x_0$ such that $f(U) \subseteq V$. Equivalently, for every $x \in U$, $|f(x) - f(x_0)| < 1$. Now, for every $x \in U$, $|f(x)| \leq |f(x) - f(x_0)| + |f(x_0)| < 1 + |f(x_0)| < \infty$. ◆

We leave the proof of the following theorem as an exercise.

**Theorem 5.3.2.** *A function $f$ from a topological space $X$ to a topological space $Y$ is continuous if and only if it is continuous at each point $x \in X$.* ∎

**Definition.** A real-valued function $f$ on a topological space $X$ is said to be **locally bounded** if, for every $x \in X$, there exists an open neighborhood $U_x$ of $x$ such that $f|_{U_x}$ is bounded in $\mathbb{R}$.

The following example follows directly from example 1 and the previous theorem.

**Example 2.** A continuous, real-valued function on a topological space is locally bounded. ◆

## Two Important Function Spaces

In section 4.8, we defined the **spaces $\mathcal{B}(X)$ of bounded functions** on an arbitrary nonempty set $X$ and, for a metric space $X$, the **spaces $\mathcal{C}(X)$ of continuous functions** on $X$ and $\mathcal{BC}(X)$ of **continuous bounded functions** on $X$. The same definitions clearly make good sense when $X$ is a topological space. Theorem 4.8.1 states that $\mathcal{B}(X)$ is a complete normed linear space under the uniform metric. The following theorem is the generalization of theorem 4.8.2.

**Theorem 5.3.3.** *If $X$ is a topological space, then the space $\mathcal{BC}(X)$ of continuous bounded functions on $X$ is a complete normed linear space.*

*Proof.* Since $\mathcal{BC}(X)$ is a subspace of $\mathcal{B}(X)$, it suffices to show that $\mathcal{BC}(X)$ is closed in $\mathcal{B}(X)$. Let $f \in \mathcal{B}(X)$ be a closure point of $\mathcal{BC}(X)$. We need to show that $f$ is continuous at each point $x_0 \in X$. For $\epsilon > 0$, there exists a function $g \in \mathcal{BC}(X)$ such that $\|f - g\|_\infty < \epsilon/3$. By the continuity of $g$ at $x_0$, there exists an open neighborhood $U$ of $x_0$ such that, for every $x \in U$, $|g(x) - g(x_0)| < \epsilon/3$. Now if $x \in U$, then $|f(x) - f(x_0)| \leq |f(x) - g(x)| + |g(x) - g(x_0)| + |g(x_0) - f(x_0)| < \epsilon$.
∎

## Homeomorphisms

**Definition.** We say that two topological spaces $(X, \mathcal{T}_X)$ and $(Y, \mathcal{T}_Y)$ are **homeomorphic** if there exists a bijection $\varphi : X \to Y$ such that both $\varphi$ and $\varphi^{-1}$ are continuous. We call such a function $\varphi$ **bicontinuous**, or **a homeomorphism**.

Intuitively speaking, two topological spaces are homeomorphic if they have identical arrangements of open sets.

**Example 3.** Any two open bounded intervals are homeomorphic. The linear function that maps $(0, 1)$ onto $(a, b)$ is clearly bicontinuous. ♦

**Example 4.** The stereographic projection is a homeomorphism from the punctured sphere onto $\mathbb{R}^2$. ♦

**Example 5.** Not every continuous bijection is a homeomorphism. The function $f(t) = (\cos t, \sin t)$ is a continuous bijection from the half-open interval $[0, 2\pi)$ onto the unit circle. ♦

**Definition.** Let $(X, \mathcal{T}_X)$ and $(Y, \mathcal{T}_Y)$ be topological spaces, let $\varphi : X \to Y$ be an injection, and let $Z = \mathfrak{R}(f)$. If $\varphi$ and $\varphi^{-1} : Z \to X$ are both continuous, we say that $\varphi$ **injects** $X$ **homeomorphically** into $Y$. We also say that $X$ is **topologically embedded** in $Y$. Here $Z$ is given the restricted topology induced by $\mathcal{T}_Y$.

**Example 6.** The inverse stereographic projection embeds $\mathbb{R}^2$ into the unit sphere. ♦

## Upper and Lower Semicontinuous Functions

**Definition.** A real-valued function $f$ on a topological space $X$ is said to be **lower semicontinuous** if, for every $a \in \mathbb{R}$, $f^{-1}((a, \infty))$ is open. We say that $f$ is **upper semicontinuous** if $f^{-1}((-\infty, b))$ is open for every $b \in \mathbb{R}$.[3]

**Theorem 5.3.4.** *Let $f$ be a real-valued function on a topological space $X$.*

*(a) $f$ is continuous if and only if it is both upper and lower semicontinuous.*

*(b) The characteristic function of an open subset is lower semicontinuous.*

---

[3] Lower semicontinuous functions played a significant role in the early development of measure theory. Upper and lower semicontinuous functions facilitate a succinct proof of Uryshon's lemma (theorem 5.11.2).

(c) The characteristic function of a closed subset is upper semicontinuous.

(d) If $\{f_\alpha\}_\alpha$ is a family of lower semicontinuous functions, then $\sup_\alpha\{f_\alpha\}$ is lower semicontinuous.

(e) If $\{f_\alpha\}_\alpha$ is a family of upper semicontinuous functions, then $\inf_\alpha\{f_\alpha\}$ is upper semicontinuous.

*Proof.* (a) *If f is both upper and lower semicontinuous, then, for all real numbers a and b, $f^{-1}(a,\infty)$ and $f^{-1}(-\infty,b)$ are open. Since intervals of the type $(a,\infty)$ and $(-\infty,b)$ form an open subbase for the usual topology on $\mathbb{R}$, f is continuous. The converse is trivial.*

(b) *If $A \subseteq X$ is open, then, for $a \in \mathbb{R}$, $\chi_A^{-1}(a,\infty)$ is open because*

$$\chi_A^{-1}(a,\infty) = \begin{cases} \varnothing & \text{if } a \geq 1, \\ A & \text{if } 0 \leq a < 1, \\ X & \text{if } a < 0. \end{cases}$$

(c) *The proof is similar to that of part (b).*

(d) *Let $f = \sup_a\{f_a\}$. Since $f^{-1}(a,\infty) = \cup_\alpha f_\alpha^{-1}(a,\infty)$, $f^{-1}(a,\infty)$ is open.*

(e) *The proof is similar to that of part (d).* ■

## Exercises

1. Let $X$, $Y$, and $Z$ be topological spaces, and let $f : X \to Y$ and $g : Y \to Z$. Prove that
   (a) if $f$ is constant, it is continuous.
   (b) if $f$ and $g$ are continuous, so is $g \circ f$;
   (c) if $A$ is a subset of $X$, then the inclusion map from $A$ to $X$ is continuous; and
   (d) if $f$ is continuous and $A \subseteq X$, then the restriction of $f$ to $A$ is continuous.

2. Let $X$ and $Y$ be topological spaces, and let $f : X \to Y$.
   (a) Prove that $f$ is continuous if and only if, for every $x \in X$ and every open neighborhood $V$ of $f(x)$ in $Y$, there exists an open neighborhood $U$ of $x$ such that $f(U) \subseteq V$.
   (b) Prove that $f$ is continuous if and only if for every subset $A$ of $X$, $f(\overline{A}) \subseteq \overline{f(A)}$.

3. Let $\mathcal{T}$ and $\mathcal{F}$ be two topologies on a set $X$. Prove that $\mathcal{T}$ is weaker that $\mathcal{F}$ if and only if the identity function $I_X : (X,\mathcal{F}) \to (X,\mathcal{T})$ is continuous. Conclude that $\mathcal{T} = \mathcal{F}$ if and only if $I_X : (X,\mathcal{F}) \to (X,\mathcal{T})$ is a homeomorphism.

4. Suppose that $f$ is a function from a topological space $X$ to a topological space $Y$ and that $X = \cup_{i=1}^n A_i$, where $A_1, \ldots, A_n$ are closed subsets of $X$. Prove that

if each $f|_{A_i}$ is continuous, then $f$ is continuous. This result is also true when each of the sets $A_i$ is open.

5. Let $f$ and $g$ be continuous real-valued functions on a topological space $X$. Prove that
   (a) $f \pm g$, $fg$ and $|f|$ are continuous,
   (b) the set $\{x \in X : f(x) \le g(x)\}$ is closed, and
   (c) the functions $h = min\{f,g\}$ and $k = max\{f,g\}$ are continuous.

6. Prove that the following subspaces of the Euclidean plane are homeomorphic:
   (a) the punctured plane $\{(x,y) : x^2 + y^2 > 0\}$
   (b) the open annulus $\{(x,y) : 1 < x^2 + y^2 < 4\}$

7. Prove that a discrete topological space $(X,\mathcal{T})$ is homeomorphic to a subspace of $\mathbb{R}$ if and only if $X$ is countable.

8. Let $a, b, c$, and $d$ be real numbers such that

$$det\begin{pmatrix} a & b \\ c & d \end{pmatrix} > 0.$$

Show that the function $f(z) = \frac{az+b}{cz+d}$ is a homeomorphism of the open upper half of the complex plane.

9. Let $X = \mathbb{R}^n - \{0\}$. Prove that the function $f(x) = \frac{x}{\|x\|_2^2}$ is continuous on $X$.

10. (a) Let $f$ be a real function on a topological space $X$. Prove that f is lower semicontinuous if and only if $-f$ is upper semicontinuous.
    (b) Prove that a subset $A$ of $X$ is open if $\chi_A$ is lower semicontinuous.
    (c) Prove that a subset $B$ of $X$ is closed if $\chi_B$ is upper semicontinuous.

11. Definition. A sequence $(x_n)$ in a topological space $X$ is said to converge to $x \in X$ if every neighborhood of $x$ contains all but finitely many terms of $(x_n)$. See problem 9 on section 4.1.

    Let $f$ be a function from a topological space $X$ to a topological space $Y$. Show that if $f$ is continuous at $x_0 \in X$, then it is sequentially continuous at $x_0$ (see theorem 4.3.2). Also give an example to show that the converse is false.

## 5.4  The Product Topology: The Finite Case

In section 4.4, we defined the product of finitely many metric spaces. In this section, we develop a construction that generalizes the concept to the case of topological spaces. Thus we define a topology on the Cartesian product of a finite number of topological spaces. Needless to say, the product topology should agree with and extend the definition of the product metric.

7. Prove that if $A_i$ is dense in $X_i$ for $1 \leq i \leq n$, then $\prod_{i=1}^{n} A_i$ is dense in the product topology.
8. Let $X$ be an infinite set, and let $\mathcal{T}$ be the co-finite topology on $X$. Prove that the product topology on $X \times X$ is not the co-finite topology.

## 5.5 Connected Spaces

Intuitively speaking, a disconnected space *comes in two pieces*. One might be tempted to define a disconnected space as the union $(X_1 \cup X_2, \mathcal{T}_1 \cup \mathcal{T}_2)$ of two topological spaces $(X_1, \mathcal{T}_1)$ and $(X_2, \mathcal{T}_2)$ where $X_1 \cap X_2 = \varnothing$. A little reflection reveals that $\mathcal{T}_1 \cup \mathcal{T}_2$ is not a topology. A topology $\mathcal{T}$ on $X = X_1 \cup X_2$ that contains $\mathcal{T}_1 \cup \mathcal{T}_2$ must contains $U_1 \cup U_2$ for any two open sets $U_1 \in \mathcal{T}_1$ and $U_2 \in \mathcal{T}_2$. In particular, $X_1 \in \mathcal{T}$ and $X_2 \in \mathcal{T}_2$. Thus $X$ is the union of two open, disjoint proper subsets of $X$. This leads us to the following definition.

**Definition.** A topological space is said to be **connected** if it is not the union of two disjoint nonempty open subsets. If $X$ is not connected, we say it is disconnected. Thus $X$ is **disconnected** if $X = P \cup Q$, where $P$ and $Q$ are open, disjoint, and $P \neq \varnothing \neq Q$. The pair $(P, Q)$ is called a **disconnection** of $X$. It is clear that $X$ is disconnected if and only if it contains a proper, nonempty subset that is both open and closed.

**Example 1.** The space $X = \{0, 1\}$ with the discrete topology is disconnected because it is the union of the open sets $\{0\}$ and $\{1\}$. We will refer to this space as the discrete space $\{0, 1\}$.

**Theorem 5.5.1.** *A topological space $X$ is disconnected if and only if there exists a continuous function from $X$ onto the discrete space $\{0, 1\}$.*

*Proof.* *Let $X$ be disconnected, and let $(P, Q)$ be a disconnection of $X$. The function $\varphi : X \to \{0, 1\}$ defined by $\varphi(P) = 0$, and $\varphi(Q) = 1$ is clearly continuous.*
   *Conversely, if $\varphi : X \to \{0, 1\}$ is a continuous surjection, then $P = \varphi^{-1}(0)$ and $Q = \varphi^{-1}(1)$ is a disconnection of $X$.* ∎

**Definition.** A subset $X$ of $\mathbb{R}$ is an **interval** if whenever $x, y \in X$ and $x < z < y$, then $z \in X$.

**Theorem 5.5.2.** *A subset $X$ of $\mathbb{R}$ is connected if and only if it is an interval. In particular, $\mathbb{R}$ is connected.*

*Proof.* Here $X$ is given the relative topology induced by the usual topology on $\mathbb{R}$. If $X$ is not an interval, then there exist two real numbers $x$ and $y$ in $X$ and a real number $z \in \mathbb{R} - X$ such that $x < z < y$. The two sets $P = X \cap (-\infty, z)$ and $Q = X \cap (z, \infty)$ form a disconnection of $X$.

Now suppose, contrary to our assertion, that $X$ is a disconnected interval, and let $\varphi : X \to \{0, 1\}$ be a continuous surjection. Since $\varphi$ is onto, there exist two real numbers $a, b \in X$ such that $\varphi(a) = 0$, and $\varphi(b) = 1$. Without loss of generality, assume that $a < b$ (otherwise, replace $\varphi$ with $1 - \varphi$). Since $X$ is an interval, the closed interval $[a, b]$ is contained in $X$. Define $P = [a, b] \cap \varphi^{-1}(0)$ and $Q = [a, b] \cap \varphi^{-1}(1)$. Clearly, $P$ and $Q$ are closed and nonempty and partition $[a, b]$. We claim that there exist sequences $a_n \in P$ and $b_n \in Q$ such that $(a_n)$ is non-decreasing, $(b_n)$ is non-increasing and $b_n - a_n = \frac{b-a}{2^{n-1}}$. This immediately leads to a contradiction because then $a = \lim_n a_n = \lim_n b_n = b \in P \cap Q = \varnothing$.

We now construct the sequences $(a_n)$ and $(b_n)$. Define $a_1 = a$, $b_1 = b$. Having found $a_2, ..., a_n$ and $b_2, ..., b_n$, let $m = \frac{a_n + b_n}{2}$. If $m \in P$, define $a_{n+1} = m$, and $b_{n+1} = b_n$. If $m \in Q$, define $a_{n+1} = a_n$, and $b_{n+1} = m$. The sequences $(a_n)$ and $(b_n)$ have the stated properties. ∎

**Theorem 5.5.3.** *The continuous image of a connected space is connected.*

*Proof.* Let $X$ be a connected space, and let $f$ be a continuous surjection of $X$ onto a topological space $Y$. If $Y$ is disconnected, there is a continuous surjection $\varphi : Y \to \{0, 1\}$. In this case, the function $\varphi \circ f$ would be a continuous surjection from $X$ onto $\{0, 1\}$. This contradicts the connectedness of $X$ and proves that $Y$ is connected. ∎

**Example 2.** The closed interval $[0, 1]$ is not homeomorphic to the circle $\mathcal{S}^1$.

Suppose there exists a homeomorphism $f : [0, 1] \to \mathcal{S}^1$. Then the restriction of $f$ to the connected subset $(0, 1)$ would be a homeomorphism. But this is a contradiction because $f((0, 1))$ is the circle with two missing points, which not connected. ◆

The following result follows directly from the last two theorems.

**Corollary 5.5.4 (the intermediate value theorem).** *If $X$ is connected and $f : X \to \mathbb{R}$ is continuous, then $f(X)$ is an interval.* ∎

**Example 3.** (a) Let $f : [a, b] \to \mathbb{R}$ be a continuous function and, say, $f(a) < f(b)$. If $k$ is between $f(a)$ and $f(b)$, then there exists a point $x \in (a, b)$ such that $f(x) = k$.

(b) Let $f : [a,b] \to [a,b]$ be continuous; then $f$ has a fixed point in $[a,b]$.

Since the range, $R$, of $f$ is connected, it is an interval. In particular, $R$ contains the interval $[f(a), f(b)]$. This proves (a). We now prove (b).

If $f(a) = a$ or $f(b) = b$, there is nothing to prove, so assume that $f(a) > a$ and $f(b) < b$. Define a function $h$ on $[a,b]$ by $h(x) = x - f(x)$. Then $h(a) < 0 < h(b)$. By (a), there is a point $x \in [a,b]$ such that $h(x) = 0$, that is, $f(x) = x$. ◆

**Theorem 5.5.5.** *If $X$ and $Y$ are connected, then the product $X \times Y$ is connected.*

*Proof.* Let $(x_0, y_0)$ and $(x_1, y_1)$ be arbitrary but fixed elements in $X \times Y$. Suppose $\varphi : X \times Y \to \{0,1\}$ is continuous. The function $i : Y \to \{x_0\} \times Y$ given by $i(y) = (x_0, y)$ is continuous; hence $\varphi \circ i$ is continuous and hence constant because $Y$ is connected. Thus $\varphi(x_0, y_0) = \varphi(x_0, y_1)$. Likewise, the function $x \mapsto \varphi(x, y_1)$ is constant, so $\varphi(x_0, y_1) = \varphi(x_1, y_1)$. Thus $\varphi(x_1, y_1) = \varphi(x_0, y_0)$, and $\varphi$ is constant. This proves that $X \times Y$ is connected. ∎

**Corollary 5.5.6.** $\mathbb{R}^n$ *is connected.*

*Proof.* Use induction, the previous theorem and the fact that $\mathbb{R}^n$ is homeomorphic to $\mathbb{R} \times \mathbb{R}^{n-1}$. ∎

**Definition.** A subset $A$ of a topological space $(X, \mathcal{T})$ is **connected** if it is a connected space with respect to the restricted topology on $A$ induced by $\mathcal{T}$.

**Example 4.** The set $X = (-1,0) \cup (0,1)$ is a disconnected subspace of $\mathbb{R}$. This is because $(-\infty, 0) \cap X = (-1, 0)$ and $(0, \infty) \cap X = (1, 0)$; hence both $(-1, 0)$ and $(1, 0)$ are open in $X$. ◆

**Theorem 5.5.7.** *Let $\{A_\alpha\}_\alpha$ be a collection of connected subsets of a topological space $X$ such that $\cap_\alpha A_\alpha \neq \emptyset$. Then $A = \cup_\alpha A_\alpha$ is connected.*

*Proof.* Let $\varphi : A \to \{0,1\}$ be continuous. Fix an element $b \in \cap_\alpha A_\alpha$. For any $a \in A$, $a \in A_\alpha$ for some $\alpha$. The restriction of $\varphi$ to $A_\alpha$ is continuous; therefore $\varphi(a) = \varphi(b)$ since $A_\alpha$ is connected. Thus $\varphi$ is not onto, and $A$ is connected. ∎

**Theorem 5.5.8.** *Let $A$ be a connected subset of a topological space $X$. If $B$ is such that $A \subseteq B \subseteq \overline{A}$, then $B$ is connected. In particular, $\overline{A}$ is connected.*

*Proof.* Let $\varphi : B \to \{0,1\}$ be continuous. Since $A$ is connected, $\varphi|_A$ is constant; say, $\varphi(A) = 0$. Now $\{0\}$ is closed in $\{0,1\}$, so $\varphi^{-1}(0)$ is closed in $B$ and contains $A$. Therefore $\varphi^{-1}(0)$ contains the closure of $A$ in $B$. But the closure of $A$ in $B$ is $\overline{A} \cap B = B$. Thus $\varphi(B) = 0$, and $\varphi$ is not onto, showing that $B$ is connected. ∎

**Definition.** Let $X$ be a topological space, and let $x, y \in X$. We say that two points $x$ and $y$ in $X$ are connected if there is a connected subset of $X$ that contains $x$ and $y$. Define a relation $\equiv$ on $X$ by $x \equiv y$ if $x$ and $y$ are connected. It is clear that $\equiv$ is an equivalence relation.

**Theorem 5.5.9.** *The equivalence classes of the relation $\equiv$ in the above definition are connected sets.*

*Proof.* Let $C$ be one of the equivalence classes and fix an element $a \in C$. For every $x \in C$, there exists a connected subset $A_x$ of $X$ containing $a$ and $x$. All the elements of $A_x$ are related; hence $A_x \subseteq C$. Since $C = \cup_{x \in C} A_x$, and $a \in \cap_{x \in C} A_x$, $C$ is connected by theorem 5.5.7. ∎

**Definition.** The equivalence classes of the relation $\equiv$ are called the **connected components of** $X$.

If $A$ is a connected subset of $X$, then all the elements of $A$ are connected (hence related). Therefore $A$ is contained in exactly one of the connected components of $X$. The summary of the above discussion is that the connected components are the maximal connected subsets of $X$. If $C$ is a connected component of $X$, then $\overline{C}$ is also connected by theorem 5.5.8. Thus $\overline{C}$ is contained in a unique connected component of $X$. Since $C \cap \overline{C} \neq \varnothing$, $\overline{C} \subseteq C$. Thus $C = \overline{C}$.

We have proved most of next result.

**Theorem 5.5.10.** *A topological space $X$ is the disjoint union of a collection $\mathcal{C}$ of disjoint, connected, closed subsets of $X$, namely, the connected components of $X$. Every connected subset of $X$ is contained in exactly one of the connected components of $X$. Every proper, nonempty subset of $X$ that is both open and closed is the union of connected components of $X$.*

*Proof.* The last assertion of the theorem is the only one we still need to prove. Let $P$ be a proper nonempty subset of $X$ that is both open and closed, and let $Q = X - P$. Then $\varnothing \neq Q \neq X$, and $Q$ is also open and closed. We show that if $C$ is a connected component of $X$, and $C \cap P \neq \varnothing$, then $C \subseteq P$. The sets $C \cap P$ and $C \cap Q$ are both open and closed in $C$. Since $C$ is connected, and $C \cap P \neq \varnothing$, $C \cap Q = \varnothing$, because otherwise the pair $(C \cap P, C \cap Q)$ would form a disconnection of $C$. This proves that $C \subseteq P$. ∎

We conclude this section with a brief excursion into path connected spaces.

**Definition.** Given two points $x$ and $y$ in a topological space $X$, a **path** from $x$ to $y$ is a continuous function $f : [0, 1] \to X$ such that $f(0) = x$, and $f(1) = y$. If there is a path from $x$ to $y$, we say that $x$ and $y$ are path connected. A topological space $X$ is **path connected** if every pair of points in $X$ are path connected.

**Example 5.** Every path connected space is connected.

Let $x$ and $y$ be points in a path connected space $X$, and let $f$ be a path from $x$ to $y$. The set $\{f(t) : t \in [0,1]\}$ is connected and contains $x$ and $y$. This shows that every two points in $X$ are connected, and hence $X$ is connected. ◆

**Example 6.** The space $X = \mathbb{R}^n - \{0\}$ is path connected.

Let $x$ and $y$ be points in $X$, and consider the line segment, $L$, that joins $x$ and $y$. If $L$ does not contain 0, then $L$ is the path we need. If $0 \in L$, take a point $z \in X$ not on $L$. The union of the two line segments that join $x$ and $y$, and then $y$ and $z$, is a path from $x$ to $z$. ◆

**Example 7.** For $n > 1$, the sphere $\mathcal{S}^{n-1}$ is path connected.

The function $f(x) = x/\|x\|_2$ maps $\mathbb{R}^n - \{0\}$ continuously onto $\mathcal{S}^{n-1}$. The result now follows from example 6 and problem 12 at the end of this section. ◆

## Exercises

1. Prove that a subset $X$ of $\mathbb{R}$ is an interval (according to the definition in this section) if and only if $X$ has one of the following types: $(-\infty, \infty)$, $(-\infty, a)$, $(-\infty, a]$, $(b, \infty)$, $[b, \infty)$, $[a, b)$, $(a, b]$, $[a, b]$, or $(a, b)$. Here $a$ and $b$ are real numbers, and $a < b$.
2. Prove that the intervals $[0, 1)$ and $(0, 1)$ are not homeomorphic. Also show that $[0, 1]$ and $[0, 1)$ are not homeomorphic.
3. Show that, for $n > 1$, $\mathbb{R}^n$ is not homeomorphic to $\mathbb{R}$.
4. Prove that a topological space $X$ is connected if and only if every nonempty proper subset of $X$ has a nonempty boundary.
5. Let $X$ be connected. Show that if there exists a continuous, nonconstant function $f : X \to \mathbb{R}$, then $X$ is uncountable.
6. Prove that if a subset $A$ of a topological space $X$ is connected, open and closed, then $A$ is a connected component of $X$.

**Definition.** A topological space $(X, \mathcal{T})$ is called **totally disconnected** if the connected components of $\mathcal{T}$ are singletons.

7. Prove that $\mathbb{Q}$ (with the usual topology) is totally disconnected. This result shows that the connected components of a topological space need not be open.

8. Prove that a topological space $X$ is totally disconnected if, for every pair of distinct points $x$ and $y$, there is a disconnection $(P, Q)$ of $X$ such that $x \in P$ and $y \in Q$.

9. Prove that if a Hausdorff space $X$ has an open base whose members are also closed, then $X$ is totally disconnected. The definition of a Hausdorff space appears in the next section.

10. Prove that the Sorgenfrey line is totally disconnected.

11. Prove the the product of two totally disconnected spaces is totally disconnected.

12. Prove that the continuous image of a path connected space is path connected.

13. Prove that the set $\{x \in \mathbb{R}^n : \|x\|_2 > 1\}$ is path connected.

**Definition.** Define a relation $\approx$ on a topological space $X$ by $x \approx y$ if $x$ and $y$ are path connected.

14. Prove that $\approx$ is an equivalence relation. The equivalence classes of $\approx$ are called the path connected components of $X$.

15. It follows from the above exercise that the path connected components partition $X$. Prove that if $A$ is a path connected subset of $X$, then $A$ is contained in exactly one of the path connected components.

16. Let $A = \{(x, sin(\frac{1}{x})) : 0 < x < 1/\pi\}$. Clearly, $A$ is path connected and hence connected. By theorem 5.5.8, the closure $\overline{A}$ of $A$ in $\mathbb{R}^2$ is also connected. Show that $\overline{A}$ is not path connected. Notice that $\overline{A} = A \cup \{(0, y) \in \mathbb{R}^2 : -1 \leq y \leq 1\}$.

## 5.6  Separation by Open Sets

Metric spaces enjoy strong separation properties, which we often take for granted. For example, two distinct points in a metric space have disjoint open neighborhoods. In chapter 4, we called this property the Hausdorff property. There is no reason to expect that the same property should hold true for an arbitrary topological space, so this property must be axiomatized. Similarly, theorem 4.2.13 shows that disjoint closed subsets of a metric space possess disjoint open neighborhoods. In the general topological setting, this property is known as normality. One important problem in topology is that of the metrizability of a topological space. Explicitly stated, under what set of conditions is a given topology induced by a metric. The fact that every metirc space is normal imposes an immediate necessary condition on a topology to be metrizable: such a topology must be normal. Of course, normality is not a sufficient condition for a space to be metrizable. In section 5.11, we prove a metrization theorem that gives a sufficient set of conditions for a topology to be metrizable. In this section, we study the three most common forms of separating points and sets in a topological space.

**Definition.** A topological space $X$ is said to be a $T_1$ **space** if, for every pair of distinct points $x$ and $y$ in $X$, there exists a neighborhood of $x$ not containing $y$ and a neighborhood of $y$ not containing $x$. The two neighborhoods may intersect.

**Definition.** A topological space $X$ is said to be **Hausdorff** (or $T_2$) if for every pair of distinct points $x$ and $y$, there is an open neighborhood $U$ of $x$ and an open neighborhood $V$ of $y$ such that $U \cap V = \varnothing$.

It is safe to say that all important topological spaces are Hausdorff. Weaker separation axioms, such as $T_1$, are used mostly to generate exercises and counterexamples.

Theorem 4.1.4 states that a metric space is Hausdorff, which supports the statement in the above paragraph since metric spaces are the most important (but not the only important) examples of topological spaces.

**Theorem 5.6.1.** *If $X$ is a Hausdorff space and $x \in X$, then $\{x\}$ is closed.*

*Proof.* We show that the set $W = X - \{x\}$ is open. For every $y \in W$, there exist open neighborhoods $U_y$ and $V_y$ of $x$ and $y$, respectively, such that $U_y \cap V_y = \varnothing$. This clearly implies that $V_y \subseteq W$ for all $y \in W$. Consequently, $W = \cup\{V_y : y \in W\}$, which is open. ∎

**Definition.** A sequence $(x_n)$ of a topological space $X$ is said to converge to a point $x \in X$ if every neighborhood of $x$ contains all but finitely many terms of the sequence.

Theorem 4.1.5 says that the limit of a convergent sequence in a metric space is unique. This is precisely because metric spaces are Hausdorff spaces.

**Example 1.** Let $X$ be a Hausdorff space, and suppose that $(x_n)$ is a convergent sequence. Then the limit is unique.
  Suppose that $\lim_n x_n = x$ and $\lim_n x_n = y$ and that $x \neq y$. Let $U$ and $V$ be disjoint open neighborhoods of $x$ and $y$, respectively. Since $\lim_n x_n = x$, there is an integer $N$ such that, for all $n > N$, $x_n \in U$. Since $U \cap V = \varnothing$, $V$ can contain only finitely many terms of $(x_n)$, which is a contradiction. ♦

**Example 2.** Let $X$ be a topological space, and let $Y$ be a Hausdorff space. If $f : X \to Y$ is continuous, then the graph of $f$, $G = \{(x, f(x)) : x \in X\}$ is closed in the product space $X \times Y$.

We will show that the complement of $G$ is open in $X \times Y$. Let $(x, y) \notin G$, thus $y \neq f(x)$. Let $U$ and $V$ be disjoint open neighborhoods of $y$ and $f(x)$, respectively. Because $f$ is continuous, there exists an open neighborhood $W$ of $x$ such that $f(W) \subseteq V$. It is easy to check that $(W \times U) \cap G = \emptyset$. ♦

**Definition.** A Hausdorff space $X$ is said to be **regular** if, for every $x \in X$ and every closed subset $F$ that does not contain $x$, there exist open sets $U$ and $V$ such that $x \in U, F \subseteq V$, and $U \cap V = \emptyset$.

Theorem 4.2.12 states that a metric space is regular.

**Example 3.** A subspace of a regular space $X$ is regular.

Let $Y$ be a subspace of $X$, let $F$ be a closed subset of $Y$ (in the restricted topology on $Y$), and let $x \in Y - F$. By theorem 5.1.6, $F = \overline{F} \cap Y$, where $\overline{F}$ denotes the closure of $F$ in $X$. Now $x \notin \overline{F}$, so by the regularity of $X$, there exist disjoint open neighborhoods $U$ and $V$ of $x$ and $\overline{F}$, respectively. The sets $U_1 = U \cap Y$ and $V_1 = V \cap Y$ are open in $Y$ and separate $x$ and $F$. ♦

The following characterization of regularity is often useful.

**Theorem 5.6.2.** *A Hausdorff space is regular if and only if for every $x \in X$ and every open neighborhood $U$ of $x$, there exists an open neighborhood $V$ of $x$ such that $\overline{V} \subseteq U$.*

*Proof. Suppose $X$ is regular, and let $x$ and $U$ be as in the statement of the theorem. By regularity, applied to $x$ and the closed set $X - U$, there exists open neighborhoods $V$ of $x$ and $W$ of $X - U$ such that $V \cap W = \emptyset$. Because $V \subseteq X - W$ and the latter set is closed, $\overline{V} \subseteq X - W$. In particular, $\overline{V} \subseteq U$.*

*Conversely, let $F$ be a closed subset of $X$ that does not contain $x$. By assumption, there exits an open neighborhood $U$ of $x$ such that $\overline{U} \subseteq X - F$. Set $V = X - \overline{U}$. The sets $U$ and $V$ are disjoint open neighborhoods of $x$ and $F$, respectively, as desired.* ∎

**Definition.** A Hausdorff space $X$ is said to be **normal** if, for every pair of disjoint closed subsets $E$ and $F$ of $X$, there exist open sets $U$ and $V$ such that $E \subseteq U$, $F \subseteq V$, and $U \cap V = \emptyset$.

Theorem 4.2.13 states that a metric space is normal.

The proof of theorem 5.6.3 mimics that of theorem 5.6.2 and is therefore omitted.

**Theorem 5.6.3.** *A Hausdorff space X is normal if and only if for every closed set E and every open neighborhood U of E, there exists an open neighborhood V of E such that $\overline{V} \subseteq U$.* ■

Products and subspaces of normal and regular spaces have dissimilar properties. For example, the product of regular spaces is regular, but the same result does not hold for the product of normal spaces. Likewise, an arbitrary subspace of a normal space need not be normal. See the exercises on section 5.7. However, the following special case is easy to prove.

**Example 4.** A closed subspace $Y$ of a normal space $X$ is normal.

Let $E$ and $F$ be closed subspaces of $Y$. Since $Y$ is closed in $X$, $E$ and $F$ are closed in $X$. By the normality of $X$, there exists disjoint open neighborhoods $U$ and $V$ of $E$ and $F$, respectively. The sets $U_1 = U \cap Y$ and $V_1 = V \cap Y$ are open in $Y$ and separate $E$ and $F$. ◆

## Exercises

1. Prove that a topological space is $T_1$ if and only if every single-point set of $X$ is closed.
2. Let $X$ be an infinite set. Prove that the co-finite topology on $X$ is $T_1$ but not Hausdorff.
3. Let $A$ be subset of a Hausdorff space $X$. Show that a point $x \in X$ is a limit point of $A$ if and only if every neighborhood of $x$ contains infinitely many points of $A$.
4. Prove that a subspace of a Hausdorff space is Hausdorff and that the product of two Hausdorff spaces is Hausdorff.
5. Prove that a topological space $X$ is Hausdorff if and only if the diagonal set $\{(x,x) : x \in X\}$ is closed in the product space $X \times X$.
6. Let $X$ and $Y$ be topological spaces, and let $f,g : X \to Y$ be continuous. Prove that if $Y$ is Hausdorff, then the set $\{x \in X : f(x) = g(x)\}$ is closed.
7. Prove that the set of fixed points of a continuous function on a Hausdorff space is closed.
8. Let $f$ and $g$ be continuous functions from a topological space $X$ to a Hausdorff space $Y$. Show that if $f$ and $g$ agree on a dense subset of $X$, then $f = g$.
9. Prove that the product of two regular spaces is regular.
10. Prove theorem 5.6.3.

11. Let $X$ be a regular space. Prove that every pair of distinct points in $X$ have neighborhoods whose closures are disjoint.

12. Let $X$ be a normal space. Prove that every pair of disjoint closed subsets of $X$ have neighborhoods whose closures are disjoint.

## 5.7  Second Countable Spaces

In this section, we study second countable, separable, and Lindelöf spaces. Theorem 4.5.1 states that all three conditions are equivalent for metric spaces. This is not true for general topological spaces, and several counterexamples are provided in this section and the section exercises to show the nonequivalence of the three conditions. However, second countability implies the other two conditions. Second countability has other pleasant consequences, especially when it is combined with normality or local compactness. The definitions in this section are identical to the those in the metric case and are included below for ease of reference.

**Definition.** A subset $A$ of a topological space $X$ is **dense** in $X$ if $\overline{A} = X$.

**Definition.** A topological space $X$ is **separable** if it contains a countable dense subset.

**Definition.** A topological space $X$ is **second countable** if the topology on $X$ contains a countable open base.

**Definition.** A topological space $X$ is said to be **a Lindelöf space** if every open cover of $X$ contains a countable subcover of $X$. The definitions of open covers and subcovers can be seen in section 4.5.

**Example 1.** Consider the **Sorgenfrey plane**, $\mathbb{R}_l^2 = \mathbb{R}_l \times \mathbb{R}_l$. In problem 11, we ask the reader to show that $\mathbb{R}_l^2$ is separable. Here we show that the subspace $L = \{(x, -x) : x \in \mathbb{R}\}$ is not separable. We claim that restriction of the topology on $\mathbb{R}_l^2$ to $L$ is the discrete topology. Since $L$ is uncountable, it is not separable. To prove our claim, let $x \in \mathbb{R}$, and consider the set $U = [x, x+1) \times [-x, -x+1)$; $U$ is open in $\mathbb{R}_l^2$, and $U \cap L = \{(x, -x)\}$. Therefore the single point $(x, -x)$ is open in $L$. ◆

**Example 2.** In problem 10, we ask the reader to show that the Sorgenfrey line $\mathbb{R}_l$ is Lindelöf. We show here that $\mathbb{R}_l^2$ is not Lindelöf. Thus the product of two Lindelöf spaces is not necessarily Lindelöf. Let $L$ be as in example 1. The line $L$ is closed in $\mathbb{R}_l^2$. Consider the open cover $\mathcal{U}$ of $\mathbb{R}_l^2$ that consists of $\{\mathbb{R}_l^2 - L\}$ and the collection $\{[x, x+1) \times [-x, -x+1) : x \in \mathbb{R}\}$. Clearly, no countable subset of $\mathcal{U}$ can cover $\mathbb{R}_l^2$. ◆

**Definition.** A collection $\mathfrak{F} = \{F_\alpha : \alpha \in I\}$ of subsets of a nonempty set $X$ is said to have the **countable intersection property** if every countable subcollection of $\mathfrak{F}$ has a nonempty intersection.

**Example 3.** A topological space $X$ is Lindelöf if and only if every collection of closed subsets of $X$ with the countable intersection property has a nonempty intersection.

Suppose $X$ is Lindelöf, and let $\mathfrak{F}$ be a collection of closed subsets with the countable intersection. If $\cap\{F_\alpha : \alpha \in I\} = \varnothing$, then $X = \cup_{\alpha \in I}(X - F_\alpha)$. Therefore $X = \cup_{n=1}^\infty (X - F_{\alpha_n})$, for some countable subset $\{\alpha_1, \alpha_2, \ldots\}$ of $I$. It follows that $\cap_{n=1}^\infty F_{\alpha_n} = \varnothing$; a contradiction.

Conversely, if $\{U_\alpha\}_{\alpha \in I}$ is an open cover of $X$ with no countable subcover, then the family $\{F_\alpha : \alpha \in I\} = \{X - U_\alpha : \alpha \in I\}$ has the countable intersection property because, for a countable subcollection $\{F_{\alpha_1}, F_{\alpha_2}, \ldots\}$ of $\mathfrak{F}$, $\cap_{n=1}^\infty F_{\alpha_n} = \cap_{n=1}^\infty (X - U_{\alpha_i}) = X - \cup_{n=1}^\infty U_{\alpha_n} \neq \varnothing$. However, $\cap_{\alpha \in I} F_\alpha = \cap_{\alpha \in I}(X - U_\alpha) = X - (\cup_{\alpha \in I} U_\alpha) = \varnothing$. ◆

**Theorem 5.7.1.** *A subset $A$ of a topological space $X$ is dense if and only if it intersects every open subset of $X$.*

*Proof.* We prove the contrapositive of each implication. If $\overline{A} \neq X$, then the set $U = X - \overline{A}$ is open, nonempty and $U \cap A = \varnothing$.

Conversely, if there exists a nonempty open set $U$ such that $U \cap A = \varnothing$, then $A \subseteq X - U$. Since $X - U$ is closed, $\overline{A} \subseteq X - U \neq X$. ∎

**Theorem 5.7.2.** *In a separable topological space $X$, every collection of pairwise disjoint open sets is countable.*

*Proof. We prove that if $X$ contains an uncountable collection of pairwise disjoint subsets, then any dense subset $A$ of $X$ is uncountable. Let $\{U_\alpha\}_{\alpha \in I}$ be an uncountable family of pairwise disjoint open subsets of $X$. By theorem 5.7.1, each $U_\alpha$ intersects $A$. Choose an element $a_\alpha \in U_\alpha \cap A$. Now $a_\alpha \neq a_\beta$ if $\alpha \neq \beta$ since $U_\alpha \cap U_\beta = \varnothing$. Hence $A$ is uncountable.* ∎

**Theorem 5.7.3.** *Let $X$ be a second countable topological space. Then*

*(a) $X$ is separable, and*
*(b) $X$ is Lindelöf.*

*Proof. Let $\{B_n\}$ be a countable open base for the topology on $X$. For each $n \in \mathbb{N}$, choose a point $a_n \in B_n$, and let $A = \{a_n : n \in \mathbb{N}\}$. If $U \neq \varnothing$ is open in $X$, then $U$*

*contains a basis element $B_n$, and hence $a_n \in A \cap U$. Theorem 5.7.1 implies that $A$
is dense. The proof of (b) is identical to that in theorem 4.5.1.* ∎

It was observed in section 5.6 that normality is a necessary condition for the
metrizability of a topological space. For second countable spaces, the normality
requirement can be relaxed, as the following theorem shows. As it turns out,
regular second countable spaces are metrizable. See the Urysohn metrization
theorem in section 5.11.

**Theorem 5.7.4.** *A regular second countable Hausdorff space $X$ is normal.*

*Proof.* Let $E$ and $F$ be disjoint closed subsets of $X$, and let $\mathfrak{B}$ be a countable open base
for the topology on $X$. For every $x \in E$, $x$ belongs to the open set $X - F$. By theorem
5.6.2, there exists an open neighborhood $W$ of $x$ such that $\overline{W} \subseteq X - F$. Choose a
basis element $B_x$ such that $x \in B_x \subseteq W$. Clearly, $E \subseteq \cup_{x \in E} B_x$. Since $\mathfrak{B}$ is countable,
the collection $\{B_x\}_{x \in E}$ can be enumerated as $\{U_n\}$. Observe that $\overline{U_n} \subseteq X - F$. A
similar argument produces a countable open cover $\{V_n\}$ of $F$ such that $V_n \in \mathfrak{B}$,
and $\overline{V_n} \subseteq X - E$.

Define $U'_n = U_n - \cup_{i=1}^{n} \overline{V_i}$, and $V'_n = V_n - \cup_{i=1}^{n} \overline{U_i}$. Notice that if $n \leq m$, then
$U'_n \cap V'_m = \varnothing$. By symmetry, if $m \leq n$, then $V'_m \cap U'_n = \varnothing$. It follows that, for
all $m, n \in \mathbb{N}$, $U'_n \cap V'_m = \varnothing$. Now define $U = \cup_{n=1}^{\infty} U'_n$ and $V = \cup_{n=1}^{\infty} V'_n$. Clearly,
$U \cap V = \varnothing$, and it is straightforward to verify that $E \subseteq U$ and $F \subseteq V$. ∎

**Example 4.** Let $X$ be a second countable topological space, and let $\mathfrak{C}$ be an open
base for the topology on $X$. Then $\mathfrak{C}$ contains a countable subset which is also
an open base for $X$.

Let $\mathfrak{B} = \{B_n : n \in \mathbb{N}\}$ be a countable open base. Let $I$ be the subset of $\mathbb{N} \times \mathbb{N}$ of
pairs $(m, n)$ for which there is a member $C \in \mathfrak{C}$ such that $B_m \subseteq C \subseteq B_n$. For each
pair $(m, n) \in I$, choose a member $C_{m,n}$ of $\mathfrak{C}$ such that $B_m \subseteq C_{n,m} \subseteq B_n$. We show
that the countable collection $\{C_{m,n} : (m, n) \in I\}$ is an open base. Let $U$ be an
open set, and let $x \in U$. Because $\mathfrak{B}$ is an open base, there is a set $B_n$ such that
$x \in B_n \subseteq U$. For the same reason, there is a member $C$ of $\mathfrak{C}$ such that $x \in C \subseteq B_n$.
Finally, there is an element $B_m$ of $\mathfrak{B}$ such that $x \in B_m \subseteq C$. Clearly, $(m, n) \in I$,
and $x \in C_{m,n} \subseteq U$. ◆

## Exercises

1. Prove that the product of two second countable spaces is second countable
   and that the product of two separable spaces is separable.

**Definition.** A point $x$ of a subset $A$ of a topological space $X$ is said to be **isolated** if $x$ has an open neighborhood $U$ such that $A \cap U = \{x\}$.

2. Prove that the set of isolated points of a second countable space $X$ is at most countable. Then show that if $X$ is uncountable, then $X$ has uncountably many limit points.

**Definition.** A topological space $X$ is said to be **first countable** if, for every $x \in X$, there is a countable collection $\{U_n\}$ of open neighborhoods of $x$ such that, for every open neighborhood $U$ of $x$, there is an integer $n$ such that $x \in U_n \subseteq U$. The collection $\{U_n\}$ is called a local base at $x$. It is sometimes convenient to have a local base $\{V_n\}$ with the additional property that $V_n \supseteq V_{n+1}$. This can be easily achieved by defining $V_n = \cap_{i=1}^n U_i$.

3. Prove that every second countable space is first countable and that every metric space is first countable.
4. Show that a subspace of a second (respectively, first) countable is second (respectively, first) countable. Also show that the product of two first countable spaces is first countable.
5. Show that a subspace of a separable space need not be separable. Hint: See problem 3 on section 5.1. For a more elaborate example, see problem 12 below.
6. Let $X$ be an uncountable set, and let $\mathcal{T}$ be the co-finite topology on $X$. Show that every infinite subset of $X$ is dense, and hence $X$ is separable. Show, however, that $X$ is not second countable. Hint: If $\{B_n\}$ is a countable collection of open subsets of $X$, then $\cap_{n=1}^{\infty} B_n$ is uncountable. Pick a point $x \in \cap_{n=1}^{\infty} B_n$, and consider the open set $U = X - \{x\}$.
7. Show that a closed subspace of a Lindelöf space if Lindelöf.
8. Let $X$ be a topological space $X$, and let $\mathfrak{B}$ be an open base for $X$. Prove that $X$ is Lindelöf if and only if every open cover of $X$ by members of $\mathfrak{B}$ has a countable subcover.
9. Show that the Sorgenfrey line is first countable, separable, but not second countable. Hint: To show that $\mathbb{R}_l$ is not second countable, let $\mathfrak{B}$ be an open base for $\mathbb{R}_l$. For every and $x \in \mathbb{R}$, there is a member $B_x \in \mathfrak{B}$ such that $x \in B_x \subseteq [x, x+1)$.
10. Prove that the Sorgenfrey line $\mathbb{R}_l$ is Lindelöf. Together with the previous problem, this problem shows that not every Lindelöf space is second countable. Hint: Use problem 8. Let $\{[a_\alpha, b_\alpha) : \alpha \in I\}$ be an open cover of $\mathbb{R}_l$ by basic open subsets of $\mathbb{R}_l$. Define $C = \cup_{\alpha \in I}(a_\alpha, b_\alpha)$. View $C$ as a subset of $\mathbb{R}$ with the usual topology; $C$ is Lindelöl because $\mathbb{R}$ is a metric space. Thus there exists a countable subset $\{\alpha_n : n \in \mathbb{N}\}$ such that $C = \cup_{n=1}^{\infty}(a_{\alpha_n}, b_{\alpha_n})$. Argue that $\mathbb{R} - C$ is countable.

11. Show that the Sorgenfrey plane $\mathbb{R}_l^2$ is separable.
12. Show that the line $L$ in example 2 is closed in $\mathbb{R}_l^2$.
13. Let $X$ be a topological space, and let $\mathfrak{B}$ be an open base for the topology on $X$. Suppose that $\mathfrak{B}$ has infinite cardinality $\aleph$. Prove that if $\mathfrak{C}$ is another open base for the topology on $X$, then $\mathfrak{C}$ contains a subset of cardinality $\leq \aleph$ that is also an open base for $X$.

## 5.8  Compact Spaces

In section 4.7, we studied compact metric spaces extensively, including several equivalent formulations of the definition of compactness. We adopt the same definition of compactness in this chapter because the other characterizations do not lend themselves easily to generalization to general topological spaces, and especially because some of the other characterizations of compact metric spaces are *false in general*. You will also see that compact spaces have pleasant separation properties. Finally, we will prove the celebrated Tychonoff theorem for the product of finitely many topological spaces. The leading theorems in this section have counterparts in section 4.7. Therefore, proofs that duplicate those in section 4.7 will be omitted.

**Definition.** A topological space $X$ is said to be **compact** if every open cover of $X$ contains a finite subcover of $X$.

**Example 1.** The co-finite topology on an infinite set $X$ is compact.

Let $\mathcal{U}$ be an open cover of $X$, and fix an element $U_1 \in \mathcal{U}$. The complement of $U_1$ is finite, say, $U_1 = X - \{x_2, \ldots, x_n\}$. Now, for each $2 \leq i \leq n$, pick an element $U_i \in \mathcal{U}$ that contains $x_i$. The finite collection $\{U_1, \ldots, U_n\}$ covers $X$. ◆

**Example 2.** A real-valued, locally bounded function $f$ on a compact space $X$ is bounded.

By the definition of local boundedness, for every $x \in X$, there exists a positive number $M_x$ and an open neighborhood $U_x$ of $x$ such that $sup_{x \in U_x} |f(x)| \leq M_x$. Clearly, $\{U_x : x \in X\}$ is an open cover of $X$. Choose points $x_1, \ldots, x_n \in X$ such that the sets $U_{x_1}, \ldots, U_{x_n}$ cover $X$, and let $M = max_{1 \leq i \leq n} M_{x_i}$. For $x \in X$, $x \in U_{x_i}$ for some $1 \leq i \leq n$, and $|f(x)| \leq M_{x_i} \leq M$. ◆

**Definition.** Let $K$ be a subset of a topological space $X$. We say that $K$ is a **compact subset** (or a compact subspace) of $X$ if it is compact in the restricted topology.

**Theorem 5.8.1.** *A subset $K$ of a topological space $X$ is compact if and only if it satisfies the following condition: if $\mathcal{U}$ is a collection of open subsets of $X$ such that*

$K \subseteq \cup\{U : U \in \mathcal{U}\}$, then there exists a finite subcollection $\{U_1, U_2, ..., U_n\}$ of $\mathcal{U}$ such that $K \subseteq \cup_{i=1}^n U_i$.
  The proof is identical to that of theorem 4.7.1. ∎

**Theorem 5.8.2.** *A closed subspace K of a compact space X is compact.*
  *The proof is identical to that of theorem 4.7.2.* ∎

**Example 3.** Every compact space has the Bolzano-Weierstrass property.
  It is sufficient to prove that if a subset $A$ of a compact space $X$ has no limit points, then it is finite. By problem 9(b) on section 5.1, $A$ is closed. By theorem 5.8.2, $A$ is compact. Every point $a \in A$ is not a limit point of $A$; hence there exists an open set $U_a$ of $A$ such that $A \cap U_a = \{a\}$. If $A$ is infinite, then the open cover $\{U_a : a \in A\}$ of $A$ would have no finite subcover. This forces $A$ to be finite, as claimed. ◆

The converse of the above example is false, but counterexamples are rather difficult.

**Theorem 5.8.3.** *A compact subspace K of a Hausdorff space X is closed.*
  *The proof is identical to that of theorem 4.7.3.* ∎

**Example 4.** Let $\{K_\alpha\}_{\alpha \in I}$ be a collection of compact subsets of a Hausdorff space $X$. If $\cap_\alpha K_\alpha = \emptyset$, then the intersection of some finite subcollection of $\{K_\alpha\}$ is empty.

Let $V_\alpha = X - K_\alpha$, and fix and element $\alpha_1 \in I$. By assumption, $\{V_\alpha\}$ covers $X$ and hence $K_{\alpha_1}$. Thus there exists a finite subset $\{\alpha_2, ..., \alpha_n\}$ of $I$ such that $K_{\alpha_1} \subseteq \cup_{i=2}^n V_{\alpha_i}$. It follows directly that $\cap_{1 \leq i \leq n} K_{\alpha_i} = \emptyset$. ◆

**Theorem 5.8.4.** *The continuous image of a compact space is compact.*
  *The proof is identical to that of theorem 4.7.4.* ∎

**Theorem 5.8.5.** *A continuous real-valued function $f : X \to \mathbb{R}$ on a compact space X is bounded and attains its maximum and minimum values.*
  *The proof is identical to that of theorem 4.7.12.* ∎

The next result follows immediately from theorem 5.3.3 and the fact that for a compact space $X$, $\mathcal{C}(X) = \mathcal{BC}(X)$.

**Theorem 5.8.6.** *Let X be a compact Hausdorff space, and let $\mathcal{C}(X)$ be the space of continuous functions on X. Then $(\mathcal{C}(X), \|.\|_\infty)$ is a complete normed linear space.* ∎

**Theorem 5.8.7.** *Let X be a compact space, and let Y be a Hausdorff space. Then a continuous bijection $\varphi : X \to Y$ is a homeomorphism from X to Y.*

*Proof. We prove that $\varphi^{-1}$ is continuous by showing that $\varphi$ is a closed mapping. Let F be a closed subset of X. By theorem 5.8.2, F is compact. By theorem 5.8.4, $\varphi(F)$ is compact in Y. Now theorem 5.8.3 implies that $\varphi(X)$ is closed, as desired.* ∎

The theorem says that when we limit our attention to compact Hausdorff spaces, a bijection $\varphi : X \to Y$ is a homeomorphism if and only if it is simply continuous. In this situation, we can show that X and Y are homeomorphic by merely showing the continuity of $\varphi$ or $\varphi^{-1}$ or by showing that $\varphi$ (or $\varphi^{-1}$) is an open (or a closed) mapping.

**Definition.** A collection $\mathfrak{F}$ of subsets of a nonempty set X is said to have the **finite intersection property** if every finite subcollection of $\mathfrak{F}$ has a nonempty intersection.

The next theorem provides a useful equivalent characterization of compactness. Its proof is left as an exercise. See example 3 on section 5.7.

**Theorem 5.8.8.** *The following are equivalent for a topological space X:*

   *(a) X is compact.*
   *(b) If $\mathfrak{F} = \{F_\alpha : \alpha \in I\}$ is a collection of closed subsets of X satisfying the finite intersection property, then $\cap\{F_\alpha : \alpha \in I\} \neq \varnothing$.* ∎

## Compactness and Separation

**Theorem 5.8.9.** *Let X be a Hausdorff space, and let F be a compact subset of X. For every $x \in X - F$, there exist disjoint open sets U and V such that $x \in U$, and $F \subseteq V$.*

*Proof. For every $y \in F$, there exist disjoint open sets $U_y$ and $V_y$ such that $x \in U_y$ and $y \in V_y$. Now $F \subseteq \cup_{y \in F} V_y$. Since F is compact, $F \subseteq \cup_{i=1}^{n} V_{y_i}$ for a finite subset $\{y_1, \ldots, y_n\}$ of F. The sets $U = \cap_{i=1}^{n} U_{y_i}$ and $V = \cup_{i=1}^{n} V_{y_i}$ have the desired properties.* ∎

**Theorem 5.8.10.** *A compact Hausdorff space is normal. Thus if E and F are disjoint closed subsets of X, then there exist disjoint open subsets U and V such that $E \subseteq U$ and $F \subseteq V$.*

*Proof.* First observe that $E$ and $F$ are compact by theorem 5.8.2. Let $x \in E$. By the previous theorem, there are disjoint open sets $U_x$ and $V_x$ such that $x \in U_x$ and $F \subseteq V_x$. Since $E \subseteq \cup_{x \in E} U_x$, and $E$ is compact, $E \subseteq \cup_{i=1}^{n} U_{x_i}$ for some finite subset $\{x_1, \dots, x_n\}$ of $E$. Set $U = \cup_{i=1}^{n} U_{x_i}$ and $V = \cap_{i=1}^{n} V_{x_i}$. The sets $U$ and $V$ have the stated properties. ∎

## Finite Products of Compact Spaces

**Lemma 5.8.11 (the tube lemma).** *Let $X$ be a topological space, and let $Y$ be a compact space. If an open subset $W$ in $X \times Y$ contains a line, $\{x\} \times Y$, then there exists a neighborhood $U$ of $x$ such that $U \times Y \subseteq W$. Here $x$ is a fixed element of $X$.*

*Proof.* For every $y \in Y$, there are open sets $U_y \subseteq X$ and $V_y \subseteq Y$ such that $(x, y) \in U_y \times V_y \subseteq W$. Thus $\{x\} \times Y \subseteq \cup_{y \in Y}(U_y \times V_y) \subseteq W$. The compactness of $\{x\} \times Y$ yields a finite subset $\{y_1, \dots, y_n\}$ such that $\{x\} \times Y \subseteq \cup_{i=1}^{n}(U_{y_i} \times V_{y_i}) \subseteq W$. Define $U = \cap_{i=1}^{n} U_{y_i}$. We claim that $U \times Y \subseteq W$. If $u \in U$, and $y \in Y$, then $y \in V_{y_i}$ for some $1 \leq i \leq n$. But $u$ belongs to $U_{y_i}$ for every $1 \leq i \leq n$. Therefore $(u, y) \in U_{y_i} \times V_{y_i} \subseteq W$. ∎

The above lemma says that if an open subset of $X \times Y$ contains a line, then it must contain a strip (or a tube, hence the name) that contains the line. Intuitively, an open subset of $X \times Y$ cannot get *arbitrarily thin* around a line. The following example illustrates the concept.

**Example 5.** The open subset $W = \{(x, y) \in \mathbb{R}^2 : x \in \mathbb{R}, |y| < \frac{1}{1+x^2}\}$ contains the $x$-axis but there is no positive number $\delta$ such that $\mathbb{R} \times (-\delta, \delta)$ is contained in $W$. ♦

**Theorem 5.8.12 (Tychonoff's theorem).** *If $X$ and $Y$ are compact spaces, so is $X \times Y$.*

*Proof.* Let $\mathcal{W}$ be an open cover of $X \times Y$. For $x \in X$, $\{x\} \times Y \subseteq \cup\{W : W \in \mathcal{W}\}$. Since $\{x\} \times Y$ is compact, there exists a finite subset $\{W_1^x, \dots, W_{n_x}^x\}$ of $\mathcal{W}$ such that $\{x\} \times Y \subseteq \cup_{i=1}^{n_x} W_i^x$. Let $W^x = \cup_{i=1}^{n_x} W_i^x$. By the previous lemma, there exists an open neighborhood $U^x$ of $x$ such that $U^x \times Y \subseteq W^x$. The collection of open sets $\{U^x\}_{x \in X}$ covers $X$; hence $X = \cup_{i=1}^{m} U^{x_i}$, for some finite subset $\{x_1, \dots, x_m\}$ of $X$. We claim that the finite collection $\{W_i^{x_j} : 1 \leq j \leq m, 1 \leq i \leq n_{x_j}\}$ covers $X \times Y$. Let $(x, y) \in X \times Y$. Then $x \in U^{x_j}$ for some $1 \leq j \leq m$. Now $(x, y) \in U^{x_j} \times Y \subseteq \cup_{i=1}^{n_{x_j}} W_i^{x_j}$. Therefore $(x, y) \in W_i^{x_j}$ for some $1 \leq i \leq n_{x_j}$. ∎

**Theorem 5.8.13 (Tychonoff's theorem).** *If $X_1, \ldots, X_n$ are compact spaces, then $\prod_{i=1}^{n} X_i$ is compact.*

*Proof. Use induction, the previous theorem, and the fact that $X_1 \times \ldots \times X_n$ is homeomorphic to $X_1 \times (X_2 \times \ldots \times X_n)$.* ∎

The following topic is included as an excursion. More properties of countably compact spaces are explored in the section exercises.

**Definition.** A topological space $X$ is said to be **countably compact** if every countable open cover of $X$ contains a finite subcover.

**Example 6.** A topological space $X$ is countably compact if and only if, for every descending sequence $F_1 \supseteq F_2 \supseteq \ldots$ of nonempty closed sets, $\cap_{n=1}^{\infty} F_n \neq \varnothing$.

Suppose $X$ is countably compact. If $\cap_{n=1}^{\infty} F_n = \varnothing$, then $X = \cup_{n=1}^{\infty}(X - F_n)$. The countable compactness assumption and the fact that the sequence $X - F_n$ is ascending imply that $X = X - F_N$ for some integer $N$. This would force $F_N = \varnothing$, which is a contradiction.

   To prove the converse, suppose that $\{U_n\}$ is an open cover of $X$, and for $n \in \mathbb{N}$, define $V_n = \cup_{i=1}^{n} U_i$. Finally define $F_n = X - V_n$. Then $F_1 \supseteq F_2 \supseteq \ldots$, and $\cap_{n=1}^{\infty} F_n = \varnothing$ because $\{V_n\}$ covers $X$. It follows that $F_n = \varnothing$ for some integer $n$, hence $X = V_n = \cup_{i=1}^{n} U_i$. ◆

## Exercises

1. Show that the union of a finite number of compact subsets of a topological space $X$ is compact.
2. Verify that the proofs of theorems 5.8.1 through 5.8.5 are those included for the corresponding theorems in section 4.7, without alteration.
3. Let $X$ be a compact Hausdorff space, and suppose there exists a countable set of continuous functions $f_n : X \to [0,1]$ such that, for every pair of distinct point $x$ and $y$ in $X$, there exists a function $f_n$ such that $f_n(x) \neq f_n(y)$. Prove that the function $d(x,y) = \sum_{n=1}^{\infty} 2^{-n}|f_n(x) - f_n(y)|$ is a metric and that it induces the topology on $X$.
4. Let $X$ be a compact space, and let $F_1 \supseteq F_2 \supseteq \ldots$ be a descending sequence of nonempty closed subsets of $X$. Prove that $\cap_{n=1}^{\infty} F_n \neq \varnothing$.

5. Prove that a compact Hausdorff space $X$ cannot be expressed as a countable union of (closed) nowhere dense subsets $\{A_n\}$.
6. Let $\mathcal{T}_1$ and $\mathcal{T}_2$ be topologies on the same set $X$ such that $\mathcal{T}_1$ is Hausdorff and $\mathcal{T}_2$ is compact. Prove that if $\mathcal{T}_1 \subseteq \mathcal{T}_2$, then $\mathcal{T}_1 = \mathcal{T}_2$. Conclude that if $\mathcal{T}$ is a compact Hausdorff topology, then any strictly larger topology than $\mathcal{T}$ is not compact, and any strictly smaller topology than $\mathcal{T}$ is not Hausdorff.[4]
7. Let $X$ be a topological space, and let $\mathfrak{B}$ be an open base for $X$. Prove that $X$ is compact if and only if every open cover of $X$ by members of $\mathfrak{B}$ has a finite subcover. The same result is true for open subbases, but it is considerably harder to prove.
8. Prove that if $X$ is a compact space and $Y$ is a Lindelöf space, then $X \times Y$ is Lindelöf.
9. Prove that any two disjoint compact subsets of a Hausdorff space have disjoint open neighborhoods.
10. Let $\mathcal{K}$ be a collection of compact subsets of a Hausdorff space $X$, and let $U$ be an open subset of $X$ such that $\cap\{K : K \in \mathcal{K}\} \subseteq U$. Prove that $U$ contains the intersection of a finite subcollection of $\mathcal{K}$.
11. Let $X$ be a Hausdorff space. Prove that if $K_1 \supseteq K_2 \supseteq \ldots$ is a sequence of descending nonempty compact subsets of $X$, then $\cap_{n=1}^{\infty} K_n \neq \emptyset$.
12. Prove that the continuous image of a countably compact space is countably compact.
13. Prove that a closed subspace of a countably compact space is countably compact.
14. Prove that a countably compact metric space is compact.
15. Prove that a second countable, countably compact space is compact.
16. Verify that the proofs included in section 4.9 for theorems 4.9.3 (the Stone-Weierstrass theorem) and 4.9.6 are valid without alteration when $X$ is a compact Hausdorff topological space.

## 5.9 Locally Compact Spaces

Without a doubt, $\mathbb{R}^n$ is the most important example of a locally compact Hausdorff space. We studied locally compact metric spaces briefly in section 4.7. In this section, we will see that locally compact Hausdorff spaces are regular (theorem 5.9.3); hence they have good separation properties. They are also *very nearly normal*. Compare theorems 5.9.2 and 5.6.3. The next section is the natural continuation of this one, where we show that every locally compact Hausdorff spaces can be embedded into a compact Hausdorff space in a special kind of way. We will take

[4] This property is sometimes described to as the *rigidity of compact Hausdorff topologies*.

another journey into locally compact spaces in section 5.11, where we establish Urysohn's theorem for locally compact Hausdorff spaces and introduce the space of continuous, compactly supported functions on such spaces.

This section is the transitional section to the remaining three sections in this chapter. It may be bypassed on the first reading of the book because locally compact metric spaces (section 4.7) are sufficient for most of the rest of the book. Locally compact Hausdorff spaces are needed only in sections 8.4 and 8.7, where frequent reference is made to the results in this section and sections 5.10 and 5.11, and where certain theorems are extended from $\mathbb{R}^n$ to locally compact Hausdorff spaces.

**Definition.** A topological space $X$ is **locally compact** if, for every $x \in X$, there exists an open set $V$ such that $x \in V$ and $\overline{V}$ is compact. Thus every point is in the interior of a compact set.

We established in section 4.7 that $\mathbb{R}^n$ is locally compact and that $l^\infty$ is not. See theorem 6.1.5 for a far-reaching result. Also in section 4.7, we showed that $\mathbb{Q}$ is not locally compact.

**Theorem 5.9.1.** *Let $X$ be a Hausdorff space. Then $X$ is locally compact if and only if, for every $x \in X$ and every open neighborhood $U$ of $x$, there exists an open neighborhood $V$ of $x$ such that $\overline{V}$ is compact and $\overline{V} \subseteq U$.*

*Proof.* Suppose $X$ is locally compact, and let $x$ and $U$ be as in the statement of the theorem. Let $K$ be a compact subset of $X$ that contains $x$ in its interior, and let $F = K - U$. As $F$ is a closed subset of the compact subset $K$, it is compact. Invoking theorem 5.8.9 yields disjoint open sets $W_1$ and $W_2$ such that $x \in W_1$ and $F \subseteq W_2$. Define $V = W_1 \cap \text{int}(K)$. Since $K$ is compact and $\overline{V} \subseteq K$, $\overline{V}$ is compact. Finally, since $V \subseteq X - W_2$, and the latter set is closed, $\overline{V} \subseteq X - W_2 \subseteq X - F$. Thus $\overline{V} \subseteq K \cap (X - F) = K - F \subseteq U$. The proof of the converse is trivial. ∎

The next result generalizes the last.

**Theorem 5.9.2.** *Let $X$ be a locally compact Hausdorff space, and let $U$ be an open neighborhood of a compact subset $K$ of $X$. Then there exists an open neighborhood $V$ of $K$ such that $\overline{V}$ is compact and $\overline{V} \subseteq U$.*

*Proof.* By theorem 5.9.1, every point $x \in K$ has an open neighborhood $V_x$ with compact closure such that $\overline{V_x} \subseteq U$. Since $K$ is compact and $K \subseteq \bigcup_{x \in K} V_x$, $K \subseteq \bigcup_{i=1}^{n} V_{x_i}$, for some finite subset $\{x_1, \ldots, x_n\}$ of $K$. The open set $V = \bigcup_{i=1}^{n} V_{x_i}$ has the desired properties. ∎

The following result is a direct consequence of theorems 5.6.2 and 5.9.1.

**Theorem 5.9.3.** *A locally compact Hausdorff space is regular.* ∎

**Theorem 5.9.4.** *Let $X$ be a second countable locally compact Hausdorff space. Then $X$ is a countable union of compact subsets of $X$.*

*Proof.* Let $\mathfrak{B}$ be a countable open base for $X$. For every $x \in X$, there is an open set $V_x$ such that $x \in V_x$ and $\overline{V}_x$ is compact. Let $B_x \in \mathfrak{B}$ be such that $x \in B_x \subseteq V_x$. Clearly, $\overline{B}_x \subseteq \overline{V}_x$; thus $\overline{B}_x$ is compact. Now $X = \cup_{x \in X} \overline{B}_x$. Since $\mathfrak{B}$ is countable, only countably many of the sets $\overline{B}_x$ can be distinct, showing that $X$ is a countable union of compact subsets of $X$. ∎

**Definition.** A topological space $X$ is said to be $\sigma$-**compact** if it is the countable union of compact subsets.

For example, $\mathbb{R}^n$ is $\sigma$-compact. More generally, the above theorem states that a second countable locally compact Hausdorff space is $\sigma$-compact.

We will use the following result in the next section to prove a simple characterization of locally compact Hausdorff spaces. The proof is left as an exercise.

**Proposition 5.9.5.** *An open subspace of a locally compact Hausdorff space is locally compact.* ∎

## Exercises

1. Prove proposition 5.9.5.
2. Prove that a closed subspace of a locally compact space is locally compact.
3. Prove that the product of two locally compact spaces is locally compact.
4. Prove that a second countable locally compact Hausdorff space is normal.
5. Let $f$ be a continuous, open mapping from a locally compact space $X$ onto a topological space $Y$. Prove that $Y$ is locally compact.
6. Prove that if $E$ and $F$ are compact subsets of a locally compact Hausdorff space $X$, then $E$ and $F$ have disjoint neighborhoods with compact closures.
7. Prove that a compact subspace of the Sorgenfrey line $\mathbb{R}_l$ is countable. Conclude that $\mathbb{R}_l$ is not locally compact. Hint: Let $K$ be compact in $\mathbb{R}_l$, and let $x \in K$. Clearly, $\mathbb{R} = \cup_{n=1}^{\infty}(-\infty, x - \frac{1}{n}) \cup [x, \infty)$. Let $n$ be the least positive integer such that $K \subseteq (-\infty, x - \frac{1}{n}) \cup [x, \infty)$. Set $a_x = x - 1/n$. Clearly, $(a_x, x] \cap K = \{x\}$. Show that if $x$ and $y$ are distinct points of $K$, then $(a_x, x] \cap (a_y, y] = \varnothing$.

## 5.10 Compactification

In this section, we show that a locally compact Hausdorff space $(X, \mathcal{T})$ can be embedded in a compact Hausdorff space $(X_\infty, \mathcal{T}_\infty)$ in the manner described in theorem 5.10.1. In that theorem, the definition of the topology $\mathcal{T}_\infty$ requires some explanation.

The prototypical and most important example of a locally compact Hausdorff space is $\mathbb{R}^n$. We focus here on $\mathbb{R}^2$, because the stereographic projection of the punctured sphere $\mathcal{S}_*^2$ onto $\mathbb{R}^2$ is easy to visualize and provides an excellent motivation for the the definition of $\mathcal{T}_\infty$. The stereographic projection has been known to mapmakers since the late sixteenth century, and it is reasonable to surmise that Alexandroff was aware of that projection when he invented the topology $\mathcal{T}_\infty$.

It is clear that a compactification of the plane (more literally, its homeomorphic image $\mathcal{S}_*^2$) is the compact sphere $\mathcal{S}^2$, which contains $\mathcal{S}_*^2$ and a single additional point $N$. Some reflection reveals that there are two types of open subsets of the compact sphere:

(a)  The open subsets of $\mathcal{S}^2$ that do not contain $N$: These are in one-to-one correspondence (through the stereographic projection) with the open subsets of the usual topology of $\mathbb{R}^2$.

(b)  The open subsets $U$ of $\mathcal{S}^2$ that contain the point $N$: The complement $K = \mathcal{S}^2 - U$ of such an open set is closed in $\mathcal{S}^2$. Since $\mathcal{S}^2$ is compact, $K$ is compact. Thus the open sets $U$ of this type are exactly the complements of compact subsets of the punctured sphere, which are in one-to-one correspondence with the compact subsets of $\mathbb{R}^2$.

The above discussion suggests that a likely construction of a compact topology that contains the usual topology on $\mathbb{R}^2$ can be obtained by adding a single point, which we call $\infty$ (this point corresponds to the point $N$ on the compact sphere), to $\mathbb{R}^2$ and define the topology on $\mathbb{R}^2 \cup \{\infty\}$ to consist of the above two types of sets. This is exactly how the topology $\mathcal{T}_\infty$ in theorem 5.10.1 is defined.

**Theorem 5.10.1.** Let $(X, \mathcal{T})$ be a locally compact Hausdorff space that is not compact. Then there exists a compact Hausdorff space $(X_\infty, \mathcal{T}_\infty)$ containing $(X, \mathcal{T})$ such that

(i)  $X_\infty - X$ is a single point,
(ii)  $\mathcal{T}$ is the restriction of $\mathcal{T}_\infty$ to $X$, and
(iii)  $X$ is dense in $(X_\infty, \mathcal{T}_\infty)$.

*Proof.* Take an object, which we give the symbol $\infty$, that does not belong to $X$, and let $X_\infty = X \cup \{\infty\}$.

We define $\mathcal{T}_\infty$ to be the collection of subsets of $X_\infty$ of one of the following two types:

(a) all the members of $\mathcal{T}$, or
(b) subsets of $X_\infty$ of the form $X_\infty - K$, where $K$ is a compact subset of $X$.

We claim that $\mathcal{T}_\infty$ is a topology that satisfies the stated properties. We leave it to the reader to verify that the intersection of two members of $\mathcal{T}_\infty$ belongs to $\mathcal{T}_\infty$. To show that the union of an arbitrary subcollection of $\mathcal{T}_\infty$ is in $\mathcal{T}_\infty$, we work out three cases:

1. Since $\mathcal{T}$ is a topology and $\mathcal{T} \subseteq \mathcal{T}_\infty$, the union of open sets of type (a) is in $\mathcal{T}_\infty$.
2. Consider the union of a collection $\{X_\infty - K_\alpha\}_\alpha$ of subsets of $X_\infty$ of type (b); $\cup_\alpha(X_\infty - K_\alpha) = X_\infty - \cap_\alpha K_\alpha$, which is in $\mathcal{T}_\infty$ because $\cap_\alpha K_\alpha$ is compact in $X$.
3. Consider the union of a subcollection $\{U_\alpha\}_\alpha$ of $\mathcal{T}$ and a subcollection $\{X - K_\beta\}_\beta$ of $\mathcal{T}_\infty$, where each $K_\beta$ is compact. By cases 1 and 2 above, $\cup_\alpha(U_\alpha) \in \mathcal{T}$ and $\cup_\beta(X_\infty - K_\beta) \in \mathcal{T}_\infty$. Write $\cup_\alpha(U_\alpha) = U$, and $\cup_\beta(X_\infty - K_\beta) = X_\infty - K$. Now $\cup_\alpha(U_\alpha) \cup \cup_\beta(X_\infty - K_\beta) = U \cup (X_\infty - K) = X_\infty - (K - U)$, which is in $\mathcal{T}_\infty$ because $K - U$ is compact in $X$. This proves that $\mathcal{T}_\infty$ is a topology.

We verify that $\mathcal{T}$ is the restriction of $\mathcal{T}_\infty$ to $X$. Given an open subset of $X_\infty$, its intersection with $X$ is in $\mathcal{T}$ since, for an open subset $U$ of $X$, $U \cap X = U$ and, for a compact subset $K$ of $X$, $(X_\infty - K) \cap X = X - K$, which is open in $X$. The converse is trivial since, for an open subset $U$ of $X$, $U \cap X_\infty = U$.

Next we show that $X$ is dense in $X_\infty$ by showing that every open neighborhood of $\infty$ intersects $X$. Such a neighborhood is of the type $X_\infty - K$, where $K$ is a compact subset of $X$. Since $X$ is not compact, $X - K \neq \varnothing$, and clearly $(X_\infty - K) \cap X \neq \varnothing$.

We now show that $\mathcal{T}_\infty$ is Hausdorff. It is sufficient to show that if $x \in X$, then $x$ and $\infty$ have disjoint open neighborhoods in $X_\infty$. Since $X$ is locally compact, there is a compact subset $K$ of $X$ such that $x \in int(K)$. Let $U = int(K)$, and $V = X_\infty - K$; $U$ and $V$ are disjoint neighborhoods of $x$ and $\infty$.

Finally, we show that $X_\infty$ is compact. Let $\mathcal{U}$ be an open cover of $X_\infty$; $\mathcal{U}$ must contain a member of the type $X_\infty - K$ for some compact subset $K$ of $X$. Let $\mathcal{U}'$ be $\mathcal{U}$ with the exclusion of $X_\infty - K$. The intersection of $X$ and the members of $\mathcal{U}'$ clearly covers $K$. Thus there exists a finite subcollection $\mathcal{U}''$ of $\mathcal{U}'$ such that $K \subseteq \cup\{X \cap W : W \in \mathcal{U}''\}$. Since $X_\infty = K \cup (X_\infty - K)$, the finite subcollection $\mathcal{U}'' \cup \{X_\infty - K\}$ of $\mathcal{U}$ covers $X_\infty$. ∎

**Definition.** The topological space $(X_\infty, \mathcal{T}_\infty)$ we constructed in theorem 5.10.1 is called the **one-point** (or **Alexandroff**) **compactification** of the locally compact Hausdorff space $X$.

**Theorem 5.10.2.** *The one-point compactification of a locally compact Hausdorff space $X$ is unique up to homeomorphism. More specifically, if $Y$ is a compact Hausdorff space and $\varphi : X \to Y$ is a topological embedding such that $Y - \varphi(X)$ is a single point, then $\varphi$ can be extended to a homeomorphism $\varphi_\infty : X_\infty \to Y$.*

*Proof.* Let $Y - \varphi(X) = \{\omega\}$, and extend $\varphi$ to a function $\varphi_\infty : X_\infty \to Y$ by defining $\varphi_\infty(\infty) = \omega$. Trivially, $\varphi_\infty$ is a bijection. Since both $X_\infty$ and $Y$ are compact Hausdorff spaces, we need only to show that $\varphi_\infty^{-1}$ is continuous; see theorem 5.8.7. Equivalently, we show that $\varphi_\infty$ is an open mapping. If $V$ is an open subset of $X$, then $\varphi_\infty(V) = \varphi(V)$, which is open by the assumption that $\varphi$ is an embedding. If $V$ contains $\infty$, then $V = X_\infty - K$ for some compact subset $K$ of $X$. Now $\varphi_\infty(V) = \varphi_\infty(X_\infty - K) = Y - \varphi_\infty(K) = Y - \varphi(K)$. The compactness of $K$ together with the continuity of $\varphi$ imply that $\varphi(K)$ is compact. By theorem 5.8.3, $\varphi(K)$ is closed, and $Y - \varphi(K)$ is open. ∎

**Example 1.** Let $\chi$ be the chordal metric on $\mathbb{R}$. Recall that $(\mathbb{R}, \chi)$ is homeomorphic to $\mathbb{R}$ with the usual topology. Therefore the one-point compactification of $\mathbb{R}$ with respect to the usual topology is homeomorphic to the one-point compactification of $(\mathbb{R}, \chi)$. By the very definition of the chordal metric, $(\mathbb{R}, \chi)$ is homeomorphic to the punctured sphere $\mathcal{S}^1 - \{N\}$. Therefore the one-point compactification of $(\mathbb{R}, \chi)$, hence $(\mathbb{R}, \mathcal{T})$, is the compactification of $\mathcal{S}^1 - N$, which is clearly $\mathcal{S}^1$. We have arrived at the following result: the one-point compactification of $\mathbb{R}$ is the circle. The same argument shows that the one-point compactification of the complex plane $\mathbb{C}$ (identified with the Euclidean plane $\mathbb{R}^2$) is the sphere. This is the reason the sphere is thought of as the extended complex plane and is often called the Riemann sphere.

**Example 2.** The one-point compactification of the open interval $(0, 1)$ is the circle $\mathcal{S}^1$. To see this, recall that the unit $(0, 1)$ is homeomorphic to the line $\mathbb{R}$. Since the one-point compactification of $\mathbb{R}$ is $\mathcal{S}^1$, the compactification of $(0, 1)$ is $\mathcal{S}^1$. ♦

**Example 3.** While the one-point compactification $X_\infty$ of a locally compact Hausdorff space $X$ is essentially unique, it is possible to embed $X$ as a dense subspace of other spaces not homeomorphic to $X_\infty$. For example, another compactification of the open unit interval $(0, 1)$ is the closed interval $[0, 1]$, which is not homeomorphic to $\mathcal{S}^1$. ♦

**Example 4.** The one-point compactification of the punctured line, $X = (-\infty, 0) \cup (0, \infty)$ is homeomorphic to the union of two externally tangent circles (a figure eight).

Each of the open half lines is homeomorphic to an open half circle, as shown in figure 5.1(a). There are several ways to see this. The stereographic projection is the easiest to visualize. The next step is to pull the two open half circles horizontally apart a distance equal to the diameter of each half circle, as shown in figure 5.1(b). Now each half circle is homeomorphic to a punctured circle. For example, function $f(e^{i\theta}) = e^{2i\theta}$ maps the half circle $\{e^{i\theta} : -\pi/2 < \theta < \pi/2\}$ onto the punctured circle $\{e^{i\theta} : -\pi < \theta < \pi\}$. Hence $X$ is homeomorphic to the union of the two tangent punctured circles shown in figure 5.1(c). If we define the point at infinity to be the missing point of tangency, we obtain the figure eight shown in figure 5.1(d). ◆

The following succinct characterization of locally compact Hausdorff spaces follows directly from theorems 5.9.5 and 5.10.1.

**Theorem 5.10.3.** *A Hausdorff space is locally compact if and only if it is an open subspace of a compact Hausdorff space.* ∎

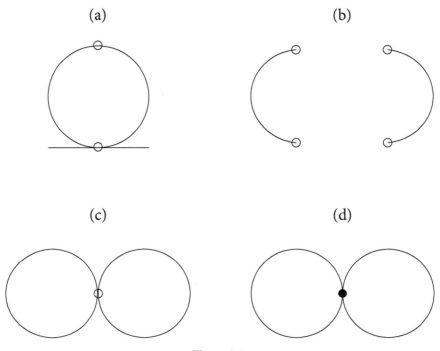

(a)        (b)

(c)        (d)

Figure 5.1

## Exercises

1. Let $X = \mathbb{N}$ with the restricted topology induced by the usual topology on $\mathbb{R}$. This topology is, in fact, the discrete topology on $\mathbb{N}$. Prove that the one-point compactification of $X$ is homeomorphic to the space $X_\infty = \{\frac{1}{n} : n \in \mathbb{N}\} \cup \{0\}$ (as a subspace of $\mathbb{R}$).

2. Prove that the one-point compactification of $\mathbb{R}^n$ is the sphere $S^n$.

3. What is the one-point compactification of the open unit disk in $\mathbb{R}^2$?

4. By generalizing the idea of example 4, make a conjecture about the one-point compactification of the union of the two open half planes $\{(x,y) \in \mathbb{R}^2 : x > 0\} \cup \{(x,y) \in \mathbb{R}^2 : x < 0\}$.

5. Prove that if a locally compact Hausdorff space $X$ is second countable, then so is $X_\infty$.

## 5.11  Metrization

We now turn to the question of which topologies are induced by a metric. Theorem 5.11.3 is the main result in this section. Although it is not the best known result, it does establish sufficient conditions for metrization. The proof techniques we develop along the path to theorem 5.11.3 are elegant and important in their own right. We first state the following definition.

**Definition.** A topological space $(X, \mathcal{T})$ is **metrizable** if there is a metric $d$ on $X$ that induces the topology $\mathcal{T}$.

**Lemma 5.11.1.** *Suppose $X$ is a normal space, and let $E$ and $F$ be disjoint closed subsets of $X$. Let $C$ be the set of rational points in the interval $[0,1]$. Then there exists a countable collection of open subsets $\{U_p : p \in C\}$ such that*

$$\text{if } p, q \in C \text{ and } p < q, \text{ then } \overline{U}_p \subseteq U_q. \tag{$*$}$$

*Additionally, for all $p \in C$, $E \subseteq U_p$, and $\overline{U}_p \subseteq X - F$.*

*Proof. Let $p_0 = 0$, and $p_1 = 1$, and let $\{p_2, p_3, p_4, \ldots\}$ be an enumeration of the rational point in $(0,1)$. Since $E \subseteq X - F$, theorem 5.6.3 yields an open set $U_1$ such that $E \subseteq U_1 \subseteq \overline{U}_1 \subseteq X - F$. Another application of theorem 5.6.3 yields an open set $U_0$ such that $E \subseteq U_0 \subseteq \overline{U}_0 \subseteq U_1$. The rest of the construction is inductive. Suppose that, for each element $p_i$ of the finite set $C_n = \{p_0, \ldots, p_n\}$, we have found an open set $U_{p_i}$ such that the sets $U_{p_1}, \ldots, U_{p_n}$ satisfy condition $(*)$ for $p, q \in C_n$. Consider the rational number $p_{n+1}$. It must fall strictly between two elements of*

$C_n$, say, $p_i < p_{n+1} < p_j$. Again by theorem 5.6.3, there exists an open set $U_{p_{n+1}}$ such that $\overline{U}_{p_i} \subseteq U_{p_{n+1}} \subseteq \overline{U}_{p_{n+1}} \subseteq U_{p_j}$. By construction, the sets $U_{p_0}, \ldots, U_{p_{n+1}}$ satisfy condition (*) for $p, q \in C_{n+1}$. Since, for every pair of points $p$ and $q$ in $C$, there is a finite set $C_n$ that contains $p$ and $q$, the proof is complete.

The inclusions $E \subseteq U_p$, and $\overline{U}_p \subseteq X - F$ for all $p \in C$ are obvious since $E \subseteq U_0$ and $\overline{U}_1 \subseteq X - F$. ∎

**Remark.** Remark. Any dense subset $C$ of $[0,1]$ containing 0 and 1 can be used in the construction of the collection $\{U_p : p \in C\}$. A commonly used such set is the set of dyadic rational numbers $D = \{0, 1, 1/2, 1/4, 3/4, 1/8, 3/8, 5/8, 7/8, \ldots\}$,[5] which is slightly more advantageous in the visualization of the construction of the sets $\{U_p\}$.

The following theorem is crucial for the proof of theorem 5.11.3. It is greatly important in its own right.

**Theorem 5.11.2 (Urysohn's lemma).** *Suppose that $X$ is a normal space and that $E$ and $F$ are disjoint closed subsets of $X$. Then there exists a continuous function $f : X \to [0,1]$ such that $f(E) = 1$, and $f(F) = 0$.*

*Proof. Let $C$ and $\{U_p : p \in C\}$ be as in lemma 5.11.1. For $p, q \in C$, define*

$$f_p(x) = \begin{cases} p & if\ x \in U_{1-p}, \\ 0 & if\ x \notin U_{1-p}, \end{cases} \quad and\ g_q(x) = \begin{cases} 1 & if\ x \in \overline{U}_{1-q}, \\ q & if\ x \notin \overline{U}_{1-q}. \end{cases}$$

*Since $f_p = p\chi_{U_{1-p}}$ and $g_q = q + (1-q)\chi_{\overline{U}_{1-q}}$, $f_p$ is lower semicontinuous and $g_q$ is upper semicontinuous by theorem 5.3.4 ($U_{1-p}$ is open, and $\overline{U}_{1-q}$ is closed). Now define*

$$f = sup_{p \in C}\{f_p\}\ and\ g = inf_{q \in C}\{g_q\}.$$

*Again theorem 5.3.4 implies that $f$ is lower semicontinuous and that $g$ is upper semicontinuous.*

*If $x \in E$, $x \in U_{1-p}$, for every $p \in C$, and hence $f(x) = sup_{p \in C}\{p\} = 1$. If $x \in F$, then $x \notin U_{1-p}$ for every $p \in C$, and hence $f(x) = 0$.*

*By theorem 5.3.4, the proof will be complete if we show that $f = g$.*

[5] $D = \{0, 1\} \cup \bigcup_{n=1}^{\infty}\{\frac{k}{2^n} : k = 1, 3, 5, \ldots, 2^n - 1\}$.

We claim that, for all $p, q \in C$, $f_p \leq g_q$. It follows immediately that $f \leq g$. If, for some $x \in X$, $g_q(x) < f_p(x)$, then $f_p(x) > 0$, and $g_q(x) < 1$. Thus

$$f_p(x) = p, \quad hence \quad x \in U_{1-p}, \quad and$$

$$g_q(x) = q, \quad hence \quad x \notin \overline{U}_{1-q}.$$

Now $p = f_p(x) > g_q(x) = q$; hence $1 - p < 1 - q$, and $\overline{U}_{1-p} \subseteq U_{1-q}$. This contradicts $x \in U_{1-p}$ and $x \notin \overline{U}_{1-q}$ and establishes the claim that $f_p \leq g_q$.

Suppose, for a contradiction, that $f(x) < g(x)$ for some $x \in X$. Because $C$ is dense in $[0, 1]$, there are points $p, q \in C$ such that $f(x) < p < q < g(x)$. Now $f(x) < p$ implies that $x \notin U_{1-p}$, and $g(x) > q$ implies $x \in \overline{U}_{1-q}$. This is a contradiction because
$1 - q < 1 - p$; hence $\overline{U}_{1-q} \subseteq U_{1-p}$. The contradiction concludes the proof. ∎

**Theorem 5.11.3 (the Urysohn metrization theorem).** *Every regular second countable topological space $(X, \mathcal{T})$ is metrizable.*

*Proof.* According to theorem 5.7.4, $X$ is actually a normal space. Let $\mathfrak{B} = \{B_1, B_2, \ldots\}$ be a countable open base for the topology. If $x \in B_1$, then, by theorem 5.6.2, and the fact that $\mathfrak{B}$ is an open base, there exists a basis element $B_k \in \mathfrak{B}$ such that $x \in B_k \subseteq \overline{B}_k \subseteq B_1$. Therefore the collection of pairs $P = \{(B_n, B_m) : \overline{B}_n \subseteq B_m\}$ is not empty. Since $P$ is countable, we enumerate $P$ as follows: $P = \{(B_{n_i}, B_{m_i}) : i \in \mathbb{N}\}$. By theorem 5.11.2, for each $i \in \mathbb{N}$, a continuous function $f_i : X \to [0, 1]$ exists such that $f_i(\overline{B}_{n_i}) = 0$, and $f_i(X - B_{m_i}) = 1$.

We now define a metric $d$ on $X$ as follows:

$$d(x, y) = \left\{ \sum_{i=1}^{\infty} \frac{|f_i(x) - f_i(y)|^2}{i^2} \right\}^{1/2}.$$

It is clear that series in the above definition converge since $|f_i(x) - f_i(y)| \leq 1$. In fact, the sequences $\varphi_x = (f_1(x), \frac{f_2(x)}{2}, \ldots, \frac{f_i(x)}{i}, \ldots)$ and $\varphi_y = (f_1(y), \frac{f_2(y)}{2}, \ldots, \frac{f_i(y)}{i}, \ldots)$ are in $l^2$ and $d(x, y)$ is nothing but the $l^2$ distance between $\varphi_x$ and $\varphi_y$. It becomes clear that $d$ is a metric once we show that the function $x \mapsto \varphi_x$ is an injection. Let $x$ and $y$ be distinct elements of $X$, and let $U$ be an open neighborhood of $x$ that excludes $y$. Choose a basis member $B_m$ such that $x \in B_m \subseteq \overline{B}_m \subseteq U$, then choose a basis member $B_n$ such that $x \in B_n \subseteq \overline{B}_n \subseteq B_m$. The pair $(B_n, B_m) \in P$, and hence $(B_n, B_m) = (B_{n_i}, B_{m_i})$ for some $i \in \mathbb{N}$. It follows that $f_i(x) = 0$ and $f_i(y) = 1$. We now show that the metric $d$ induces the topology $\mathcal{T}$.

We prove that every d-open subset $U$ of $X$ is $\mathcal{T}$-open. Let $x \in U$, and let $r > 0$ be such that $B(x,r) \subseteq U$. Here $B(x,r)$ is the d-ball of radius $r$ centered at $x$. We will show that there exists a $\mathcal{T}$-open set $V$ such that $x \in V \subseteq B(x,r)$. First choose an integer $N$ such that $\sum_{i=N+1}^{\infty} \frac{1}{i^2} < \frac{r^2}{2}$. Since each $f_i$ is continuous, there exists an open neighborhood $V_i$ of $x$ such that, for every $y \in V_i$, $|f_i(x) - f_i(y)|^2 < \frac{r^2}{2N}$. We claim that the set $V = \cap_{i=1}^{N} V_i$ is the set we seek. If $y \in V$, then $[d(x,y)]^2 = \sum_{i=1}^{\infty} \frac{|f_i(x)-f_i(y)|^2}{i^2} = \sum_{i=1}^{N} \frac{|f_i(x)-f_i(y)|^2}{i^2} + \sum_{i=N+1}^{\infty} \frac{|f_i(x)-f_i(y)|^2}{i^2} < \frac{r^2}{2N} \sum_{i=1}^{N} \frac{1}{i^2} + \sum_{i=N+1}^{\infty} \frac{1}{i^2} < \frac{r^2}{2} + \frac{r^2}{2}$.

To show that every $\mathcal{T}$-open set is d-open, it is sufficient to show that every basic open set $B_m$ is d-open. Let $x \in B_m$. We need to show that there exists $r > 0$ such that $B(x,r) \subseteq B_m$. By theorem 5.6.2, there exists a basis element $B_n$ such that $x \in B_n \subseteq \overline{B_n} \subseteq B_m$. Now $(B_n, B_m) \in P$, say, $(B_n, B_m) = (B_{n_i}, B_{m_i})$. Let $r = \frac{1}{2i}$. If $y \in B(x,r)$, then $\sum_{j=1}^{\infty} \frac{|f_j(x)-f_j(y)|^2}{j^2} < \frac{1}{4i^2}$. In particular, $\frac{|f_i(x)-f_i(y)|^2}{i^2} < \frac{1}{4i^2}$. Thus $|f_i(x) - f_i(y)| < \frac{1}{2}$. Because $f_i(x) = 0$, $|f_i(y)| < \frac{1}{2}$. Since $f_i(X - B_{m_i}) = 1$, $y \in B_{m_i} = B_m$. ∎

The conditions of theorem 5.11.3 are not necessary for a space to be metrizable. For example, the space $l^{\infty}$ is metrizable but not second countable. However, if we limit ourselves to compact Hausdorff spaces, the conditions of theorem 5.11.3 are necessary as well as sufficient, as the next theorem shows.

**Theorem 5.11.4.** *A compact Hausdorff space is metrizable if and only if it is second countable.*

*Proof.* A compact Hausdorff space $X$ is normal by theorem 5.8.10. Therefore if $X$ is second countable, it is metrizable by the previous theorem. To prove the converse, recall that a compact metric space is second countable (see problem 7 on section 4.7 and theorem 4.5.1). ∎

We now venture back into locally compact Hausdorff spaces. The following theorem is the closest analog of theorem 5.11.2 for locally compact Hausdorff spaces, which need not be normal. It is sometimes referred to as **Urysohn's theorem for locally compact spaces**.

**Theorem 5.11.5.** *Let $X$ be a locally compact Hausdorff space, and let $K$ and $F$ be disjoint subsets of $X$ such that $K$ is compact and $F$ is closed. Then there exists a continuous function $f : X \to [0,1]$ such that $f(K) = 1$, and $f(F) = 0$.*

*Proof.* Applying theorem 5.9.2 to the compact set $K$ and the open set $X - F$, there exists an open subset $V$ with compact closure such that $K \subseteq V \subseteq \overline{V} \subseteq X - F$. Applying theorem 5.9.2 again to the compact set $K$ and the open set $V$, we can find an open set $U$ with compact closure such that $K \subseteq U \subseteq \overline{U} \subseteq V$. Now the subspace $\overline{V}$ with the restricted topology is a compact Hausdorff space and is therefore normal by theorem 5.8.10. Applying theorem 5.11.2 to the closed subsets $K$ and $\overline{V} - U$ of $\overline{V}$, there is a continuous function $f : \overline{V} \to [0,1]$ such that $f(K) = 1$, and $f(\overline{V} - U) = 0$. Extend $f$ to a continuous function $f : X \to [0,1]$ by defining $f(x) = 0$, for all $x \in X - U$. Problem 4 on section 5.3 is relevant here to show the continuity of the extended $f$. ∎

**Definition.** Let $f$ be a complex-valued function on a topological space $X$. The **support** of $f$, written $supp(f)$, is the closure of the set $\{x \in X : f(x) \neq 0\}$. A function $f : X \to \mathbb{C}$ is said to have **compact support** if $supp(f)$ is compact. The set $\mathcal{C}_c(X)$ of **continuous, compactly supported functions** is clearly a subspace of $\mathcal{BC}(X)$.

The following corollary is of pivotal importance in studying a certain class of measures on locally compact Hausdorff spaces. In section 8.4, we present the main examples of such a measure: Lebesgue and Radon measures.

**Corollary 5.11.6.** *Suppose that $X$ is a locally compact Hausdorff space, that $K$ is a compact subspace of $X$, and that $V$ is an open neighborhood of $K$. Then there exists a continuous function of compact support, $f : X \to [0,1]$, such that $f(K) = 1$, and $supp(f) \subseteq V$.*

*Proof.* Apply theorem 5.9.2 to find an open set $U$ with compact closure such that $K \subseteq U \subseteq \overline{U} \subseteq V$. Now apply theorem 5.11.5 to the sets $K$ and $F = X - U$ to find a function $f : X \to [0,1]$ such that $F(K) = 1$ and $f(X - U) = 0$. Observe that $supp(f) \subseteq \overline{U}$, which is compact. ∎

**Definition.** Let $X$ be a locally compact Hausdorff space. A continuous, scalar-valued function $f$ is said to **vanish** at $\infty$ if, for every $\epsilon > 0$, there exists a compact subset $K$ of $X$ such that $|f(x)| < \epsilon$ for every $x \in X - K$. The set $\mathcal{C}_0(X)$ of all scalar-valued functions on $X$ that vanish at $\infty$ is clearly a vector space.

**Theorem 5.11.7.** *A function $f \in \mathcal{C}_0(X)$ is bounded and the space $\mathcal{C}_0(X)$ is a complete normed linear space under the supremum norm.*

*Proof.* We leave it to the reader to show that $\mathcal{C}_0(X) \subseteq \mathcal{BC}(X)$. We prove that $\mathcal{C}_0(X)$ is closed in $\mathcal{BC}(X)$. Let $f \in \mathcal{BC}(X)$ be a closure point of $\mathcal{C}_0(X)$, and let $\epsilon > 0$. There exists a function $g \in \mathcal{C}_0(X)$ such that $\|f - g\|_\infty < \epsilon/2$. Let $K$ be a compact subset *of X such that $|g(x)| < \epsilon/2$ whenever $x \notin K$. Now if $x \notin K$, then $|f(x)| \leq |f(x) - g(x)| + |g(x)| < \epsilon/2 + \epsilon/2$.* ∎

**Theorem 5.11.8.** *Suppose X is a locally compact Hausdorff space. Then $\mathcal{C}_c(X)$ is dense in $\mathcal{C}_0(X)$.*

*Proof.* Let $g \in \mathcal{C}_0(X)$, let $\epsilon > 0$, and let $K$ be a compact subset of $X$ such that $|g(x)| < \epsilon$ for $x \in X - K$. By theorem 5.9.2, there is an open subset $V$ with compact closure such that $K \subseteq V$. By corollary 5.11.6, there exists a function $f \in \mathcal{C}_c(X)$ such that $f(K) = 1$, $0 \leq f(x) \leq 1$, and $\mathrm{supp}(f) \subseteq V$. The function $fg$ is in $\mathcal{C}_c(X)$ and $\|g - fg\|_\infty < \epsilon$. ∎

## Exercises

In the problems below, $X$ is a locally compact Hausdorff space.

1. Prove that $\mathcal{C}_0(X) \subseteq \mathcal{BC}(X)$.
2. Prove that $f \in \mathcal{C}_0(X)$ if and only if, for every $\epsilon > 0$, the set $\{x \in X : |f(x)| \geq \epsilon\}$ is compact.
3. Prove that $f \in \mathcal{C}_0(X)$ if and only if $f$ is the restriction to $X$ of a function $g \in \mathcal{C}(X_\infty)$ such that $g(\infty) = 0$. Here $X_\infty$ is the one-point compactification of $X$.

## 5.12  The Product of Infinitely Many Spaces

This section generalizes section 5.4. First we review some terminology and notation.

Let $\{X_\alpha\}_{\alpha \in I}$ be an arbitrary collection of nonempty sets. The Cartesian product $X = \prod_{\alpha \in I} X_\alpha$ is the set of all functions $x : I \to \cup_{\alpha \in I} X_\alpha$ such that, for every $\alpha \in I$, $x(\alpha) \in X_\alpha$. We write $x_\alpha$ instead of $x(\alpha)$, and we denote an element of $X$ by $x = (x_\alpha)_{\alpha \in I}$, or simply $x = (x_\alpha)$. For a fixed $\alpha \in I$, the projection of $X$ onto the factor set $X_\alpha$ is the function $\pi_\alpha(x) = x_\alpha$.

Let $\{(X_\alpha, \mathcal{T}_\alpha)\}_{\alpha \in I}$ be a collection of topological spaces, and let $X = \prod_\alpha X_\alpha$ be the Cartesian product of the underlying sets. As in the definition of the product topology in section 5.4, we would like the product topology to guarantee the continuity of all the projections $\pi_\alpha : X \to X_\alpha$. One might be tempted to adopt the following simple generalization of the product of finitely many spaces. Consider the topology $\mathcal{T}$, which has the following subbase:

$$\left\{ \prod_{\alpha \in I} U_\alpha : \alpha \in I, U_\alpha \in \mathcal{T}_\alpha \right\}.$$

It would be a hasty decision to define $\mathcal{T}$ to be the product topology. Although $\mathcal{T}$ certainly guarantees the continuity of all the projections, it is too wasteful because, in order to guarantee the continuity of $\pi_\alpha$, we only need the openness of sets of the form $\pi_\alpha^{-1}(U_\alpha)$, where $U_\alpha \in \mathcal{T}_\alpha$. A little reflection shows that

$$\pi_\alpha^{-1}(U_\alpha) = U_\alpha \times \prod_{\beta \neq \alpha} X_\beta.\,^6$$

Therefore the smallest topology which guarantees the continuity of all the projections is the topology whose subbase is the collection $\{\pi_\alpha^{-1}(U_\alpha) : \alpha \in I, U_\alpha \in \mathcal{T}_\alpha\}$.

We now formalize the above motivation to define the product topology

$$\mathfrak{S} = \{\pi_\alpha^{-1}(U_\alpha) : \alpha \in I, U_\alpha \in \mathcal{T}_\alpha\}.$$

Since $\cup\{S : S \in \mathfrak{S}\} = X$, theorem 5.2.3 applies, and the following definition is meaningful.

**Definition.** The **product topology** on $X$ is the weakest topology that contains $\mathfrak{S}$.

By construction, $\mathfrak{S}$ is a subbase for the product topology (theorem 5.2.3).

An open base $\mathfrak{B}$ for the product topology on $X$ consists of finite intersections of members of $\mathfrak{S}$. Thus a typical member of $\mathfrak{B}$ is a set of the form

$$\cap_{i=1}^n \pi_{\alpha_i}^{-1}(U_{\alpha_i}) = U_{\alpha_1} \times \dots \times U_{\alpha_n} \times \prod_{\alpha \neq \alpha_i} X_\alpha,$$

where $\{\alpha_1, \dots, \alpha_n\}$ is a finite subset of $I$.

To reiterate, the above set is the set of all $x \in X$ such that $\pi_{\alpha_i}(x) \in U_i$ for all $1 \leq i \leq n$.

The following theorem is a restatement of the definition of the product topology. See the proof of theorem 5.4.1.

**Theorem 5.12.1.** *The product topology is the weakest topology relative to which all the projections $\pi_\alpha : X \to X_\alpha$ are continuous.* ∎

---

6 The set $\pi_\alpha^{-1}(U_\alpha)$ is the set of all elements $x$ in $X$ such that $x_\alpha \in U_\alpha$ and the other coordinates, $x_\beta$, of $x$ are unrestricted elements of $X_\beta$. This is exactly the set $U_\alpha \times \prod_{\beta \neq \alpha} X_\beta$.

**Example 1.** The product of an arbitrary collection $\{X_\alpha : \alpha \in I\}$ of Hausdorff spaces is Hausdorff.

Let $x = (x_\alpha)$ and $y = (y_\alpha)$ be distinct elements of $\prod_\alpha X_\alpha$. Fix an element $\alpha \in I$ for which $x_\alpha \neq y_\alpha$. Since $X_\alpha$ is Hausdorff, there exist open neighborhoods $U_\alpha$ and $V_\alpha$ of $x_\alpha$ and $y_\alpha$, respectively, such that $U_\alpha \cap V_\alpha = \emptyset$. Now $\pi_\alpha^{-1}(U_\alpha)$ and $\pi_\alpha^{-1}(V_\alpha)$ are disjoint open neighborhoods of $x$ and $y$, respectively. ◆

**Example 2.** Let $\{X_\alpha : \alpha \in I\}$ be a collection of connected spaces, let $X = \prod_{\alpha \in I} X_\alpha$, and let $\varphi : X \to \{0,1\}$ be continuous. If $x, y \in X$ are such that $x_\alpha = y_\alpha$ except for a finite subset $F$ of $I$, then $\varphi(x) = \varphi(y)$.

Define a function $j : \prod_{\alpha \in F} X_\alpha \to X$ as follows : $z \mapsto j_z$, where

$$j_z(\alpha) = \begin{cases} z_\alpha & \text{if } \alpha \in F, \\ y_\alpha & \text{if } \alpha \notin F. \end{cases}$$

For each $\beta \in I$, the function $\pi_\beta \circ j : \prod_{\alpha \in F} X_\alpha \to X_\beta$ is equal to $\pi_\beta$ if $\beta \in F$, and it is constant if $\beta \notin F$. By problem 6 at the end of this section, the function $j$ is continuous. Now $\prod_{\alpha \in F} X_\alpha$ is connected by theorem 5.5.5; hence $j(\prod_{\alpha \in F} X_\alpha)$ is connected by theorem 5.5.3. Since $x$ and $y$ are in $j(\prod_{\alpha \in F} X_\alpha)$, $\varphi(x) = \varphi(y)$. ◆

**Example 3.** If $\{X_\alpha : \alpha \in I\}$ is a collection of connected spaces, then $X = \prod_{\alpha \in I} X_\alpha$ is connected.

Let $\varphi : X \to \{0,1\}$ be a continuous function, and suppose that $U = \varphi^{-1}(0) \neq \emptyset$. We show that $\varphi(X) = \{0\}$. Fix an element $a = (a_\alpha) \in U$, and let $x \in X$ be arbitrary. Since $U$ is open in $X$, there is a basis element $B = \cap_{i=1}^n \pi_{\alpha_i}^{-1}(U_{\alpha_i})$ such that $a \in B \subseteq U$. Define an element $y \in X$ as follows:

$$y_\alpha = \begin{cases} a_\alpha & \text{if } \alpha = \alpha_i, \\ x_\alpha & \text{if } \alpha \neq \alpha_i. \end{cases}$$

Then $y \in B \subseteq U$ and $\varphi(y) = 0$. Since $x_\alpha = y_\alpha$ for all $\alpha \neq \alpha_i$, $\varphi(x) = \varphi(y) = 0$ by example 2. ◆

**Theorem 5.12.2.** *If, for each $\alpha$, $F_\alpha$ is closed in $X_\alpha$, then $\prod_\alpha F_\alpha$ is closed in $X$.*

*Proof.* Let $U_\alpha = X_\alpha - F_\alpha$. We claim that $\prod_{\alpha \in I} F_\alpha = X - \cup_\alpha \pi_\alpha^{-1}(U_\alpha)$. The result follows since $\cup_\alpha \pi_\alpha^{-1}(U_\alpha)$ is open in $X$. Let $x = (x_\alpha)$ be an element of $X$. Now $x \in X - \prod_\alpha F_\alpha$ if and only if $x \notin \prod_\alpha F_\alpha$, if and only if $x_\alpha \notin F_\alpha$ for some $\alpha \in I$ if and only if $x_\alpha \in U_\alpha$ for some $\alpha \in I$, if and only if $x \in \cup_\alpha \pi_\alpha^{-1}(U_\alpha)$. ■

**Theorem 5.12.3.** *For each $\alpha \in I$, let $\mathfrak{B}_\alpha$ be an open base for $X_\alpha$. Then the family of subsets of X of the form $\cap_{\alpha \in F} \pi_\alpha^{-1}(B_\alpha)$, where F ranges over finite subsets of I and $B_\alpha \in \mathfrak{B}_\alpha$, is an open base for the product topology.*
*The proof is left as an exercise.* ∎

**Example 4.** The product of a countable collection of second countable spaces is second countable.

This follows directly from the previous theorem. Indeed, when $I$ is countable and, for each $\alpha \in I$, $\mathfrak{B}_\alpha$ is countable, then the family $\cap_{\alpha \in F} \pi_\alpha^{-1}(B_\alpha)$, where $F$ ranges over finite subsets of $I$ and $B_\alpha \in \mathfrak{B}_\alpha$, is countable. ◆

We need the following lemma before we tackle Tychonoff's theorem.

**Lemma 5.12.4.** *Let $\{X_\alpha\}_\alpha$ be a collection of topological spaces, and let X be the product space. If $\mathfrak{F}$ is a collection of closed subsets of X possessing the finite intersection property, then there exists a family $\mathfrak{F}^*$ of subsets of X, not necessarily closed, which is maximal subject to the following conditions:*

*(a) $\mathfrak{F}^*$ has the finite intersection property, and*
*(b) $\mathfrak{F} \subseteq \mathfrak{F}^*$.*

*Furthermore, $\mathfrak{F}^*$ is closed under the formation of finite intersections.*

*Proof.* *Consider the family $\mathfrak{D}$ of subsets of X containing $\mathfrak{F}$ and having the finite intersection property. Order $\mathfrak{D}$ by set inclusion, and let $\mathfrak{C}$ be a chain in $\mathfrak{D}$. We will verify that $\cup\{\mathcal{C} : \mathcal{C} \in \mathfrak{C}\}$ is an upper bound on $\mathfrak{C}$. Let $F_1, \ldots, F_n$ be members of $\cup\{\mathcal{C} : \mathcal{C} \in \mathfrak{C}\}$. Then there are members $\mathcal{C}_1, \ldots, \mathcal{C}_n$ of $\mathfrak{C}$ such that $F_i \in \mathcal{C}_i$. Since $\mathfrak{C}$ is a chain, one of the families $\mathcal{C}_1, \ldots, \mathcal{C}_n$, say, $\mathcal{C}_1$, contains all the others. Now all the sets $F_1, \ldots, F_n$ are in $\mathcal{C}_1$; hence $\cap_{i=1}^{n} F_i \neq \emptyset$, and $\cup\{\mathcal{C} : \mathcal{C} \in \mathfrak{C}\}$ has the finite intersection property. Clearly, $\mathcal{C}$ contains $\mathfrak{F}$. By Zorn's lemma, $\mathfrak{D}$ contains a maximal member $\mathfrak{F}^*$. If there are sets $F_1$ and $F_2$ in $\mathfrak{F}^*$ such that $F_1 \cap F_2 \notin \mathfrak{F}^*$, then $\mathfrak{F}^* \cup \{F_1 \cap F_2\}$ would have properties (a) and (b), which contradicts the maximality of $\mathfrak{F}^*$. This proves that the intersection of two (hence any finite number of) sets in $\mathfrak{F}^*$ is in $\mathfrak{F}^*$.* ∎

**Theorem 5.12.5 (Tychonoff's theorem).** *Let $\{(X_\alpha, \mathcal{T}_\alpha)\}_{\alpha \in I}$ be a collection of compact topological spaces. Then the product space X is compact.*

*Proof.* *Let $\mathfrak{F}$ be a collection of closed subsets of X that has the finite intersection property. We prove that $\cap\{F : F \in \mathfrak{F}\} \neq \emptyset$. By theorem 5.8.8, X is compact. Let $\mathfrak{F}^*$ be a collection of subsets of X having the properties described in lemma 5.12.4.*

We will show that $\cap\{\overline{F} : F \in \mathfrak{F}^*\} \neq \varnothing$. This will establish the theorem because the members of $\mathfrak{F}$ are closed and $\cap\{\overline{F} : F \in \mathfrak{F}^*\} \subseteq \cap\{\overline{F} : F \in \mathfrak{F}\} = \cap\{F : F \in \mathfrak{F}\}$.

For a fixed $\alpha \in I$, consider the following collection of sets:

$$\{\overline{\pi_\alpha(F)} : F \in \mathfrak{F}^*\}.$$

This is a family of closed subsets of $X_\alpha$, and it has the finite intersection property because if $F_1, \ldots, F_n$ are in $\mathfrak{F}^*$, then $\cap_{i=1}^n \overline{\pi_\alpha(F_i)} \supseteq \overline{\cap_{i=1}^n \pi_\alpha(F_i)} \supseteq \overline{\pi_\alpha(\cap_{i=1}^n F_i)} \neq \varnothing$. Since each $X_\alpha$ is compact, there is an element $x_\alpha \in \cap\{\overline{\pi_\alpha(F)} : F \in \mathfrak{F}^*\}$ (theorem 5.8.8). Let $x = (x_\alpha)$. We will show that $x \in \cap\{\overline{F} : F \in \mathfrak{F}^*\}$. Let $U = \cap_{i=1}^n \pi_{\alpha_i}^{-1}(U_{\alpha_i})$ be an arbitrary basic open neighborhood of $x$. We claim that $U$ intersects every $F \in \mathfrak{F}^*$. This will show that $x \in \overline{F}$, and the proof will be complete. Since $x_{\alpha_i} \in U_{\alpha_i}$, and $x_{\alpha_i} \in \overline{\pi_{\alpha_i}(F)}$ for every $F \in \mathfrak{F}^*$, $U_{\alpha_i} \cap \pi_{\alpha_i}(F) \neq \varnothing$ for every $F \in \mathfrak{F}^*$. Thus $\pi_{\alpha_i}^{-1}(U_{\alpha_i}) \cap F \neq \varnothing$ for every $F \in \mathfrak{F}^*$. By the maximality of $\mathfrak{F}^*$, it must be the case that $\pi_{\alpha_i}^{-1}(U_{\alpha_i}) \in \mathfrak{F}^*$. Since $\mathfrak{F}^*$ is closed under the formation of finite intersections, $U = \cap_{i=1}^n \pi_{\alpha_i}^{-1}(U_{\alpha_i}) \in \mathfrak{F}^*$. In particular, $U \cap F \neq \varnothing$ for every $F \in \mathfrak{F}^*$. ∎

**Example 5.** (the **box topology**). Let $\{(X_\alpha, \mathcal{T}_\alpha)\}_{\alpha \in I}$ be a collection of topological spaces, and let $X = \prod_{\alpha \in I} X_\alpha$ be the Cartesian product of the underlying sets. Consider the topology $\mathcal{T}$ whose open subbase is

$$\mathfrak{S} = \left\{ \prod_{\alpha \in I} U_\alpha : U_\alpha \in \mathcal{T}_\alpha \right\}.$$

We alluded to this topology in the section preamble. It is a well-known topology, although it is more intellectually curious than practically important. When each of the spaces $X_\alpha$ is compact and Hausdorff, the box topology is Hausdorff, because it contains the product topology, which is Hausdorff by example 1. However, the box topology is not compact, by problem 6 on section 5.8. In fact, the product topology has the optimality feature of being the smallest Hausdorff topology on $X$ that admits the continuity of the projections, and the largest topology on $X$ that admits Tychonoff's theorem. ◆

We conclude this section by showing that not every compact Hausdorff space is metrizable.

**Example 6.** Let $I$ be an uncountable set and, for each $\alpha \in I$, let $X_\alpha = [0,1]$. The space $X = [0,1]^I = \prod_{\alpha \in I} X_\alpha$ is compact by Tychonoff's theorem, and Hausdorff by example 1. We show that $X$ is not metrizable.

Let 0 denote the zero function from $I$ to $[0,1]$, and let $A$ consist of all elements $x = (x_\alpha) \in X$ such that $x_\alpha = 0$ for finitely many $\alpha \in I$, and $x_\alpha = 1$ otherwise. We show that $0 \in \overline{A}$. Suppose $B = \cap_{i=1}^{n} \pi_{\alpha_i}^{-1}(U_{\alpha_i})$ is a basic open neighborhood of 0. The element $x = (x_\alpha)$ defined below is in $A \cap B$, and hence $A \cap B \neq \emptyset$:

$$x_\alpha = \begin{cases} 0 & \text{if } \alpha = \alpha_i, \\ 1 & \text{if } \alpha \neq \alpha_i. \end{cases}$$

We show that, for any sequence $x^{(n)} = (x_\alpha^n)$ in $A$, $\lim_n x^{(n)} \neq 0$. The proof will be complete by theorem 4.2.5. Let $I_n$ be the subset of elements $\alpha \in I$ for which $x_\alpha^{(n)} = 0$. The set $J = \cup_{n=1}^{\infty} I_n$ is countable since each of the sets $I_n$ is finite. Because $I$ is uncountable, $I - J \neq \emptyset$. Pick an element $\beta \in I - J$. By construction, $x_\beta^{(n)} = 1$ for all $n \in \mathbb{N}$. Now consider the open set $V = \pi_\beta^{-1}([0, 1/2))$; $V$ is a neighborhood of 0 that contains no terms of the sequence $(x^{(n)})$. ◆

## Exercises

1. Let $\{X_\alpha\}_\alpha$ be a collection of topological spaces, and let $A_\alpha \subseteq X_\alpha$. Prove that $\prod_\alpha A_\alpha = \prod_\alpha \overline{A}_\alpha$.

2. Prove theorem 5.12.3.

3. Prove that the product of an arbitrary collection of regular spaces is regular.

4. Prove that the product of a countable collection of separable spaces is separable.

5. Let $\{X_\alpha\}_\alpha$ be a family of topological spaces, and let $X = \prod_\alpha X_\alpha$. Prove that a sequence $(x^{(n)}) \in X$ converges to $x \in X$ if and only if, for each $\alpha$, $\pi_\alpha(x^{(n)})$ converges to $\pi_\alpha(x)$.

6. Let $\{X_\alpha\}_\alpha$ be a collection of topological spaces, and let $f$ be a function from a topological space $Y$ to the product space $\prod_\alpha X_\alpha$. Prove that $f$ is continuous if and only if each of the compositions $\pi_\alpha \circ f : Y \to X_\alpha$ is continuous.

7. Let $\{X_\alpha\}_{\alpha \in I}$ and $\{Y_\alpha\}_{\alpha \in I}$ be two collections of topological spaces, and, for each $\alpha \in I$, let $f_\alpha : X_\alpha \to Y_\alpha$ be continuous. Define a function $F : \prod_\alpha X_\alpha \to \prod_\alpha Y_\alpha$ by $F(x) = (f_\alpha(x_\alpha))_{\alpha \in I}$. Prove that $F$ is continuous.

8. Prove that the product of a countable collection of metrizable spaces is metrizable. Hint: Let $\{(X_n, d_n)\}$ be a countable collection of metric spaces. Without loss of generality, assume that each of the metrics $d_n$ is bounded by 1 (see theorem 4.3.9). Prove that the metric

$$D(x,y) = \sup_{n \in \mathbb{N}} \frac{d_n(x_n, y_n)}{n}$$

induces the product topology on $\prod_{n=1}^{\infty} X_n$. Here $x = (x_n)$, and $y = (y_n)$, are elements of $\prod_{n=1}^{\infty} X_n$.

9. In the notation of the previous exercise, prove that the metric

$$d(x,y) = \sum_{n=1}^{\infty} 2^{-n} d_n(x_n, y_n)$$

also induces the product topology on $\prod_{n=1}^{\infty} X_n$.

10. In the notation of problem 8, prove that if $d_n$ is a complete metric for every $n \in \mathbb{N}$, then $D$ is a complete metric.

# 6

# Banach Spaces

*Mathematics is the most beautiful and most powerful creation of the human spirit.*

Stefan Banach

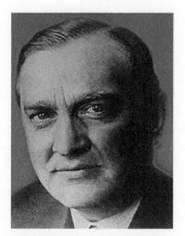

Stefan Banach. 1892–1945

In 1902 Banach began his secondary education at the Henryk Sienkiewicz Gymnasium in Kraków,[1] where he graduated in 1910. He then went to Lvov where he studied engineering at Lvov Technical University, graduating in 1914, shortly before World War I broke out in August. With the outbreak of the war, the Russian troops occupied the city of Lvov. Having poor vision in his left eye, Banach was not physically fit for army service. During the war, he worked building roads but also spent time in Kraków, where he earned money by teaching in the local schools. He also attended mathematics lectures at the Jagiellonian University in Kraków.

A life-changing event occurred in the spring of 1916 when Banach met Steinhaus, who was living in Kraków, waiting to take up a post at the Jan Kazimierz University in Lvov. Steinhaus and Banach wrote a joint paper, which was published in *The Bulletin of the Kraków Academy* after the war ended in 1918. From that time,

---

[1] A European secondary school that prepares students for the university.

*Fundamentals of Mathematical Analysis.* Adel N. Boules, Oxford University Press (2021). © Adel N. Boules.
DOI: 10.1093/oso/9780198868781.003.0006

Banach started to produce important mathematics papers at a rapid rate. On Steinhaus's initiative, the Mathematical Society of Kraków was set up in 1919. The society later became the Polish Mathematical Society in 1920. It was also through Steinhaus that Banach met his future wife, Lucja Braus, whom he married in 1920.

Banach was offered an assistantship to Lomnicki at Lvov Technical University in 1920. He lectured there in mathematics and submitted a dissertation. This was, of course, not the standard route to a doctorate, for Banach had no university mathematics qualifications. However, an exception was made to allow him to submit his thesis "On Operations on Abstract Sets and their Application to Integral Equations." This thesis is sometimes said to mark the birth of functional analysis. In his dissertation, Banach defined axiomatically what today is called a *Banach space*, a term which was coined later by Fréchet. The importance of Banach's work is that he developed a systematic theory of functional analysis, where before there had only been isolated results, which were later seen to fit into the new theory.

In 1922 the Jan Kazimierz University in Lvov awarded Banach his qualification to become a university professor, and in 1924 Banach was promoted to full professor. The years between the wars were extremely busy for Banach. As well as continuing to produce a stream of important papers, he wrote arithmetic, geometry, and algebra texts for high schools. In 1929, together with Steinhaus, he started a new journal, *Studia Mathematica*, and Banach and Steinhaus became the first editors. Another important publishing venture, begun in 1931, was a new series of mathematical monographs. These were set up under the editorship of Banach and Steinhaus, from Lvov, and Knaster, Kuratowski, Mazurkiewicz, and Sierpiński from Warsaw. The first volume in the series, *Théorie des opérations linéaires*, was written by Banach and appeared in 1932. It was a French version of a volume he originally published in Polish in 1931 and quickly became a classic. Another important influence on Banach was the fact that Kuratowski was appointed to the Lvov Technical University in 1927 and worked there until 1934. Banach collaborated with Kuratowski, and they wrote some joint papers during this period. Banach proved a number of fundamental results on normed linear spaces, including the Hahn-Banach theorem, the Banach-Steinhaus theorem, the Banach-Alaoglu theorem, Banach's open mapping theorem, and the Banach fixed point theorem. In addition, he contributed to measure theory, integration, topological vector spaces, and set theory.

In 1939, just before the start of World War II, Banach was elected as President of the Polish Mathematical Society. At the beginning of the war, Soviet troops occupied Lvov. Banach had been on good terms with the Soviet mathematicians before the war started, visiting Moscow several times, and he was treated well by the new Soviet administration. He was allowed to continue to hold his chair at the university, and he became Dean of the Faculty of Science at the university, now renamed the Ivan Franko University. Life at this stage was little changed for Banach, who continued his research, his textbook writing, lecturing, and holding sessions in cafés. Sobolev and Alexandroff visited Banach in Lvov in 1940, and Banach attended conferences in the Soviet Union. He was in Kiev when Germany invaded the Soviet Union, and he returned immediately to his family in Lvov.

The Nazi occupation of Lvov in June 1941 meant that Banach lived under very difficult conditions. He was arrested under suspicion of trafficking in German currency but was released after a few weeks. As soon as the Soviet troops retook Lvov, Banach renewed his contacts. He met Sobolev outside Moscow, but by this time he was seriously ill. Sobolev, giving an address at a memorial conference for Banach, said of this meeting[2]

Despite heavy traces of the war years under German occupation, and despite the grave illness that was undercutting his strength, Banach's eyes were still lively. He remained the same sociable, cheerful, and extraordinarily well-meaning and charming Stefan Banach whom I had seen in Lvov before the war. That is how he remains in my memory: with a great sense of humor, an energetic human being, a beautiful soul, and a great talent.

Banach had planned to go to Kraków after the war to take up the chair of mathematics at the Jagiellonian University, but he died in Lvov in 1945 of lung cancer.

## 6.1 Finite vs. Infinite-Dimensional Spaces

This section draws some sharp distinctions between finite and infinite-dimensional spaces. Although some of the results in this section have intrinsic importance and will be used later in the book, they are collected here to convince the reader that infinite-dimensional spaces are truly vast compared to finite-dimensional ones and that a very different set of tools is needed for studying them. Among other results, we will see that local compactness characterizes finite-dimensional normed linear spaces, and that an infinite-dimensional Banach space cannot have a countable linear basis.

**Definition.** A **Banach space** is a complete normed linear space.
   Examples of Banach spaces include $(\mathbb{K}^n, \|.\|_2)$, $(\mathcal{C}[0, 1], \|.\|_\infty)$, and all $l^p$ spaces.

[2] J. J. O'Connor and E. F. Robertson, "Stefan Banach," in *MacTutor History of Mathematics*, (St Andrews: University of St Andrews, 1998), http://mathshistory.st-andrews.ac.uk/Biographies/Banac/, accessed Nov. 1, 2020.

**Lemma 6.1.1.** *Let $X$ be an $n$-dimensional vector space. Then there exists a norm $\|.\|^*$ on $X$ such that $(X, \|.\|^*)$ is isometric to $(\mathbb{K}^n, \|.\|_\infty)$. In particular, $(X, \|.\|^*)$ is complete and locally compact.*

*Proof.* Fix a basis $\{x_1, ..., x_n\}$ of $X$, and define $\|x\|^* = max_{1 \leq i \leq n}|a_i|$, where $x = \sum_{i=1}^{n} a_i x_i$ is the unique representation of $x$ as a linear combination of the basis elements. The mapping $T : x \mapsto (a_1, ..., a_n)$ is clearly a linear isometry from $(X, \|.\|^*)$ onto $(\mathbb{K}^n, \|.\|_\infty)$. ∎

**Theorem 6.1.2.** *Let $(X, \|.\|)$ be an $n$-dimensional normed linear space, and let $\|.\|^*$ be the norm on $X$ defined in lemma 6.1.1. Then there exist positive constants $\alpha$ and $\beta$ such that, for all $x \in X, \beta\|x\|^* \leq \|x\| \leq \alpha\|x\|^*$.*

*Proof.* We continue to use the notation of the proof of the previous lemma. Let $\alpha = n \, max_{1 \leq i \leq n}\|x_i\|$. Then

$$\|x\| = \|\sum_{i=1}^{n} a_i x_i\| \leq \sum_{i=1}^{n} |a_i|\|x_i\| \leq max_{1 \leq i \leq n}\|x_i\|\sum_{i=1}^{n} |a_i|$$

$$\leq n \, max_{1 \leq i \leq n}\|x_i\| \, max_{1 \leq i \leq n}|a_i| = \alpha\|x\|^*.$$

To prove the other inequality, define a function $\lambda : (X, \|.\|^*) \to \mathbb{R}$ by $\lambda(x) = \|x\|$. Now $\lambda$ is continuous because if $\lim_n \|x_n - x\|^* = 0$, then $|\lambda(x_n) - \lambda(x)| = |\|x_n\| - \|x\|| \leq \|x_n - x\| \leq \alpha\|x_n - x\|^*$. Hence $\lambda(x_n) \to \lambda(x)$. By lemma 6.1.1, $(X, \|.\|^*)$ is locally compact; hence the closed unit sphere $S = \{x \in X : \|x\|^* = 1\}$ in $(X, \|.\|^*)$ is compact (see problem 9 on section 4.7). Thus the restriction of $\lambda$ to $S$ assumes a minimum value $\beta = \lambda(x_0)$ at some point $x_0 \in S$. The constant $\beta$ must be positive since, otherwise, $\lambda(x_0) = \|x_0\| = 0$, and hence $x_0 = 0$, which is not possible. We have shown that, for every $x \in S, \|x\| \geq \beta$. Now, for a nonzero vector $x \in X, \frac{x}{\|x\|^*} \in S$; hence $\|\frac{x}{\|x\|^*}\| \geq \beta$, and $\|x\| \geq \beta\|x\|^*$. ∎

**Corollary 6.1.3.** *All norms on a finite-dimensional normed linear space are equivalent.*

*Proof.* By theorem 6.1.2, an arbitrary norm on a finite-dimensional space is equivalent to $\|.\|^*$; hence any two norms are equivalent. ∎

**Theorem 6.1.4.** *A finite-dimensional proper subspace $F$ of a normed linear space $X$ is closed and nowhere dense.*

*Proof.* By lemma 6.1.1, F is complete and hence closed in X. To show that X is nowhere dense, let $\{x_1, \ldots, x_n\}$ be a basis for F, and let $x \in X - F$. If F contains a ball of radius $\delta$, then it would contain the ball B of radius $\delta$ and centered at the origin. But then B (hence F) would contain a multiple of x, namely, $\frac{\delta x}{2\|x\|}$. This contradiction shows that F is nowhere dense in X. ∎

**Example 1.** Let F be a finite-dimensional subspace of a normed linear space X, and let $x \in X - F$. Then there exists a point $z \in F$ such that $\|x - z\| = dist(x, F)$.

Let $B = B[x, r]$ be a closed ball centered at x, and assume r is large enough so that $B \cap F \neq \varnothing$. The set $K = B \cap F$ is a closed and bounded subset of F. By the Heine-Borel theorem (see problem 6 at the end of this section), K is compact. The function $f : K \to \mathbb{R}$ given by $f(y) = \|x - y\|$ is continuous and positive. Therefore $d = min\{f(y) : y \in K\}$ is positive and is attained at some point $z \in K$. Since $d \leq r$ and $\|x - y\| > r$ for every vector $y \in F - B$, $\|x - z\| = dist(x, F)$. ◆

**Example 2.** The following is a direct application of the previous example. Take $X = \mathcal{C}[0, 1]$, and $F = \mathbb{P}_n$. For any function $f \in X$, there is a polynomial $p_n^* \in \mathbb{P}_n$ such that $\|f - p_n^*\|_\infty = dist(f, \mathbb{P}_n)$. ◆

The polynomial $p_n^*$ is the best approximation of f in $\mathbb{P}_n$. It can be shown that $p_n^*$ is unique. Observe that $p_n^*$ can have degree less than n.

**Example 3.** For a function $f \in \mathcal{C}[0, 1]$, the sequence of best approximations $p_n^*$ converges uniformly to f.

Let $\epsilon > 0$. By the Weierstrass polynomial approximation theorem, there exists a polynomial q such that $\|f - q\|_\infty < \epsilon$. Let N be the degree of q. Then, for every $n > N$, $q \in \mathbb{P}_n$. Since $p_n^*$ is the best approximation of f in $\mathbb{P}_n$, $\|f - p_n^*\|_\infty \leq \|f - q\|_\infty < \epsilon$. This shows that $lim_n \|f - p_n^*\|_\infty = 0$. ◆

The following theorem establishes the fact that local compactness is exclusively a property of finite-dimensional spaces.

**Theorem 6.1.5.** *A normed linear space X is locally compact if and only if it is finite dimensional.*

*Proof.* Finite-dimensional spaces are locally compact by lemma 6.1.1 and corollary 6.1.3. Now suppose X is a locally compact normed linear space. Thus the closed unit ball $B = \{x \in X : \|x\| \leq 1\}$ is compact. Since $B \subseteq \cup_{x \in B} B(x, 1/2)$, $B \subseteq \cup_{i=1}^n B(x_i, 1/2)$ for a finite subset $\{x_1, \ldots, x_n\}$ of B. We will show that $\{x_1, \ldots, x_n\}$ spans X. Let $F = Span\{x_1, \ldots, x_n\}$, and suppose, for a contradiction, that there *is a vector $x \in X - F$. By example 1, there is an element $z \in F$ which is closest to $x$, and $d = dist(x, F) = \|x - z\| > 0$. Now $\frac{x-z}{\|x-z\|} \in B$, so there is an element $x_i \in B$ such that $\|\frac{x-z}{\|x-z\|} - x_i\| < 1/2$, and $\|x - z - \|x - z\|x_i\| \leq \|x - z\|/2 = d/2$. But $z + \|x - z\|x_i \in F$, so $\|x - z - \|x - z\|x_i\| \geq d$. This contradiction concludes the proof.* ∎

**Remark.** The last theorem implies that compact subsets of an infinite-dimensional space have empty interiors and are therefore thought of as *rather thin and scarce sets*. However, compact sets continue to play an important role in the study of infinite-dimensional spaces. The Hilbert cube is an example of a rather exotic compact set.

**Example 4.** The **Hilbert cube** $\mathcal{H}$ is a compact, convex, nowhere dense subset of $l^2$. Furthermore, $Span(\mathcal{H})$ is dense in $l^2$.

It is simple to show that $\mathcal{H}$ is closed and convex. Once we show that $\mathcal{H}$ is compact, the above remark implies that it has an empty interior.
Let $(x^k)_{k=1}^\infty$ be a sequence in $\mathcal{H}$, and write $x^k = (x_1^k, x_2^k, ...)$. The sequence $(x_1^k)_{k=1}^\infty$ is bounded by 1, so there exists a strictly increasing sequence $(k_p^1)_{p=1}^\infty$ of positive integers such that $x_1 = \lim_{p\to\infty} x_1^{k_p^1}$ exists. Since the sequence $(x_2^{k_p^1})$ is bounded by 1/2, it contains a convergent subsequence $(x_2^{k_p^2})_{p=1}^\infty$. Let $x_2 = \lim_p x_2^{k_p^2}$. Continue inductively to find sequences $(k_p^n)_{p=1}^\infty$ of positive integers such that, for each $n \geq 1$, $(k_p^{n+1})$ is a subsequence of $(k_p^n)$ such that $x_n = \lim_p x_n^{k_p^n}$ exists. Consider the diagonal sequence $(k_p^p)_{p=1}^\infty$, and observe that it is a subsequence of $(k_p^n)_{p=1}^\infty$ for every $n \geq 1$, thus $\lim_p x_n^{k_p^p} = x_n$. Let $x = (x_n)$, and observe that $|x_n| = \lim_p |x_n^{k_p^p}| \leq 1/n$; thus $x \in \mathcal{H}$. We claim that $\lim_p x^{k_p^p} = x$ in $l^2$. For simplicity of notation, write $y^p = x^{k_p^p}$. Let $\epsilon > 0$, and choose an integer $N$ such that $\sum_{n=N+1}^\infty \frac{1}{n^2} < \epsilon^2/8$. For $n = 1, 2, ..., N$, $\lim_p y_n^p = x_n$; thus there exists an integer $P$ such that, for $n = 1, ..., N$, $p > P$ implies that $|y_n^p - x_n|^2 < \frac{\epsilon^2}{2N}$. Hence, for $p > P$,

$$\sum_{n=1}^N |y_n^p - x_n|^2 < \epsilon^2/2. \text{ Also,}$$

$$\sum_{n=N+1}^\infty |y_n^p - x_n|^2 < \sum_{n=N+1}^\infty \left(\frac{2}{n}\right)^2 = 4\sum_{n=N+1}^\infty \frac{1}{n^2} < \epsilon^2/2.$$

These last two inequalities imply that, for $p > P$, $\|y^p - x\|_2 < \epsilon$.

Finally, we show that $Span(\mathcal{H})$ is dense in $l^2$. Let $x = (x_n) \in l^2$, and let $\epsilon > 0$. Choose an integer $N$ such that $\sum_{n=N+1}^{\infty} |x_n|^2 < \epsilon^2$. Define $y_1 = (1,0,0,...)$, $y_2 = (0, 1/2, 0, 0, ...)$, ... , $y_N = (0, 0, ..., 0, 1/N, 0, ...)$, and set $h = \sum_{n=1}^{N} a_n y_n$, where $a_n = n x_n$. Clearly, $h \in Span(\mathcal{H})$ and $\|x - h\|_2 < \epsilon$. ◆

Two very useful tools for studying finite-dimensional spaces are local compactness and the existence of a finite linear basis. We already saw in theorem 6.1.5 that infinite-dimensional normed linear spaces are never locally compact. By definition, an infinite-dimensional space cannot have a finite Hamel basis. The following theorem should thoroughly convince the reader that a Hamel basis is of no practical use as a tool for studying infinite-dimensional Banach spaces. However, see the concept of a Schauder basis in the exercises following this section and the next section.

**Theorem 6.1.6.** *An infinite-dimensional Banach space does not have a countable Hamel basis.*

*Proof. Let $X$ be an infinite-dimensional Banach space, and let $B = \{x_1, x_2, ..\}$ be a countably infinite independent subset of $X$. The finite-dimensional spaces $F_n = Span\{x_1, ..., x_n\}$ are closed and nowhere dense in $X$. Baire's theorem implies that $\cup_{n=1}^{\infty} F_n \neq X$. Since $Span(B) = \cup_{n=1}^{\infty} F_n$, $Span(B) \neq X$. Therefore no countable subset of $X$ spans $X$. ∎*

The following result will be used frequently later in the book. The motivation for the theorem is provided below.

Let $M$ be a proper subspace of $\mathbb{R}^n$. It is an elementary fact of linear algebra (see problem 7 on section 3.7) that there is a unit vector $x$ orthogonal to $M$. In this case, $dist(x, M) = 1$.

Generalizing this result to Banach space is more challenging because we lack the concept of orthogonality, which is a available only for inner product spaces. The result below provides the next best alternative to the desirable property that $dist(x, M) = 1$; we can pick a unit vector $x$ whose distance from $M$ is arbitrarily close to 1.

**Lemma 6.1.7 (Riesz's lemma).** *Let $M$ be a closed proper subspace of a normed linear space $X$, and let $0 < \theta < 1$. Then there exists a unit vector $x$ such that $dist(x, M) > \theta$.*

*Proof. Let $v \in X - M$, and let $\delta = dist(v, M)$. Since $\theta < 1$, there exists $y_0 \in M$ such that $\delta \le \|v - y_0\| < \delta/\theta$. Define $x = \dfrac{v - y_0}{\|v - y_0\|}$. For $y \in M$, $y_0 + \|v - y_0\| y \in M$ and $\|v - (y_0 + \|v - y_0\| y)\| \ge \delta$. Now*

$$\|x - y\| = \left\|\frac{v - y_0}{\|v - y_0\|} - y\right\| = \frac{1}{\|v - y_0\|}\|v - y_0 - \|v - y_0\|y\|$$

$$= \frac{1}{\|v - y_0\|}\|v - (y_0 + \|v - y_0\|y)\| \geq \frac{\delta}{\|v - y_0\|} > \frac{\delta}{\delta/\theta} = \theta. \ \blacksquare$$

## Exercises

1. (a) Prove that a sequence $(x_n)$ in a normed linear space $X$ is Cauchy if and only if $\lim_n(x_{p_n} - x_{q_n}) = 0$ for every pair $(p_n)$ and $(q_n)$ of increasing sequences of positive integers.

   (b) Show that if $\lim_n x_n = x$, then $\lim_n \frac{1}{n}\sum_{i=1}^n x_i = x$.

2. Let $w$ be a fixed positive function in $\mathcal{C}[0,1]$. For $f \in \mathcal{C}[0,1]$, define $\|f\| = \|fw\|_\infty$. Prove that $\|.\|$ is a norm, and determine if it is equivalent to the uniform norm on $\mathcal{C}[0,1]$.

3. Let $X$ be a normed linear space. Prove that $X$ is a Banach space if and only if the closed unit ball in $X$ is complete.

4. Let $X$ be a normed linear space. Prove that $X$ is separable if and only if the closed unit sphere in $X$ is separable.

5. Let $(x_n)$ be a sequence in a Banach space $X$ such that, for every $\epsilon > 0$, there exists a convergent sequence $(y_n)$ in X such that $\|x_n - y_n\| < \epsilon$ for all $n \in \mathbb{N}$. Prove that $(x_n)$ is convergent.

6. **The Heine-Borel theorem.** Let $V$ be a finite-dimensional subspace of a normed linear space $X$. Show that a subset $K$ of $V$ is compact if and only if it is closed and bounded.

7. Let $X$ be an infinite-dimensional normed linear space. Show that $X$ contains a compact countable subset that is not contained in any finite-dimensional subspace of $X$. Hint: Let $\{x_1, x_2, ...\}$ be an infinite independent subset of $X$, and let $\xi_n = \frac{x_n}{n\|x_n\|}$. Consider the set $\{\xi_n\} \cup \{0\}$.

8. Prove that, for $1 \leq p < \infty$, the linear dimension of $l^p$ is $\mathfrak{c}$. Hint: For each $0 < \lambda < 1$, let $x_\lambda = (\lambda, \lambda^2, \lambda^3, ...)$. Show that the set $\{x_\lambda\}$ is independent, then use example 7 on section 4.5.

**Definition.** Let $(x_n)$ be a sequence in a normed linear space $X$. We say the the series $\sum_{n=1}^\infty x_n$ converges if the sequence of partial sums $S_n = \sum_{i=1}^n x_i$ converges to an element $x \in X$. We write $x = \sum_{n=1}^\infty x_n$ for $\lim_n S_n$ and say the $x$ is the sum of the series. We say that $\sum_{n=1}^\infty x_n$ is **absolutely convergent** if $\sum_{n=1}^\infty \|x_n\| < \infty$.

9. Prove that if $X$ is a Banach space, then every absolutely convergent series is convergent.

10. Prove that if every absolutely convergent series in a normed linear space $X$ is convergent, then $X$ is a Banach space. The proof outline is as follows. Let $(x_n)$ be a Cauchy sequence in $X$. It is enough to show that $(x_n)$ contains a convergent subsequence. Choose a subsequence $(x_{n_k})$ of $(x_n)$ such that $\|x_{n_{k+1}} - x_{n_k}\| < 2^{-k}$. Define $y_k = x_{n_{k+1}} - x_{n_k}$. Show that $\sum_{k=1}^{\infty} \|y_k\| < \infty$. By assumption, $\sum_{k=1}^{\infty} y_k$ converges. But $\sum_{k=1}^{\infty} y_k = -x_{n_1} + \lim_k x_{n_k}$.

**Definition.** Let $X$ be a normed linear space. A countable subset $\{u_n\}$ of $X$ is a **Schauder basis** for $X$ if every element $x \in X$ can be expressed uniquely as $x = \sum_{n=1}^{\infty} a_n u_n$ where $a_n \in \mathbb{K}$.

11. Prove that a Schauder basis is independent.
12. Prove that if $X$ has a Schauder basis then it is separable.
13. Find a Schauder basis for $l^p, 1 \leq p < \infty$.
14. Prove that if $M$ is a subspace of a normed linear space $X$, then $\overline{M}$ is also a subspace of $X$.
15. Let $M$ be a closed subspace of a normed linear space $X$, and let $x \in X$. Prove that $dist(x, M) \leq \|x\|$.
16. Use Riesz's theorem to produce another proof that an infinite-dimensional normed linear space is not locally compact.

## 6.2  Bounded Linear Mappings

The boundedness of a linear transformation on a normed linear space and its continuity are used synonymously. Every linear transformation on a finite-dimensional space is continuous. The picture is far more complicated for linear transformations on infinite-dimensional spaces. In this chapter and the next, we study continuous linear transformations exclusively because nonlinear transformations and discontinuous linear transformations fall outside the realm of beginning linear functional analysis.

In this section, we study the various equivalent characterizations of boundedness, the space of bounded linear transformations on a normed linear space, and the dual space in particular. The section concludes with a typical representation theorem, which gives a concrete description of the dual of a normed linear space. Throughout this section, $X$ and $Y$ are normed linear spaces.

**Definition.** A linear mapping $T : X \to Y$ is said to be **bounded** if there exists a constant $M > 0$ such that for every $x \in X$,

$$\|T(x)\| \leq M\|x\|.$$

**Theorem 6.2.1.** *Let $T : X \to Y$ be linear. The following are equivalent:*

*(a) $T$ is continuous.*
*(b) $T$ is continuous at one point $x_0 \in X$.*
*(c) $T$ is continuous at 0.*
*(d) $T$ is bounded.*

*Proof. (a) implies (b), obviously.*

*(b) implies (c). Let $x_n \to 0$ in $X$. Then $x_n + x_0$ converges to $x_0$. By assumption, $\lim_n T(x_n + x_0) = T(x_0)$. But $\lim_n T(x_n + x_0) = \lim_n T(x_n) + T(x_0)$; hence $\lim_n T(x_n) = 0$.*

*(c) implies (d). Suppose $T$ is not bounded. Then, for every $n \in \mathbb{N}$, there exists $x_n \in X$ such that $\|T(x_n)\| > n\|x_n\|$. Let $\xi_n = \frac{x_n}{n\|x_n\|}$. Then $\lim_n \xi_n = 0$ in $X$, but $\|T(\xi_n)\| = \frac{\|T(x_n)\|}{n\|x_n\|} > 1$. Thus, $\lim_n T(\xi_n) \neq 0$ in $Y$, and $T$ is not continuous at 0.*

*(d) implies (a). Suppose that there is a constant $M > 0$ such that, for every $x \in X$, $\|T(x)\| \leq M\|x\|$, and $\lim_n x_n = x$ in $X$. Then*

$$\lim_n \|T(x_n) - T(x)\| = \lim_n \|T(x_n - x)\| \leq \lim_n M\|x_n - x\| = 0. \quad \blacksquare$$

Let $T : X \to Y$ be a bounded linear mapping. The **norm** of $T$ is

$$\|T\| = \sup_{x \neq 0} \frac{\|T(x)\|}{\|x\|}.$$

Notice that since $T$ is bounded, there is a constant $M > 0$ such that $\|T(x)\| \leq M\|x\|$ for all $x \in X$; therefore $\frac{\|T(x)\|}{\|x\|} \leq M$, and hence $\|T\|$ is finite. It also follows directly from the definition that $\|T(x)\| \leq \|T\|\|x\|$.

**Example 1.** Let $(c_n)$ be a bounded sequence and, for a sequence $x = (x_1, x_2, \ldots) \in l^2$, define $T(x) = (c_1 x_1, c_2 x_2, \ldots)$. We claim that $T$ is a bounded linear mapping on $l^2$. Indeed, $\|T(x)\|_2^2 = \sum_{n=1}^{\infty} |c_n x_n|^2 \leq \|c\|_\infty^2 \sum_{n=1}^{\infty} |x_n|^2 = \|c\|_\infty^2 \|x\|_2^2$. This estimate shows that $T(x) \in l^2$ and that $\|T\| \leq \|c\|_\infty$. The linearity of $T$ is obvious. $\blacklozenge$

**Example 2.** A bounded linear mapping $T : X \to Y$ maps bounded sets into bounded sets.
Let $A$ be a bounded subset of $X$ and suppose that $\|x\| \leq r$ for every $x \in A$. Then $\|T(x)\| \leq \|T\|\|x\| \leq \|T\|r < \infty$. $\blacklozenge$

**Example 3.** If $X$ is finite dimensional, then every linear mapping $T : X \to Y$ is bounded.

Let $\{x_1, \dots, x_n\}$ be a linear basis for $X$, and let $M = max_{1 \leq i \leq n} \|T(x_i)\|$. Since all norms on $X$ are equivalent, we may assume that the norm on $X$ is the 1-norm. Thus if $x = \sum_{i=1}^n a_i x_i$, then $\|x\| = \sum_{i=1}^n |a_i|$. Now $\|T(x)\| = \|T(\sum_{i=1}^n a_i x_i)\| \leq \sum_{i=1}^n |a_i| \|T(x_i)\| \leq M \sum_{i=1}^n |a_i| = M\|x\|$. ◆

**Example 4.** If $X$ is infinite dimensional, then there exists a linear unbounded mapping from $X$ to $Y$.

Fix a nonzero element $y \in Y$, and let $S_1 = \{x_1, x_2, \dots\}$ be an infinite independent set of unit vectors in $X$. Let $S_2 \subseteq X$ be such that $S = S_1 \cup S_2$ is a linear basis for $X$. Define a function $T : S \to Y$ as follows:

$$T(x) = \begin{cases} ny & \text{if } x = x_n, \\ 0 & \text{if } x \in S_2. \end{cases}$$

Extend $T$ to a linear mapping $T : X \to Y$. See theorem 3.4.4. Since $S_1$ is bounded but $T(S_1)$ is not, $T$ is not bounded by example 2. ◆

**Theorem 6.2.2.** *Let $T : X \to Y$ be a bounded linear mapping. Then*

$$\|T\| = sup_{\|x\| \leq 1} \|T(x)\| = sup_{\|x\|=1} \|T(x)\|.$$

*Proof.* Let $M = sup_{\|x\| \leq 1} \|T(x)\|$. For every $x \in X, x \neq 0, \|T(\frac{x}{\|x\|})\| \leq M$, hence $\frac{\|T(x)\|}{\|x\|} \leq M$. Thus $\|T\| \leq M$. To prove that $M \leq \|T\|$, fix a vector $x \in X$ such that $0 < \|x\| \leq 1$. By definition of $\|T\|$, $\|T\| \geq \frac{\|Tx\|}{\|x\|} \geq \|Tx\|$. Since $x$ is arbitrary, it follows that $M = sup_{\|x\| \leq 1} \|Tx\| \leq \|T\|$, as desired.

The proof that $\|T\| = sup_{\|x\|=1} \|T(x)\|$ is similar. ■

**Theorem 6.2.3.** *Let $X$ and $Y$ be normed linear spaces, and let $\mathcal{L}(X, Y)$ be the set of all bounded linear mappings from $X$ to $Y$. Then $\mathcal{L}(X, Y)$ is a normed linear space with the operations $(T_1 + T_2)(x) = T_1(x) + T_2(x)$, $(aT)(x) = aT(x)$, and the norm $\|T\| = sup_{x \neq 0} \frac{\|T(x)\|}{\|x\|}$. Furthermore, if $Y$ is a Banach space, then so is $\mathcal{L}(X, Y)$.*

*Proof.* We first show that a linear combination of bounded linear mappings is bounded. Let $a$ and $b$ be scalars, and let $T_1, T_2 \in \mathcal{L}(X, Y)$. Then, for every $x \in X$, $\|aT_1(x) + bT_2(x)\| \leq |a| \|T_1(x)\| + |b| \|T_2(x)\| \leq |a| \|T_1\| \|x\| + |b| \|T_2\| \|x\| = (|a| \|T_1\| + |b| \|T_2\|) \|x\|$. This shows that $\|aT_1 + bT_2\| \leq |a| \|T_1\| + |b| \|T_2\|$ and *that $\mathcal{L}(X,Y)$ is closed under addition and scalar multiplication. Verifying the rest of the axioms for a vector space is routine.*

*The above inequality can also be used to verify the defining properties of a norm. For example, taking $a = b = 1$ gives $\|T_1 + T_2\| \le \|T_1\| + \|T_2\|$. The identity $\|aT\| = |a|\|T\|$ is obvious.*

*It remains to show that $\mathcal{L}(X,Y)$ is complete if $Y$ is complete. Suppose $(T_n)$ is a Cauchy sequence in $\mathcal{L}(X,Y)$, and let $\epsilon > 0$. By assumption, there exists a positive integer $N$ such that, for $m,n > N$, $\|T_n - T_m\| < \epsilon$. For all $x \in X$, $\|T_n(x) - T_m(x)\| = \|(T_n - T_m)(x)\| \le \|T_n - T_m\|\|x\| < \epsilon\|x\|$. Thus $(T_n(x))$ is a Cauchy sequence in $Y$, and hence $\lim_n T(x_n)$ exists for every $x \in X$. Define $T(x) = \lim_n T_n(x)$. We show that $T \in \mathcal{L}(X,Y)$. The linearity of $T$ is straightforward; if $x, y \in X$, and $a$ and $b$ are scalars, then $T(ax + by) = \lim_n T_n(ax + by) = \lim_n aT_n(x) + bT_n(y) = aT(x) + bT(y)$. To show that $T$ is bounded, let $\epsilon = 1$. There is a positive integer $N$ such that, for $m, n \ge N, \|T_n - T_m\| \le 1$. In particular, for all $n \ge N, \|T_n(x) - T_N(x)\| \le \|x\|$. Hence, for all $x \in X$, and all $n \ge N$, $\|T_n(x)\| \le \|T_n(x) - T_N(x)\| + \|T_N(x)\| \le \|x\| + \|T_N\|\|x\| = (1 + \|T_N\|)\|x\|$. Taking the limit as $n \to \infty$, we obtain $\|T(x)\| \le (1 + \|T_N\|)\|x\|$. Thus $\|T\| \le (1 + \|T_N\|)$. Finally, we show that $\lim_n \|T_n - T\| = 0$. Let $\epsilon > 0$, and let $N$ be such that $\|T_n - T_m\| < \epsilon$ for all $m, n > N$. For all $x \in X$, $\|T_n(x) - T_m(x)\| \le \epsilon\|x\|$. Taking the limit as $m \to \infty$, we have $\|T_n(x) - T(x)\| \le \epsilon\|x\|$ for all $x \in X$, and all $n > N$. Thus $\|T_n - T\| \le \epsilon$ for all $n > N$; hence $\lim_n T_n = T$.* ∎

An important special case of theorem 6.2.3 is the space $\mathcal{L}(X, \mathbb{K})$ of all bounded linear functionals from a normed linear space $X$ to the base field. This space is known as the **dual space** of $X$, and is denoted by $X^*$. Since $\mathbb{K}$ is complete, $X^*$ is a Banach space, even when $X$ is not complete.

Another important special case of theorem 6.2.3 is the space $\mathcal{L}(X) = \mathcal{L}(X,X)$ of bounded linear transformations on a Banach space $X$. Elements of $\mathcal{L}(X)$ are also called bounded operators on $X$. The norm $\|T\| = sup_{x \ne 0} \frac{\|T(x)\|}{\|x\|}$ is called, not surprisingly, the **operator norm** on $\mathcal{L}(X)$.

**Example 5.** Let $\|.\|$ and $\|.\|'$ be norms on a vector space $X$. Then $\|.\|$ and $\|.\|'$ are equivalent if and only if there exist positive constants $k_1$ and $k_2$ such that, for every $x \in X$, $k_1\|x\| \le \|x\|' \le k_2\|x\|$. Note the contrast between this result and exercise 12 on section 4.3.

If $k_1$ and $k_2$ exist, the two norms are equivalent by theorem 4.3.9. Conversely, the equivalence of the two norms implies the bi-continuity of the identity mapping $I : (X, \|.\|) \to (X, \|.\|')$. The continuity of $I$ implies the existence of a positive constant $k_2$ (namely, the norm of $I$ relative to the given norms) such that $\|x\|' = \|I(x)\|' \le k_2\|x\|$. The existence of $k_1$ is established in a similar way. ◆

**Example 6.** It is easy to verify that the function $\|f\| = \|f\|_\infty + |f(0)| + |f(1)|$ defines a norm on $\mathcal{C}[0,1]$. This norm is equivalent to the uniform norm on $\mathcal{C}[0,1]$ by the previous example since $\|f\|_\infty \le \|f\| \le 3\|f\|_\infty$. ◆

**Example 7.** Let $L : l^2 \to l^2$ be the operator defined by $L(x_1, x_2, ...) = (x_2, x_3, ...)$, and let $T_n = L^n$. Thus $T_n(x_1, x_2, ...) = (x_{n+1}, x_{n+2}, ...)$. Since $\|T_n(x)\|_2 \le \|x\|_2$ for every $x \in l^2$, $\|T_n\| \le 1$. In fact, $\|T_n\| = 1$ because if we take $x = e_{n+1}$, then $T_n(x) = e_1$, and $\|T_n(x)\|_2 = 1 = \|x\|_2$. Next we show that $\lim_n T_n(x) = 0$ for every $x \in l^2$. Indeed, $\|T_n(x)\|_2^2 = \sum_{k=n+1}^{\infty} |x_k|^2$. Being the tail end of a convergent series, the last quantity approaches 0 as $n \to \infty$. Observe that the sequence $(T_n)$ does not converge in the operator norm because if it did, it should converge to the zero operator. This is obviously false because $\|T_n\| = 1$. ◆

**Definition.** A bounded linear operator $T$ on a Banach space $X$ is said to be **bounded away from zero** if there is a constant $c > 0$ such that $\|Tx\| \ge c\|x\|$ for all $x \in X$.

**Example 8.** Let $X$ be a Banach space, and let $T \in \mathcal{L}(X)$ be bounded away from zero. Then $T$ is one-to-one, $\mathfrak{R}(T)$ is closed in $X$, and $T^{-1} : \mathfrak{R}(T) \to X$ is bounded.

If $Tx = 0$, then $\|x\| \le \|Tx\|/c = 0$; hence $x = 0$, and $T$ is one-to-one.

To prove that $\mathfrak{R}(T)$ is closed, we show that if $(x_n)$ is such that $\lim_n Tx_n = y$, then $y \in \mathfrak{R}(T)$. It is enough to show that $(x_n)$ is Cauchy because if we set $x = \lim_n x_n$, then $y = \lim_n Tx_n = Tx$. Since $(Tx_n)$ is a Cauchy sequence, $\|x_n - x_m\| \le (1/c)\|Tx_n - Tx_m\| \to 0$ as $n, m \to \infty$. Thus $(x_n)$ is a Cauchy sequence, as desired. Finally, the inequality $\|Tx\| \ge c\|x\|$ implies that $\|T^{-1}\| \le 1/c$. ◆

We conclude this section with an example of a representation theorem. The result below gives a concrete characterization of the dual of the sequence spaces $l^p$.

**Theorem 6.2.4.** *Let* $1 \le p < \infty$. *The dual* $(l^p)^*$ *of* $l^p$ *is isometrically isomorphic to* $l^q$, *where* $\dfrac{1}{p} + \dfrac{1}{q} = 1$.

*Proof.* Fix a real number $p > 1$. For $y \in l^q$, define a functional $\lambda_y : l^p \to \mathbb{K}$ by $\lambda_y(x) = \sum_{n=1}^{\infty} x_n y_n$. By Hölder's inequality, $|\lambda_y(x)| \le \sum_{n=1}^{\infty} |x_n y_n| \le \|x\|_p \|y\|_q$. Thus $\lambda_y$ is a bounded linear functional on $l^p$, and $\|\lambda_y\| \le \|y\|_q$. The linearity of $\lambda_y$ is clear. To show that $\|\lambda_y\| = \|y\|_q$, define a sequence $(x_n)$ as follows $x_n = 0$

*if $y_n = 0$, and $x_n = |y_n|^q/y_n$ if $y_n \neq 0$. Note that $\|x\|_p = \|y\|_q^{q/p} = \|y\|_q^{q-1}$; hence*
*$x \in l^p$. Now $\lambda_y(x) = \sum_{n=1}^{\infty} x_n y_n = \sum_{n=1}^{\infty} |y_n|^q = \|y\|_q^q = \|y\|_q^{q-1} \|y\|_q = \|x\|_p \|y\|_q$.*
*Thus $\|\lambda_y\| = \|y\|_q$.*

*The mapping $\Lambda : l^q \to (l^p)^*$ given by $y \mapsto \lambda_y$ is clearly linear, and the fact that*
*$\|\lambda_y\| = \|y\|_q$ makes $\Lambda$ an isometry. It remains to show that $\Lambda$ maps $l^q$ onto $(l^p)^*$.*
*Let $\lambda$ be a bounded linear functional on $l^p$. We need to show that $\lambda = \lambda_y$ for*
*some $y \in l^q$. Let $e_n$ be the canonical vectors in $\mathbb{K}(\mathbb{N})$. For $x = (x_1, x_2, ...) \in l^p$, the*
*sequence $\xi_n = \sum_{i=1}^{n} x_i e_i = (x_1, x_2, ..., x_n, 0, 0, 0, ...)$ converges to $x$ in $l^p$. Let $y_n = \lambda(e_n)$, and let $y = (y_n)$. First we show that $y \in l^q$. Let $\eta_n = (y_1, y_2, ..., y_n, 0, 0, 0...)$,*
*and define $\lambda_n(x) = \sum_{i=1}^{n} x_i y_i$. By the part of the theorem we already estab-*
*lished, $\lambda_n \in (l^p)^*$, and $\|\lambda_n\| = \|\eta_n\|_q = (\sum_{i=1}^{n} |y_i|^q)^{1/q}$. Now $|\lambda_n(x)| = |\lambda(\xi_n)| \leq \|\lambda\| \|\xi_n\|_p \leq \|\lambda\| \|x\|_p$; hence $\|\lambda_n\| \leq \|\lambda\|$. Therefore $(\sum_{i=1}^{n} |y_i|^q)^{1/q}$ is bounded by*
*$\|\lambda\|$; hence $\sum_{n=1}^{\infty} |y_n|^q < \infty$, that is, $y \in l^q$. Finally, we show that $\lambda = \lambda_y$:*

$$\lambda(x) = \lambda(\lim_n \xi_n) = \lim_n \lambda(\xi_n) = \lim_n \lambda(\sum_{i=1}^{n} x_i e_i)$$

$$= \lim_n \sum_{i=1}^{n} x_i \lambda(e_i) = \lim_n \sum_{i=1}^{n} x_i y_i = \sum_{n=1}^{\infty} x_n y_n = \lambda_y(x). \blacksquare$$

We sometimes summarize the above result by saying that *the dual of $l^p$ is $l^q$* instead
of saying that the dual of $l^p$ is isometrically isomorphic to $l^q$. This slight abuse of
language is common.

## Exercises

1. Prove that if a linear function $T : X \to Y$ maps bounded sets into bounded
   sets, then $T$ is bounded.
2. Let $T : X \to Y$ be a bounded linear mapping. Show that the closed ball
   $\{y \in Y : \|y\| \leq \|T\|\}$ is the smallest ball in $Y$ that contains the image of the
   closed unit ball, $\{x \in X : \|x\| \leq 1\}$ in $X$.
3. Let $T : X \to Y$ be a bounded linear mapping. Show that $\|T\| = \sup_{\|x\|<1} \|T(x)\|$.
4. Show that a bounded linear mapping $T : X \to Y$ is uniformly continuous.
5. Prove that every linear functional on $X$ is bounded if and only if
   $dim(X) < \infty$.
6. Let $\lambda \in X^*$. Prove that $\lambda$ is an open mapping.
7. Let $T : X \to X$ be a linear mapping such that whenever $x_n \to 0$, then $\{T(x_n)\}$
   is bounded. Prove that $T$ is bounded. Hint: If $T$ is unbounded, then there

exists a sequence $(x_n)$ such that $x_n \to 0$ but $\|T(x_n)\|$ is bounded away from 0. Consider the sequence $\xi_n = x_n / \sqrt{\|x_n\|}$.

8. Suppose $X$ is an $n$-dimensional normed linear space. Prove that $dim(X^*) = n$.

9. Let $T : X \to Y$ be a bounded linear injection. Prove that the following conditions are equivalent:

   (a) $T$ is an isometry from $X$ onto $Y$.
   (b) $T(S_X) = S_Y$.
   (c) $T(B_X) = B_Y$.

   Here $B_X$ and $B_Y$ are the closed unit balls in $X$ and $Y$, respectively, and $S_X$ and $S_Y$ are the unit spheres in $X$ and $Y$, respectively.

10. We know that if $1 \le p < q \le \infty$, then $l^p \subset l^q$; see problem 3 on section 3.6. Let $i : l^p \to l^q$ be the inclusion map. Find $\|i\|$.

11. Define a linear operator $T \in \mathcal{L}(c_0)$ as follows: for $x = (x_n)$, $T(x) = (\frac{x_n}{n})$. Find $\|T\|$ and show that $\mathfrak{R}(T)$ is dense in $c_0$.

12. In connection with example 1, show that $\|T\| = \|c\|_\infty$.

13. Let $X$ be the space of polynomials equipped with the norm $\|f\| = sup_{0 \le x \le 1} |f(x)|$. Prove that differentiation is an unbounded operator on $X$.

14. Define a function $\|.\|'$ on the space of null sequences $c_0$ by $\|x\|' = \sum_{n=1}^{\infty} 2^{-n}|x_n|$. Here $x = (x_n)$. Prove that the given function is a norm and that it is not equivalent to the infinity norm on $c_0$. Hint: The sequence $(1, 1, \ldots, 1, 0, 0, 0, \ldots)$ is Cauchy in $\|.\|'$.

15. Let $\|.\|_1$ and $\|.\|_2$ be equivalent norms on a Banach space $X$. Prove that the closed unit balls $B_1 = \{x \in X : \|x\|_1 \le 1\}$ and $B_2 = \{x \in X : \|x\|_2 \le 1\}$ are homeomorphic. Hint: Consider the function $\varphi : B_1 \to B_2$ defined by

$$\varphi(x) = \begin{cases} \frac{\|x\|_1}{\|x\|_2} x & \text{if } x \neq 0, \\ 0 & \text{if } x = 0. \end{cases}$$

16. Prove that $c_0^* = l^1$ and $(l^1)^* = l^\infty$.

17. Let $\lambda : X \to \mathbb{K}$ be a nonzero linear functional on a vector space $X$. Prove that there exists a one-dimensional subspace $M$ of $X$ such that $X = Ker(\lambda) \oplus M$.

18. Let $\lambda : X \to \mathbb{K}$ be a nonzero linear functional on a normed linear space $X$. Prove that the following are equivalent:

   (a) $\lambda$ is bounded.
   (b) $Ker(\lambda)$ is closed.
   (c) $Ker(\lambda)$ is not dense in $X$.

   Conclude that $\lambda$ is unbounded if and only if $Ker(\lambda)$ is dense in $X$.

19. Let $\lambda : X \to \mathbb{K}$ be a nonzero linear functional on a normed linear space $X$. Prove that the following are equivalent:

(a) $\lambda$ is unbounded.

(b) There is a sequence $(x_n)$ in $X$ such that $||x_n|| = 1$ and $\lim_n |\lambda(x_n)| = \infty$.

(c) There is a sequence $(x_n)$ in $X$ such that $\lim_n x_n = 0$ and $\lambda(x_n) = 1$.

20. Let $T : \mathcal{C}[0,1] \to \mathcal{C}[0,1]$ be the linear operator $(Tf)(x) = \int_0^x f(t)dt$. Show that $T$ is bounded, and find its norm.

21. Let $\lambda : \mathcal{C}[0,1] \to \mathbb{K}$ be the linear functional $\lambda(f) = \int_0^1 f(t)dt$. Show that $\lambda$ is bounded, and find its norm.

22. Define a linear operator $P_n$ on the space of convergent sequences $c$ by
$$P_n(x) = (x_1, x_2, \ldots, x_n, x_n, x_n, \ldots).$$

(a) Prove that $||P_n|| = 1$ and that $\lim P_n(x) = x$ for all $x \in X$.

(b) Prove that $u_1 = (1, 1, 1\ldots), u_2 = (0, 1, 1, 1, \ldots), u_3 = (0, 0, 1, 1, 1, \ldots), \ldots$, is a Schauder basis for $c$.

23. Let $\{t_n\}$ be a countable dense subset of $[0,1]$, where $t_1 = 0, t_2 = 1$. For $n \in \mathbb{N}$, define an operator $P_n$ on $\mathcal{C}[0,1]$ as follows: $P_n f$ is the continuous, piecewise linear function with nodes $t_1, \ldots, t_n$ such that $(P_n f)(t_i) = f(t_i)$ for $1 \le i \le n$. Show that $||P_n|| = 1$ for all $n \in \mathbb{N}$ and that, for every $f \in \mathcal{C}[0,1]$, $\lim_n ||P_n f - f||_\infty = 0$.

24. This is a continuation of the previous exercise. Define $u_1(x) = 1$, and, for $n \ge 2$, define $u_n$ to be the continuous, piecewise linear function such that $u_n(t_n) = 1$, and $u_n(t_i) = 0$ for $1 \le i \le n - 1$. Prove that $\{u_i\}_{i=1}^n$ is a basis for the range of $P_n$, and hence conclude that $\{u_n\}_{n=1}^\infty$ is a Schuader basis for $\mathcal{C}[0,1]$.

**Definition.** Let $\{u_n\}$ be a Schauder basis for a Banach space $X$. Thus every $x \in X$ has a unique representation $x = \sum_{n=1}^\infty a_n(x)u_n$. Define the **canonical projections** $P_n : X \to Span\{u_1, \ldots, u_n\}$ by $P_n(x) = \sum_{i=1}^n a_i(x)u_i$. Notice that the last three problems include examples of canonical projections. We assume, without proof, the fact that the set $\{P_n\}$ is uniformly bounded, that is, $sup_n ||P_n|| < \infty$.

25. Let $\{u_n\}$ be a Schauder basis for a Banach space $X$, and consider the series representation $x = \sum_{n=1}^\infty a_n u_n$ of an element $x \in X$. Each of the coefficients $a_n$ is clearly a linear functional on $X$. Prove that $a_n \in X^*$. Hint: $a_n(x)u_n = P_n(x) - P_{n-1}(x)$.

# 6.3 Three Fundamental Theorems

In addition to the Hahn-Banach theorem, the three theorems we present in this section are of fundamental importance. All three theorems require completeness; hence they apply only to Banach spaces.

In chapter 4 (see problem 5 on section 4.8), we encountered an example where a family of pointwise bounded functions on a complete metric space is, in fact, uniformly bounded on a ball. Lemma 6.3.1 is similar in spirit, and its proof demonstrates the centrality of Baire's theorem in this section. Because the boundedness of a linear function on a ball implies its boundedness, it must not be surprising that when $X$ is a Banach space, pointwise boundedness implies uniform boundedness. This is the uniform boundedness principle.

The open mapping theorem is a central theorem in functional analysis, and one cannot exaggerate its importance. Lemma 6.3.3 is critical to the proof of the open mapping theorem, and, again, completeness is crucial. The closed graph theorem comes in quite handy in certain applications to prove the boundedness of a linear function. It follows rather easily from the open mapping theorem. Later in the book, you will see many applications of the three theorems, as well as the Hahn-Banach theorem.

In this section, $X$ and $Y$ are normed linear spaces.

A family of bounded linear functions $\{T_\alpha\}_{\alpha \in I}$ from $X$ to $Y$ such that, for each $x \in X$, $sup_{\alpha \in I}\{\|T_\alpha(x)\|\} < \infty$ is said to be **pointwise bounded**. If $sup_{\alpha \in I}\|T_\alpha\| < \infty$, we say that the family $\{T_\alpha\}$ is **uniformly bounded**.

**Example 1.** Let $X$ and $Y$ be normed linear spaces, and suppose that $dim(X) < \infty$. If a family of linear transformations $\{T_\alpha\}_{\alpha \in I}$ from $X$ to $Y$ is pointwise bounded, then $sup_{\alpha \in I}\|T_\alpha\| < \infty$. To see this, fix a basis $\{x_1, ..., x_n\}$ for $X$, and use the 1-norm on $X$. Thus if $x = \sum_{i=1}^{n} a_i x_i$, then $\|x\| = \sum_{i=1}^{n} |a_i|$. Define $M_i = sup_{\alpha \in I}\|T_\alpha(x_i)\|$, and let $M = max_{1 \leq i \leq n} M_i$. For any $\alpha \in I$, we have

$$\|T_\alpha(x)\| = \left\| T_\alpha \left( \sum_{i=1}^{n} a_i x_i \right) \right\| \leq \sum_{i=1}^{n} |a_i| \|T_\alpha(x_i)\| \leq \sum_{i=1}^{n} M_i |a_i| \leq M\|x\|. \blacklozenge$$

The uniform boundedness principle generalizes example 1 to the infinite-dimensional case.

**Lemma 6.3.1.** *Let $X$ be a Banach space, and let $Y$ be a normed linear space. Suppose $\{T_\alpha\}_{\alpha \in I}$ is a family of bounded linear functions $X \to Y$ such that, for each $x \in X$, $sup_{\alpha \in I}\{\|T_\alpha(x)\|\} < \infty$. Then there exists a ball $B(x_0, \delta)$ such that $sup\{\|T_\alpha(x)\| : x \in B(x_0, \delta), \alpha \in I\} < \infty$.*

*Proof.* For each $n \in \mathbb{N}$, let $F_n = \cap_{\alpha \in I}\{x \in X : \|T_\alpha(x)\| \leq n\}$. Note that each $F_n$ is closed and that $X = \cup_{n=1}^{\infty} F_n$. Since $X$ is complete, Baire's theorem forces at least one set $F_N$ to have a nonempty interior. Thus there exists a ball $B = B(x_0, \delta) \subseteq int(F_N) \subseteq F_N$. Now, for every $x \in B$ and every $\alpha \in I$, $\|T_\alpha(x)\| \leq N$. $\blacksquare$

**Theorem 6.3.2 (the uniform boundedness principle).** *Under the assumptions of lemma 6.3.1, the family $\{T_\alpha\}$ is a bounded subset of $\mathcal{L}(X, Y)$, that is, $\sup_\alpha \|T_\alpha\| < \infty$.*

*Proof.* We continue to use the notation of lemma 6.3.1.

Let $\beta = \sup_\alpha \|T_\alpha(x_0)\|$. We claim that $\sup\{\|T_\alpha(x)\| : \|x\| \le 1, \alpha \in I\} \le \frac{2}{\delta}(N + \beta)$. This will show that $\sup\{\|T_\alpha\| : \alpha \in I\} \le \frac{2}{\delta}(N + \beta)$. Let $\alpha \in I$, and $\|x\| \le 1$. Then $x_0 + \frac{\delta x}{2} \in B$, and

$$\|T_\alpha(x)\| = \frac{2}{\delta}\left\|T_\alpha\left(\frac{\delta x}{2}\right)\right\| = \frac{2}{\delta}\left\|T_\alpha\left(x_0 + \frac{\delta x}{2}\right) - T_\alpha(x_0)\right\|$$

$$\le \frac{2}{\delta}\left\{\left\|T_\alpha\left(x_0 + \frac{\delta x}{2}\right)\right\| + \|T_\alpha(x_0)\|\right\} \le \frac{\delta}{2}(N + \beta). \blacksquare$$

The above theorem fails when $X$ is not complete. See problem 1 at the end of this section.

In the lemma below, we use the notation $B_\delta$ to denote an open ball in $X$ of radius $\delta$ centered at 0. We use the same notation, in addition to the prime character, to indicate an open ball in $Y$. Thus $B'_r$ denotes an open ball in $Y$ of radius $r$ and centered at 0.

**Lemma 6.3.3.** *Suppose that $X$ and $Y$ are Banach spaces and that $T$ is a bounded linear mapping from $X$ to $Y$. If, for some $r > 0, B'_r \subseteq \overline{T(B_1)}$, then $B'_r \subseteq T(B_3)$. Equivalently, $B'_{r/3} \subseteq T(B_1)$.*

*Proof.* First observe that $B'_r \subseteq \overline{T(B_1)}$ implies that $B'_{r/2^i} \subseteq \overline{T(B_{1/2^i})}$, for every $i \in \mathbb{N}$. Pick $y \in B'_r$. There exists $x_1 \in B_1$ such that $\|y - T(x_1)\| < r/2$. Now $y - T(x_1) \in B'_{r/2} \subseteq \overline{T(B_{1/2})}$, so there exists $x_2 \in B_{1/2}$ such that $\|y - T(x_1) - T(x_2)\| < r/4$. Continuing in this manner, we can construct a sequence $(x_n)$ in $X$ such that $x_n \in B_{1/2^{n-1}}$ (i.e., $\|x_n\| < 1/2^{n-1}$), and $\|y - T(x_1) - T(x_2) - ... - T(x_n)\| < r/2^n$. Because $\|x_n\| < 1/2^{n-1}$, the sequence $S_n = x_1 + ... + x_n$ is a Cauchy sequence in $X$; hence $x = \lim_n S_n$ exists. Now $T(x) = T(\lim_n S_n) = \lim_n T(S_n) = y$, and $\|x\| = \lim_n \|S_n\| = \lim_{n \to \infty} \|x_1 + ... + x_n\| \le \lim_n \sum_{i=1}^n \|x_i\| \le \sum_{i=1}^\infty 1/2^{i-1} = 2 < 3$. We have shown that every $y \in B'_r$ is the image of an element $x \in B_3$. This proves the result. $\blacksquare$

**Theorem 6.3.4 (the open mapping theorem).** *Suppose that $X$ and $Y$ are Banach spaces and that $T : X \to Y$ is a bounded linear mapping from $X$ onto $Y$. Then there exists a number $\delta > 0$ such that $B'_\delta \subseteq T(B_1)$.*

*Proof.* Since $T$ is onto, $Y = \cup_{n=1}^{\infty} \overline{T(B_n)}$. Baire's theorem implies that $\overline{T(B_N)}$ has a nonempty interior for some positive integer N. Thus there exists an element $y_0 \in Y$ and a positive number $r$ such that $B(y_0, r) \subseteq \overline{T(B_N)}$. We claim that $B'_r \subseteq \overline{T(B_{2N})}$. Let $y \in Y$ be such that $\|y\| < r$, and let $\epsilon > 0$. Both $y_0$ and $y_0 + y$ are in $B(y_0, r)$, so there are vectors $\xi$ and $\eta$ in $B_N$ such that $\|y + y_0 - T(\xi)\| < \epsilon/2$, and $\|y_0 - T(\eta)\| < \epsilon/2$. Let $x = \xi - \eta$. Then $\|x\| \leq \|\xi\| + \|\eta\| < 2N$, and $\|y - T(x)\| = \|y - T(\xi - \eta) - y_0 + y_0\| \leq \|y + y_0 - T(\xi)\| + \|T(\eta) - y_0\| < \epsilon/2 + \epsilon/2 = \epsilon$. This proves that $B'_r \subseteq \overline{T(B_{2N})}$, which establishes our claim and implies that $B'_{r/2N} \subseteq \overline{T(B_1)}$. By lemma 6.3.3, $B'_\delta \subseteq T(B_1)$, where $\delta = \frac{r}{6N}$. ∎

**Corollary 6.3.5.** *Under the assumptions of the open mapping theorem, given $r > 0$, there exists $\delta > 0$ such that $B'_\delta \subseteq T(B_r)$.* ∎

The following theorem justifies the name of the open mapping theorem.

**Theorem 6.3.6.** *Under the assumptions of the open mapping theorem, $T$ is an open mapping.*

*Proof.* Let $U$ be an open subset of $X$. We need to show that $T(U)$ is open in $Y$. Let $y = T(x) \in T(U)$. Since $U$ is open, there exists $r > 0$ such that $B(x, r) \subseteq U$. Corollary 6.3.5 implies that there is a positive number $\delta$ such that $B'_\delta \subseteq T(B_r)$. Now $T(B(x, r)) = T(x + B_r) = T(x) + T(B_r) \supseteq y + B'_\delta = B(y, \delta)$. This concludes the proof. ∎

The continuity of a function does not imply its openness. For example, the function $f(x) = \sin x$ is continuous but not open, since the image of interval $(0, \pi)$ is $(0, 1]$.

**Example 2.** Under the assumptions of the open mapping theorem, there exists a constant $M > 0$ such that, for every $y \in Y$, there is an element $x \in T^{-1}(y)$ such that $\|x\| \leq M\|y\|$. By the open mapping theorem, there exists a positive number $\delta$ such that $B'_\delta \subseteq T(B_1)$. For a nonzero vector $y \in Y$, $\frac{\delta y}{2\|y\|} \in B'_\delta$; hence there is a vector $x_1 \in X$ such that $\|x_1\| \leq 1$ and $T(x_1) = \frac{\delta y}{2\|y\|}$. Define $x = \frac{2x_1\|y\|}{\delta}$. One can see that $T(x) = y$, and $\|x\| \leq \frac{2\|y\|}{\delta}$. The constant we seek is $M = \frac{2}{\delta}$. ◆

The following results represent a small sample of applications of the open mapping theorem.

**Theorem 6.3.7.** *Let $X$ and $Y$ be Banach spaces, and let $T : X \to Y$ be a bounded bijection. Then $T$ is a homeomorphism.*

*Proof.* By theorem 6.3.6, $T$ is an open mapping; hence $T^{-1}$ is continuous, and $T$ is a homeomorphism. ∎

The above theorem also follows from example 2. If $T$ is injective, then $\|T^{-1}\| \leq M$.

The following theorem states that it is not possible for a complete norm to be strictly stronger than another.

**Theorem 6.3.8.** *Let $X$ be a Banach space under each of the norms $\|.\|$ and $\|.\|'$. If there exits a constant $\alpha > 0$ such that $\|x\| \leq \alpha\|x\|'$ for every $x \in X$, then there exists a constant $\beta > 0$ such that $\|x\|' \leq \beta\|x\|$ for every $x \in X$.*

*Proof.* Consider the identity mapping $I_X : (X, \|.\|') \to (X, \|.\|)$. The assumption $\|x\| \leq \alpha\|x\|'$ is equivalent to the boundedness of $I_X$. By theorem 6.3.7, the inverse of $I_X$ is also continuous. Thus $I_X : (X, \|.\|) \to (X, \|.\|')$ is bounded. Thus there exists a positive constant $\beta$ such that, for all $x \in X$, $\|x\|' \leq \beta\|x\|$. ∎

**Definition.** Let $(X, d)$ and $(Y, \rho)$ be metric spaces, and let $T : X \to Y$. The graph of $T$ is the subset $G = \{(x, T(x)) : x \in X\}$ of $X \times Y$. We say that the graph of $T$ is **closed** if $G$ is closed in the product metric on $X \times Y$.

Recall that a sequence $(x_n, y_n) \in X \times Y$ converges to $(x, y)$ if and only if $x_n \to x$ and $y_n \to y$. Thus the graph of $T$ is closed if whenever $x_n \to x$, and $T(x_n) \to y$, then $(x, y) \in G$, or simply $y = T(x)$. It is a simple exercise to verify that if $T$ is continuous, then the graph of $T$ is closed in $X \times Y$. For Banach spaces and linear mappings, the converse is true.

**Theorem 6.3.9 (the closed graph theorem).** *Let $X$ and $Y$ be Banach spaces, and let $T$ be a linear mapping from $X$ to $Y$. If the graph of $T$ is closed, then $T$ is bounded.*

*Proof.* Define a norm on $X$ as follows: $\|x\|' = \|x\| + \|T(x)\|$. We first show that $\|.\|'$ is complete, and hence $(X, \|.\|')$ is a Banach space. If $(x_n)$ is a Cauchy sequence in $\|.\|'$, then, for $\epsilon > 0$, there is a natural number $N$ such that, for $m, n > N$, $\|x_n - x_m\|' < \epsilon$. In particular, both $(x_n)$ and $(T(x_n))$ are Cauchy sequences in $X$ and $Y$, respectively. The completeness of $X$ and $Y$ guarantees that both sequences converge, say, $x = \lim_n x_n$, and $y = \lim_n T(x_n)$. The assumption that the graph of graph of $T$ is closed implies that $y = T(x)$. Now $\|x_n - x\|' = \|x_n - x\| + \|T(x_n) - T(x)\| = \|x_n - x\| + \|T(x_n) - y\| \to 0$ as $n \to \infty$. This demonstrates the completeness of $\|.\|'$. Now $\|x\| \leq \|x\| + \|T(x)\| = \|x\|'$. By theorem 6.3.8, the two norms $\|.\|$ and $\|.\|'$ are equivalent; thus the boundedness of $T$ in one norm is equivalent to its *boundedness in the other. But the boundedness of T in the $\|.\|'$ norm is immediate from the inequality $\|T(x)\| \leq \|T(x)\| + \|x\| = \|x\|'$.* ∎

The following examples show that both the linearity of the $T$ and the completeness of the spaces are needed for the closed graph theorem to hold.

**Example 3.** The function $f : \mathbb{R} \to \mathbb{R}$, defined below, is discontinuous but its graph is closed:

$$f(x) = \begin{cases} \frac{1}{x} & \text{if } x \neq 0, \\ 0 & \text{if } x = 0. \end{cases} \blacklozenge$$

**Example 4.** Let $X = \mathcal{C}[0,1]$ be equipped with the 1-norm, and let $Y = \mathcal{C}[0,1]$ be equipped with the uniform norm. The identity function $I : X \to Y$ is discontinuous by example 7 on section 3.6. However, the graph of $I$ is closed. Suppose $\lim_n \|f_n - f\|_1 = 0$ and $\lim_n \|I(f_n) - g\|_\infty = \lim_n \|f_n - g\|_\infty = 0$. Since convergence in the uniform norm implies convergence in the 1-norm, $\lim_n \|f_n - g\|_1 = 0$. Now the uniqueness of limits forces $f = g$. ◆

## Exercises

1. Let $\lambda_n : \mathbb{K}(\mathbb{N}) \to \mathbb{K}$ be the functional defined by $\lambda_n(x) = nx_n$. Prove that the set $\{\lambda_n\}$ is pointwise bounded but not uniformly bounded. Here $x = (x_n)$, and $\mathbb{K}(\mathbb{N})$ is given the supremum norm.
2. **The Banach-Steinhaus theorem.** Let $X$ and $Y$ be Banach spaces, and let $(T_n)$ be a sequence of bounded linear mappings from $X$ to $Y$ such that, for every $x \in X$, $T(x) = \lim_n T_n(x)$ exists. Prove that $T$ is bounded and that $\|T\| \leq \liminf_n \|T_n\|$. Is it necessarily true that $\lim_n T_n = T$ in $\mathcal{L}(X, Y)$?
3. Let $(y_n)$ be a sequence such that $\sum_{n=1}^\infty x_n y_n < \infty$ for all sequences $(y_n) \in l^q$. Prove that $(x_n) \in l^p$. Here $p$ and $q$ are conjugate Hölder exponents with $p > 1$.
4. Let $(y_n)$ be a sequence such that $\sum_{n=1}^\infty x_n y_n < \infty$ for all sequences $(y_n)$ that converge to 0. Prove that $\sum_{n=1}^\infty |y_n| < \infty$.
5. Let $X$ be a Banach space, and suppose that the sequence $\lambda_n \in X^*$ is pointwise bounded. Prove that $\lambda_n$ is equicontinuous.
6. Let $M$ and $N$ be closed subspaces of a Banach space $(X, \|.\|)$ such that $X = M \oplus N$. Thus every $x \in X$ can be written uniquely as $x = y + z$, where $y \in M, z \in N$. Define a norm on $X$ by $\|x\|' = \|y\| + \|z\|$. Prove that $\|.\|'$ is equivalent to $\|.\|$.

## 6.4 The Hahn-Banach Theorem

The importance of the Hahn-Banach theorem cannot be overstated. The results following theorem 6.4.4 represent only a sample of the wide range of applications of the Hahn-Banach theorem. Unlike the three major theorems of the previous section, the Hahn-Banach theorem does not require completeness.

The Hahn-Banach theorem has many guises, and one of them is an extension theorem. The following example shows that, from the purely algebraic perspective, extending a linear functional on a subspace $M$ of a vector space $X$ is a trivial task. Compare the following example to theorem 6.4.4.

**Example 1.** Let $M$ be a subspace of a vector space $X$, and let $\lambda$ be a linear functional on $M$. Then $\lambda$ can be extended to a linear functional on $X$.

Let $S_1$ be a basis for $M$, and choose a subset $S_2$ of $X$ such that $S_1 \cup S_2$ is a basis for $X$. Define a function $\Lambda : S \to \mathbb{C}$ as follows:

$$\Lambda(x) = \begin{cases} \lambda(x) & \text{if } x \in S_1, \\ 0 & \text{if } x \in S_2. \end{cases}$$

Extend the function $\Lambda$ by linearity to a functional $\Lambda$ on $X$. The restriction of $\Lambda$ to $M$ is clearly $\lambda$. ◆

One of the corollaries of the Hahn-Banach theorem (theorem 6.4.5) is a powerful separation theorem. Earlier in the book, we saw examples of separation theorems by linear functionals, albeit in a slightly different context. See example 10 in section 4.7. The following example shows, once again, that, from the algebraic point of view, the problem of separating a subspace from a point outside it is a simple one. Compare the result below to theorem 6.4.5.

**Example 2.** Let $M$ be a proper subspace of a vector space $X$, and let $x_0 \in X - M$. There exists a linear functional $\lambda$ on $X$ such that $\lambda(M) = 0$ and $\lambda(x_0) \neq 0$.

Choose a basis $S_1$ for $M$. Since $S_1 \cup \{x_0\}$ is independent, there is a subset $S_2$ of $X$ such that $S_1 \cup \{x_0\} \cup S_2$ is a basis for $X$. Define a function $\lambda : S_1 \cup \{x_0\} \cup S_2 \to \mathbb{C}$ by

$$\lambda(x) = \begin{cases} 1 & \text{if } x = x_0, \\ 0 & \text{if } x \in S_2 \cup S_2. \end{cases}$$

Extend $\lambda$ by linearity to a linear functional $\lambda$ on $X$. Clearly, $\lambda(M) = 0$. ◆

**Definition.** Let $X$ be a complex normed linear space. A real functional $u$ on $X$ is said to be a bounded real-valued functional if

(a) $u(x+y) = u(x) + u(y)$ for all $x, y \in X$,
(b) $u(ax) = au(x)$ for all $x \in X$ and $a \in \mathbb{R}$, and
(c) $\|u\| = \sup_{x \neq 0} \frac{|u(x)|}{\|x\|} < \infty.$

**Lemma 6.4.1.** *Let $\lambda$ be a bounded complex functional on a normed linear space $X$, and let $u$ be the real part of $\lambda$. Then*

(a) $\lambda(x) = u(x) - iu(ix)$, *and*
(b) $u$ *is a bounded real functional on $X$ and $\|u\| = \|\lambda\|$.*
    *Conversely, if $u$ is a bounded real functional on $X$, then $\lambda(x) = u(x) - iu(ix)$ is a bounded complex functional on $X$.*

*Proof.* Write $\lambda(x) = u(x) + iv(x)$. On the one hand, $\lambda(ix) = i\lambda(x) = iu(x) - v(x)$. On the other hand, $\lambda(ix) = u(ix) + iv(ix)$. Equating the right-hand sides of the above identities yields $v(x) = -u(ix)$, and hence (a). Since, for any complex number $z$, $|Re(z)| \leq |z|$, $|u(x)| \leq |\lambda(x)|$; hence $\|u\| \leq \|\lambda\|$. If $\lambda(x) \neq 0$, let $\alpha = \frac{|\lambda(x)|}{\lambda(x)}$. Then $|\lambda(x)| = \alpha\lambda(x) = \lambda(\alpha x) = u(\alpha x) \leq \|u\|\|\alpha x\| = |\alpha|\|u\|\|x\| = \|u\|\|x\|$. Thus $\|\lambda\| \leq \|u\|$, and this establishes (b).

Conversely, if $u$ is a bounded real functional on $X$ and $\lambda(x) = u(x) - iu(ix)$, then the additivity of $\lambda$ is straightforward. Now $\lambda(ix) = u(ix) - iu(-x) = u(ix) + iu(x) = i[u(x) - iu(ix)] = i\lambda(x)$. Hence $\lambda((a+ib)x) = \lambda(ax) + \lambda(ibx) = a\lambda(x) + i\lambda(bx) = a\lambda(x) + ib\lambda(x) = (a+ib)\lambda(x)$. Thus $\lambda$ is complex linear. The boundedness of $\lambda$ follows from the proof of part (b). ∎

**Lemma 6.4.2.** *Let $M$ be a a subspace of a real normed linear space $X$, and let $x_0 \in X - M$. If $u$ is a bounded real functional on $M$, then $u$ has an extension $U$ to a bounded real functional on $N = M \oplus Span\{x_0\}$ such that $\|U\| = \|u\|$.*

*Proof. Without loss of generality, assume that $\|u\| = 1$. Every element of $N$ can be written uniquely as $x + \alpha x_0$, where $x \in M$ and $\alpha \in \mathbb{R}$. Define $U : N \to \mathbb{R}$ by $U(x + \alpha x_0) = u(x) + \alpha b$, where $b$ is a constant to be determined later in the proof. The linearity of $U$ is obvious, and since $U$ extends $u$, $\|u\| \leq \|U\|$. It remains to show that $\|U\| \leq 1$. It suffices to show that a constant $b$ exists such that $|u(x) - b| \leq \|x - x_0\|$ for every $x \in M$, because then $|U(x)| = |u(x) + \alpha b| = |-\alpha||\frac{u(x)}{-\alpha} - b| \leq |-\alpha|\|\frac{x}{-\alpha} - x_0\| = \|x + \alpha x_0\|$; hence $\|U\| \leq 1$. We now show that a constant $b$ exists such that $|u(x) - b| \leq \|x - x_0\|$ for every $x \in M$. For $x, y \in X, u(x) - u(y) = u(x - y) \leq \|u\|\|x - y\| = \|x - y\| \leq \|x - x_0\| + \|y - x_0\|$. Therefore $u(x) - \|x - x_0\| \leq u(y) + \|y - x_0\|$, and $b_1 = \sup_{x \in M}\{u(x) - \|x -$*

$x_0\|\} \leq \inf_{y \in M}\{u(y) + \|y - x_0\|\} = b_2$. Any constant $b$ such that $b_1 \leq b \leq b_2$ satisfies $u(x) - \|x - x_0\| \leq b \leq u(x) + \|x - x_0\|$ for every $x \in M$. For such a constant $b$, $|u(x) - b| \leq \|x - x_0\|$. The existence of $b$ concludes the proof. ∎

**Lemma 6.4.3.** *Let $M$ be a subspace of a real normed linear space $X$, and let $u$ be a bounded real functional on $M$. Then $u$ has a bounded real extension, $U$, on $X$ such that $\|U\| = \|u\|$.*

*Proof.* Consider the family $\mathfrak{B} = \{(M_\alpha, U_\alpha) : \alpha \in I\}$ of extensions of $u$ that satisfy the conclusion of the theorem. Thus, for each $\alpha \in I$, $M$ is a subspace of $M_\alpha$, $U_\alpha$ extends $u$, and $\|U_\alpha\| = \|u\|$. Order $\mathfrak{B}$ by set and function inclusion: $(M_\alpha, U_\alpha) \subseteq (M_\beta, U_\beta)$ means, by definition, that $M_\alpha \subseteq M_\beta$, and $U_\beta$ extends $U_\alpha$. If $\{\mathfrak{C} = (M_\alpha, U_\alpha) : \alpha \in J\}$ is a chain in $\mathfrak{B}$, let $(N, U) = (\cup_{\alpha \in J} M_\alpha, \cup_{\alpha \in J} U_\alpha)$. It is easy to verify that $N$ is a subspace of $X$, that $U$ is well defined and, linear and that $\|U\| = \|u\|$. All the properties follow from the fact that $\mathfrak{C}$ is a chain. Thus $(N, U)$ is an upper bound of $\mathfrak{C}$. By Zorn's lemma, $\mathfrak{B}$ has a maximal member, $(M^*, U^*)$. It must be the case that $M^* = X$ because otherwise we can pick an element $x_0 \in X - M$ and use lemma 6.4.2 to extend $U^*$ to $M^* \oplus Span\{x_0\}$, which would contradict the maximality of $(M^*, U^*)$. ∎

**Theorem 6.4.4 (the Hahn-Banach theorem).** *Let $\lambda$ be a bounded linear functional on a subspace $M$ of a complex normed linear space $X$. Then $\lambda$ has an extension to a bounded linear functional, $\Lambda$, on $X$ such that $\|\Lambda\| = \|\lambda\|$.*

*Proof.* Consider $X$ as a real normed linear space simply by limiting the scalar field to $\mathbb{R}$, and let $u$ be the real part of $\lambda$. By lemma 6.4.1, $u$ is a bounded real functional on $M$, and $\|u\| = \|\lambda\|$. By lemma 6.4.3, $u$ has an extension, $U$, to $X$ such that $\|U\| = \|u\|$. Define $\Lambda : X \to \mathbb{C}$ by $\Lambda(x) = U(x) - iU(ix)$. By lemma 6.4.1, $\Lambda$ is a bounded linear functional on $X$, and $\|\Lambda\| = \|U\| = \|u\| = \|\lambda\|$. ∎

We now look at some applications of the Hahn-Banach theorem. The results below are important in their own right.

**Theorem 6.4.5.** *Let $M$ be a subspace of a normed linear space $X$, and let $x_0 \in X$. Then $x_0 \in \overline{M}$ if and only if there does not exist a bounded linear functional $\lambda$ on $X$ such that $\lambda(M) = 0$ and $\lambda(x_0) \neq 0$.*

*Proof.* We show that if $x_0 \notin \overline{M}$, then there exists a functional $\Lambda \in X^*$ such that $\Lambda(M) = 0$, and $\Lambda(x_0) = 1$. Since $x_0 \notin \overline{M}$, there exists a number $\delta > 0$ such that $\|x - x_0\| \geq \delta$ for every $x \in M$. Let $N = M \oplus Span\{x_0\}$, and define a function $\lambda : N \to \mathbb{C}$ by $\lambda(x + \alpha x_0) = \alpha$; $\lambda$ is clearly linear on $N$, $\lambda(M) = 0$, and

$\lambda(x_0) = 1$. Now, for any $x \in M, \alpha \neq 0, \|\frac{x}{-\alpha} - x_0\| \geq \delta$. Thus $\delta|\lambda(x + \alpha x_0)| = \delta|\alpha| \leq |\alpha|\|\frac{x}{-\alpha} - x_0\| = \|x + \alpha x_0\|$. Thus $\lambda$ is bounded on $N$ ($\|\lambda\| \leq 1/\delta$). Extend $\lambda$ to a bounded linear functional $\Lambda$ on $X$. The functional $\Lambda$ has the desired properties. Conversely, if $x_0 \in \overline{M}$ and $\lambda \in X^*$ is such that $\lambda(M) = 0$, then there exists a sequence of vectors $(x_n)$ in $M$ such that $\lim_n x_n = x$. Now $\lambda(x_0) = \lambda(\lim_n x_n) = \lim_n \lambda(x_n) = 0$. ∎

The following result can be used to prove certain approximation theorems. It follows immediately from the previous theorem.

**Corollary 6.4.6.** *Let $M$ be a subspace of a normed linear space $X$. If, for $\lambda \in X^*, \lambda(M) = 0$ implies that $\lambda = 0$, then $M$ is dense in $X$.* ∎

**Example 3.** Let $A$ be a dense subset of $[-\pi, \pi]$. For a fixed $t \in A$, the sequence $\xi_t = (\frac{e^{int}}{n})_{n=1}^{\infty}$ is in $l^2$. We claim that the subspace $M = Span\{\xi_t : t \in A\}$ is dense in $l^2$. We use the above corollary and show that, for a bounded linear functional $\lambda$ on $l^2$, $\lambda(M) = 0$ is possible only if $\lambda = 0$. By theorem 6.2.4, there exists a sequence $(y_n) \in l^2$ such that, for every sequence $x = (x_n) \in l^2, \lambda(x) = \sum_{n=1}^{\infty} x_n y_n$. For every $t \in [-\pi, \pi], \sum_{n=1}^{\infty} |\frac{y_n}{n} e^{int}| = \sum_{n=1}^{\infty} \frac{|y_n|}{n} \leq \|y\|_2 \{\sum_{n=1}^{\infty} \frac{1}{n^2}\}^{1/2} < \infty$, and the series $\sum_{n=1}^{\infty} \frac{y_n}{n} e^{int}$ converges absolutely and uniformly on $[-\pi, \pi]$ to a continuous function $F(t)$. By assumption, $F$ vanishes on a dense subset of $[-\pi, \pi]$, so F is identically equal to the zero function. Theorem 4.10.5 implies that $\frac{y_n}{n} = 0$ for all $n \in \mathbb{N}$. Thus $y_n = 0$, and $\lambda = 0$. ◆

Another corollary of the Hahn-Banach theorem is the following separation theorem.

**Corollary 6.4.7.** *Let $X$ be a normed linear space, and let $x_0 \in X, x_0 \neq 0$. Then there exists a bounded linear functional $\lambda$ on $X$ such that $\lambda(x_0) = \|x_0\|$, and $\|\lambda\| = 1$. In particular, if $y \in X$ and $\lambda(y) = 0$ for all $\lambda \in X^*$, then $y = 0$.*

*Proof.* Let $M = Span\{x_0\}$, and define a functional $\lambda : M \to \mathbb{C}$ by $\lambda(\alpha x_0) = \alpha\|x_0\|$. Clearly, $\lambda(x_0) = \|x_0\|$, and $\|\lambda\| = 1$. By the Hahn-Banach theorem, $\lambda$ has a norm-preserving extension to $X$. ∎

The following important construction relies heavily on the above corollary. As we established in section 6.2, the dual $X^*$ of a normed linear space $X$ is a Banach space with the norm $\|\lambda\| = sup_{x \neq 0} \frac{|\lambda(x)|}{\|x\|}$; $X^*$, in turn, had a dual, $X^{**}$, which is a Banach space known as the **second dual** of $X$. Now $X$ can be linearly and isometrically embedded into $X^{**}$ as follows. For an element $x \in X$, define an element $\hat{x}$ of $X^{**}$ by $\hat{x}(\lambda) = \lambda(x)$. The linearity of $\hat{x}$, as well as that of the mapping $X \to X^{**}$ defined by $x \mapsto \hat{x}$, are obvious. We now show that $\|\hat{x}\| = \|x\|$. Since $|\hat{x}(\lambda)| = |\lambda(x)| \leq \|\lambda\|\|x\|$, $\frac{|\hat{x}(\lambda)|}{\|\lambda\|} \leq \|x\|$. Hence $\|\hat{x}\| \leq \|x\|$. We now show that $\|\hat{x}\| = \|x\|$. By corollary 6.4.7, there exists $\lambda \in X^*$ such that $\|\lambda\| = 1$, and $\lambda(x) = \|x\|$. Now $|\hat{x}(\lambda)| = |\lambda(x)| = \|x\|$. Therefore $\|\hat{x}\| \geq \|x\|$ and $\|\hat{x}\| = \|x\|$. We have proved the following result.

**Theorem 6.4.8.** *Let $X$ be a normed linear space, and let $\varphi : X \to X^{**}$ be the function $\varphi(x) = \hat{x}$. Then $\varphi$ is a linear isometry.* ∎

The function $\varphi$ in the above theorem is known as the **natural embedding** of $X$ into $X^{**}$. We use the notation $\hat{X}$ to denote the range of $\varphi$. Thus $\hat{X} = \{\hat{x} : x \in X\}$.

The above theorem provides the neatest construction of the **completion of a normed linear space**.

**Theorem 6.4.9.** *Let $X$ be a normed linear space. Then $X$ can be linearly and isometrically embedded as a dense subspace of a Banach space. Thus every normed linear space has a completion.*

*Proof.* We know that $X^{**}$ is a Banach space. Let $\hat{X}$ be the image of $X$ under the natural embedding $\varphi$ in theorem 6.4.8. The desired completion of $X$ is the closure of $\hat{X}$ in $X^{**}$. ∎

**Definition.** A Banach space $X$ is **reflexive** if $\hat{X} = X^{**}$. Thus $X$ is reflexive if every member of $X^{**}$ is of the form $\hat{x}$ for some $x \in X$.

**Example 4.** The $l^p$ spaces are reflexive for $1 < p < \infty$. This follows directly from theorem 6.2.4. ♦

The result below is important in its own right, but it also helps us decide whether certain spaces are reflexive.

**Example 5.** Let $X$ be a normed linear space. If $X^*$ is separable, then $X$ is separable.

Let $\{\lambda_n\}$ be a countable dense subset of $X^*$. Since $\|\lambda_n\| = sup_{\|x\|=1}|\lambda_n(x)|$, there exist unit vectors $x_n \in X$ such that $|\lambda_n(x_n)| \geq \|\lambda_n\|/2$. Let $M = Span\{x_1, x_2, ...\}$. We employ theorem 6.4.6. Suppose that $\lambda \in X^*$ is such that $\lambda(M) = 0$. Let $\epsilon > 0$, and pick a positive integer $n$ such that $\|\lambda_n - \lambda\| < \epsilon$. By the definition of $x_n$, and the fact that $\lambda(x_n) = 0$, we have $\|\lambda_n\|/2 \leq |\lambda_n(x_n)| = |\lambda_n(x_n) - \lambda(x_n)| = |(\lambda_n - \lambda)(x_n)| \leq \|\lambda_n - \lambda\| < \epsilon$. Therefore $\|\lambda\| \leq \|\lambda - \lambda_n\| + \|\lambda_n\| < \epsilon + 2\epsilon = 3\epsilon$. This means that $\lambda = 0$, and, by corollary 6.4.6, $M$ is dense in $X$. Now the countable set $\{\sum_{i=1}^{n} a_i x_i : n \in \mathbb{N}, a_i \in \mathbb{Q} + i\mathbb{Q}\}$ is dense in $X$. ♦

**Definition.** A closed subspace $M$ of a Banach space $X$ is said to be **complemented** if there exists a closed subspace $N$ of $X$ such that $X = M \oplus N$.

The definition of the algebraic complement of an arbitrary subspace of a vector space was introduced in section 3.4. The current definition requires both $M$ and $N$ to be closed subspaces of $X$. The direct sum of two closed subspaces of a Banach space is sometimes referred to as the topological direct sum of the two subspaces.

The very definition suggests that not every closed subspace of a Banach space has a closed complement. However, the following examples identify two important special cases where closed complements are guaranteed.

**Example 6.** If $M$ is a finite-dimensional subspace of a Banach space $X$, then $M$ is complemented.

Let $\{x_1, \ldots, x_n\}$ be a basis for $M$. For $x \in M, x = \sum_{i=1}^{n} a_i(x)x_i$ for a unique set of coefficients $a_1(x), \ldots, a_n(x)$. Each $a_i$ is a continuous linear functional on $M$. By the Hahn-Banach theorem, each $a_i$ has an extension to a functional $\lambda_i \in X^*$. Define an operator $P : X \to X$ by $P(x) = \sum_{i=1}^{n} \lambda_i(x)x_i$. It is easy to see that $P \in \mathcal{L}(X)$, that $P(x) = x$ for every $x \in M$, and that $P^2 = P$. Let $N = Ker(P) = \cap_{i=1}^{n} Ker(\lambda_i)$. Clearly, $N$ is a closed subspace of $X$. For $x \in X, x = x - P(x) + P(x)$. By the above, $P(x) \in M$, and $x - P(x) \in N$, since $P(x - P(x)) = P(x) - P^2(x) = P(x) - P(x) = 0$. This shows that $M + N = X$. If $x \in M \cap N$, then $\lambda_i(x) = 0$ for every $1 \leq i \leq n$; hence $x = \sum_{i=1}^{n} a_i(x)x_i = \sum_{i=1}^{n} \lambda_i(x)x_i = 0$. We have shown that $M \oplus N = X$. ♦

**Example 7.** If $N$ is a closed, finite co-dimensional subspace of $X$, then $N$ is complemented.

Recall that the co-dimension of $N$ is the dimension of the quotient space $X/N$. Pick vectors $x_1, \ldots, x_n$ such that $\{\tilde{x}_1, \ldots, \tilde{x}_n\}$ is a basis for $X/N$, where $\tilde{x}_i = x_i + N$, and let $M = Span\{x_1, \ldots, x_n\}$. We claim that $M \oplus N = X$. For $x \in X$, $x + N = \sum_{i=1}^{n} a_i(x_i + N) = (\sum_{i=1}^{n} a_i x_i) + N$. Therefore $y = x - \sum_{i=1}^{n} a_i x_i \in N$, and $x = \sum_{i=1}^{n} a_i x_i + y \in M + N$. If $x \in M \cap N$, then $x = \sum_{i=1}^{n} a_i x_i \in N$. Thus $\sum_{i=1}^{n} a_i \tilde{x}_i = 0$; hence $a_i = 0$ for every $1 \leq i \leq n$ by the independence of $\{\tilde{x}_1, \ldots, \tilde{x}_n\}$, and $x = 0$. ♦

## Exercises

1. Let $M$ be a closed maximal subspace of a normed linear space $X$. Prove that there exists a functional $\lambda \in X^*$ such that $Ker(\lambda) = M$.

2. Let $A$ be a subset of a normed linear space $X$. Prove that $A$ is bounded in $X$ if and only if, for every $\lambda \in X^*$, $\lambda(A)$ is bounded in $\mathbb{C}$.
3. Prove that if $\{x_1, \ldots, x_n\}$ is an independent subset of a normed linear space $X$, and $\{\alpha_1, \ldots, \alpha_n\}$ is an arbitrary set of complex numbers, then there exists $\lambda \in X^*$ such that $\lambda(x_i) = \alpha_i$ for all $1 \leq i \leq n$.
4. Let $M$ be a closed subspace of a normed linear space $X$, and let $x_0 \notin M$. Prove there exists $\lambda \in X^*$ such that $\lambda(M) = 0, \|\lambda\| \leq 1$, and $|\lambda(x_0)| = dist(x_0, M)$.
5. Prove that, for an element $x$ of a normed linear space $X$, $\|x\| = sup\{|\lambda(x)| : \lambda \in X^*, \|\lambda\| = 1\}$.

**Definition.** A sequence $(x_n)$ in a normed linear space $X$ is said to **converge weakly** to an element $x \in X$ if, for every $\lambda \in X^*$, $\lim_n \lambda(x_n) = \lambda(x)$. We use the notation $x_n \to^w x$ to indicate the weak convergence of $(x_n)$ to $x$.

6. Prove that if $(x_n)$ is weakly convergent, then $(\|x_n\|)$ is bounded.
7. Prove that if $x_n \to^w x$ and $y_n \to^w y$, then for any scalars $a$ and $b$, $ax_n + by_n \to^w ax + by$.
8. Prove that the weak limit of a sequence, if it exists, is unique.
9. Prove that $l^1$ is not reflexive.

**Definition.** A bounded operator $P$ on a Banach space $X$ is called a **bounded projection** if $P^2 = P$. Equivalently, if $Px = x$ for every $x \in M = \mathfrak{R}(P)$. See problems 13 and 14 on section 3.4 for the general properties of the projection of a vector space onto a subspace.

10. Let $M$ be a closed subspace of a Banach space $X$. Prove that $M$ is complemented if and only if there exists a bounded projection $P$ on $X$ such that $M = \mathfrak{R}(P)$. Hint: Suppose $X = M \oplus N$, where $M$ and $N$ are closed subspaces of $X$, and let $P : X \to X$ be the projection of $X$ onto $M$. Use the closed graph theorem to prove the boundedness of $P$.
11. Suppose that $M$ and $N$ are closed, complementary subspaces of a Banach space $X$, and let $T_1 : M \to X$ and $T_2 : N \to X$ be bounded linear mappings. Define $T : X \to X$ by $T(x) = T_1(y) + T_2(z)$, where $x = y + z, y \in M, z \in N$. Prove that $T$ is bounded.

## 6.5 The Spectrum of an Operator

The spectrum of a square matrix $A$ is simply its set of eigenvalues, and the eigenvalues of $A$ are easy to characterize. They are exactly the complex numbers $\lambda$

for which the matrix $A - \lambda I$ is not invertible. We recall the simple fact that $A - \lambda I$ is not invertible if and only if the linear operator $T$ it generates on $\mathbb{K}^n$ is not one-to-one, and this is the case if and only if $T$ in not onto.

The definition of the spectrum of an operator $T$ on an infinite-dimensional space is exactly the same as it is for a matrix. The stark distinction here is that not every point in the spectrum of an operator on an infinite-dimensional space is an eigenvalue. This is because such an operator may be one-to-one but not onto or conversely. See example 1. Thus the spectrum consists of two main parts: the complex numbers $\lambda$ for which $T - \lambda I$ is not one-to-one (the eigenvalues) and those for which $T - \lambda I$ is one to one but not onto. The spectrum of an operator $T$ often carries valuable information about $T$, and, in some cases, the eigenvalues of an operator and the corresponding eigenvectors completely define the operator.

**Definition.** A **Banach algebra** is a Banach space $X$ that is also an algebra with a multiplicative identity $I$ such that the norm satisfies the following additional assumptions:

(a) $\|I\| = 1$, and
(b) $\|ST\| \leq \|S\|\|T\|$ for all $S$ and $T$ in $X$.

We know that the set $\mathcal{L}(X)$ of bounded linear operators on a Banach space $X$ is a Banach space. In fact, $\mathcal{L}(X)$ is a Banach algebra with the composition of operators as the multiplication operation. The composition of two operators $S$ and $T$ is usually denoted by $ST$ rather than $SoT$. Property (a) is obvious, and property (b) follows from the inequalities $\|(ST)(x)\| = \|S(T(x))\| \leq \|S\|\|T(x)\| \leq \|S\|\|T\|\|x\|$.

For the convenience of the reader, we list below the properties that make $\mathcal{L}(X)$ a Banach algebra: for operators $T, S, U \in \mathcal{L}(X)$ and all $a, b \in \mathbb{K}$,

(a) $(ST)U = S(TU)$,
(b) $(ab)T = a(bT)$,
(c) $(T + S)U = TU + SU$ and $U(T + S) = UT + US$,
(d) $\|I\| = 1$, and
(e) $\|ST\| \leq \|S\|\|T\|$.

The algebra $\mathcal{L}(X)$ is called the **operator algebra** on $X$.

**Definition.** An operator $T \in \mathcal{L}(X)$ is called **invertible** if there exists an operator $S \in \mathcal{L}(X)$ such that $ST = TS = I$.

If $T$ is a bounded linear bijection of a Banach space $X$, then its inverse is bounded by theorem 6.3.7. Thus a bounded operator $T$ fails to be invertible if

(a) $T$ is not one-to-one, that is, $Ker(T) \neq \{0\}$, or

(b) $T$ is not onto, that is, $\mathfrak{R}(T) \neq X$.

We point out here an important distinction between operators on finite vs. infinite-dimensional spaces. Every linear operator on a finite-dimensional space is bounded, and such an operator is one-to-one if and only if it is onto. This is not the case in infinite dimensions, as the following example illustrates.

**Example 1.** The **right shift operator** $R$ and the **left shift operator** $L$ on $l^2$ are, respectively,

$$R(x_1, x_2, ...) = (0, x_1, x_2, ...),$$
$$L(x_1, x_2, ...) = (x_2, x_3, ...).$$

It is clear that $R$ is one-to-one but not onto, while $L$ is onto but not one-to-one. ♦

**Definition.** The **spectrum**, $\sigma(T)$, of an operator $T \in \mathcal{L}(X)$ is the set of all complex numbers $\lambda$ such that $T - \lambda I$ is not invertible. It follows that there are two types of points in the spectrum:

(a) Complex numbers $\lambda$ such that $Ker(T - \lambda I) \neq \{0\}$: Such a number $\lambda$ is called an **eigenvalue** of $T$. Specifically, $\lambda$ is an eigenvalue of $T$ if there exists a nonzero vector $x$ such that $Tx = \lambda x$. In this case, we say that $x$ is an **eigenvector** of $T$ corresponding (or belonging) to the eigenvalue $\lambda$. The set of eigenvalues of $T$ is known as the **point spectrum** of $T$. The set $Ker(T - \lambda I)$ is called the **eigenspace** of $T$ corresponding to the eigenvalue $\lambda$.

(b) Complex numbers $\lambda$ such that $T - \lambda I$ is one-to-one but not onto, that is, $\mathfrak{R}(T - \lambda I) \neq X$. We will not dwell on this part of the spectrum, since the eigenvalues are the only important part of the spectrum for our purposes.

The complement of the spectrum of $T$ in the complex plane is called the **resolvent set** of $T$ and is denoted $\rho(T)$. Thus $\lambda \in \rho(T)$ if and only if $(T - \lambda I)^{-1}$ exists. If $\lambda \in \rho(T)$, we use the notation $T_\lambda$ to denote $(T - \lambda I)^{-1}$.

**Example 2.** Define an operator $T$ on $\mathcal{C}[0, 1]$ as follows: for $f \in \mathcal{C}[0, 1]$, $(Tf)(x) = xf(x)$. The reader can easily verify that $T$ has no eigenvalues. Thus the spectrum consists only of complex numbers $\lambda$ for which $T - \lambda I$ is not onto. For $\lambda \in \mathbb{C}$ and $g \in \mathcal{C}[0, 1]$, if there exists a function $f \in \mathcal{C}[0, 1]$ such that $(T - \lambda I)f = g$, then $f(x) = \frac{g(x)}{x - \lambda}$. Therefore the spectrum is the interval $[0, 1]$.

**Example 3.** Every complex number $\lambda$ in the open unit disk is an eigenvalue of the left shift operator on $l^2$.

If $0 \neq \lambda \in \mathbb{C}$ and $|\lambda| < 1$, then the vector $x_\lambda = (\lambda, \lambda^2, \lambda^3, ...)$ is clearly in $l^2$ and $L(x_\lambda) = \lambda x_\lambda$. Also, $\lambda = 0$ is an eigenvalue of $L$ because $L(e_1) = 0$, where $e_1 = (1, 0, 0, ...)$. ♦

**Lemma 6.5.1.** If $T \in \mathcal{L}(X)$, and $\|T\| < 1$, then $(I - T)^{-1}$ exists, $(I - T)^{-1} = \sum_{n=0}^{\infty} T^n$, and $\|(I - T)^{-1}\| \leq \frac{1}{1 - \|T\|}$.

*Proof.* First observe that $\sum_{n=0}^{\infty} \|T^n\| \leq \sum_{n=0}^{\infty} \|T\|^n = \frac{1}{1 - \|T\|}$. Thus the series $\sum_{n=0}^{\infty} T^n$ converges to an operator $S \in \mathcal{L}(X)$. Now $(I - T) \sum_{j=0}^{n} T^j = I - T^{n+1}$. Taking the limit as $n \to \infty$, $(I - T)S = I$. Similarly, $S(I - T) = I$; hence $(I - T)^{-1} = S$. ∎

**Theorem 6.5.2.** Let $T \in \mathcal{L}(X)$. If $\lambda \in \mathbb{C}$ and $|\lambda| > \|T\|$, then $\lambda \in \rho(T)$.

*Proof.* Since $\|T\| < |\lambda|, \|T/\lambda\| < 1$. By lemma 6.5.1, $(I - T/\lambda)^{-1}$ exists. Thus $T - \lambda I$ is invertible since $(T - \lambda I)^{-1} = -\lambda^{-1}(I - T/\lambda)^{-1}$. Notice that, in this case, $T_\lambda = (T - \lambda I)^{-1} = \frac{-1}{\lambda} \sum_{n=0}^{\infty} \frac{T^n}{\lambda^n}$. ∎

**Corollary 6.5.3.** The spectrum $\sigma(T)$ of an operator $T \in \mathcal{L}(X)$ is bounded.

*Proof.* By theorem 6.5.2, $\sigma(T)$ is contained in the closed disk $\{z \in \mathbb{C} : |z| \leq \|T\|\}$. ∎

**Theorem 6.5.4.** The spectrum $\sigma(T)$ of an operator $T \in \mathcal{L}(X)$ is a closed, hence compact, subset of $\mathbb{C}$.

*Proof.* We show that $\rho(T)$ is an open subset of $\mathbb{C}$. Let $\lambda_0 \in \rho(T)$, and let $\lambda \in \mathbb{C}$. Recall the notation $T_\lambda = (T - \lambda I)^{-1}$. Now

$$T - \lambda I = (T - \lambda_0 I) - (\lambda - \lambda_0)I = (T - \lambda_0 I)[I - (\lambda - \lambda_0)T_{\lambda_0}].$$

If $|\lambda - \lambda_0| < 1/\|T_{\lambda_0}\|$, then, by lemma 6.5.1, $I - (\lambda - \lambda_0)T_{\lambda_0}$ is invertible, and hence $T - \lambda I$ is also invertible, being the composition of invertible operators. This shows that $\rho(T)$ contains the disk in the complex plane centered at $\lambda_0$ of radius $1/\|T_{\lambda_0}\|$, and therefore $\rho(T)$ is open in $\mathbb{C}$. ∎

**Definition.** The **spectral radius** of an operator $T$ is the number

$$r(T) = \sup\{|\lambda| : \lambda \in \sigma(T)\}.$$

Thus $r(T)$ is the radius of the smallest closed disk in the complex plane that contains $\sigma(T)$. By theorem 6.5.3, $r(T) \le \|T\|$. It is possible that $r(T) < \|T\|$. See problem 5 on section 7.4.

**Example 4.** Let $L : l^2 \to l^2$ be the left shift operator. It is clear that $\|L(x)\|_2 \le \|x\|_2$, and since $\|L(e_2)\|_2 = \|e_1\|_2 = 1 = \|e_2\|_2$, $\|L\| = 1$. Therefore the spectrum of $R$ is contained in the closed unit disk, $D$. It follows directly from this and example 3 that $\sigma(L) = D$. Thus, $\|L\| = r(L) = 1$. ◆

The last conclusion of the previous example is true for the right shift operator. We derive it without directly computing $\sigma(R)$.

**Example 5.** For the right shift operator $R$, $\|R\| = r(R) = 1$.

Since $R$ is an isometry, $\|R\| = 1$. The result follows if we prove that $\lambda = 1 \in \sigma(R)$. We show below that $R - I$ is not onto.

We formally compute the inverse image, $x = (x_n)$, under $R - I$ of an element $y = (y_n) \in l^2$. If $(R - I)(x) = y$, then $(-x_1, x_1 - x_2, x_2 - x_3, ...) = (y_1, y_2, ...)$. Equating the corresponding terms, we have $-x_1 = y_1, x_1 - x_2 = y_2, ..., x_n - x_{n+1} = y_{n+1}, ...$.

Solving for $x$, we have $x_1 = -y_1, x_2 = -y_1 - y_2, ..., x_n = -y_1 - y_2 - ... - y_n$. Now if $y = (\frac{1}{n}) \in l^2$, then there is no $x \in l^2$ such that $(R - I)(x) = y$ since $x_n = -\sum_{i=1}^{n} \frac{1}{i} \to -\infty$. ◆

Before we show that the spectrum of a bounded linear operator on a Banach space is not empty, we need to establish the following identity: for $\lambda$ and $\mu \in \rho(T)$,

$$T_\lambda - T_\mu = (\lambda - \mu)T_\lambda T_\mu, \tag{1}$$

$$T_\lambda = (T - \lambda I)^{-1} = T_\lambda(T - \mu I)T_\mu = T_\lambda[T - \lambda I + (\lambda - \mu)I]T_\mu$$
$$= [I + (\lambda - \mu)T_\lambda]T_\mu = T_\mu + (\lambda - \mu)T_\lambda T_\mu.$$

We need the following result from complex analysis, which we state without proof.

**Lemma 6.5.5 (Liouville's theorem).** *If $F(z)$ is a bounded differentiable complex function defined on the entire complex plane, then $F$ is constant.* ∎

**Theorem 6.5.6.** *The spectrum of a bounded linear operator $T$ on a Banach space $X$ is nonempty.*

*Proof.* Suppose, contrary to the above statement, that $\sigma(T) = \varnothing$. Thus $\rho(T) = \mathbb{C}$. For an arbitrary but fixed functional $g \in (\mathcal{L}(X))^*$, define a function $F : \mathbb{C} \to \mathbb{C}$ by $F(\lambda) = g(T_\lambda)$. By identity (1), $\frac{F(\lambda) - F(\mu)}{\lambda - \mu} = g(T_\lambda T_\mu)$.

As $\mu \to \lambda$, $g(T_\lambda T_\mu) \to g(T_\lambda^2)$. Therefore $F'(\lambda) = \lim_{\mu \to \lambda} \frac{F(\lambda) - F(\mu)}{\lambda - \mu} = g(T_\lambda^2)$, and $F$ is differentiable at every point of the complex plane. If $|\lambda| \geq 1 + \|T\|$, then, by lemma 6.5.1,

$$\|T_\lambda\| = \frac{1}{|\lambda|}\|(\frac{T}{\lambda} - I)^{-1}\| \leq \frac{1}{|\lambda|}\frac{1}{1 - \|T\|/|\lambda|} = \frac{1}{|\lambda| - \|T\|} \leq 1. \qquad (2)$$

*Thus* $\|T_\lambda\|$ *is bounded by 1 outside the closed disk, D, of radius* $1 + \|T\|$. *Therefore, outside the disk D,* $|F(\lambda)| = |g(T_\lambda)| \leq \|g\|$. *Because F is continuous on D, it is bounded on D; hence F is a bounded differentiable function on the entire complex plane. By lemma 6.5.5, $F(\lambda)$ is constant. If $\epsilon > 0$, there exists a positive constant R such that $\|T_\lambda\| < \epsilon$ for $|\lambda| \geq R$ (see inequality (2) above). Consequently, for such $\lambda$, $|F(\lambda)| \leq \|g\|\epsilon$. Since $\epsilon$ is arbitrary, and F is constant, $F(\lambda) = 0$ for all $\lambda \in \mathbb{C}$. Now since g is an arbitrary element of $(\mathcal{L}(X))^*$, $T_\lambda = 0$ (see corollary 6.4.7). This is impossible because $T_\lambda$ is invertible.* ∎

The following formula for the spectral radius is well known.

**Theorem 6.5.7 (Gelfand's theorem).** *Let T be a bounded operator on a Banach space X. Then $r(T) = \lim_n \|T^n\|^{1/n}$.*

*Proof.* By problem 9 at the end of this section, $r(T^n) = [r(T)]^n$. Therefore $r(T) = [r(T^n)]^{1/n} \leq \|T^n\|^{1/n}$, and $r(T) \leq \liminf_n \|T^n\|^{1/n}$. The proof will be complete if we show that $\limsup_n \|T^n\|^{1/n} \leq r(T)$.

Let $\lambda \in \mathbb{C}$ be such that $|\lambda| > \|T\|$. By theorem 6.5.2, $T_\lambda = (T - \lambda I)^{-1} = \frac{-1}{\lambda}\sum_{n=0}^{\infty}\frac{T^n}{\lambda^n}$. If $g \in (\mathcal{L}(X))^*$, then $g(T_\lambda) = \frac{-1}{\lambda}\sum_{n=0}^{\infty}\frac{g(T^n)}{\lambda^n}$. By the proof of theorem 6.5.6, the function $F(\lambda) = g(T_\lambda)$ is differentiable for all $\lambda \in \rho(T)$; thus the function $F(\lambda)$ extends the series $\frac{-1}{\lambda}\sum_{n=0}^{\infty}\frac{g(T^n)}{\lambda^n}$ to the set $\{z \in \mathbb{C} : |\lambda| > r(T)\}$. Therefore the series expansion $\frac{-1}{\lambda}\sum_{n=0}^{\infty}\frac{g(T^n)}{\lambda^n}$ is valid for all complex numbers $\lambda$ such that $|\lambda| > r(T).$[3] Now, for an arbitrary real number $a > r(T)$, the series $\frac{-1}{a}\sum_{n=0}^{\infty}\frac{g(T^n)}{a^n}$ is convergent; hence the sequence $g(\frac{T^n}{a^n})$ is bounded. Since $g \in (\mathcal{L}(X))^*$ is arbitrary, $T^n/a^n$ is bounded in $\mathcal{L}(X)$. Let $K > 0$ be such that $\|T^n/a^n\| \leq K$. Then $\|T^n\|^{1/n} \leq K^{1/n}a$, and $\limsup_n \|T^n\|^{1/n} \leq a$. Since a is an arbitrary number greater than $r(T)$, $\limsup_n \|T^n\|^{1/n} \leq r(T).$ ∎

---

[3] The series involved here are Laurent series.

## Exercises

1. Show that the composition of invertible operators is invertible.
2. Show that if $\lim_n T_n = T$, and $\lim_n S_n = S$ in $\mathcal{L}(X)$, then $\lim_n T_n S_n = TS$.
3. Let $\lambda_1, \ldots, \lambda_n$ be distinct eigenvalues of a bounded operator $T$, and let $u_1, \ldots, u_n$ be eigenvectors that correspond to $\lambda_1, \ldots, \lambda_n$, respectively. Prove that $u_1, \ldots, u_n$ are independent. Hint: Use induction on $n$.
4. Prove the following version of lemma 6.5.1. If $T \in \mathcal{L}(X)$ is such that $\|I - T\| < 1$, then $T$ is invertible, $T^{-1} = \sum_{n=0}^{\infty} (I - T)^n$, and $\|T^{-1}\| \leq \frac{1}{1 - \|I - T\|}$.
5. Let $T \in \mathcal{L}(X)$ be an invertible operator, and let $S \in \mathcal{L}(X)$. Prove that if $\|S - T\| < \frac{1}{\|T^{-1}\|}$, then $S$ is invertible.
6. Let $T, S \in \mathcal{L}(X)$ be invertible operators, such that $\|S - T\| < \frac{1}{2\|T^{-1}\|}$. Prove that $\|S^{-1} - T^{-1}\| < 2\|T^{-1}\|^2 \|S - T\|$. Hint: First show that $\|I - T^{-1}S\| < 1/2$, then use the identity $S^{-1} = [I - (I - T^{-1}S)]^{-1} T^{-1}$ to show that $S^{-1} - T^{-1} = \sum_{n=1}^{\infty} (I - T^{-1}S)^n T^{-1}$.
7. Let $U$ be the set of all invertible operators in $\mathcal{L}(X)$. Prove that $U$ is open in $\mathcal{L}(X)$ and that inversion is a homeomorphism on $U$.
8. Let $T$ and $S$ be commuting bounded linear operators on a Banach space $X$. Prove that if $ST$ is invertible, then $S$ and $T$ are invertible. Also give an example of two singular operators whose composition is invertible.
9. Prove that, for $T \in \mathcal{L}(X)$, $\sigma(T^n) = \{\mu^n : \mu \in \sigma(T)\}$. Conclude that $r(T^n) = [r(T)]^n$. Hint: Let $\lambda \in \mathbb{C}$, and let $t^n - \lambda = (t - \mu_1) \ldots (t - \mu_n)$. Then $T^n - \lambda I = (T - \mu_1 I) \ldots (T - \mu_n I)$.
10. For a fixed function $w \in \mathcal{C}[0,1]$, define an operator $T$ on $\mathcal{C}[0,1]$ by $(Tf)(x) = f(x)w(x)$. Show that $T$ is a bounded operator and that $\|T\| = \|w\|_\infty$. Also give a sufficient condition for $T$ to be invertible.

## 6.6 Adjoint Operators and Quotient Spaces

In section 3.7, we defined the adjoint of an operator on a finite-dimensional inner product space, and, in chapter 7, we will study adjoints of operators on a Hilbert space. The definition of the adjoints on Banach spaces $X$ is more complicated. In fact, the adjoint of a bounded operator on a Banach space $X$ is a bounded operator on the dual space $X^*$. Among other results, we prove that an operator $T$ and its adjoint, $T^*$ have the same norm, the same spectrum, and the same spectral radius. We also study annihilators and quotient spaces. Little subsequent material rests on this section, and it is possible to study the remainder of the book independently of this section.

Notation. The **duality bracket**: Let $X$ be a Banach space. For $x \in X$ and $\lambda \in X^*$, we write $\langle x, \lambda \rangle$ for $\lambda(x)$. This is a notational convenience that also facilitates certain computations. In addition, the notation equalizes the roles of $X$ and $X^*$. We already saw that $X$ acts on $X^*$ in much the same way $X^*$ acts on $X$. See, for example, the construction leading up to theorem 6.4.8. Observe that $|\langle x, \lambda \rangle| \leq \|x\|\|\lambda\|$, reminiscent of the Cauchy-Schwarz inequality. We revert to the traditional notation $\lambda(x)$ when convenient.

**Theorem 6.6.1.** *Let $X$ be a Banach space, let $x \in X$, $\lambda \in X^*$, and let $T \in \mathcal{L}(X)$. Then*

*(a)* $\|\lambda\| = sup\{|\langle x, \lambda \rangle| : x \in X, \|x\| \leq 1\}$,
*(b)* $\|x\| = \|\hat{x}\| = sup\{|\langle x, \lambda \rangle| : \lambda \in X^*, \|\lambda\| \leq 1\}$, *and*
*(c)* $\|T\| = sup\{|\langle Tx, \lambda \rangle| : x \in X, \lambda \in X^*, \|x\| \leq 1, \|\lambda\| \leq 1\}$.

*Proof.* *(a) and (b) are previously established facts in new notation. To prove (c),*

$$\|T\| = sup\{\|Tx\| : \|x\| \leq 1\} = sup_{\|x\| \leq 1} sup_{\|\lambda\| \leq 1} |\langle Tx, \lambda \rangle|$$
$$= sup\{|\langle Tx, \lambda \rangle| : \|x\| \leq 1, \|\lambda\| \leq 1\}. \blacksquare$$

**Definition.** Let $T \in \mathcal{L}(X)$. We define the **adjoint operator** $T^*$ on $X^*$ by the requirement that for all $x \in X$,

$$\langle Tx, \lambda \rangle = \langle x, T^*(\lambda) \rangle.$$

Using conventional notation rather than duality brackets, the requirement in the above definition can be written as $\lambda(Tx) = (T^*(\lambda))(x)$ for every $x \in X$. This simply means that $T^*(\lambda) = \lambda o T$, which can well be taken as the definition of the operator $T^*$. It is obvious that $T^* \in \mathcal{L}(X^*)$.

**Example 1.** In this example, we use theorem 6.2.4 and identify $(l^1)^*$ with $l^\infty$. For elements $x = (x_n) \in l^1$ and $\lambda = (\lambda_n) \in l^\infty$, define $T(x) = (x_2, x_3, ...)$ and $S(\lambda) = (0, \lambda_1, \lambda_2, ...)$. Clearly, $T \in \mathcal{L}(l^1)$ and $S \in \mathcal{L}(l^\infty)$. We claim that $S = T^*$. We need to verify that $\lambda o T = S(\lambda)$, which is straightforward since, for $x \in l^1$, $\lambda(T(x)) = \sum_{n=1}^{\infty} \lambda_n x_{n+1} = (S(\lambda))(x)$. $\blacklozenge$

**Theorem 6.6.2.** $\|T^*\| = \|T\|$.

*Proof.* *By theorem 6.6.1,*

$$\|T\| = sup\{|\langle Tx, \lambda \rangle| : \|x\| \leq 1, \|\lambda\| \leq 1\}$$
$$= sup\{|\langle x, T^*\lambda \rangle| : \|x\| \leq 1, \|\lambda\| \leq 1\}$$
$$= sup\{\|T^*\lambda\| : \|\lambda\| \leq 1\} = \|T^*\|. \blacksquare$$

**Example 2.** $r(T) = r(T^*)$.

It is straightforward to show that, for $n \in \mathbb{N}$, $(T^n)^* = (T^*)^n$. It follows that $\|(T^*)^n\| = \|(T^n)^*\| = \|T^n\|$. Employing theorem 6.5.7, we have

$$r(T^*) = \lim_n \|(T^*)^n\|^{1/n} = \lim_n \|T^n\|^{1/n} = r(T). \; \blacklozenge$$

The next example utilizes several of the ideas of sections 6.3 and 6.4.

**Example 3.** An operator $T \in \mathcal{L}(X)$ is invertible if and only if $T^*$ is invertible.

If $T$ is invertible, then $TT^{-1} = T^{-1}T = I_X$. Equating the adjoints of the operators in the previous identities and using problem 2 at the end of this section, we have $(T^{-1})^* T^* = T^* (T^{-1})^* = I_{X^*}$. Thus $(T^*)^{-1} = (T^{-1})^*$.

Now suppose $T^*$ is invertible, and write $S = (T^*)^{-1}$.

We show that $T$ is bounded away from zero. It then follows from example 8 on section 6.2 that $\mathfrak{R}(T)$ is closed and that $T$ is injective. Now

$$\|x\| = sup\{|\langle x, \lambda \rangle| : \lambda \in X^*, \|\lambda\| \leq 1\}$$
$$= sup\{|\langle x, T^* S\lambda \rangle| : \lambda \in X^*, \|\lambda\| \leq 1\}$$
$$= sup\{|\langle Tx, S\lambda \rangle| : \lambda \in X^*, \|\lambda\| \leq 1\}$$
$$\leq \|Tx\| sup\{\|S\lambda\| : \lambda \in X^*, \|\lambda\| \leq 1\} = \|Tx\|\|S\|.$$

Thus $\|Tx\| \geq c\|x\|$, where $c = 1/\|S\|$.

If we show that $T$ is surjective, then $T$ is invertible by theorem 6.3.7, and the proof will be complete. Suppose there is an element $y \in X - \mathfrak{R}(T)$. Since $\mathfrak{R}(T)$ is closed, theorem 6.4.5 yields an element $\lambda \in X^*$ such that $\lambda(y) \neq 0$ and $\lambda(Tx) = 0$ for every $x \in X$. Now $(T^*\lambda)(x) = \lambda(Tx) = 0$ for every $x \in X$; hence $T^*\lambda = 0$. Since $T^*$ is injective, $\lambda = 0$, and this is a contradiction. $\blacklozenge$

**Example 4.** $\sigma(T) = \sigma(T^*)$.

For any $\lambda \in \mathbb{C}$, $(T - \lambda I)^* = T^* - \lambda I$. By the above example, $T - \lambda I_X$ is invertible if and only if $T^* - \lambda I_{X^*}$ is invertible, and the result follows. Observe that the result of example 2 is a trivial consequence of this result. $\blacklozenge$

**Theorem 6.6.3.** Let $T \in \mathcal{L}(X)$. Then, for every $x \in X$, $(Tx)\hat{} = T^{**}(\hat{x})$.

*Proof.* For $\lambda \in X^*$, $(Tx)\hat{}(\lambda) = \lambda(Tx) = (\lambda \circ T)(x) = (T^*\lambda)(x) = \hat{x}(T^*\lambda) = (\hat{x} \circ T^*)(\lambda) = (T^{**}(\hat{x}))(\lambda)$. Thus $(Tx)\hat{} = T^{**}(\hat{x})$. $\blacksquare$

Loosely interpreted, the above theorem says that $T$ is the restriction of $T^{**}$ to $X$.

**Definition.** Let $M$ be a subspace of a Banach space $X$, and let $N$ be a subspace of $X^*$. The **annihilator** $M^\perp$ of $M$ consists of all the functionals in $X^*$ that vanish on $M$. Symbolically, $M^\perp = \{\lambda \in X^* : \lambda(M) = 0\}$. Similarly, the annihilator of $N$ is $N_\perp = \{x \in X : \lambda(x) = 0\ \forall \lambda \in N\}$.

**Example 5.** Let $M$ be the set of all sequences in $l^1$ where every even term is 0. We claim that $M^\perp$ is the set $S$ of all the sequences in $l^\infty$ where every odd term is 0. It is clear that if $\lambda = (\lambda_n) \in S$, then, for every $x = (x_n) \in M$, $\sum_{n=1}^\infty x_n \lambda_n = 0$; hence $S \subseteq M^\perp$. Conversely, if $\lambda = (\lambda_n) \in M^\perp$, then $\lambda(e_{2n+1}) = 0$ for every positive integer $n$. This means that $\lambda_{2n+1} = 0$, and $\lambda \in S$. Observe that $M^\perp$ is a closed subspace $l^\infty$, consistent with the theorem below. ◆

**Theorem 6.6.4.** $M^\perp$ is a closed subspace of $X^*$, and $N_\perp$ is a closed subspace of $X$.

*Proof.* For $x \in M$, $x^\perp = \{\lambda \in X^* : \lambda(x) = 0\}$ is a closed subspace of $X^*$ because $x^\perp = \mathrm{Ker}(\hat{x})$. Consequently, $M^\perp = \cap_{x \in M} x^\perp$ is a closed subspace of $X^*$. Similarly, $N_\perp = \cap_{\lambda \in N} \mathrm{Ker}(\lambda)$ is a closed subspace of $X$. ∎

**Theorem 6.6.5.** *Let $T \in \mathcal{L}(X)$, let $\mathcal{N}(T)$ and $\mathfrak{R}(T)$ be the kernel and range of $T$, respectively, and let $\mathcal{N}(T^*)$, and $\mathfrak{R}(T^*)$ be the kernel and range of $T^*$. Then*

*(a) $\mathcal{N}(T) = \mathfrak{R}(T^*)_\perp$.*
*(b) $\mathcal{N}(T^*) = \mathfrak{R}(T)^\perp$.*

*Proof.* (a) $x \in \mathcal{N}(T)$ if and only if $Tx = 0$, if and only if $\langle Tx, \lambda \rangle = 0$ for all $\lambda \in X^*$, if and only if $\langle x, T^*\lambda \rangle = 0$ for all $\lambda \in X^*$, if and only if $x \in \mathfrak{R}(T^*)_\perp$.

(b) $\lambda \in \mathcal{N}(T^*)$ if and only if $T^*\lambda = 0$, if and only if $\langle x, T^*\lambda \rangle = 0$ for all $x \in X$ if and only if $\langle Tx, \lambda \rangle = 0$ for every $x \in X$, if and only if $\lambda \in \mathfrak{R}(T)^\perp$. ∎

## Quotient Spaces

Let $M$ be a closed subspace of a normed linear space $X$. We define a norm on the quotient space $X/M$ as follows: for $\tilde{x} = x + M \in X/M$, $\|\tilde{x}\| = \inf\{\|x - y\| : y \in M\} = dist(x, M)$. We leave it to the reader to verify that the norm we just defined on $X/M$ is well defined and that $\|\tilde{x}\| = 0$ if and only if $\tilde{x} = 0$. The triangle inequality is the only norm property yet to be verified. Let $x_1, x_2 \in X$, and $y_1, y_2 \in M$. Since $y_1 + y_2 \in M$, $dist(x_1 + x_2, M) \le \|(x_1 + x_2) - (y_1 + y_2)\| \le \|x_1 - y_1\| + \|x_2 - y_2\|$. Because the last inequality is valid for arbitrary elements $y_1$ and $y_2$ of $M$, $dist(x_1 + x_2, M) \le dist(x_1, M) + dist(x_2, M)$, that is, $\|\tilde{x}_1 + \tilde{x}_2\| \le \|\tilde{x}_1\| + \|\tilde{x}_2\|$.

**Remarks.**  1. If $\|\tilde{x}\| < \delta$, then there exists $\xi \in \tilde{x}$ such that $\|\xi\| < \delta$. This is because if $\|\tilde{x}\| < \delta$, then there exists $y \in M$ such that $\|x - y\| < \delta$. Set $\xi = x - y$.
   2. For $x \in X$, $\|\tilde{x}\| \leq \|x\|$. This is because $0 \in M$; hence $\|x\| = \|x - 0\| \geq \|\tilde{x}\|$.
   3. It follows directly from remark 2 that if $(x_n)$ converges to $x$ in X, then $(\tilde{x}_n)$ converges to $\tilde{x}$ in $X/M$. In particular, if the series $\sum_{n=1}^{\infty} \xi_n$ converges in X, then $\sum_{n=1}^{\infty} \tilde{\xi}_n$ converges in $X/M$.

**Theorem 6.6.6.** *Let M be a closed subspace of a Banach space X. Then $X/M$ is a Banach space.*

*Proof.* We use the result of problem 10 on section 6.1. Suppose $\sum_{n=1}^{\infty} \|\tilde{x}_n\| < \infty$. We prove that $\sum_{n=1}^{\infty} \tilde{x}_n$ converges in $X/M$. By remark 1, for every $n \in \mathbb{N}$, there exists an element $\xi_n \in \tilde{x}_n$ such that $\|\xi_n\| < \|\tilde{x}_n\| + 1/2^n$. Now $\sum_{n=1}^{\infty} \|\xi_n\| \leq \sum_{n=1}^{\infty} [\|\tilde{x}_n\| + 1/2^n] = 1 + \sum_{n=1}^{\infty} \|\tilde{x}_n\| < \infty$. By the completeness of X, the series $\sum_{n=1}^{\infty} \xi_n$ converges in X, and, by remark 3, $\sum_{n=1}^{\infty} \tilde{\xi}_n = \sum_{n=1}^{\infty} \tilde{x}_n$ converges in $X/M$. ∎

**Theorem 6.6.7.** *Let M be a closed subspace of a Banach space X. Then*

   (a) $(X/M)^*$ *is isometrically isomorphic to $M^{\perp}$.*
   (b) $X^*/M^{\perp}$ *is isometrically isomorphic to $M^*$.*

*Proof.* (a) Define a map $\delta : M^{\perp} \to (X/M)^*$ by $\delta : \lambda \mapsto \delta_{\lambda}$, where $\delta_{\lambda}(x + M) = \delta_{\lambda}(\tilde{x}) = \lambda(x)$; $\delta$ is onto since if $\mu \in (X/M)^*$, define a functional $\lambda : X \to \mathbb{C}$ by $\lambda(x) = \mu(\tilde{x})$. It is easy to see that $\lambda \in X^*$, $\lambda \in M^{\perp}$, and $\delta_{\lambda} = \mu$. To show that $\delta$ is an isometry, first notice that $\|\tilde{x}\| \leq \|x\|$ and, by remark 1, if $\|\tilde{x}\| < 1$, there exists an element $x \in \tilde{x}$ such that $\|x\| < 1$. Therefore $\|\delta_{\lambda}\| = sup_{\|\tilde{x}\| < 1} |\delta_{\lambda}(\tilde{x})| = sup_{\|x\| < 1} |\lambda(x)| = \|\lambda\|$.

(b) Let $\mu \in M^*$. By the Hahn-Banach theorem, $\mu$ has an extension $\lambda \in X^*$. Define a mapping $\sigma : M^* \to X^*/M^{\perp}$ by $\sigma_{\mu} = \lambda + M^{\perp}$; $\sigma$ is well defined because if $\lambda$ and $\lambda'$ are bounded extensions of $\mu$, then $(\lambda - \lambda')(M) = 0$; hence $\lambda - \lambda' \in M^{\perp}$, and $\lambda + M^{\perp} = \lambda' + M^{\perp}$. The linearity of $\sigma$ is obvious and since the restriction of any $\lambda \in X^*$ to M is in $M^*$, $\sigma$ is onto. It remains to show that $\|\sigma_{\mu}\| = \|\mu\|$. Observe that $\sigma_{\mu}$ is the collection of all bounded extensions of $\mu$. Thus, by the definition of the quotient norm, $\|\sigma_{\mu}\| = inf\{\|\lambda\|\}$, where $\lambda$ is a bounded extension of $\mu$. Since, for any such $\lambda$, $\|\mu\| \leq \|\lambda\|$, it follows that $\|\mu\| \leq \|\sigma_{\mu}\| \leq \|\lambda\|$. The Hahn-Banach theorem also guarantees an extension $\lambda$ for which $\|\lambda\| = \|\mu\|$. Thus $\|\sigma_{\mu}\| = \|\mu\|$. ∎

## Exercises

1. Show that if $T, S \in \mathcal{L}(X)$ and $a$, and $b$ are scalars, then $(aT + bS)^* = aT^* + bS^*$. Conclude that if $X$ is reflexive, then the correspondence $\Psi : T \mapsto T^*$ is an isometric isomorphism from $\mathcal{L}(X)$ to $\mathcal{L}(X^*)$.

2. If $T$ and $S$ are as in the above exercise, show that $(ST)^* = T^*S^*$.

3. Let $X$ be a Banach space. Prove that $X^\perp = \{0\}, \{0\}^\perp = X^*$. State and prove the corresponding statements for $X^*$.

4. Let $M$ be a subspace of a Banach space $X$. Prove that $(M^\perp)_\perp = \overline{M}$. Hint: Use theorem 6.4.5 to show that if $x \notin \overline{M}$, then $x \notin (M^\perp)_\perp$.

5. Let $X$ be a Banach space, and let $T \in \mathcal{L}(X)$. Prove that $\overline{\mathfrak{R}(T)} = \mathcal{N}(T^*)_\perp$. Conclude that $\mathfrak{R}(T)$ is dense if and only if $T^*$ is one-to-one.

6. Let $X$ be a Banach space, and let $T \in \mathcal{L}(X)$. Show that if $x_n \to^w x$, then $Tx_n \to^w Tx$.

7. Let $S$ and $T$ be commuting bounded linear operators on $X$. Prove that the eigenspaces of $T$ are $S$-invariant.

8. Let $T \in \mathcal{L}(X)$, and suppose $M$ is a $T$-invariant subspace of $X$. Prove that $M^\perp$ is invariant under $T^*$.

9. Verify the details of the proof that the norm defined on $X/M$ is indeed a norm.

10. Show that if $M$ is a closed subspace of a Banach space $X$, then the quotient map $\pi : X \to X/M$ is continuous. Also prove that if $N$ is a finite-dimensional subspace of $X$, then $\pi(N)$ is a finite-dimensional subspace of $X/M$.

11. In the quotient space $l^\infty/c_0$, prove that $\|\tilde{x}\| = \limsup_n |x_n|$. Hint: For $\epsilon > 0$, there are finitely many $n \in \mathbb{N}$ such that $|x_n| > \limsup_n |x_n| + \epsilon$.

12. Let $X$ be a Banach space, and let $T \in \mathcal{L}(X)$. Define $\overline{T} : X/Ker(T) \to \mathfrak{R}(T)$ by $\overline{T}(\tilde{x}) = T(x)$. Prove that $\overline{T}$ is a bounded isomorphism. Hint: To show the continuity of $\overline{T}$, suppose $\tilde{x}_n \to 0$, and choose $x_n \in \tilde{x}_n$ such that $\|x_n\| < \|\tilde{x}_n\| + 1/n$.

13. Let $R$ be the right shift operator on $l^2$, let $M_1$ be the range of $R$, and let $M_2$ be the range of $R^2$. Determine the quotient spaces $l^2/M_1$ and $l^2/M_2$. Conclude that if $M_1$ and $M_2$ are isomorphic closed subspaces of a Banach space $X$, then it is not necessarily true that $X/M_1$ and $X/M_2$ are isomorphic.

14. Prove that if $M$ is a closed subspace of a separable Banach space $X$, then $X/M$ is separable.

15. Let $M$ be a closed subspace of a Banach space $X$. Prove that if $X^*$ is separable, then so is $M^*$.

16. Let $M$ be a closed subspace of a Banach space $X$. Prove that if $M$ and $X/M$ are separable, then $X$ is separable. Hint: Let $\{x_n\} \subseteq X$ be such that $\{\tilde{x}_n\}$ is dense in $X/M$, and let $\{y_m\} \subseteq M$ be dense in $M$.

17. Let $X$ be a normed linear space, and let $M$ be a closed subspace of $X$. Prove that if $M$ and $X/M$ are Banach spaces, then so is $X$.
18. Let $M$ be a closed subspace of a Banach space $X$, and let $N$ be a finite-dimensional subspace of $X$. Show that $M+N$ is closed. Hint: Consider $\pi^{-1}(\pi(N))$, where $\pi : X \to X/M$ is the quotient map.
19. Let $M$ be a complemented subspace of a Banach space $X$. Show that $M^{\perp}$ is complemented in $X^*$. Hint: Let $P$ be the projection of $X$ onto $M$, and let $N = Ker(P)$. Show that $M^{\perp} = Ker(P^*)$ and that $N^{\perp} = \Re(P^*)$.
20. Define a linear operator $T \in \mathcal{L}(c_0)$ as follows: for $x = (x_n)$, $T(x) = (\frac{x_n}{n})$. Describe $T^*$. Recall the result of problem 16 on section 6.2.

## 6.7  Weak Topologies

The weak topologies are defined in much the same way the product topology is defined. They are designed to guarantee the continuity of a certain class of functions. We urge the reader to look up theorem 5.4.1, the definition of the product topology in section 5.12, and theorem 5.12.1. This section is terminal and may be omitted without loss of continuity.

**Definition.** Let $X$ be a normed linear space. The **weak topology** on $X$ is the smallest topology relative to which all the bounded linear functionals on $X$ are continuous. We use the abbreviation $w$-topology for the weak topology on $X$.

**Definition.** Let $X$ be a normed linear space, and let $X^*$ be its dual. The **weak\* topology** on $X^*$ is the smallest topology on $X^*$ relative to which the functionals $\hat{x}$ are continuous. Here $\hat{x}$ is the image of $x \in X$ under the natural embedding of $X$ into $X^{**}$. We use the abbreviation $w^*$-topology for the weak\* topology on $X^*$.

Notice that the definitions of the $w$- and $w^*$-topologies are asymmetric. Only the functional on $X^*$ of the form $\hat{x}$ is admitted in the definition of the $w^*$-topology on $X^*$. Thus if $X$ is not reflexive, then the functionals in $X^{**} - \hat{X}$ are not guaranteed to be continuous in the $w^*$-topology, and indeed they are not. See theorem 6.7.6.

In order to eliminate any potential confusion, we specifically refer to the topology generated by the norm on a space $X$ (or its dual $X^*$) as the **norm topology on $X$** (or $X^*$). The **norm topology** is also referred to as the **strong topology**. We denote the closed unit balls of a normed linear space $X$ and its dual $X^*$ by $B$ and $B^*$, respectively. We use notation such as $(B^*, w^*)$ to indicate the closed unit ball of $X^*$, when it is endowed with the $w^*$-topology.

It follows directly from the definitions that an open base for the $w$-topology is the collection of sets of the form $\cap_{i=1}^{n}\{x \in X : |\lambda_i(x) - \lambda_i(x_0)| < r\}$, where $r > 0$, $x_0 \in X$, and $\{\lambda_1, ..., \lambda_n\}$ is a finite subset of $X^*$. Similarly, an open base for the $w^*$-topology is the collection of all sets of the type $\cap_{i=1}^{n}\{\lambda \in X^* : |\lambda(x_i) - \lambda_0(x_i)| < r\}$, where $r > 0$, $\lambda_0 \in X^*$, and $\{x_1, ..., x_n\}$ is a finite subset of $X$.

In the exercises in section 6.4, we introduced the notion of a weakly convergent sequence in a normed linear space. We now reconcile this concept with the definition of the weak topologies. First recall the definition of a convergent sequence in a topological space.

**Definition.** A sequence $(x_n)$ in a topological space $X$ is said to converge to a point $x \in X$ if every open neighborhood of $x$ contains all but finitely many terms of the sequence $(x_n)$. Thus if $U$ is an open neighborhood of $x$, then there exists a natural number $N$ such that $x_n \in U$ for every $n \geq N$.

**Theorem 6.7.1.**

(a) A sequence $(x_n)$ converges to $x$ in the $w$-topology on a normed linear space $X$ if and only if $\lim_n \lambda(x_n) = \lambda(x)$ for every $\lambda \in X^*$.

(b) A sequence $(\lambda_n)$ converges to $\lambda$ in the $w^*$-topology if and only if $\lim_n \lambda_n(x) = \lambda(x)$ for every $x \in X$.

*Proof. We prove part (b). Let $\lambda_n$ and $\lambda_0$ be such that $\lim_n \lambda_n(x) = \lambda_0(x)$ for every $x \in X$. We show that $\lambda_n$ converges to $\lambda_0$ in the $w^*$-topology. If $U$ is a $w^*$-open neighborhood of $\lambda_0$, then there exists $r > 0$ and a finite subset $\{x_1, ..., x_m\}$ of $X$ such that $\cap_{i=1}^{m}\{\lambda \in X^* : |\lambda(x_i) - \lambda_0(x_i)| < r\} \subseteq U$. Since for all $1 \leq i \leq m$, $\lim_n \lambda_n(x_i) = \lambda_0(x_i)$, there is a natural number $N$ such that $|\lambda_n(x_i) - \lambda_0(x_i)| < r$ for all $n > N$ and all $1 \leq i \leq m$. This means that $\lambda_n \in \cap_{i=1}^{m}\{\lambda \in X^* : |\lambda(x_i) - \lambda_0(x_i)| < r\} \subseteq U$, for every $n > N$. The proof of the converse is a partial reversal of the above argument.* ∎

**Theorem 6.7.2.** *Let $X$ be a finite-dimensional normed linear space. Then*

(a) *The $w$-topology and the norm topology on $X$ coincide.*

(b) *The $w^*$-topology and the norm topology on $X^*$ coincide.*

*Proof. We prove part (b). We show that if $U = \{\lambda \in X^* : \|\lambda - \lambda_0\| < r\}$, then $U$ contains a $w^*$-neighborhood $V$ of $\lambda_0$. Let $\{e_1, ..., e_n\}$ be a basis for $X$, and define a norm on $X^*$ by $\|\lambda\|' = max_{1 \leq i \leq n}|\lambda(e_i)|$. Since all norms on $X^*$ are equivalent, there exists $\delta > 0$ such that $\|\lambda\|' < \delta$ implies that $\|\lambda\| < r$. The $w^*$-open neighborhood $V = \cap_{i=1}^{n}\{\lambda \in X^* : |\lambda(e_i) - \lambda_0(e_i)| < \delta\}$ of $\lambda_0$ is contained in $U$.* ∎

Before we can prove that the $w$-topology is different from the norm topology on an infinite-dimensional normed linear space, we need a fact from general vector space theory.

**Lemma 6.7.3.** *Let $U$, $V$, and $W$ be vector spaces, and let $\pi : U \to V$ and $\varphi : U \to W$ be linear mappings such that $\mathrm{Ker}(\pi) \subseteq \mathrm{Ker}(\varphi)$. Then there exists a linear mapping $\psi : V \to W$ such that $\psi o \pi = \varphi$.*

*Proof.* Let $V_1 = \Re(\pi)$, and define $\psi : V_1 \to W$ by $\psi(\pi(x)) = \varphi(x)$. The condition $\mathrm{Ker}(\pi) \subseteq \mathrm{Ker}(\varphi)$ guarantees that $\psi$ is well defined. Let $V_2$ be an algebraic complement of $V_1$ in $V$, and extend the definition of $\psi$ to $V$ by $\psi(v) = v_1$, where $v = v_1 + v_2$ and $v_i \in V_i$. By construction, $\psi o \pi = \varphi$. ∎

**Lemma 6.7.4.** *Let $X$ be a vector space, and let $\lambda$ and $\lambda_1, \dots, \lambda_n$ be linear functionals on $X$. If $\cap_{i=1}^n \mathrm{Ker}(\lambda_i) \subseteq \mathrm{Ker}(\lambda)$, then $\lambda$ is a linear combination of $\lambda_1, \dots, \lambda_n$.*

*Proof.* Define $\pi : X \to \mathbb{K}^n$ by $\pi(x) = (\lambda_1(x), \dots, \lambda_n(x))$. The condition $\cap_{i=1}^n \mathrm{Ker}(\lambda_i) \subseteq \mathrm{Ker}(\lambda)$ implies that $\mathrm{Ker}(\pi) \subseteq \mathrm{Ker}(\lambda)$. The previous lemma produces a functional $\psi : \mathbb{K}^n \to \mathbb{K}$ such that $\psi o \pi = \lambda$. Because $\psi$ is linear, there exist scalars $a_1, \dots, a_n$ such that for $(v_1, \dots, v_n) \in \mathbb{K}^n$, $\psi(v_1, \dots, v_n) = \sum_{i=a}^n a_i v_i$. Now, for $x \in X$, $\lambda(x) = (\psi o \pi)(x) = \psi(\lambda_1(x), \dots, \lambda_n(x)) = \sum_{i=1}^n a_i \lambda_i(x)$. ∎

**Theorem 6.7.5.** *A weakly open subset $U$ of an infinite-dimensional normed linear space $X$ is unbounded.*

*Proof.* Without loss of generality, we assume that $0 \in U$. Then there is $r > 0$ and a finite subset $\{\lambda_1, \dots, \lambda_n\}$ of $X^*$ such that $\cap_{i=1}^n \{x \in X : |\lambda_i(x)| < r\} \subseteq U$. The set $N = \cap_{i=1}^n \mathrm{Ker}(\lambda_i)$ is clearly contained in $U$. If $N = \{0\}$, then, for every $\lambda \in X^*$, $N \subseteq \mathrm{Ker}(\lambda)$. By lemma 6.7.4, every $\lambda \in X^*$ would be a linear combination of $\lambda_1, \dots, \lambda_n$, contradicting the assumption that $X$, hence $X^*$, is infinite dimensional. Thus $N \neq \{0\}$, and, for any nonzero $x \in N$, the line $\{cx : c \in \mathbb{R}\} \subseteq N$; hence $U$ is unbounded. ∎

The above theorem implies that the weak and norm topologies on an infinite-dimensional space $X$ are distinct since no open bounded subset of $X$ can be weakly open.

Weak topologies are generally intricate, and good caution must be exercised when formulating arguments involving them. In metric topologies, when one speaks of an open neighborhood of a point $x$, one instinctively thinks of an open ball centered at $x$. A $w$-open neighborhood of a point looks nothing like an open ball since bounded subsets of $X$ are never $w$-open.

We now prove that the $w^*$-topology is very tight in the sense that it admits the continuity of no linear functionals other than the functionals $\hat{x}$ used in the definition of the $w^*$-topology.

**Theorem 6.7.6.** *Let $X$ be a Banach space, and let $F$ be a $w^*$-continuous linear functional on $X^*$. Then $F = \hat{x}$ for some $x \in X$.*

*Proof.* Let $D$ be the open unit disk in the complex plane. By the $w^*$-continuity of $F$, $F^{-1}(D)$ is $w^*$-open; hence it contains a $w^*$-neighborhood of $0$ of the form $U = \cap_{i=1}^n \{\lambda \in X^* : |\lambda(x_i)| < r\}$ for some $r > 0$, and some finite subset $\{x_1, \dots, x_n\}$ of $X$. In particular, $F(U)$ is a bounded subset of the complex plane. We show that $\cap_{i=1}^n Ker(\hat{x}_i) \subseteq Ker(F)$. If $\lambda \in \cap_{i=1}^n Ker(\hat{x}_i)$, then, clearly, $\lambda \in U$ and $c\lambda \in U$, for all $c \in \mathbb{R}$; hence $|c||F(\lambda)| = |F(c\lambda)|$ is bounded for all $c \in \mathbb{R}$. This forces $F(\lambda) = 0$. Therefore $\cap_{i=1}^n Ker(\hat{x}_i) \subseteq Ker(F)$. By lemma 6.7.4, $F$ is a linear combination of $\hat{x}_1, \dots, \hat{x}_n$; hence $F \in \hat{X}$. ∎

**Theorem 6.7.7 (the Banach-Alaoglu theorem).** *Let $X$ be a normed linear space. Then $B^* = \{\lambda \in X^* : \|\lambda\| \leq 1\}$ is compact in the $w^*$-topology.*

*Proof.* For each $x \in X$, let $D_x = \{z \in \mathbb{C} : |z| \leq \|x\|\}$, and let $D = \prod_{x \in X} D_x$. By Tychonoff's theorem, $D$ is compact. For each $\lambda \in B^*$, define $f_\lambda \in D$ by $f_\lambda(x) = \lambda(x)$. Since, for $x \in X$, $|f_\lambda(x)| \leq \|\lambda\|\|x\| \leq \|x\|$, $f_\lambda(x) \in D_x$, and, indeed, $f_\lambda \in D$. The function $f : B^* \to D$ given by $\lambda \mapsto f_\lambda$ clearly injects $B^*$ into $D$. For the rest of the proof, we identify $\lambda$ and $f_\lambda$ and consider $B^*$ as a subset of $D$. The $w^*$-topology on $B^*$ is the restriction of the product topology on $D$ to $B^*$. The proof will be complete if we show that $B^*$ is closed in $D$. Let $\mu \in D$ be a closure point of $B^*$. We need to show that $\mu \in B^*$. Fix a pair of points $x, y \in X$, and let $\epsilon > 0$. The $D$-open set $\{g \in D : |g(x) - \mu(x)| < \epsilon/3, |g(y) - \mu(y)| < \epsilon/3, |g(x+y) - \mu(x+y)| < \epsilon/3\}$ intersects $B^*$, so there exists an element $\lambda \in B^*$ such that $|\lambda(x) - \mu(x)| < \epsilon/3, |\lambda(y) - \mu(y)| < \epsilon/3$, and $|\lambda(x+y) - \mu(x+y)| < \epsilon/3$. Since $\lambda(x+y) - \lambda(x) - \lambda(y) = 0$,

$$|\mu(x+y) - \mu(x) - \mu(y)| = |\mu(x+y) - \mu(x) - \mu(y) - \lambda(x+y) + \lambda(x) + \lambda(y)|$$
$$\leq |\lambda(x+y) - \mu(x+y)| + |\lambda(x) - \mu(x)| + |\lambda(y) - \mu(y)| < \epsilon.$$

Since $\epsilon$ is arbitrary, $\mu(x+y) = \mu(x) + \mu(y)$. The homogeneity of $\mu$ is proved using a similar argument. Finally, $\mu \in B^*$ because $\mu(x) \in D_x$, so $|\mu(x)| \leq \|x\|$, which means that $\mu$ is bounded and $\|\mu\| \leq 1$. ∎

The following theorem is curious if not very practical.

**Corollary 6.7.8.** *Every Banach space X is isometrically isomorphic to a closed subspace of $\mathcal{C}(K)$ for some compact Hausdorff space K.*

*Proof.* By the Banach-Alaoglu theorem, the space $K = (B^*, w^*)$ is compact. Define a function $F : X \to \mathcal{C}(K)$ by $F(x) = \hat{x}$; $F$ is a linear isometry since $\|F(x)\|_\infty = \sup\{|\hat{x}(\lambda)| : \lambda \in B^*\} = \|\hat{x}\| = \|x\|$. (Note: The norm on $\mathcal{C}(K)$ is the supremum norm.) Thus X is isometrically isomorphic to $F(X)$, which is a closed subspace of $\mathcal{C}(K)$. ∎

**Theorem 6.7.9.** *Let X be a separable Banach space. Then $(B^*, w^*)$ is metrizable.*

*Proof.* Let $\{x_n\}$ be a dense subset of B. For $\lambda, \mu \in B^*$, define

$$d(\lambda, \mu) = \sum_{n=1}^{\infty} 2^{-n} |\lambda(x_n) - \mu(x_n)|.$$

If $d(\lambda, \mu) = 0$, then $(\lambda - \mu)(x_n) = 0$ for all n. The density of $\{x_n\}$ forces $\lambda = \mu$. The other defining properties of a metric are easy to verify. We show that the identity function $I : (B^*, w^*) \to (B^*, d)$ is a homeomorphism. It follows that $(B^*, w^*)$ is metrizable. Since $(B^*, w^*)$ is compact and $(B^*, d)$ is Hausdorff, it suffices to show that I is continuous. See theorem 5.8.7. To this end, we prove that a d-open ball $U = \{\lambda \in B^* : d(\lambda, \lambda_0) < r\}$ contains a $w^*$-open neighborhood V of $\lambda_0$. Because $|\lambda(x_i) - \lambda_0(x_i)| \leq 2$ for all $\lambda \in B^*$ and all $i \in \mathbb{N}$, there exists a positive integer N such that such that $\sum_{i=N+1}^{\infty} 2^{-i} |\lambda(x_i) - \lambda_0(x_i)| < r/2$ for all $\lambda \in B^*$. For every $\lambda \in B^*$ satisfying $|\lambda(x_i) - \lambda_0(x_i)| < r/2$ for all $1 \leq i \leq N$, we have $d(\lambda, \lambda_0) = \sum_{i=1}^{N} 2^{-i} |\lambda(x_i) - \lambda_0(x_i)| + \sum_{i=N+1}^{\infty} 2^{-i} |\lambda(x_i) - \lambda_0(x_i)| < r/2 + r/2$. Thus the $w^*$-neighborhood $V = \cap_{i=1}^{N} \{\lambda \in B^* : |\lambda(x_i) - \lambda_0(x_i)| < r/2\}$ of $\lambda_0$ is contained in U. ∎

**Corollary 6.7.10.** *If X is separable, so is $(X^*, w^*)$.*

*Proof.* By theorems 6.7.7 and 6.7.9, $(B^*, w^*)$ is compact and metrizable; hence, it is separable. Since $X^* = \cup_{n=1}^{\infty} nB^*$, $X^*$ is separable in the $w^*$-topology. ∎

The converse of theorem 6.7.9 is also true. Recall (see theorem 4.9.10) that if K is a compact metric space, then $\mathcal{C}(K)$ is separable.

**Theorem 6.7.11.** *Let X be a Banach space. Then $(B^*, w^*)$ is metrizable if and only if X is separable.*

*Proof.* Suppose $(B^*, w^*)$ is metrizable. Theorem 6.7.7 implies that $K = (B^*, w^*)$ is a compact metrizable space; hence $\mathcal{C}(K)$ is separable. The mapping $F : X \to \mathcal{C}(K)$ defined by $F(x) = \hat{x}$ is an isometry. Hence X is separable because it is isometric to $F(X)$, which is separable, being a subspace of a separable metric space. ∎

# Exercises

1. Prove that the $w$- and the $w^*$-topologies are Hausdorff.
2. Complete the proof of theorem 6.7.1.
3. Complete the proof of theorem 6.7.2.
4. Let $X$ be an infinite-dimensional normed linear space. Prove that every $w$-neighborhood of 0 contains an infinite-dimensional subspace of $X$. Hint: Examine the proof of theorem 6.7.5, and see problem 19 on section 3.4.
5. In connection with corollary 6.7.8, prove that $F(X)$ is a closed subspace of $\mathcal{C}(K)$.
6. Let $M$ be a subspace of a normed linear space $X$. Prove that the $w$-closure of $M$ is a subspace of $X$.
7. Prove that every norm-closed subspace $M$ of a Banach space $X$ is $w$-closed. Conclude that the $w$-closure and the norm-closure of a subspace of $X$ coincide.
8. Prove that a Banach space is separable if it is weakly separable.
9. Prove that if $X$ a Banach space such that $X^*$ is separable, then $(B, w)$ is metrizable.
10. Let $K$ be a compact subset of a Banach space $X$. Prove that the weak and norm topologies on $K$ coincide. Hint: See problem 6 on section 5.8.

# 7

# Hilbert Spaces

*Wir müssen wissen.*
*Wir werden wissen.*

David Hilbert. 1862–1943

David Hilbert. 1862–1943

Upon graduation from the Wilhelm Gymnasium, where he spent his final year of schooling, Hilbert enrolled at the University of Königsberg in the autumn of 1880. He received his Ph.D. from Königsberg in 1885, remained there as a member of staff from 1886 to 1895, and was promoted to the rank of professor in 1893. In 1895 Hilbert was appointed to the chair of mathematics at the University of Göttingen, where he spent the rest of his career. Among Hilbert's numerous students were Hermann Weyl, Felix Bernstein, Otto Blumenthal, Richard Courant, Alfred Haar, and Hugo Steinhaus.

Hilbert contributed to many branches of mathematics, including geometry, algebraic number fields, functional analysis, integral equations, mathematical physics, and the calculus of variations. Hilbert's work in geometry had the greatest influence in that area after Euclid. A systematic study of the axioms of Euclidean geometry led Hilbert to propose twenty-one such axioms, and he analyzed their

*Fundamentals of Mathematical Analysis.* Adel N. Boules, Oxford University Press (2021). © Adel N. Boules.
DOI: 10.1093/oso/9780198868781.003.0007

significance. He published *Grundlagen der Geometrie* in 1899, putting geometry in a formal axiomatic setting. Hilbert is most remembered for studying infinite-dimensional Euclidean spaces, which are now known as Hilbert spaces.

Hilbert's famous twenty-three Paris problems challenged (and still today challenge) mathematicians to solve fundamental questions. Hilbert's famous speech *The Problems of Mathematics* was delivered to the Second International Congress of Mathematicians in Paris. It was a speech full of optimism for mathematics in the coming century, and he felt that open problems were the sign of vitality in the subject. Hilbert's problems included the continuum hypothesis, Goldbach's conjecture, and the Riemann hypothesis.

Hilbert's mathematical abilities were nicely summed up by Otto Blumenthal, his first student:

In the analysis of mathematical talent one has to differentiate between the ability to create new concepts that generate new types of thought structures and the gift for sensing deeper connections and underlying unity. In Hilbert's case, his greatness lies in an immensely powerful insight that penetrates into the depths of a question. All of his works contain examples from far-flung fields in which only he was able to discern an interrelatedness and connection with the problem at hand. From these, the synthesis, his work of art, was ultimately created. Insofar as the creation of new ideas is concerned, I would place Minkowski higher, and of the classical great ones, Gauss, Galois, and Riemann. But when it comes to penetrating insight, only a few of the very greatest were the equal of Hilbert.

Hilbert retired in 1930, and the city of Königsberg made him an honorary citizen. He gave an address which ended with famous words that now appear on his epitaph:

Wir müssen wissen, wir werden wissen: We must know, we shall know.[1]

## 7.1 Definitions and Basic Properties

Let $\{u_1, u_2, ...\}$ be an infinite orthonormal sequence of vectors in an inner product space $H$, and let $x \in H$. In the introduction to section 4.10, we posed the following problem. Under what conditions does the sequence of orthogonal projections, $S_n x = \sum_{i=1}^{n} \langle x, u_i \rangle u_i = \sum_{i=1}^{n} \hat{x}_i u_i$, of $x$ on the finite-dimensional space

---

[1] Perhaps as a rebuttal of Du Bois-Raymond's statement *"we do not know and will not know,"* reflecting the idea that scientific knowledge is unknown and unknowable.

$M_n = Span(\{u_1, \dots, u_n\})$, converge to $x$. Regardless of whether $S_n x$ converges to $x$, it is a Cauchy sequence. To see this, recall the result of problem 5 on section 3.7 (also see theorem 7.2.6,) which states that $\sum_{n=1}^{\infty} |\hat{x}_n|^2 < \infty$. Now, for $m > n$, $\|S_m x - S_n x\|^2 \le \sum_{i=n+1}^{m} |\hat{x}_i|^2$. The sum in the last expression tends to 0 as $n \to \infty$ because it is the middle section of the convergent series $\sum_{i=1}^{\infty} |\hat{x}_i|^2$. Thus we have a sufficient condition for the convergence of the sequence $S_n x$: the completeness of $H$. This is exactly the definition of a Hilbert space. The completeness of $H$ merely guarantees the convergence of $S_n x$. It does not guarantee that $\lim_n S_n x = x$, as the following situation illustrates. If $u \in H$ is unit vector orthogonal to each $u_n$, then $S_n u = 0$ for all $n \in \mathbb{N}$; hence $\lim_n S_n u = 0 \ne u$. To remedy this situation, one may want to impose the condition that no such vector $u$ exists. Equivalently, this means that the sequence $\{u_1, u_2, \dots\}$ is a maximal orthonormal subset of $H$, and this is precisely the definition of a countable orthonormal basis for $H$. Hilbert spaces and orthonormal bases are the subject of our study in this section and the next. The question about the smallest Hilbert space $H$ in which trigonometric series of functions in $H$ converge will be settled in section 8.9, together with related questions pertaining orthogonal polynomials. It is strongly recommended that you study sections 3.7 and 4.10 before you tackle this chapter.

**Definition.** A **Hilbert space** is a complete inner product space.

**Example 1.** The spaces $\mathbb{K}^n$ and $l^2$ are Hilbert spaces. ◆

**Example 2.** The space $(\mathbb{K}(\mathbb{N}), \|.\|_2)$ is not a Hilbert space. We use the fact that a subspace of $l^2$ is complete if and only if it is closed. Now $\mathbb{K}(\mathbb{N})$ is not closed in $l^2$ because it contains the sequence $x_1 = (1, 0, 0, \dots), x_2 = (1, 1/2, 0, 0, \dots), \dots,$ $x_n = (1, 1/2, 1/3, \dots, 1/n, 0, 0, \dots)$ The limit of the sequence $(x_n)$ is the harmonic sequence $x = (1, 1/2, \dots, 1/n, \dots)$ because $\|x_n - x\|_2^2 = \sum_{j=n+1}^{\infty} |x_j|^2 \to 0$ as $n \to \infty$. Clearly, $x \notin \mathbb{K}(\mathbb{N})$. ◆

For ease of reference, we state, without proof, a few results from section 3.7. We urge the reader to look up the proofs and the basic definitions in section 3.7.

**Theorem 7.1.1 (the Cauchy-Schwarz inequality).** *If $H$ is an inner product space, then, for all $x, y \in H$, $|\langle x, y \rangle| \le \|x\|\|y\|$. Equality holds if and only if $x$ and $y$ are linearly dependent.* ∎

**Corollary 7.1.2.** *In an inner product space $H$, $\|x + y\| \le \|x\| + \|y\|$. Here $\|x\| = \langle x, x \rangle^{1/2}$ is the norm on $H$ induced by the inner product.* ∎

**Theorem 7.1.3.** *Let $x$ and $y$ be elements of an inner product space $H$.*

(a) **The Pythagorean theorem.** *If $x$ and $y$ are orthogonal, then*

$$\|x+y\|^2 = \|x\|^2 + \|y\|^2.$$

(b) **The Parallelogram law.**

$$\|x+y\|^2 + \|x-y\|^2 = 2\|x\|^2 + 2\|y\|^2.$$

(c) **The Polarization identity.**

$$\|x+y\|^2 - \|x-y\|^2 + i\|x+iy\|^2 - i\|x-iy\|^2 = 4\langle x,y\rangle.$$

*Proof. We leave the proof as an exercise.* ∎

Not all norms are induced by an inner product. However, we have the following result, which we limit to real normed linear spaces for simplicity.

**Example 3.** Suppose that $(X, \|.\|)$ is a real normed linear space and that the norm satisfies the parallelogram identity. Then the function

$$\langle x,y\rangle = \tfrac{1}{4}\left[\|x+y\|^2 - \|x-y\|^2\right]$$

is an inner product that generates the norm.

It is clear that $\langle x,x\rangle = \|x\|^2$, thus establishing the positivity of the function $\langle .,.\rangle$ and that it generates the norm. The symmetry of $\langle .,.\rangle$ is obvious. Next we establish the linearity of $\langle .,.\rangle$.

We leave it to the reader to use the parallelogram identity to show that

$$\|x+y+z\|^2 + \|x\|^2 + \|y\|^2 + \|z\|^2 = \|x+y\|^2 + \|y+z\|^2 + \|x+z\|^2. \qquad (1)$$

Replacing $z$ with $-z$ in identity (1), we have

$$\|x+y-z\|^2 + \|x\|^2 + \|y\|^2 + \|z\|^2 = \|x+y\|^2 + \|y-z\|^2 + \|x-z\|^2. \qquad (2)$$

Subtracting (2) from (1), we obtain

$$\|x+y+z\|^2 - \|x+y-z\|^2 = \|x+z\|^2 - \|x-z\|^2 + \|y+z\|^2 - \|y-z\|^2,$$

which is equivalent to $\langle x+y,z\rangle = \langle x,z\rangle + \langle y,z\rangle$.

Using the linearity property we just established and induction, it follows that, for $m \in \mathbb{N}$, $\langle mx,y\rangle = m\langle x,y\rangle$. Since $\langle -x,y\rangle = -\langle x,y\rangle$, the identity

$\langle mx, y \rangle = m \langle x, y \rangle$ holds for all $m \in \mathbb{Z}$. Using this, for all $n \in \mathbb{N}$, $\langle x, y \rangle = \langle n \frac{1}{n} x, y \rangle = n \langle \frac{1}{n} x, y \rangle$. Equivalently, $\langle \frac{1}{n} x, y \rangle = \frac{1}{n} \langle x, y \rangle$.

We have shown that, for all $q \in \mathbb{Q}$, $\langle qx, y \rangle = q \langle x, y \rangle$. It is easy to see that if $\lim_n x_n = x$, then $\lim_n \langle x_n, y \rangle = \langle x, y \rangle$. Now the homogeneity property, $\langle \alpha x, y \rangle = \alpha \langle x, y \rangle$ for $\alpha \in \mathbb{R}$, is immediate because $\mathbb{Q}$ is dense in $\mathbb{R}$. ◆

**Definition.** For a subset $A$ of an inner product space $H$, the **annihilator of** $A$ is the set of all vectors that are orthogonal to every element in $A$. Symbolically, $A^\perp = \{x \in H : \langle x, a \rangle = 0 \; \forall a \in A\}$.

**Example 4.** Consider the space $\mathcal{C}[-\pi, \pi]$ with the inner product $\langle f, g \rangle = \int_{\pi}^{-\pi} f(x) \overline{g(x)} dx$. Let $M$ be the set of functions in $\mathcal{C}[-\pi, \pi]$ that vanish on the interval $[-\pi, 0]$, and let $N$ be the set of all functions in $\mathcal{C}[-\pi, \pi]$ that vanish on the interval $[0, \pi]$. Every function in $M$ is orthogonal to every function in $N$. Thus $N \subseteq M^\perp$ and $M \subseteq N^\perp$. ◆

**Theorem 7.1.4.** *For subsets $A$ and $B$ of an inner product space $H$,*

(a) $A \subseteq A^{\perp\perp}$;
(b) *if $A \subseteq B$, then $A^\perp \supseteq B^\perp$;*
(c) $A^\perp$ *is a closed subspace of $H$; and*
(d) $A^\perp = M^\perp$, *where $M = \mathrm{Span}(A)$.*

*Proof.* (a) *and* (b) *are obvious.*

(c) *Since $A^\perp = \cap_{a \in A} a^\perp$, it is enough to prove that $a^\perp$ is a closed subspace of $H$. If $\alpha, \beta \in \mathbb{K}$ and $x, y \in a^\perp$, then $\langle \alpha x + \beta y, a \rangle = \alpha \langle x, a \rangle + \beta \langle y, a \rangle = 0$. Thus $a^\perp$ is a subspace of $H$. If $x_n \in a^\perp$ and $\lim_n x_n = x$, then $\langle x, a \rangle = \langle \lim_n x_n, a \rangle = \lim_n \langle x_n, a \rangle = 0$. The continuity of the inner product in its arguments has been used here. The proof of part (d) is left as an exercise.* ■

**Definition.** If $M$ is a closed subspace of a Hilbert space $H$, the closed subspace $M^\perp$ is called the **orthogonal complement** of $M$ (rather than the annihilator of $M$.) The reason for the above terminology will become apparent in theorem 7.1.7.

Example 8 in section 4.7 is a very special case of the theorem below. Observe that the completeness of $H$ is crucial here.

**Theorem 7.1.5.** *Let $C$ be a closed convex subset of a Hilbert space $H$, and let $x \in H$. Then there exists a unique element $y \in C$ such that $\|x - y\| = \mathrm{dist}(x, C) = \inf_{z \in C} \|x - z\|$.*

*Proof.* If $x \in C$, take $y = x$. If $x \notin C$, then $\delta = \text{dist}(x, C) > 0$ because $C$ is closed. There exists a sequence $(y_n)$ in $C$ such that $\lim_n \|x - y_n\| = \delta$. By the parallelogram law,

$$\|y_n - y_m\|^2 = \|(y_n - x) - (y_m - x)\|^2 = 2\|y_n - x\|^2 + 2\|y_m - x\|^2$$
$$- \|y_n - x + y_m - x\|^2.$$

Now $\|y_n - x - y_m - x\|^2 = 4\|x - (\frac{y_n + y_m}{2})\|^2 \geq 4\delta^2$. The last inequality is true because $\frac{y_n + y_m}{2} \in C$ due to the convexity of $C$. Thus

$$\|y_n - y_m\|^2 \leq 2\|y_n - x\|^2 + 2\|y_m - x\|^2 - 4\delta^2 \to 0 \text{ as } m, n \to \infty.$$

This shows that $(y_n)$ is a Cauchy sequence, and hence $y = \lim_n y_n$ exists. Since $C$ is closed, $y \in C$. Now $\delta = \lim_n \|x - y_n\| = \|x - y\|$, and $y$ is one of the closest points in $C$ to $x$. To show that $y$ is unique, suppose $z \in C$ is such that $\|x - z\| = \delta$. By the parallelogram law, and as in the calculation above,

$$\|y - z\|^2 = 2\|y - x\|^2 + 2\|x - z\|^2 - \|y + z - 2x\|^2$$
$$= 2\delta^2 + 2\delta^2 - 4\|x - (\frac{y + z}{2})\|^2 \leq 2\delta^2 + 2\delta^2 - 4\delta^2 = 0. \blacksquare$$

**Corollary 7.1.6.** *If $C$ is a closed convex subset of a Hilbert space $H$, then $C$ contains a unique element of smallest norm.*

*Proof.* Apply the above theorem with $x = 0$. $\blacksquare$

**Theorem 7.1.7 (the projection theorem).** *Let $M$ be a closed subspace of a Hilbert space $H$. Then $H = M \oplus M^{\perp}$, where $M^{\perp}$ is the orthogonal complement of $M$.*

*Proof.* Let $x \in H$, let $y$ be the closest element of $M$ to $x$, and let $z = x - y$. Write $\delta = \text{dist}(x, M) = \|z\|$. We show that $z \in M^{\perp}$. Let $w \in M$ and, without loss of generality, assume that $\|w\| = 1$. For any $\alpha \in \mathbb{K}, y + \alpha w \in M$, so

$$\delta^2 \leq \|x - y - \alpha w\|^2 = \|z - \alpha w\|^2 = \langle z - \alpha w, z - \alpha w \rangle$$
$$= \|z\|^2 - \alpha \langle w, z \rangle - \overline{\alpha} \langle z, w \rangle + |\alpha|^2 \|w\|^2$$
$$= \delta^2 - 2\text{Re}(\alpha \langle w, z \rangle) + |\alpha|^2.$$

Therefore $2\text{Re}(\alpha \langle w, z \rangle) \leq |\alpha|^2$. Since the above is true for an arbitrary $\alpha$, choose $\alpha = \langle z, w \rangle$. We now have $2|\langle w, z \rangle|^2 \leq |\langle w, z \rangle|^2$; hence $\langle w, z \rangle = 0$. The proof is now complete because $M \cap M^{\perp} = \{0\}$. $\blacksquare$

As an immediate consequence of the above theorem, every element $x \in H$ can be written uniquely as $x = y + z$, where $y \in M$ and $z \in M^\perp$.

**Example 5.** Let $M$ be the set of all sequences in $l^2$ whose even terms are zero, and let $N$ be the set of all sequences in $l^2$ whose odd terms are zero. It is easy to see that $M$ and $N$ are closed subspaces of $l^2$ and that $N = M^\perp$. Trivially, every vector $x = (x_n) \in l^2$ can be written as $x = (x_1, 0, x_3, 0, ...) + (0, x_2, 0, x_4, ...) \in M \oplus N.$ ◆

**Definition.** The element $y$ in theorem 7.1.7 is called the **orthogonal projection** of $x$ on $M$. It is worth reiterating that $y$ is the closest element in $M$ to $x$. The mapping $P_M : H \to H$ defined by $P_M(x) = y$ is called the **projection operator** (or simply the projection) of $H$ onto $M$.

**Theorem 7.1.8.** *Let $M$ be a closed subspace of a Hilbert space $H$, and let $P = P_M$ be the projection of $H$ on $M$. Then*

(a) *$P$ is bounded and $\|P\| = 1$,*
(b) *$\mathfrak{R}(P) = M$, and*
(c) *$P^2 = P$.*

*Proof.* (a) Let $x, x' \in H$, and let $x = y + z$, and $x' = y' + z'$, where $y, y' \in M$ and $z, z' \in M^\perp$. Then $x + x' = (y + y') + (z + z')$. Since $y + y' \in M$ and $z + z' \in M^\perp$, the uniqueness of the orthogonal projection of $x + x'$ on $M$ forces $P(x + x') = y + y' = P(x) + P(x')$. The proof that $P(ax) = aP(x)$ is similar. Now $\|x\|^2 = \|y\|^2 + \|z\|^2$; thus $\|P(x)\| = \|y\| \leq \|x\|$, so $\|P\| \leq 1$. Since $P(x) = x$ for every $x \in M$, $\|P\| = 1$. This proves (a). Parts (b) and (c) are obvious. ∎

The following theorem gives a complete and simple characterization of the dual of a Hilbert space. The Riesz representation theorem basically says that a Hilbert space is isometrically isomorphic to itself in a very natural way.

**Theorem 7.1.9 (the Riesz representation theorem).** *Let $\lambda$ be a bounded linear functional on a Hilbert space $H$. Then there exists a unique element $y \in H$ such that $\lambda(x) = \langle x, y \rangle$ for all $x \in H$. Furthermore, $\|\lambda\| = \|y\|$.*

*Proof.* If $\lambda = 0$, take $y = 0$. Otherwise, let $M = \text{Ker}(\lambda)$; $M$ is a closed subspace of $H$ because $M = \lambda^{-1}(0)$, and $M \neq H$ because $\lambda \neq 0$. By the projection theorem, $H = M \oplus M^\perp$. Pick a nonzero element $z \in M^\perp$. Then $\lambda(z) \neq 0$, and, by replacing $z$ with $z/\lambda(z)$, we may assume that $\lambda(z) = 1$. For $x \in H$, $x = x - \lambda(x)z + \lambda(x)z$. It is easy to verify that $w = x - \lambda(x)z \in M$.

*Observe that $\langle x, z \rangle = \langle w, z \rangle + \langle \lambda(x)z, z \rangle = \lambda(x)\|z\|^2$. Define $y = z/\|z\|^2$. Then, by the above identity, $\lambda(x) = \frac{\langle x, z \rangle}{\|z\|^2} = \langle x, y \rangle$. To prove that $y$ is unique, suppose that there is another element $y_1 \in H$ such that $\langle x, y \rangle = \langle x, y_1 \rangle$ for all $x \in H$. Then $\langle x, y - y_1 \rangle = 0$ for all $x \in H$. Choose $x = y - y_1$. Then $\|y - y_1\|^2 = 0$; hence $y = y_1$. Finally, $|\lambda(x)| = |\langle x, y \rangle| \leq \|x\|\|y\|$. Thus $\|\lambda\| \leq \|y\|$.*

*Also $|\lambda(y)| = |\langle y, y \rangle| = \|y\|^2 = \|y\|\|y\|$. This shows that $\|\lambda\| \geq \|y\|$ and that $\|\lambda\| = \|y\|$.* ∎

Recall that a hyperplane in $\mathbb{R}^n$ is nothing other than the translation of the null-space of a linear functional on $\mathbb{R}^n$, that all linear functionals on $\mathbb{R}^n$ are continuous, and that all maximal subspaces are closed. In infinite dimensions, the null-space of a linear functional $\lambda$ is closed if and only if $\lambda$ is continuous. The following result is the exact analog of example 10 in section 4.7.

**Example 6.** Let $C$ be a closed convex subset of a real Hilbert space $H$, and let $a \in H - C$. Then there exists a bounded functional $\lambda$ on $H$ and a constant $b$ such that $\lambda(y) < b$ for every $y \in C$, and $\lambda(a) > b$.

The obtuse angle criterion extends to the current situation, and the proof is identical to that in example 9 in section 4.7. Thus if $z$ is the closest element in $C$ to $a$, then, for every $y \in C$, $\langle a - z, y - z \rangle \leq 0$. Let $m = (a + z)/2$, and define $n = a - z$, $\lambda(x) = \langle x, n \rangle$, and $b = \lambda(m)$. As in example 10 in section 4.7, we may assume that $m = 0$; hence $b = 0$. It is easy to verify that $\lambda(y) < 0$ for all $y \in C$ and that $\lambda(a) > 0$. ◆

## The Completion of an Inner Product Space.

**Example 7.** If $(x_n)$ and $(y_n)$ are Cauchy sequences in an inner product space, then $\lim_n \langle x_n, y_n \rangle$ exists.

We prove that the sequence $\langle x_n, y_n \rangle$ is Cauchy in $\mathbb{C}$; hence the limit in question exists. Recall that Cauchy sequences are bounded. Now

$$|\langle x_n, y_n \rangle - \langle x_m, y_m \rangle| = |\langle x_n - x_m, y_n \rangle + \langle x_m, y_n - y_m \rangle|$$
$$\leq \|x_n - x_m\|\|y_n\| + \|x_m\|\|y_n - y_m\| \to 0 \text{ as } m, n \to \infty. ◆$$

**Theorem 7.1.10.** *Let $(X, \langle ., . \rangle)$ be an incomplete inner product space. Then there exists a Hilbert space $H$ that contains $X$ as a dense subspace such that the inner product on $X$ is the restriction of the inner product on $H$. If $X$ is separable, so is $H$.*

*Proof.* Let $\|.\|$ be the norm on $X$ induced by the inner product, and let $H$ be the completion of $X$ with respect to the norm on $X$ (theorem 6.4.9). Refer to the extended norm by $\|.\|'$. For $x, y \in H$, choose sequences $(x_n)$ and $(y_n)$ in $X$ such that $x_n \to x$ and $y_n \to y$, and extend the definition of the inner product to $H$ by $\langle x, y \rangle' = \lim_n \langle x_n, y_n \rangle$. We leave it to the reader to verify that the inner product we just defined is well defined and that it is indeed an inner product. Clearly, the inner product on $H$ extends that on $X$. Finally, we prove that that the extended inner product induces the extended norm on $H$. For a sequence $(x_n)$ converging to $x \in H$, $\langle x, x \rangle' = \lim_n \langle x_n, x_n \rangle = \lim_n \|x_n\|^2 = \lim_n (\|x_n\|')^2 = (\|x\|')^2$. ∎

## Exercises

1. Prove that the norm on $\mathcal{C}[0,1]$ generated by the inner product

$$\langle f, g \rangle = \int_0^1 f(x)\overline{g(x)}dx$$

is not complete.

2. Prove the parallelogram law and the polarization identity.

3. Let $x$ and $y$ be nonzero vectors in an inner product space. Prove that there exists a unique number $0 \le \theta \le \pi$ such that $\cos \theta = \frac{Re\langle x,y \rangle}{\|x\|\|y\|}$. Conclude that $\|x+y\|^2 = \|x\|^2 + \|y\|^2 + 2\|x\|\|y\| \cos \theta$.

4. Prove the **Apollonius identity**: For vectors $x, y$, and $z$ in an inner product space, $\|z-x\|^2 + \|z-y\|^2 = \frac{1}{2}\|x-y\|^2 + 2\|z - \frac{x+y}{2}\|^2$.

5. Let $A$ be a subset of a Hilbert space $H$, and let $M = Span(A)$. Prove that $A^\perp = M^\perp$.

6. Let $M$ be a closed subspace of a Hilbert space. Prove that $M = M^{\perp\perp}$. Give an example to show that the result fails if $M$ is not closed. More generally, show that $M^{\perp\perp} = \overline{M}$.

7. Show that if $A$ is a subset of a Hilbert space $H$, then $A^{\perp\perp}$ is the smallest closed subspace of $H$ containing $A$.

8. Let $(x_n)$ and $(y_n)$ be sequences in an inner product space. Prove that
   (a) if $\lim_n x_n = 0$, and $(y_n)$ is bounded, than $\lim_n \langle x_n, y_n \rangle = 0$; and
   (b) if $y \perp x_n$ for each $n \in \mathbb{N}$, and $\sum_{n=1}^\infty x_n$ is convergent, then $y \perp \sum_{n=1}^\infty x_n$.

9. Prove that if an element $x$ in a Hilbert space is orthogonal to every vector in a dense subset of $H$, then $x = 0$.

10. Let $(x_n)$ be a sequence of mutually orthogonal vectors in a Hilbert space $H$. Prove that $\sum_{n=1}^\infty x_n$ converges in $H$ if and only if $\sum_{n=1}^\infty \|x_n\|^2 < \infty$.

11. Use Hilbert space methods to provide an easy proof of the Hahn-Banach theorem for Hilbert spaces.

12. Let $M = \{x = (x_1, \ldots, x_n) \in \mathbb{R}^n : \sum_{i=1}^{n} x_i = 1\}$. Show that $M$ is closed and convex, and find the element in $M$ closest to the origin.

13. Let $C$ be a closed convex subset of a Hilbert space $H$, let $x \in H - C$, and let $y$ be the closest element of $C$ to $x$. Prove that, for every $z \in C$, $Re\langle x - y, z - y \rangle \leq 0$.

14. Let $\delta_n$ be a positive sequence, and let $C = \{x \in l^2 : |x_n| \leq \delta_n\}$. Show that $C$ is compact if and only if $\sum_{n=1}^{\infty} \delta_n^2 < \infty$.

## 7.2  Orthonormal Bases and Fourier Series

In the introduction to section 7.1, we made the case for the existence of a maximal orthonormal sequence $\{u_1, u_2, \ldots\}$ in a Hilbert space $H$. As you will see in this section, some Hilbert spaces do not admit countable maximal orthonormal subsets. Perhaps we must first tackle the problem of the existence of a maximal orthogonal subset of $H$, then examine the problem of which Hilbert spaces possess a countable such subset. In this section, we provide solutions to both problems and reveal the basic structure of a Hilbert space, hence paving the way to answer the problems posed in section 4.10.

The proof of the following theorem can be seen in section 3.7.

**Theorem 7.2.1.** *An orthogonal subset S of a Hilbert space H is independent.* ■

**Definition.** An **orthonormal basis** for a Hilbert space $H$ is a maximal orthonormal subset of $H$. An orthonormal subset of $H$ is maximal if it is not properly contained in another orthonormal subset of $H$.

**Example 1.** We show that $\{e_n : n \in \mathbb{N}\}$ is an orthonormal basis for $l^2$.
It is clear that $S$ is orthonormal. If $x = (x_n) \in l^2$ is orthogonal to $S$, then, for every $n \in \mathbb{N}$, $x_n = \langle x, e_n \rangle = 0$, and hence $x = 0$. ◆

In the theorem below, we prove a little more than the existence of an orthonormal basis for an arbitrary Hilbert space.

**Theorem 7.2.2.** *Every orthonormal subset A of a Hilbert space H is contained in an orthonormal basis for H. In particular, every Hilbert space contains an orthonormal basis.*

*Proof.* Let $\mathfrak{B}$ be the collection of all orthonormal subsets of $H$ that contain $A$; $\mathfrak{B}$ is not empty since $A$ is one of its members. Order $\mathfrak{B}$ by set inclusion. It is rather straightforward to show that the union of the members of a chain in $\mathfrak{B}$ is an orthonormal subset of $H$ that contains $A$ and is therefore an upper bound of the chain. By Zorn's lemma, $\mathfrak{B}$ has a maximal member, that is, an orthonormal basis of $H$ containing $A$. To prove that every Hilbert space possesses an orthonormal basis, apply the result we just proved with $A = \{x\}$, where $x$ is a unit vector. ∎

The goal of this section is to represent an arbitrary element of a Hilbert space $H$ in terms of a basis of some kind. If $dim(H) < \infty$, the goal is too trivial, and if $dim(H) = \infty$, the goal is unrealistic if one insists on looking at a Hamel basis because any such basis is uncountable and hence too big to be useful. The only realistic expectation is to hope to express an arbitrary element of $H$ as a series of the basis elements, as was achieved in section 4.10 for trigonometric series of continuous functions. This means that $H$ has a Schauder basis, which immediately suggests that we investigate separable Hilbert spaces (see problem 12 on section 6.1). The following theorem is the happy coincidence we hope for.

**Theorem 7.2.3.** *A Hilbert space $H$ is separable if and only if every orthonormal basis of $H$ is countable.*

*Proof.* If $H$ is separable, then $H$ contains a countable dense subset $\{x_1, x_2, ...\}$ and, clearly, $H = \cup_{n \in \mathbb{N}} B(x_n, 1/2)$. If $S = \{u_\alpha\}_{\alpha \in I}$ is an orthonormal basis for $H$, then, for $\alpha, \beta \in I$, $\|u_\alpha - u_\beta\| = \sqrt{2}$. Since the diameter of each of the balls $B(x_n, 1/2)$ is 1, no such ball can contain more that one member of $S$. Therefore $S$ is at most countable.

Conversely, if $H$ possesses a countable orthonormal basis $S = \{u_n : n \in \mathbb{N}\}$, let $A$ be the collection of all finite linear combinations of element in $S$ with coefficients in $\mathbb{Q} + i\mathbb{Q}$. We claim that $A$ is dense in $H$. This will conclude the proof because $A$ is countable. To prove the claim, let $M$ be the closure of $A$. To show that $M$ is a subspace of $H$, let $x, y \in M$, and let $a, b \in \mathbb{K}$. Then there exist sequences $(x_n)$ and $(y_n)$ in $A$, and sequences $a_n, b_n \in \mathbb{Q} + i\mathbb{Q}$ such that $\lim_n x_n = x, \lim_n y_n = y, \lim_n a_n = a$, and $\lim_n b_n = b$. The sequence $(a_n x_n + b_n y_n)$ is in $A$, and $\lim_n a_n x_n + b_n y_n = ax + by$. Therefore $ax + by \in M$. We now show that $M = H$. If not, then $H = M \oplus M^\perp$, and $M^\perp \neq \{0\}$. Pick a unit vector $z \in M^\perp$. Then $S \cup \{z\}$ is an orthonormal subset of $H$ that properly contains $S$. This contradicts the maximality of $S$ and completes the proof. ∎

**Example 2.** It is possible for a separable inner product space (hence for a separable Hilbert space) to contain uncountably many pairs of orthogonal vectors. Consider the space $\mathcal{C}[-\pi, \pi]$ with the inner product $\langle f, g \rangle = \frac{1}{2\pi} \int_{-\pi}^{\pi} f(x)\overline{g(x)}$;

$\mathcal{C}[-\pi, \pi]$ is separable.[2] In the notation of example 4 on section 7.1, every pair of functions $(f, g) \in M \times N$ is orthogonal. Since both $M$ and $N$ are uncountable, we have proved our assertion. ◆

We focus mostly but not exclusively on separable Hilbert spaces. The existence of inseparable Hilbert spaces of arbitrary Hilbert dimension will be presented in the excursion at the end of this section. Many of the results we develop in this chapter are valid for inseparable Hilbert spaces. Examples include the projection theorem, the Riesz representation theorem, and the next three theorems. Also, in the definition below, the set $I$ need not be countable; hence $H$ is not assumed to be separable.

**Definition.** Let $S = \{u_\alpha : \alpha \in I\}$ be an orthonormal subset of a Hilbert space $H$. It is not assumed that $S$ is an orthonormal basis. For an element $x \in H$, the scalars $\hat{x}_\alpha = \langle x, u_\alpha \rangle$ are called the **Fourier coefficients** of $x$ relative to $S$.

**Theorem 7.2.4.** Let $S = \{u_1, \dots, u_n\}$ be an orthonormal subset of $H$, and let $x \in \mathrm{Span}(S)$. Then $x = \sum_{i=1}^{n} \hat{x}_i u_i$, and $\|x\|^2 = \sum_{i=1}^{n} |\hat{x}_i|^2$.

*Proof.* See the proof of theorem 3.7.5. ∎

**Theorem 7.2.5.** *Suppose* $S = \{u_1, \dots, u_n\}$ *be an orthonormal subset of* $H$, *and let* $M = \mathrm{Span}(S)$. *For a vector* $x \in H$, *the vector* $y = \sum_{i=1}^{n} \hat{x}_i u_i$ *is the orthogonal projection of* $x$ *on* $M$. *In particular, for all scalars* $a_1, \dots, a_n$, $\|x - \sum_{i=1}^{n} \hat{x}_i u_i\| \leq \|x - \sum_{i=1}^{n} a_i u_i\|$. *Furthermore* $\sum_{i=1}^{n} |\hat{x}_i|^2 \leq \|x\|^2$.

*Proof.* We only need to show that the vector $z = x - y = x - \sum_{i=1}^{n} \hat{x}_i u_i$ is in $M^\perp$. The rest of the assertions follow from the projection theorem and theorem 7.2.4. Now, for a fixed $1 \leq j \leq n$,

$$\langle z, u_j \rangle = \langle x, u_j \rangle - \sum_{i=1}^{n} \hat{x}_i \langle u_i, u_j \rangle = \langle x, u_j \rangle - \hat{x}_j = 0. \quad \blacksquare$$

**Theorem 7.2.6 (Bessel's inequality).** *Let* $\{u_n\}$ *be an orthonormal subset of a Hilbert space* $H$. *Then, for* $x \in H$, $\sum_{n=1}^{\infty} |\hat{x}_n|^2 \leq \|x\|^2$.

*Proof.* By theorem 7.2.5, $\sum_{i=1}^{n} |\hat{x}_i|^2 \leq \|x\|^2$ for each $n \in \mathbb{N}$. Taking the limit as $n \to \infty$ yields Bessel's inequality. ∎

---

[2] The set of trigonometric polynomials with rational coefficients is dense in $\mathcal{C}[-\pi, \pi]$. See corollary 4.10.3.

**Theorem 7.2.7.** *Let H be a separable Hilbert space, and let $S = \{u_n : n \in \mathbb{N}\}$ be an orthonormal subset of H. Then the following are equivalent:*

(a) *S is an orthonormal basis for H.*
(b) *For every $x \in H$, $x = \sum_{n=1}^{\infty} \hat{x}_n u_n$.*
(c) *Span(S) is dense in H.*
(d) *For every $x \in H$, $\|x\|^2 = \sum_{n=1}^{\infty} |\hat{x}|^2$.*
(e) **Parseval's identity.** *For every $x, y \in H$, $\langle x, y \rangle = \sum_{n=1}^{\infty} \hat{x}_n \overline{\hat{y}}_n$.*

*Proof.* (a) implies (b). Let $y_n = \sum_{k=1}^{n} \hat{x}_k u_k$.

For $n < m$, $\|y_m - y_n\|^2 = \|\sum_{k=n+1}^{m} \hat{x}_k u_k\|^2 = \sum_{k=n+1}^{m} |\hat{x}_k|^2 \to 0$ as $m, n \to \infty$, because $\sum_{n=1}^{\infty} |\hat{x}_n|^2 < \infty$ (Bessel's inequality). This shows that $(y_n)$ is a Cauchy sequence in H; hence it converges to, say y. Thus $y = \sum_{n=1}^{\infty} \hat{x}_n u_n$. We need to show that $y = x$. For a fixed $k \in \mathbb{N}$, $\langle y, u_k \rangle = \lim_{n \to \infty} \langle y_n, u_k \rangle = \langle x, u_k \rangle$. Thus $y - x$ is orthogonal to each $u_k$. If $y - x \neq 0$, then $S \cup \{\frac{y-x}{\|y-x\|}\}$ would be an orthonormal set that properly contains S. The maximality of S forces $y = x$.

That (b) implies (c) is obvious, since $x = \lim_{n \to \infty} \sum_{k=1}^{n} \hat{x}_k u_k$, and each $\sum_{k=1}^{n} \hat{x}_k u_k$ is in Span(S).

(c) implies (d). Suppose, for some $x \in H$, $\|x\|^2 > \sum_{n=1}^{\infty} |\hat{x}_n|^2$, and let $\delta^2 = \|x\|^2 - \sum_{n=1}^{\infty} |\hat{x}_n|^2$. We show that the ball $B(x, \delta)$ contains no finite linear combination of S. This will show that Span(S) is not dense in H. If $\sum_{k=1}^{n} a_k u_k \in$ Span(S), then, by theorem 7.2.5, $\|x - \sum_{k=1}^{n} a_k u_k\|^2 \geq \|x - \sum_{k=1}^{n} \hat{x}_k u_k\|^2 = \|x\|^2 - \|\sum_{k=1}^{n} \hat{x}_k u_k\|^2 = \|x\|^2 - \sum_{k=1}^{n} |\hat{x}_k|^2 = \delta^2$.

(d) implies (e). The identity $\|x\|^2 = \sum_{n=1}^{\infty} |\hat{x}_n|^2$ can be written as $\|x\|^2 = \|\hat{x}\|_2^2$, where $\hat{x} = (\hat{x}_n) \in l^2$, and $\|\hat{x}\|_2$ is the $l^2$-norm of $\hat{x}$. Now, assuming (d) is true, then, for every $\alpha \in \mathbb{K}$, $\langle x + \alpha y, x + \alpha y \rangle = \langle \hat{x} + \alpha \hat{y}, \hat{x} + \alpha \hat{y} \rangle$. Equivalently, $\alpha \langle y, x \rangle + \overline{\alpha} \langle x, y \rangle = \alpha \langle \hat{y}, \hat{x} \rangle + \overline{\alpha} \langle \hat{x}, \hat{y} \rangle$. Setting $\alpha = 1/2$, we obtain $\text{Re}(\langle x, y \rangle) = \text{Re}(\langle \hat{x}, \hat{y} \rangle)$. Setting $\alpha = 1/2i$ yields $\text{Im}\langle x, y \rangle = \text{Im}\langle \hat{x}, \hat{y} \rangle$. This proves that $\langle x, y \rangle = \langle \hat{x}, \hat{y} \rangle$, which is equivalent to $\langle x, y \rangle = \sum_{n=1}^{\infty} \hat{x}_n \overline{\hat{y}}_n$.

(e) implies (a). Suppose there exists a unit vector u such that $S \cup \{u\}$ is orthonormal. Then $\hat{u}_k = \langle u, u_k \rangle = 0$ for all $k \in \mathbb{N}$, and $1 = \langle u, u \rangle = \sum_{k=1}^{\infty} \hat{u}_k \overline{\hat{u}}_k = 0$. This contradiction shows that (a) is true. ∎

**Example 3.** Every element in $l^2$ can be written as a series $x = \sum_{n=1}^{\infty} x_n e_n$.

Consider the vectors $y_n = x - \sum_{i=1}^{n} x_i e_i = (0, 0, \ldots, 0, x_{n+1}, x_{n+2}, \ldots)$. Since $\lim_n \|y_n\|^2 = \lim_n \sum_{i=n+1}^{\infty} |x_i|^2 = 0$, $x = \sum_{n=1}^{\infty} x_n e_n$. ◆

**Example 4.** Consider the countable set of functions $S = \{e^{int} : n \in \mathbb{Z}\}$. By corollary 4.10.3, $Span(S)$ is dense in $(\mathcal{C}[-\pi, \pi], \|.\|_2)$. Therefore $Span(S)$ is dense in the completion, $H$, of $(\mathcal{C}[-\pi, \pi], \|.\|_2)$. Thus the set $\{e^{int} : n \in \mathbb{Z}\}$ is an orthonormal basis for $H$. We will see in section 8.9 that $H$ is the space $\mathfrak{L}^2(-\pi, \pi)$ of (Lebesgue) square integrable functions on $(-\pi, \pi)$.

For exactly the same reason, (see theorem 4.10.8), the set of normalized Legendre polynomials $\{\tilde{P}_n : n \in \mathbb{N}\}$ is an orthonormal basis for $\mathfrak{L}^2(-1, 1)$. ◆

**Definition.** Two Hilbert spaces $H_1$ and $H_2$ are isomorphic (as Hilbert spaces) if there exists an isomorphism $T : H_1 \to H_2$ such that, for all $x, y \in H_1$,

$$\langle x, y \rangle = \langle Tx, Ty \rangle.$$

It follows directly that such an isomorphism is also an isometry because

$$\|x\|^2 = \langle x, x \rangle = \langle Tx, Tx \rangle = \|Tx\|^2.$$

**Theorem 7.2.8 (the Riesz-Fisher theorem).** *Let $H$ be a separable Hilbert space.*

(a) *If $dim(H) = n$, then $H$ is isomorphic to $\mathbb{K}^n$.*
(b) *If $dim(H) = \infty$, then $H$ is isomorphic to $l^2$.*

*Proof. We only prove the second statement. The proof of the first statement is simpler. Let $\{u_n\}$ be an orthonormal basis for $H$. For $x \in H$, let $T(x) = (\hat{x}_n)_{n=1}^\infty = \hat{x}$; $T : H \to l^2$ is linear since $(ax + by)\hat{} = a\hat{x} + b\hat{y}$. The fact that $\langle x, y \rangle = \langle Tx, Ty \rangle$ is Parseval's identity in theorem 7.2.7. To verify that $T$ is one-to-one, suppose that $\hat{x} = \hat{y}$. Then $(x - y)\hat{} = 0$, and $\sum_{n=1}^\infty |\hat{x}_n - \hat{y}_n|^2 = 0$. Therefore $\hat{x}_n = \hat{y}_n$. Hence, by theorem 7.2.7, $x = \sum_{n=1}^\infty \hat{x}_n u_n = \sum_{n=1}^\infty \hat{y}_n u_n = y$; $T$ is onto because if $(a_n) \in l^2$, then the series $\sum_{n=1}^\infty a_n u_n$ converges to a vector $x \in H$ such that $\hat{x} = (a_n)$. See problem 3 at the end of this section.* ∎

We offer a few observations on some crucial differences between Banach and Hilbert spaces. This will hopefully explain why Hilbert spaces have such an elegant and uncluttered structure compared to a general Banach space.

The closest point property and the projection theorem (theorems 7.1.5 and 7.1.7, respectively) are at the heart of the constructions of this chapter. An examination of the proof of theorem 7.1.5 reveals that the parallelogram law delivers both the existence and the uniqueness of the closest point to a closed convex set. The parallelogram law is a direct result of the fact that the norm on a Hilbert space is induced by an inner product, which is what sets Hilbert spaces apart from general Banach spaces, where the closest point property fails as does the conclusion of the projection theorem. The following simple example illustrates one of the discussion points.

**Example 5.** Consider the space $\mathbb{R}^2$ with the norm $\|x\|_\infty = max\{|x_1|, |x_2|\}$. The set $M = \{x = (x_1, x_2) \in \mathbb{R}^2 : |x_1| \le 1, |x_2| \le 1\}$ is closed and convex. Every point on the line segment $\{(1, y) : |y| \le 1\}$ has distance 1 from the point $x = (2, 0)$, and $dist(x, M) = 1$. There are examples where the very existence of a closest point is not guaranteed. See problems 6–8 at the end of this section for a slight expansion of this discussion. ♦

It was mentioned in section 6.4 that not every closed subspace of a Banach space is complemented. Theorem 7.1.7 guarantees that every closed subspace of a Hilbert space is complemented. Projections in Banach spaces play a similar role to orthogonal projections in proving that certain closed subspaces are complemented. See problem 10 on section 6.4 for necessary and sufficient conditions for a closed subspace of a Banach space to be complemented. Also examine example 6 in section 6.4.

## Excursion: Inseparable Hilbert Spaces

Inseparable Hilbert spaces do exist. They are mostly a curiosity and do not have much practical use. We include the discussion below for the satisfaction of the inquisitive reader.

The motivation for the definition below and the construction in theorem 7.2.9 is provided by the following example.

**Example 6.** Let $S = \{u_\alpha : \alpha \in I\}$ be an uncountable orthonormal subset of a Hilbert space $H$. For a vector $x \in H$, consider the set of Fourier coefficients $\{\hat{x}_\alpha : \alpha \in I\}$. We claim that $\hat{x}_\alpha = 0$ for all but countably many $\alpha \in I$.

Let $\{u_{\alpha_1}, ..., u_{\alpha_n}\}$ be a finite subset of $S$. By theorem 7.2.5, $\sum_{i=1}^n |\hat{x}_{\alpha_i}|^2 \le \|x\|^2 < \infty$. It follows that $\sum_{\alpha \in I} |\hat{x}_\alpha|^2 < \infty$ (see example 1 in section 4.10 and the definition preceding it); hence the set $\{\alpha \in I : \hat{x}_\alpha \ne 0\}$ is countable. ♦

The above example strongly suggests the following definition.

**Definition.** Let $I$ be an infinite set, and let $\aleph = Card(I)$. Define $l^2(\aleph)$ to be the set of all functions $x : I \to \mathbb{C}$ such that $x_\alpha = 0$ for all but countably many $\alpha \in I$ and $\|x\| = (\sum_{\alpha \in I} |x_\alpha|^2)^{1/2} < \infty$. To eliminate any danger of ambiguity, let $I_x = \{\alpha_1, \alpha_2, ...\}$ be the subset of $I$ for which $x_\alpha \ne 0$. The notation $\sum_{\alpha \in I} |x_\alpha|^2$ means $\sum_{i=1}^\infty |x_{\alpha_i}|^2$. We will continue to employ this notation for the remainder of this discussion.

**Theorem 7.2.9.** *The set $l^2(\aleph)$ is a Hilbert space with the operations defined within the proof.*

*Proof. Let $x = (x_\alpha)$, and $y = (y_\alpha) \in l^2(\aleph)$. We show that $x + y \in l^2(\aleph)$ and that $\|x + y\| \leq \|x\| + \|y\|$. Let $I_x = \{\alpha \in I : x_\alpha \neq 0\}$, and $I_y = \{\alpha \in I : y_\alpha \neq 0\}$, and let $J = I_x \cup I_y$. Since $J$ is countable, we can write $J = \{\alpha_1, \alpha_2, ...\}$. Note that $\hat{x} = (x_{\alpha_i})$ and $\hat{y} = (y_{\alpha_i})$ are in $l^2$; hence $\|\hat{x} + \hat{y}\|_2 \leq \|\hat{x}\|_2 + \|\hat{y}\|_2$. But every $\alpha$ for which $x_\alpha + y_\alpha \neq 0$ is in $J$; hence $\|x + y\| = (\sum_{i=1}^{\infty} |x_{\alpha_i} + y_{\alpha_i}|^2)^{1/2} = \|\hat{x} + \hat{y}\|_2 \leq \|\hat{x}\|_2 + \|\hat{y}\|_2 = \|x\| + \|y\|$. The fact that $\|ax\| = |a| \|x\|$ for all $x \in l^2(\aleph)$ and all scalars $a$ requires an even simpler argument. The rest of the properties of a normed linear space are easily verifiable. Thus $l^2(\aleph)$ is a normed linear space.*

*Define an inner product on $l^2(\aleph)$ as follows: $\langle x, y \rangle = \langle \hat{x}, \hat{y} \rangle = \sum_{i=1}^{\infty} x_{\alpha_i} \overline{y}_{\alpha_i}$. This inner product induces the norm on $l^2(\aleph)$ we defined earlier. We now show the completeness of $l^2(\aleph)$. Suppose $(x^{(n)})$ is a Cauchy sequence in $l^2(\aleph)$, let $I_n = \{\alpha \in I : x_\alpha^{(n)} \neq 0\}$, and let $J = \cup_{n \in \mathbb{N}} I_n$. Then $J$ is a countable subset of $I$, and we can write $J = \{\alpha_1, \alpha_2, ...\}$. Since $\|\hat{x}^{(m)} - \hat{y}^{(n)}\| = \|x^{(n)} - y^{(n)}\|$, $(\hat{x}^{(n)})$ is a Cauchy sequence in $l^2$ and is therefore convergent to an element $\hat{x} = (x_1, x_2, ...) \in l^2$. Define $x \in l^2(\aleph)$ by*

$$x_\alpha = \begin{cases} x_i & \text{if } \alpha = \alpha_i, \\ 0 & \text{otherwise.} \end{cases}$$

*Clearly, $x^{(n)}$ converges to $x$ in $l^2(\aleph)$.* ∎

The reader can now anticipate the theorem that must be stated next: the set $\{e_\alpha\}_{\alpha \in I}$ is an orthonormal basis for $l^2(\aleph)$, where $e_\alpha(\beta) = \delta_{\alpha, \beta}$. Thus, for any cardinal number $\aleph$, we have constructed a Hilbert space whose orthonormal basis has cardinality $\aleph$. Such a space is also unique up to Hilbert space isomorphism in the sense that it depends only on $\aleph$ and not on the particular set $I$ in the above construction. We leave it to the interested reader to reflect on the details.

The cardinality of an orthonormal basis of a Hilbert space $H$ is known as the **Hilbert dimension** of H.

## Exercises

1. Let $\{u_n\}$ be an orthonormal basis for a separable Hilbert space $H$, and let $\{v_n\}$ be an orthonormal set in $H$ such that $\sum_{n=1}^{\infty} \|u_n - v_n\|^2 < 1$. Prove that $\{v_n\}$ is an orthonormal basis for $H$.

2. Let $S = \{v_1, v_2, ...\}$ be an orthonormal subset of a separable Hilbert space $H$ (not necessarily an orthonormal basis), and let $M = \overline{Span(S)}$. Prove that if $P$ is the projection of $H$ onto $M$, then $Px = \sum_{i=1}^{\infty} \langle x, v_i \rangle v_i$.

3. Let $\{u_n\}$ be an orthonormal basis for a separable Hilbert space $H$, and let $(a_n) \in l^2$. Prove that there is an element $x \in H$ such that $\hat{x}_n = a_n$.

4. Use theorems 6.2.4 and 7.2.8 to provide an alternative proof of the Riesz representation theorem for separable Hilbert spaces.

5. Let $\{u_n\}$ be an orthonormal basis for a separable Hilbert space $H$. Define a function $\|.\|' : H \to \mathbb{R}$ as follows: $\|x\|' = \sum_{n=1}^{\infty} 2^{-n}|\hat{x}_n|$. Show that $\|.\|'$ is a norm on $H$ and that it is not equivalent to the original norm on $H$.

6. Let $X = \mathcal{C}[0,1]$ endowed with the uniform norm, and let $M$ be the subset of $X$ consisting of all functions $f$ such that $f(0) = 0, f(1) = 1, f \geq 0$, and $\int_0^1 f(x)dx = 1$. Prove that $M$ is closed and convex and that $dist(0, M) = 1$. Also show that, for every $f \in M$, $\|f\|_\infty > 1$, and hence $M$ contains no element of smallest norm.

**Definition.** A Banach space $X$ is strictly convex if whenever $x \neq y$, and $\|x\| = \|y\|$, then $\|\frac{x+y}{2}\| < \|x\|$. Geometrically, strict convexity means that if $x$ and $y$ are equidistant from the origin, then the midpoint is strictly closer to the origin than $x$ (and y).

7. Prove that a Hilbert space is strictly convex.

8. Let $X$ be a strictly convex Banach space, let $M$ be a closed convex subset, and let $x \in X$. Show that if there is a point $y \in M$ such that $\|x - y\| = dist(x, M)$, then $y$ is unique.

**Definition.** A sequence $(x_n)$ in a Hilbert space is said to **converge weakly** to $x \in H$ if, for every $y \in H$, $\lim_n \langle x_n, y \rangle = \langle x, y \rangle$. In light of the Riesz representation theorem, this definition is consistent with the corresponding definition for Banach spaces introduced in the problem set in section 6.4. Some of the exercises below are repetitive of problems on section 6.4, but the proofs can be significantly simplified in the context of Hilbert spaces.

9. Prove that if $\{u_n\}$ is an orthonormal basis for a separable Hilbert space, then $(u_n)$ converges weakly to 0.

10. Show that a norm convergent sequence is weakly convergent but not conversely.

11. Show that the weak limit of a sequence in a Hilbert space, if it exists, is unique.

12. Show that if $\lim_n \langle x_n, y \rangle = \langle x, y \rangle$ for every $y$ in a dense subset of $H$, then $x_n \to^w x$.

13. Let $\{u_n\}$ be an orthonormal basis for a separable Hilbert space $H$. Prove the $x_n \to^w x$ if and only if $\lim_n \langle x_n, u_j \rangle = \langle x, u_j \rangle$ for every $j \in \mathbb{N}$.

14. Show that if $x_n \to^w x$, and $\lim_n \|x_n\| = \|x\|$, then $x_n$ is norm convergent to $x$.

15. Prove that if $x_n \to^w x$, then $\{\|x_n\|\}$ is bounded and $\|x\| \leq \liminf_n \|x_n\|$. Hint: The linear functionals $\lambda_n(\xi) = \langle \xi, x_n \rangle$ are pointwise bounded.

16. Let $x_n \to^w x$. Prove that there exists a subsequence $x_{n_k}$ of $(x_n)$ such that $\frac{1}{N}\sum_{k=1}^{N} x_{n_k}$ is strongly convergent to $x$. Hint: Without loss of generality, assume that $x = 0$. Inductively define a subsequence $x_{n_k}$ of $(x_n)$ such that $|\langle x_{n_i}, x_{n_k} \rangle| < 2^{-k}$ for $i = 1, \ldots, k-1$. Now

$$\|\frac{1}{N}\sum_{k=1}^{N} x_{n_k}\|^2 = \frac{1}{N^2}\sum_{k=1}^{N} \|x_{n_k}\|^2 + \frac{1}{N^2}\sum_{k=2}^{N}\sum_{i=1}^{k-1} 2Re\langle x_{n_i}, x_{n_k} \rangle.$$

This is a version of the **Banach-Saks theorem**.

## 7.3  Self-Adjoint Operators

This section establishes the broad characteristics of self-adjoint operators. We also study projection operators and prove a theorem (7.3.11) which produces ample examples of self-adjoint operators. Self-adjoitness (more generally, normality) is not nearly sufficient to produce a result resembling the spectral theorem for normal operators on finite-dimensional inner product spaces. The complete picture emerges in the next section.

In the finite-dimensional case, the definition of the adjoint is straightforward, owing to the simplicity of the characterization of linear functional on a finite-dimensional inner product spaces (see example 9 in section 3.7). The definition below in the infinite-dimensional case requires the full power of the Riesz representation theorem. Let $H$ be a separable Hilbert space, and let $T \in \mathcal{L}(H)$. For a fixed element $y \in H$, define a functional $\lambda_y$ by $\lambda_y(x) = \langle Tx, y \rangle$. It is clear that $\lambda_y$ is linear. In fact, $\lambda_y$ is bounded since $|\lambda_y(x)| = |\langle Tx, y \rangle| \leq \|Tx\|\|y\| \leq \|T\|\|x\|\|y\|$. By the Riesz representation theorem, there exists a unique element $T^*y \in H$ such that, for all $x \in H$, $\lambda_y(x) = \langle x, T^*y \rangle$. We therefore have a function $T^* : H \to H$ defined by the requirement that

$$\langle Tx, y \rangle = \langle x, T^*y \rangle$$

for all $x, y \in H$.

The above equation is the defining property of the **adjoint operator** $T^*$ of $T$. The reader can easily see that the definition is consistent with the definition of the adjoint operator on a Banach space that was introduced in section 6.6.

It is easy to verify that $T^*$ is linear. For example,

$$\langle x, T^*(y_1 + y_2)\rangle = \langle Tx, y_1 + y_2\rangle = \langle Tx, y_1\rangle + \langle Tx, y_2\rangle$$
$$= \langle x, T^*y_1\rangle + \langle x, T^*y_2\rangle = \langle x, T^*y_1 + T^*y_2\rangle.$$

This shows that $T^*(y_1 + y_2) = T^*(y_1) + T^*(y_2)$. We now show that $T^* \in \mathcal{L}(H)$.
$\|T^*y\|^2 = \langle T^*y, T^*y\rangle = \langle T(T^*y), y\rangle \le \|T(T^*y)\|\|y\| \le \|T\|\|T^*y\|\|y\|$. Thus $\|T^*y\| \le \|T\|\|y\|$ for every $y \in H$. Hence $T^* \in \mathcal{L}(H)$, and $\|T^*\| \le \|T\|$.

**Theorem 7.3.1.** *For $T, T_1, T_2 \in \mathcal{L}(H)$ and $\alpha \in \mathbb{K}$,*

(a) $(T_1 + T_2)^* = T_1^* + T_2^*$.
(b) $(\alpha T)^* = \bar{\alpha} T^*$.
(c) $(T_1 T_2)^* = T_2^* T_1^*$. *Consequently, for every $n \in \mathbb{N}$, $(T^*)^n = (T^n)^*$.*
(d) $T^{**} = T$.
(e) $\|T^*\| = \|T\|$.
(f) $\|T^* T\| = \|T\|^2$.

*Proof.* The computations needed to prove parts (a)–(d) are simple. As an example, we establish part (c): $\langle T_1 T_2 x, y\rangle = \langle T_2 x, T_1^* y\rangle \langle x, T_2^* T_1^* y\rangle$, which, by definition, means that $(T_1 T_2)^* = T_2^* T_1^*$. We already saw that $\|T^*\| \le \|T\|$. Applying the same fact to $T^*$ and using part (d), we have $\|T^*\| \le \|T^{**}\| = \|T\|$, thus proving (e). To prove (f), $\|T^* T\| \le \|T^*\|\|T\| = \|T\|^2$. Also,

$$\|Tx\|^2 = \langle Tx, Tx\rangle = \langle x, T^* Tx\rangle \le \|x\|\|T^* Tx\| \le \|x\|\|T^* T\|\|x\| = \|T^* T\|\|x\|^2,$$

which implies that $\|T\|^2 \le \|T^* T\|$, and the proof of (f) is complete. ∎

**Definition.** An operator $T \in \mathcal{L}(H)$ is called **self-adjoint** if $T^* = T$. Thus $T$ is self-adjoint if and only if, for all $x, y \in H$, $\langle Tx, y\rangle = \langle x, Ty\rangle$.

**Example 1.** Let $\{u_n : n \in \mathbb{N}\}$ be an orthonormal basis for a separable Hilbert space $H$, and fix a positive integer $N$. The projection operator $P : H \to H$ defined by $Px = \sum_{n=1}^{N} \hat{x}_n u_n$ is self-adjoint. For all vectors $x$ and $y$ in $H$,

$$\langle Px, y\rangle = \langle \sum_{n=1}^{N} \hat{x}_n u_n, y\rangle = \sum_{n=1}^{N} \hat{x}_n \bar{\hat{y}}_n,$$

while

$$\langle x, Py\rangle = \sum_{n=1}^{N} \langle x, \hat{y}_n u_n\rangle = \sum_{n=1}^{N} \bar{\hat{y}}_n \langle x, u_n\rangle = \sum_{n=1}^{N} \hat{x}_n \bar{\hat{y}}_n. \blacklozenge$$

Projection operators are the simplest self-adjoint operators, and, in a way, are building blocks we can use to generate more examples of self-adjoint operators.

**Theorem 7.3.2.** *(a) The sum of self-adjoint operators is self-adjoint.*
*(b) If T is self-adjoint and $\alpha \in \mathbb{R}$, then $\alpha T$ is self-adjoint.*
*(c) The composition of two self-adjoint operators $T_1$ and $T_2$ is self-adjoint if and only if $T_1 T_2 = T_2 T_1$.*
*(d) The set of self-adjoint operators is closed in $\mathcal{L}(H)$.*

*Proof.* We leave the proof of (a)–(c) to the reader. To prove (d), let $(T_n)$ be a sequence of self-adjoint operators that converges in $\mathcal{L}(H)$ to T. We show that $T^* = T$. Then

$$\|T - T^*\| \leq \|T - T_n\| + \|T_n - T_n^*\| + \|T_n^* - T^*\|$$
$$= \|T - T_n\| + \|(T_n - T)^*\| = 2\|T - T_n\| \to 0 \text{ as } n \to \infty. \blacksquare$$

**Theorem 7.3.3.** *The eigenvalues of a self-adjoint operator T are real.*

*Proof.* Let $\lambda$ be an eigenvalue of T with eigenvector u. $\lambda\langle u, u \rangle = \langle \lambda u, u \rangle = \langle Tu, u \rangle = \langle u, Tu \rangle = \langle u, \lambda u \rangle = \bar{\lambda}\langle u, u \rangle$. Since $\langle u, u \rangle \neq 0, \bar{\lambda} = \lambda$, and $\lambda$ is real. $\blacksquare$

**Theorem 7.3.4.** *If T is a self-adjoint operator, then eigenvectors of T corresponding to distinct eigenvalues are orthogonal.*

*Proof.* Let $Tu_1 = \lambda_1 u_1, Tu_2 = \lambda_2 u_2$, where $u_1 \neq 0 \neq u_2$, and $\lambda_1 \neq \lambda_2$. Then $\lambda_1\langle u_1, u_2 \rangle = \langle \lambda_1 u_1, u_2 \rangle = \langle Tu_1, u_2 \rangle = \langle u_1, Tu_2 \rangle = \langle u_1, \lambda_2 u_2 \rangle = \lambda_2\langle u_1, u_2 \rangle$. Thus $(\lambda_1 - \lambda_2)\langle u_1, u_2 \rangle = 0$, and $\langle u_1, u_2 \rangle = 0$. $\blacksquare$

**Example 2.** The set of eigenvalues of a self-adjoint operator on a separable Hilbert space is at most countable.

Since a separable Hilbert space cannot contain an uncountable subset of orthogonal vectors and since eigenvectors corresponding to distinct eigenvalues are orthogonal, the set of eigenvalues is at most countable. ◆

**Lemma 7.3.5.** *(a) Let H be a complex Hilbert space, and let $T \in \mathcal{L}(H)$. If $\langle Tx, x \rangle = 0$ for all $x \in H$, then $T = 0$.*
*(b) Let H be a real Hilbert space, and let $T \in \mathcal{L}(H)$ be self-adjoint. If $\langle Tx, x \rangle = 0$ for all $x \in H$, then $T = 0$.*

*Proof.* It is sufficient to show that $\langle Tx, y \rangle = 0$ for all $x, y \in H$, because, in that case, $\langle Tx, Tx \rangle = 0$, so $\|Tx\|^2 = 0$, and $T = 0$. For $x, y \in H$, and scalars $\alpha$ and $\beta$, $0 = \langle T(\alpha x + \beta y), \alpha x + \beta y \rangle = \alpha\bar{\beta}\langle Tx, y \rangle + \bar{\alpha}\beta\langle Ty, x \rangle$. If we take $\alpha = \beta = 1$, then

$\langle Tx, y \rangle + \langle Ty, x \rangle = 0$. *If we take* $\alpha = i, \beta = 1$, *then* $i\langle Tx, y \rangle - i\langle Ty, x \rangle = 0$. *The above two identities imply that* $\langle Tx, y \rangle = 0$.

*The proof of part (b) is a straightforward specialization of the proof of (a).* ∎

**Remark.** Part (b) of the above theorem is false if $T$ is not self-adjoint. For example, if $T : \mathbb{R}^2 \to \mathbb{R}^2$ is the 90° rotation of the plane, then $\langle Tx, x \rangle = 0$ for all $x \in \mathbb{R}^2$.

**Theorem 7.3.6.** *Let $H$ be a complex Hilbert space, and let $T \in \mathcal{L}(H)$. Then $T$ is self-adjoint if and only if $\langle Tx, x \rangle$ is real for all $x \in H$.*

*Proof.* If $T$ is self-adjoint, then $\langle Tx, x \rangle = \langle x, T^*x \rangle = \langle x, Tx \rangle = \overline{\langle Tx, x \rangle}$. Thus $\langle Tx, x \rangle$ is real. Conversely, if $\langle Tx, x \rangle$ is real for all $x \in H$, then $\langle Tx, x \rangle = \overline{\langle Tx, x \rangle} = \langle x, T^*x \rangle = \langle T^*x, x \rangle$. Thus $\langle (T^* - T)x, x \rangle = 0$ for all $x$; hence $T^* - T = 0$, by the previous lemma. ∎

**Theorem 7.3.7.** *Let $T \in \mathcal{L}(H)$ be self-adjoint. Then*

$$\|T\| = \sup\{|\langle Tx, x \rangle| : \|x\| = 1\}.$$

*Proof.* Let $M = \sup\{|\langle Tx, x \rangle| : \|x\| = 1\}$. If $\|x\| = 1$, then

$$|\langle Tx, x \rangle| \le \|T\|\|x\|^2 = \|T\|.$$

*Thus* $M \le \|T\|$.

It follows from the definition of $M$ that $|\langle Tx, x \rangle| \le M\|x\|^2$ for all $x \in H$. The following identities are easy to verify:

$$\langle T(x+y), x+y \rangle - \langle T(x-y), x-y \rangle = 2\langle Tx, y \rangle + 2\langle Ty, x \rangle = 2\langle Tx, y \rangle + 2\langle y, Tx \rangle$$
$$= 2\langle Tx, y \rangle + 2\overline{\langle Tx, y \rangle} = 4Re(\langle Tx, y \rangle).$$

Thus

$$|Re\langle Tx, y \rangle| \le \frac{1}{4}|\langle T(x+y), x+y \rangle| + \frac{1}{4}|\langle T(x-y), x-y \rangle|$$
$$\le \frac{M}{4}\{\|x+y\|^2 + \|x-y\|^2\} = \frac{M}{2}\{\|x\|^2 + \|y\|^2\}.$$

The summary of the above calculations is that

$$\text{for all } x, y \in H, |Re\langle Tx, y \rangle| \le \frac{M}{2}\{\|x\|^2 + \|y\|^2\} \tag{3}$$

*If $\|x\| = 1$, and $Tx \neq 0$, let $y = \frac{Tx}{\|Tx\|}$. Then*

$$Re\left\langle Tx, \frac{Tx}{\|Tx\|}\right\rangle = \frac{1}{\|Tx\|}\langle Tx, Tx\rangle = \|Tx\|.$$

*This and inequality (3) yield*

$$\|Tx\| \leq \frac{M}{2}\left\{\|x\|^2 + \|\frac{Tx}{\|Tx\|}\|^2\right\} = M.$$

*This completes the proof.* ∎

**Theorem 7.3.8.** *If $T$ is a self-adjoint operator, then $r(T) = \|T\|$.*

*Proof.* Since $|\lambda| \leq \|T\|$ for all $\lambda \in \sigma(T)$, it is sufficient to find an element $\lambda \in \sigma(T)$ such that $|\lambda| = \|T\|$. By the previous theorem, there exists a sequence of unit vectors $(x_n)$ such that $\lim_n |\langle Tx_n, x_n\rangle| = \|T\|$. Thus there exists a subsequence $(y_n)$ of $(x_n)$ such that $\lim_n \langle Ty_n, y_n\rangle = \|T\|$, or $\lim_n \langle Ty_n, y_n\rangle = -\|T\|$. Therefore there exists a real number $\lambda$ such that $|\lambda| = \|T\|$ and $\lim_n \langle Ty_n, y_n\rangle = \lambda$. Now

$$\|Ty_n - \lambda y_n\|^2 = \|Ty_n\|^2 - 2\lambda\langle Ty_n, y_n\rangle + \lambda^2\|y_n\|^2 \leq \|T\|^2 - 2\lambda\langle Ty_n, y_n\rangle + \lambda^2$$
$$= 2\lambda^2 - 2\lambda\langle Ty_n, y_n\rangle = 2\lambda(\lambda - \langle Ty_n, y_n\rangle) \to 0.$$

*If $T - \lambda I$ is invertible, then $1 = \|y_n\| = \|(T - \lambda I)^{-1}(T - \lambda I)y_n\| \leq \|(T - \lambda I)^{-1}\|\|Ty_n - \lambda y_n\| \to 0$. This contradiction shows that $\lambda \in \sigma(T)$.* ∎

**Definition.** A bounded operator $P \in \mathcal{L}(H)$ is a **projection** if, for some closed subspace $M$ of $H$, $P$ is the (orthogonal) projection of $H$ onto $M$. See the projection theorem. We remind the reader that we use the notation $P_M$ to denote the projection of $H$ onto its closed subspace $M$

**Theorem 7.3.9.** *A bounded operator $P$ is a projection if and only if it is idempotent and self-adjoint.*

*Proof. Suppose $P$ is the projection of $H$ onto a closed subspace $M$. The fact that $P^2 = P$ has been established in theorem 7.1.8, We show that $P$ is self-adjoint. First observe that, for all $x, y \in H$, $Px \in M$ and $Py - y \in M^\perp$; hence $\langle Px, Py - y\rangle = 0$.*
*Now for $x, y \in H$,*

$$\langle Px, y\rangle = \langle Px, y - Py\rangle + \langle Px, Py\rangle = \langle Px, Py\rangle = \langle Px - x, Py\rangle + \langle x, Py\rangle = \langle x, Py\rangle.$$

Conversely, suppose that $P$ is self-adjoint and idempotent. Let $M = \{x \in H : Px = x\}$. Being the kernel of the bounded operator $P - I$, $M$ is a closed subspace of $H$. We show that $P$ is the projection of $H$ onto $M$ by showing that, for $x \in H$, $y = Px \in M$, and $z = x - Px \in M^{\perp}$. Because $Py = P(Px) = P^2 x = Px = y$, $y \in M$. Now if $w \in M$, then $\langle z, w \rangle = \langle x - Px, w \rangle = \langle x, w \rangle - \langle Px, w \rangle = \langle x, w \rangle - \langle x, P^* w \rangle = \langle x, w \rangle - \langle x, Pw \rangle = \langle x, w \rangle - \langle x, w \rangle = 0.$ ∎

**Definition.** Two projections $P_M$ and $P_N$ are said to be orthogonal if $M \perp N$. Notice that, in this case, $M + N = M \oplus N$.

**Theorem 7.3.10.** *The sum of two orthogonal projections $P_M$ and $P_N$ is a projection and, in this case, $P_M + P_N = P_{M \oplus N}$. Consequently, the sum of a finite set of pairwise orthogonal projections is a projection.*

*Proof.* It is easy to verify that $P_M P_N = P_N P_M = 0$. For example, $P_M(P_N x) = 0$ because $P_N x \in N$, and $N \subseteq M^{\perp}$. Now theorem 7.3.9 implies that $P_M + P_N$ is a projection since the sum of self-adjoint operators is self-adjoint and $(P_M + P_N)^2 = P_M^2 + P_M P_N + P_N P_M + P_N^2 = P_M^2 + P_N^2 = P_M + P_N$. We now show that $\mathfrak{R}(P_M + P_N) = M \oplus N$. Clearly, $\mathfrak{R}(P_M + P_N) \subseteq M \oplus N$. Conversely, if $x = y + z \in M \oplus N$, where $y \in M$, and $z \in N$, then $(P_M + P_N)(x) = P_M(y) + P_N(y) + P_M(z) + P_N(z) = P_M(y) + P_N(z) = y + z = x.$ ∎

**Example 3.** Let $M$ and $N$ be the following closed subspaces of $l^2$: $M = \overline{Span(\{e_{2n} : n \in \mathbb{N}\})}$ and $N = \overline{Span(\{e_{2n+1} : n \in \mathbb{N}\})}$. Since $M \oplus N = l^2$, $P_M + P_N = I.$ ♦

The following construction produces an abundance of examples of self-adjoint operators. This is also the first step toward understanding the structure of compact self-adjoint operators.

**Theorem 7.3.11.** *Let $(P_n)$ be a sequence of pairwise orthogonal projections, and let $\lambda_n$ be a sequence of nonzero complex numbers such that $\lim_n \lambda_n = 0$. Define $T : H \to H$ by $Tx = \sum_{n=1}^{\infty} \lambda_n P_n x$. Then*

(a) *$T$ is a bounded operator;*
(b) *$T^* = \sum_{n=1}^{\infty} \bar{\lambda}_n P_n$; therefore if each $\lambda_n \in \mathbb{R}$, then $T$ is self-adjoint; and*
(c) *$\{\lambda_n\}$ is the set of nonzero eigenvalues of $T$.*

*Proof.* We show that the sequence of operators $S_n = \sum_{i=1}^{n} \lambda_i P_i$ is a Cauchy sequence in $\mathcal{L}(H)$. The previous theorem shows that, for $n < m$, $\sum_{k=n}^{m} P_k$ is a projection; hence $\| \sum_{k=n}^{m} P_k \| = 1$. (See theorem 7.1.8.) Observe that the mutual orthogonality of the projections $P_n$ implies that, for every $x \in H$, the vectors $P_n x$ are orthogonal.

Let $\epsilon > 0$. Since $\lambda_n \to 0$, there exists a positive integer $N$ such that, for all $n > N, |\lambda_n| < \epsilon$. Now, for $m > n > N$, and an arbitrary $x \in H$, $\|\sum_{k=n}^{m} \lambda_k P_k x\|^2 = \sum_{k=n}^{m} |\lambda_k|^2 \|P_k x\|^2 \le \epsilon^2 \sum_{k=n}^{m} \|P_k x\|^2 = \epsilon^2 \|\sum_{k=n}^{m} P_k x\|^2 \le \epsilon^2 \|\sum_{k=n}^{m} P_k\|^2 \|x\|^2 = \epsilon^2 \|x\|^2$. This shows that the sequence $S_n$ of partial sums of the series defining $T$ is Cauchy; hence the series converges, and $T \in \mathcal{L}(H)$. This proves part (a).

Now $\langle Tx, y \rangle = \langle \sum_{n=1}^{\infty} \lambda_n P_n x, y \rangle = \sum_{n=1}^{\infty} \lambda_n \langle P_n x, y \rangle = \sum_{n=1}^{\infty} \lambda_n \langle x, P_n y \rangle = \langle x, \sum_{n=1}^{\infty} \overline{\lambda_n} P_n y \rangle$; hence we obtain the stated formula for $T^*$.

Fix a positive integer $m$. For a nonzero element $x \in M_m = \mathfrak{R}(P_m)$, $P_m x = x$, and $P_n x = 0$ for all $n \ne m$. Thus $Tx = \sum_{n=1}^{\infty} \lambda_n P_n x = \lambda_m P_m x = \lambda_m x$. Hence $\lambda_m$ is an eigenvalue of $T$. To show that $\{\lambda_n\}$ is the entire set of nonzero eigenvalues of $T$, let $u \in H$ and $0 \ne \mu \in \mathbb{C}$ be such that $\mu \ne \lambda_n$ and $Tu = \mu u$. Since $u = \frac{1}{\mu} Tu$, $u \in \mathfrak{R}(T)$. We show below that $u \in \mathfrak{R}(T)^\perp$; hence $u = 0$, which will establish (c). For $x \in M_n$, $Tx = \lambda_n x$, $T^* x = \overline{\lambda_n} x$, so

$$\mu \langle u, Tx \rangle = \langle \mu u, Tx \rangle = \langle Tu, \lambda_n x \rangle = \langle u, T^*(\lambda_n x) \rangle$$
$$= \langle u, \lambda_n \overline{\lambda_n} x \rangle = \lambda_n \langle u, \lambda_n x \rangle = \lambda_n \langle u, Tx \rangle.$$

Thus $(\mu - \lambda_n)\langle u, Tx \rangle = 0$. Since $\mu \ne \lambda_n$, $\langle u, Tx \rangle = 0$. We have shown that $u \in M_n^\perp$ for every $n \in \mathbb{N}$. Therefore $u \perp S = Span\{\cup_{n \in \mathbb{N}} M_n\}$; hence $u \perp \bar{S}$. Clearly, $\mathfrak{R}(T) \subseteq \bar{S}$; hence $u \in \mathfrak{R}(T)^\perp$. ■

**Example 4.** Consider the operator $T : l^2 \to l^2$ defined by $Tx = \sum_{n=1}^{\infty} \frac{\hat{x}_n}{n} e_n$. Here $P_n$ is the projection of $l^2$ on the one-dimensional subspace spanned by $e_n$. By the above theorem, $T$ is self-adjoint, and the set $\{\lambda_n = \frac{1}{n} : n \in \mathbb{N}\}$ is the entire set of nonzero eigenvalues. Since the spectrum of $T$ is closed, $\lambda = 0$ is in $\sigma(T)$. However, since $T$ is injective, $\lambda = 0$ is not an eigenvalue of $T$. We now show that the set $S = \{\frac{1}{n} : n \in \mathbb{N}\} \cup \{0\}$ is the entire spectrum of $T$. If $\lambda \in \mathbb{C} - S$, then $\delta = dist(\lambda, S) > 0$. Now $(T - \lambda I)x = \sum_{n=1}^{\infty} (\frac{1}{n} - \lambda)\hat{x}_n e_n$; hence $\|((T - \lambda I)(x)\|^2 = \sum_{n=1}^{\infty} |\frac{1}{n} - \lambda|^2 |\hat{x}_n|^2 \ge \delta^2 \sum_{n=1}^{\infty} |\hat{x}_n|^2 = \delta^2 \|x\|^2$. Thus $T - \lambda I$ is bounded away from zero. In the same manner, the adjoint of $T - \lambda I$, namely, $T - \overline{\lambda} I$, is bounded away from zero. Hence $T$ is invertible by problem 11 at the end of this section. ◆

**Theorem 7.3.12.** Let $T \in \mathcal{L}(H)$. Then

(a) $\mathfrak{R}(T)^\perp = \mathcal{N}(T^*)$, and
(b) $\mathcal{N}(T^*)^\perp = \overline{\mathfrak{R}(T)}$.

*Proof.* $y \in \mathfrak{R}(T)^{\perp}$ *if and only if* $\langle y, Tx \rangle = 0$ *for all* $x \in H$ *if and only if* $\langle T^*y, x \rangle = 0$ *for all* $x \in H$ *if and only if* $T^*y = 0$ *if and only if* $y \in \mathcal{N}(T^*)$. *Part (b) follows from (a) and* $\mathcal{N}(T^*)^{\perp} = \mathfrak{R}(T)^{\perp\perp} = \overline{\mathfrak{R}(T)}$. ∎

The above theorem has many applications. Here is a simple example.

**Example 5.** Suppose $T$ is a self-adjoint operator. If $\lambda$ is not an eigenvalue of $T$, then the range of $T - \lambda I$ is dense.

If $\lambda$ is not an eigenvalue of $T$, $\overline{\lambda}$ is not an eigenvalue of $T$. Applying the previous theorem, we have $\overline{\mathfrak{R}(T - \lambda I)} = \mathcal{N}(T^* - \overline{\lambda}I)^{\perp} = \mathcal{N}(T - \overline{\lambda}I)^{\perp} = \{0\}^{\perp} = H$. ◆

The following example shows that the entire spectrum, not just the eigenvalues, of a self-adjoint operator is contained in $\mathbb{R}$. Problem 16 at the end of this section provides a sharper result.

**Example 6.** Let $T$ be a self-adjoint operator on $H$. Then $\sigma(T) \subseteq \mathbb{R}$.

It is enough to show that if $\lambda \in \mathbb{C}$ and $\mu = Im(\lambda) \neq 0$, then $\lambda \in \rho(T)$. Let $x \in H$. Using theorem 7.3.6, we have $\mu \|x\|^2 = -Im(\langle (T - \lambda I)x, x \rangle)$. Thus $|\mu| \|x\|^2 \leq |\langle (T - \lambda I)x, x \rangle| \leq \|(T - \lambda I)x\| \|x\|$. Hence $|\mu| \|x\| \leq \|(T - \lambda I)x\|$. This proves that $T - \lambda I$ is bounded away from zero. In particular, by example 8 in section 6.2, $T - \lambda I$ is injective, and its range is closed. By the previous example, $\mathfrak{R}(T - \lambda I) = H$. This shows that $T - \lambda I$ is invertible. ◆

The next example illustrates that theorem 7.3.11 is not the only way to construct self-adjoint operators and that we have wide control over the design of the spectrum.

**Example 7.** Let $\{q_n : n \in \mathbb{N}\}$ be an enumeration of the rational numbers in $[0, 1]$, and let $\{u_n : n \in \mathbb{N}\}$ be an orthonormal basis for a Hilbert space $H$. Define an operator $T$ as follows: $T(x) = \sum_{n=1}^{\infty} q_n \hat{x}_n u_n$. We show that $\sigma(T) = [0, 1]$.

The assumptions of theorem 7.3.11 are clearly not satisfied. However, $T$ is bounded because $\|Tx\|^2 = \sum_{n=1}^{\infty} q_n^2 |\hat{x}_n|^2 \leq \sum_{n=1}^{\infty} |\hat{x}_n|^2 = \|x\|^2$. Thus $\|T\| \leq 1$. The verification that $T$ is self-adjoint is similar to example 1, and we leave it to the reader. Since $Tu_n = q_n u_n$, each $q_n$ is an eigenvalue of $T$. Since $\sigma(T)$ is closed, $[0, 1] \subseteq \sigma(T)$. By the above example and corollary 6.5.3 $\sigma(T) \subseteq [-1, 1]$. It is easy to see that if $\lambda \in [-1, 0)$, then $|q_n - \lambda| \geq |\lambda|$ for every $n \in \mathbb{N}$. A calculation similar to that in example 4 shows that $\|(T - \lambda I)x\| \geq |\lambda| \|x\|$, and hence $T - \lambda I$ is bounded away from zero. Since $T - \lambda I$ is self-adjoint, it is invertible by problem 11 at the end of this section. ◆

## Normal and Unitary Operators

We briefly discuss two classes of bounded operators. The section exercises extend the discussion. The definition of a normal operator is the same as that in the finite-dimensional case discussed in section 3.7.

**Definition.** A bounded operator $T$ on a Hilbert space is **normal** if $TT^* = T^*T$.

Observe that every self-adjoint operator is normal and that, for an arbitrary operator $T$, $TT^*$ and $T^*T$ are self-adjoint.

**Example 8.** Let $T$ be a normal operator. Then, for $n \in \mathbb{N}$, $\|T^n\| = \|T\|^n$.
  We apply the result of problem 8 in the section exercises to the self-adjoint operator $TT^*$:

$$\|T\|^{2n} = \left(\|T\|^2\right)^n = \|T^*T\|^n = \|(T^*T)^n\| = \|(T^n)^*(T^n)\| = \|T^n\|^2.$$

Now equate the square roots of the extreme quantities of the last string. ♦

**Example 9.** A bounded operator $T$ is normal if and only if, for every $x \in H$, $\|Tx\| = \|T^*x\|$. Consequently, $\mathcal{N}(T) = \mathcal{N}(T^*)$.
  If $T$ is normal, then $\|Tx\|^2 - \|T^*x\|^2 = \langle Tx, Tx \rangle - \langle T^*x, T^*x \rangle = \langle x, T^*Tx \rangle - \langle x, TT^*x \rangle = \langle x, (T^*T - TT^*)x \rangle = \langle x, 0 \rangle = 0$.
  If $\|Tx\| = \|T^*x\|$, then, by the above calculation, $\langle x, (T^*T - TT^*)x \rangle = 0$ for all $x \in H$. Since $T^*T - TT^*$ is self-adjoint, it is 0, by lemma 7.3.5. ♦

**Example 10.** Let $\mu$ be an arbitrary complex number, and define $T(x) = \mu x$. It is clear that $T^*x = \overline{\mu}x$. By the previous example, $T$ is normal because $\|Tx\| = |\mu|\|x\| = |\overline{\mu}|\|x\| = \|\overline{\mu}x\| = \|T^*x\|$. ♦

**Definition.** A bounded operator $U$ is **unitary** if $UU^* = U^*U = I$.

Observe that $U$ is unitary if and only if $U^{-1} = U^*$ and that every unitary operator is normal.

**Example 11.** $U$ is unitary if and only if $\|Ux\| = \|x\|$ for every $x \in H$.
  If $U$ is unitary, then $\|Ux\|^2 = \langle Ux, Ux \rangle = \langle x, U^*Ux \rangle = \langle x, x \rangle = \|x\|^2$. Conversely, if $\langle Ux, Ux \rangle = \langle x, x \rangle$ for all $x$, then $\langle x, U^*Ux \rangle - \langle x, x \rangle = 0$ for all $x \in H$; hence $U^*U = I$ by lemma 7.3.5. ♦

**Example 12.** For $\theta \in [0, 2\pi)$, the operator $U(x) = e^{i\theta}x$ is unitary, by the previous example. ♦

The results of the last two examples are consistent with the fact the unitary matrices resemble rotations of the plane. Also see the last three problems in the section exercises.

## Exercises

1. Let $T$ be a linear operator on $H$ such that, for all $x, y \in H$,

$$\langle Tx, y \rangle = \langle x, Ty \rangle.$$

   Prove that $T$ is bounded. Hint: Use the closed graph theorem.
2. Let $T \in \mathcal{L}(H)$, and let $S$ be an invertible operator. Show that $T$ and $S^{-1}TS$ have the same eigenvalues.
3. Show that if $T$ is invertible, then $(T^{-1})^* = (T^*)^{-1}$.
4. Prove that $\lambda \in \sigma(T)$ if and only if $\bar{\lambda} \in \sigma(T^*)$.
5. Complete the proof of theorem 7.3.1.
6. Complete the proof of theorem 7.3.2.
7. Let $T$ be a linear operator on a complex Hilbert space $H$. Prove that, for all $x, y \in H$,

$$\langle T(x+y), x+y \rangle - \langle T(x-y), x-y \rangle + i\langle T(x+iy), x+iy \rangle.$$
$$- i\langle T(x-iy), x-iy \rangle = 4\langle Tx, y \rangle$$

   For a real Hilbert space, prove that if $T$ is self-adjoint, then

$$\langle T(x+y), x+y \rangle - \langle T(x-y), x-y \rangle = 4\langle Tx, y \rangle.$$

   Use the above identities to provide another proof of lemma 7.3.5.
8. Let $T$ be a self-adjoint operator on a separable Hilbert space $H$.
   (a) Prove that $\|T\|^2 = \|T^2\|$. By induction, $\|T\|^{2^k} = \|T^{2^k}\|$ for every positive integer $k$.
   (b) Prove that, for every positive integer $n$, $\|T^n\| = \|T\|^n$. Hint: Choose an integer $k$ such that $1 \leq n \leq 2^k$; $\|T\|^{2^k} = \|T^{2^k}\| = \|T^n T^{2^k-n}\|$.
9. Prove that if $x_n \to^w x$, and $T \in \mathcal{L}(H)$, then $Tx_n \to^w Tx$.
10. Let $T \in \mathcal{L}(H)$. Prove there are unique self-adjoint operators $A$ and $B$ such that $T = A + iB$ and $T^* = A - iB$.
11. Prove that a bounded operator on a Hilbert space is invertible if and only if both $T$ and $T^*$ are bounded away from zero. Consequently, if $T$ is self-adjoint, the mere assumption that $T$ is bounded away from zero implies the invertibility of $T$.

12. Let $\{u_n\}$ be an orthonormal basis for $H$, and let $\{\lambda_n\} \in \mathbb{C}$ be such that $\lim_n \lambda_n = \lambda$ and $\lambda_n \neq \lambda$ for all $n$. Prove that the function $Tx = \sum_{n=1}^{\infty} \lambda_n \langle x, u_n \rangle u_n$ is a bounded linear operator on $H$. Prove also that each $\lambda_n$ is an eigenvalue of $T$, but $\lambda$ is not.

13. Let $T \in \mathcal{L}(H)$. Show that if a subspace $M$ is invariant under $T$, then $M^{\perp}$ is invariant under $T^*$.

14. Let $R$ and $L$ be the right and left shift operators on $l^2$, respectively.
    (a) Show that $R^* = L$.
    (b) Describe the eigenvalues of each operator.
    (c) Prove that $\sigma(R) = \sigma(L) = $ the closed unit disk in the complex plane.

15. Let $T$ be a self-adjoint operator on $H$, and let $\lambda$ be a complex number. Prove that $\lambda \in \sigma(T)$ if and only if $inf_{\|x\|=1}\|(T-\lambda I)(x)\| = 0$. Hint: If there exists a constant $\delta > 0$ such that $\|(T-\lambda I)(x)\| \geq \delta\|x\|$ for every $x \in H$, then, by example 8 in section 6.2, $\mathfrak{R}(T-\lambda I)$ is closed. Show that it is also dense in $H$. To prove the converse, examine the proof of theorem 7.3.8. Observe that this result is false if $T$ is not self-adjoint. The right shift on $l^2$ satisfies $\|Rx\| = \|x\|$ but $0 \in \sigma(R)$.

16. Let $T$ be a self-adjoint operator on a separable Hilbert space $H$, let $m = inf_{\|x\|=1}\langle Tx, x \rangle$, and let $M = sup_{\|x\|=1}\langle Tx, x \rangle$. Prove that $\sigma(T) \subseteq [m, M]$ and that both $m$ and $M$ are in $\sigma(T)$. Hint: Since $\sigma(T+\mu I) = \sigma(T) + \mu$, we may assume (by considering $T + \mu I$ for a sufficiently large positive constant $\mu$) that $0 \leq m \leq M$. By theorem 7.3.7, $\|T\| = M$. Thus $\sigma(T) \subseteq [-M, M]$. Let $\delta > 0$, and let $\lambda = m - \delta$. For every unit vector $u$, $\|Tu - \lambda u\| \geq \langle Tu - \lambda u, u \rangle$. Show that $\langle Tu - \lambda u, u \rangle \geq \delta$. By the previous problem, $m - \delta \notin \sigma(T)$. This proves that $\sigma(T) \subseteq [m, M]$. To show that $M \in \sigma(T)$, use theorem 7.3.7 to find a sequence of unit vectors $(u_n)$ such that $\lim_n \langle Tu_n, u_n \rangle = M$. Show that $\lim_n Tu_n - Mu_n = 0$, and again use problem 15. To show that $m \in \sigma(T)$, assume (by considering $T - \mu I$ for a sufficiently large positive constant $\mu$) that $m \leq M \leq 0$. Apply the result you just obtained to the operator $S = -T$ to conclude that $-m \in \sigma(S)$.

17. Let $T$ be a self-adjoint operator on $H$, let $M$ be a closed, $T$-invariant subspace of $H$, and let $N = M^{\perp}$. If $T_1$ and $T_2$ are the restrictions of $T$ to $M$ and $N$, respectively, prove that $\mathfrak{R}(T) = \mathfrak{R}(T_1) \oplus \mathfrak{R}(T_2)$ and that $\sigma(T) = \sigma(T_1) \cup \sigma(T_2)$. Hint: Use problem 15 to show that if $\lambda \notin \sigma(T_1) \cup \sigma(T_2)$, then $\lambda \notin \sigma(T)$.

18. Prove that if $P$ is the projection on a closed subspace $M$, then $I - P$ is the projection of $H$ on $M^{\perp}$.

19. Let $P$ be a projection. Prove that $0$ and $1$ are the only eigenvalues of $P$. What is $\sigma(P)$?

20. Let $P$ be a projection. Show that, for all $x \in H$, $\langle Px, x \rangle = \|Px\|^2$.

21. Show that the composition $P_M P_N$ of two projections is a projection if and only if $P_M$ and $P_N$ commute. In this case, $P_M P_N = P_{M \cap N}$.

**Definition.** A bounded operator $T \in \mathcal{L}(H)$ is **positive** if $\langle Tx, x \rangle \geq 0$ for every $x \in H$. If $\langle Tx, x \rangle > 0$ for all nonzero vectors $x \in H, T$ is said to be strictly positive.

22. Prove that the eigenvalues of a positive operator are nonnegative. If $T$ is strictly positive, prove that the eigenvalues of $T$ are positive.
23. Prove that a bounded operator $T$ on a Hilbert space is invertible if there exists a positive constant $c$ such that $T - cI$ is a positive operator.
24. Show that, for $T \in \mathcal{L}(T)$, $T^*T$ is a positive operator.
25. (a) Prove that if each $\lambda_n \geq 0$, then the operator $T$ defined in theorem 7.3.11 is positive.
    (b) For $\alpha > 0$, define $T^\alpha(x) = \sum_{n=1}^{\infty} \lambda_n^\alpha P_n(x)$. Prove that $T^\alpha T^\beta = T^{\alpha+\beta}$.
26. Let $S$ and $T$ be commuting self-adjoint operators. Prove that the operator $S + \alpha T$ is normal for every $\alpha \in \mathbb{C}$.
27. Let $T$ be a normal operator. Prove that $r(T) = \|T\|$. Conclude that if $T \neq 0$, then $\sigma(T)$ contains at least one nonzero point. Hint: Use example 8 and theorem 6.5.7. Observe that this result generalizes theorem 7.3.8 and provides an alternative proof of it.
28. Let $T$ be a normal operator. Prove that if $\lambda$ is an eigenvalue of $T$ and $u$ is a corresponding eigenvector, then $\bar{\lambda}$ is an eigenvalue of $T^*$ and $u$ is a corresponding eigenvector.
29. Prove that eigenvectors of a normal operator corresponding to distinct eigenvalues are orthogonal.
30. Prove that a bounded operator $U$ is unitary if and only if $\langle Ux, Uy \rangle = \langle x, y \rangle$ for all $x, y \in H$.
31. If $U$ is a unitary operator and $\{u_n\}$ is an orthonormal basis for $H$, prove that $\{Uu_n\}$ is an orthonormal basis for $H$.
32. Prove that if $\lambda$ is an eigenvalue of a unitary operator, then $|\lambda| = 1$.

## 7.4 Compact Operators

In section 3.7. we established the spectral theorem for normal operators on finite-dimensional inner product spaces. The question now is how much of the finite-dimensional theory can be generalized to self-adjoint (generally, normal) operators on a separable Hilbert space.

Self-adjoint operators on an infinite-dimensional separable Hilbert space share some of the properties of Hermitian matrices. For example, the eigenvalues of such an operator are real, and eigenvalues corresponding to distinct eigenvalues are orthogonal. However, the spectral theorem does not extend to self-adjoint operators for a simple reason: A self-adjoint operator on an infinite-dimensional separable Hilbert space may not have any eigenvalues. The following example assumes familiarity with the space $\mathfrak{L}^2(0,1)$ of (Lebesgue) square integrable functions on $[0,1]$, with the inner product $\langle f,g \rangle = \int_0^1 f(x)\overline{g(x)}dx$. The unfamiliar reader can think of $\mathfrak{L}^2(0,1)$ as the completion of $\mathcal{C}[0,1]$ with respect to the given inner product. See theorem 7.1.10 and example 4 in section 7.2.

**Example 1.** The operator $T$ on $\mathfrak{L}^2(0,1)$ defined by $(Tu)(x) = xu(x)$ is clearly self-adjoint and has no eigenvalues. ◆

In this section, we study compact operators in some depth. The culmination of the section is the spectral theorem for compact self-adjoint operators.

**Definition.** A linear operator $T$ on a separable Hilbert space $H$ is **compact** if it maps bounded sets into relatively compact sets. Thus $T$ is compact if whenever $A$ is a bounded subset of $H$, then $\overline{T(A)}$ is compact.

**Example 2.** (a) Compact operators are clearly bounded.
  (b) The identity operator, $I$, on an infinite-dimensional Hilbert space is never compact. The image of the unit ball, which is bounded, is itself. But, in infinite-dimensional space, no ball is relatively compact, so $I$ is not compact.
  (c) Define $T : l^2 \to l^2$ as follows: for $x = (x_n) \in l^2$, $T(x) = (x_1,0,x_3,0,x_5,...)$. The set $A = \{e_{2n-1} : n \in \mathbb{N}\}$ is bounded, but its image $T(A) = A$ is not relatively compact. Hence $T$ is not a compact operator. ◆

**Example 3.** A bounded operator on a Hilbert space $H$ is compact if and only if $\overline{T(B)}$ is compact, where $B$ is the unit ball in $H$
  Suppose $\overline{T(B)}$ is compact, and let $A$ be a bounded subset of $H$. Then $A$ is contained in a ball $B_r$ of radius $r$ and centered at the origin. Because $\overline{T(B_r)} = r\overline{T(B)}$, $\overline{T(B_r)}$ is compact; hence $\overline{T(A)}$ is compact. The converse is trivial. ◆

**Example 4.** Define $D : l^2 \to l^2$ as follows: for an element $x = (x_n) \in l^2$, $D(x) = (x_1, x_2/2, x_3/3, ...)$. Being a closed subset of the Hilbert cube, $\overline{D(B)}$ is compact. Thus $D$ is compact, by the previous example. ◆

**Theorem 7.4.1.** *An operator $T \in \mathcal{L}(H)$ is compact if and only if, for every bounded sequence $(x_n)$ in $H$, $(T(x_n))$ contains a convergent subsequence.*

*Proof.* Suppose $T$ is compact, and let $(x_n)$ be a bounded sequence in $H$, say, $\|x_n\| \leq r$. By assumption, $\overline{T(B(0,r))}$ is compact and contains $T(x_n)$. By the sequential compactness of $\overline{T(B(0,r))}$, $(T(x_n))$ contains a convergent subsequence.

Conversely, if $T$ is not compact, then there exists a bounded subset $A$ such that $\overline{T(A)}$ is not compact. In particular, $T(A)$ is not totally bounded. Thus there exists a positive number $\epsilon$ and a sequence $(x_n)$ in $A$ such that $\|T(x_n) - T(x_m)\| \geq \epsilon$ for all $m, n \in \mathbb{N}$. See the proof of theorem 4.7.6. We have constructed a bounded sequence $(x_n)$ for which $T(x_n)$ contains no convergent subsequence. ∎

**Example 5.** A compact operator $T$ maps weakly convergent sequences into (norm) convergent sequences. The converse is also true. See problem 2 at the end of this section.

Let $(x_n)$ be a sequence in $H$ such that $x_n \to^w x$. We show that every subsequence of $(T(x_n))$ contains a subsequence that converges to $T(x)$. Pick a subsequence $y_k = x_{n_k}$ of $(x_n)$. Since $y_k \to^w x$ and since weak sequences are bounded (problem 15 on section 7.2), the previous theorem yields a subsequence $z_p = y_{k_p}$ of $(y_k)$ such that $T(z_p)$ is convergent. Set $\lim_p T(z_p) = z$. In particular, $T(z_p) \to^w z$. Now $z_p \to^w x$, so, by problem 6 on section 6.6, $T(z_p) \to^w T(x)$. By the uniqueness of weak limits (problem 11 on section 7.2), $z = T(x)$. ◆

**Theorem 7.4.2.** *The set $\mathcal{K}$ of compact operators on a separable Hilbert space $H$ is a closed subspace of $\mathcal{L}(H)$.*

*Proof.* We leave it to the reader to verify that $\mathcal{K}$ is a vector space. To prove that $\mathcal{K}$ is closed, let $T \in \mathcal{L}(H)$ be in the closure of $\mathcal{K}$. Let $(x_n)$ be a bounded sequence in $H$, and suppose that $\|x_n\| \leq r$. If $\epsilon > 0$, there exists a compact operator $K$ such that $\|T - K\| < \epsilon$. Since $K$ is compact, a subsequence $(y_n)$ of $(x_n)$ exists such that $K(y_n)$ is convergent. In particular, $K(y_n)$ is a Cauchy sequence, so there exists a positive integer $N$ such that, for $m, n > N$, $\|Ky_n - Ky_m\| < \epsilon$. Now, for $m, n > N$, $\|Ty_n - Ty_m\| \leq \|Ty_n - Ky_n\| + \|Ky_n - Ky_m\| + \|Ky_m - Ty_m\| \leq \|T-K\|\|y_n\| + \epsilon + \|K-T\|\|y_m\| < r\epsilon + \epsilon + r\epsilon$. Thus $Ty_n$ is Cauchy; hence it is convergent. ∎

**Theorem 7.4.3.** *(a) If $T$ is compact and $S \in \mathcal{L}(H)$, then $ST$ and $TS$ are compact.*
*(b) If $T$ is compact, and $H$ is infinite dimensional, then $0 \in \sigma(T)$.*

*Proof.* The proof of part (a) is a straightforward application of theorem 7.4.1 and the fact that a bounded operator maps bounded sequences into bounded sequences and convergent sequences into convergent sequences. To prove (b), suppose $0 \notin \sigma(T)$. Then $T$ is invertible, so there exists a bounded operator $S$ such that $ST = I$. By part (a), $ST$ would be compact, so $I$ would be compact, which is false by example 2. ∎

**Definition.** An operator $T \in \mathcal{L}(H)$ is said to be of **finite rank** if $\mathfrak{R}(T)$ is finite dimensional.

**Theorem 7.4.4.**
 (a) *A bounded, finite-rank operator T is compact.*
 (b) *If T is compact and $\mathfrak{R}(T)$ is closed, then T is of finite rank.*

*Proof.* (a) *Suppose $dim(\mathfrak{R}(T)) < \infty$. The continuity of T implies that $T(A)$ is a bounded subset of $\mathfrak{R}(T)$ for every bounded subset A of H. But bounded subsets of a finite-dimensional space are relatively compact by the Heine-Borel theorem. This proves that T is compact.*

(b) *If $\mathfrak{R}(T)$ is closed, then it is a Banach space, and T maps H onto $\mathfrak{R}(T)$. The open mapping theorem implies that T is an open mapping. Coupled with the compactness of T, this implies that the image $T(B)$ of the unit ball, B, in H is relatively compact and contains a ball $B' = \{x \in \mathfrak{R}(T) : \|x\| < \delta\}$. In particular, the closed ball $\overline{B'}$ in $\mathfrak{R}(T)$ is compact. This cannot happen unless $\mathfrak{R}(T)$ is finite dimensional.* ∎

**Example 6.** Let $(a_{ij})$ be an infinite matrix such that $\sum_{i=1}^{\infty} \sum_{j=1}^{\infty} |a_{ij}|^2 < \infty$, and define an operator $T$ on $l^2$ as follows: for $x = (x_n) \in l^2$, $T(x) = \sum_{j=1}^{\infty} a_{ij} x_j$. We claim that $T$ is compact. Observe that the assumptions imply that $| \sum_{j=1}^{\infty} a_{ij} x_j |^2 \le \sum_{j=1}^{\infty} |a_{ij}|^2 \sum_{j=1}^{\infty} |x_j|^2 = \|x\| \sum_{j=1}^{\infty} |a_{ij}|^2$. Also, $\lim_{n \to \infty} \sum_{i=n+1}^{\infty} \sum_{j=1}^{\infty} |a_{ij}|^2 = 0$.
 For $n \in \mathbb{N}$ let $P_n$ be the projection of $l^2$ onto the finite-dimensional subspace $Span(\{e_1, ..., e_n\})$. Thus, for $x = (x_n) \in l^2$, $P_n(x) = (x_1, ..., x_n, 0, 0, 0, ...)$. Since $P_n$ is compact, $P_n T$ is compact by theorem 7.4.3. If we show that $\lim_n \|T - P_n T\| = 0$, the proof will be complete by theorem 7.4.2. Now $\|(T - P_n T)x\|^2 = \sum_{i=n+1}^{\infty} |\sum_{j=1}^{\infty} a_{ij} x_j|^2 \le \sum_{i=n+1}^{\infty} \|x\|^2 \sum_{j=1}^{\infty} |a_{ij}|^2$. This shows that $\|T - P_n T\| \le \sum_{i=n+1}^{\infty} \sum_{j=1}^{\infty} |a_{ij}|^2$. Since $\lim_n \sum_{i=n+1}^{\infty} \sum_{j=1}^{\infty} |a_{ij}|^2 = 0$, we are done. ♦

Not every compact operator is of finite rank. The following theorem provides the next best result.

**Theorem 7.4.5.** *Every compact operator T on a separable Hilbert space H is the limit of a sequence of finite-rank operators.*

*Proof. Let B be the closed unit ball in H. Since $T(B)$ is relatively compact, for every $n \in \mathbb{N}$, there exists a finite subset $F_n$ of H such that $T(B) \subseteq \cup_{y \in F_n} B(y, 1/n)$. Let $M_n = Span\{F_n\}$, and let $P_n$ be the projection of H onto $M_n$. Finally, let $T_n = P_n T$. Note that $\mathfrak{R}(T_n)$ has finite dimension because it is contained in $M_n$. Thus each $T_n$ if a finite-rank operator. We now show that, for $x \in B$, $\|T_n x - Tx\| < 2/n$. This will prove that $\lim_n T_n = T$. Fix $n \in \mathbb{N}$ and write $F_n = \{y_1, ..., y_N\}$. If $x \in B$,*

$Tx \in T(B) \subseteq \cup_{i=1}^{N} B(y_i, 1/n)$. Thus, for some $1 \leq i \leq N$, $\|Tx - y_i\| < 1/n$. Now $\|T_n x - y_i\| = \|P_n(Tx) - P_n(y_i)\| = \|P_n(Tx - y_i)\| \leq \|P_n\| \|Tx - y_i\| = \|Tx - y_i\| < 1/n$. Finally, $\|T_n x - Tx\| \leq \|T_n x - y_i\| + \|y_i - Tx\| < 1/n + 1/n = 2/n$. ∎

**Theorem 7.4.6.** *Let $T \in \mathcal{L}(T)$. Then $T$ is compact if and only if $T^*$ is compact.*

*Proof.* Since $T^{**} = T$, it is enough to show that if $T$ is compact, then $T^*$ is compact. Let $(y_n)$ be a sequence in the closed unit ball $B$ of $H$. We show that $T^* y_n$ contains a convergent subsequence. For each $n \in \mathbb{N}$, define $\lambda_n \in H^*$ by $\lambda_n(x) = \langle x, y_n \rangle$. For $x, x' \in H$,

$$|\lambda_n(x) - \lambda_n(x')| = |\langle x - x', y_n \rangle| \leq \|x - x'\|.$$

It follows that the sequence $\lambda_n$ is equicontinuous on $H$ and, in particular, on $\overline{T(B)}$, which is compact by assumption. Ascoli's theorem guarantees a subsequence $\lambda_{n_k}$ that converges uniformly on $\overline{T(B)}$.

Now

$$\|T^* y_{n_i} - T^* y_{n_j}\| = sup_{x \in B} |\langle x, T^*(y_{n_i} - y_{n_j}) \rangle| = sup_{x \in B} |\langle Tx, y_{n_i} - y_{n_j} \rangle|$$
$$= sup_{x \in B} |\langle Tx, y_{n_i} \rangle - \langle Tx, y_{n_j} \rangle|$$
$$= sup_{x \in B} |\lambda_{n_i}(Tx) - \lambda_{n_j}(Tx)|.$$

The uniform convergence of $\lambda_{n_k}$ on $T(B)$ guarantees that the last quantity can be made less than $\epsilon$ for sufficiently large integers $i$ and $j$. Thus $T^*(y_{n_k})$ is Cauchy and hence convergent. ∎

## The Eigenvalues of a Compact Operator

**Theorem 7.4.7 (the Riesz-Schauder theorem).** *Let $T$ be compact, and let $r > 0$. Then the set of eigenvalues $\lambda$ of $T$ such that $|\lambda| > r$ is finite.*

*Proof.* Suppose there exist infinitely many eigenvalues $\lambda_n$ of $T$ such that $|\lambda_n| > r$. For each eigenvalue $\lambda_n$, choose an eigenvector $x_n$, and let $M_n = \text{Span}\{x_1, \ldots, x_n\}$. Note that $M_n$ is properly contained in $M_{n+1}$ and that $T(M_n) \subseteq M_n$. By Riesz's lemma, for every $n \geq 2$, there exists a unit vector $y_n \in M_n$ such that $\text{dist}(y_n, M_{n-1}) \geq 1/2$. It is easy to verify that $(T - \lambda_m I) y_m \in M_{m-1}$. Now if $n < m$, then $Ty_n - (T - \lambda_m I) y_m \in M_{m-1}$, so $\frac{1}{\lambda_m} [Ty_n - (T - \lambda_m I) y_m] \in M_{m-1}$, and $\|Ty_n - Ty_m\| = |\lambda_m| \|\frac{1}{\lambda_m} [Ty_n - (T - \lambda_m I) y_m] - y_m\| \geq |\lambda_m| \text{ dist}(y_m, M_{m-1}) \geq r/2$. Thus $(Ty_n)$ contains no convergent subsequence, contradicting the compactness of $T$. ∎

**Theorem 7.4.8.** *For a compact operator T on a separable Hilbert space H,*

(a) *the set of eigenvalues of T is at most countable and can be arranged as follows:* $|\lambda_1| \geq |\lambda_2| \geq ...$; *and the set* $\{\lambda_n\}$ *of eigenvalues of T has no nonzero limit points,*

(b) *If T has infinitely many eigenvalues, then* $\lim_n \lambda_n = 0$; *and*

(c) *if* $0 \neq \lambda \in \mathbb{C}$, *and* $L = T - \lambda I$, *then* $dim(Ker(L)) < \infty$.

*Proof.* Let $\Lambda$ be the set of nonzero eigenvalues of $T$, and assume $\Lambda$ is infinite. Let $r$ be the spectral radius of $T$, and let $r_n = r/n$ $(n \in \mathbb{N})$. If $U_n$ is the complement in the complex plane of the closed disk of radius $r_n$ centered at 0, then $\mathbb{C} - \{0\} = \cup_{n=1}^{\infty} U_n$, and $\Lambda = \cup_{n=1}^{\infty} \Lambda \cap U_n$. Since each of the sets $\Lambda_n = \Lambda \cap U_n$ is finite by theorem 7.4.7, $\Lambda$ is countable. If $\Lambda$ has a nonzero limit point, $z$, then $z \in U_n$ for some positive integer n. Because $U_n$ is open, it contains a disk centered at $z$, and such a disk would contains infinitely many points of $\Lambda$. This would contradict the finiteness of $\Lambda_n$, so no such point $z$ exists. Next let $n \in \mathbb{N}$ be such that $\Lambda \cap U_n \neq \emptyset$. Since $\Lambda \cap U_n$ is finite, the eigenvalues $\lambda$ such that $|\lambda| > r_n$ can be enumerated such that $|\lambda_1| \geq |\lambda_2| \geq ... |\lambda_{N_1}|$. Since $\Lambda$ is infinite, there exists an integer $m > n$ such that $\Lambda \cap U_m$ properly contains $\Lambda \cap U_n$. Arrange the eigenvalues in $(U_m - U_n) \cap \Lambda$ in such a way that $|\lambda_{N_1+1}| \geq |\lambda_{N_1+2}| \geq ... \geq |\lambda_{N_2}|$. Continuing in this manner, we can enumerate all the eigenvalues in the desired fashion.

(b) Any disk centered at 0 contains all but finitely many of the points $\lambda_n$. This proves part (b).

(c) Write $N_L$ for $Ker(L)$. Note that $T(N_L) \subseteq N_L$; hence the restriction of $T$ to $N_L$ is compact. Since $Tx = \lambda x$ for all $x \in N_L$, $I = \frac{1}{\lambda} T$ on $N_L$. Thus the identity operator on $N_L$ is compact, so $N_L$ is finite dimensional. ∎

Now that we have established enough of the basic properties of compact operators, we give examples of how compact operators can be constructed. We hope this will help motivate some of the results we discuss later in the section. The following builds on the constructions of theorem 7.3.11.

**Example 7.** Let $P_n$ be a sequence of pairwise orthogonal projections, and let $\lambda_n$ be a sequence of nonzero complex numbers such that $\lim_n \lambda_n = 0$. Define $T : H \to H$ by $Tx = \sum_{n=1}^{\infty} \lambda_n P_n x$. By theorem 7.3.11, $T \in \mathcal{L}(H)$. If, in addition, the rank of each $P_n$ is finite (i.e., $P_n$ projects $H$ onto a finite-dimensional subspace), then $T$ is compact, by theorems 7.4.4 and 7.4.2. By theorem 7.3.11, $\lambda_n$ are all the nonzero eigenvalues of $T$. The importance of theorem 7.3.11 and this example is that they not only produce an abundance of examples of self-adjoint and compact operators but also illustrate that we have wide control over tailoring the spectrum, as the examples below illustrate. Also see problem 4 at the end of this section for an example of the ultimate tailoring of the spectrum of a bounded operator. ♦

**Example 8.** Let $\{u_n\}$ be an orthonormal basis for $H$ and define $T(x) = \sum_{n=1}^{\infty} \lambda_n \langle x, u_n \rangle u_n$, where $(\lambda_n)$ is a sequence of nonzero complex numbers such that $\lim_n \lambda_n = 0$. This is a special case of the previous example; each $P_n$ is the projection on the one-dimensional subspace spanned by $u_n$. In this example, $\lambda = 0$ is not an eigenvalue of $T$, as the reader can easily verify. ♦

**Example 9.** Let $\{u_n\}$ be an orthonormal basis for $H$ and define $T(x) = \sum_{n=1}^{\infty} \lambda_n \langle x, u_{2n} \rangle u_{2n}$, where $(\lambda_n)$ is a sequence of nonzero complex numbers such that $\lim_n \lambda_n = 0$. In this case, $\lambda = 0$ is an eigenvalue of $T$. In fact, $dim(\mathcal{N}(T)) = \infty$ since $T(u_{2n+1}) = 0$ for all positive integers $n$. ♦

The following subsection is independent of the subsequent subsections and can be bypassed without loss of continuity.

## The Fredholm Theory

We will adopt the following standing assumptions for the remainder of this section: $T$ is a compact operator on a separable Hilbert space $H$, and $\lambda$ is a nonzero complex number. We also use the following notation: $L = T - \lambda I$, $L^* = T^* - \bar{\lambda} I$; $N_L = Ker(L)$; $R_L = \mathfrak{R}(L)$; $N_{L^*} = Ker(L^*)$; $R_{L^*} = \mathfrak{R}(L^*)$.

In the calculations in the rest of this section, we repeatedly use the fact that $T$ commutes with the powers of $L$. This is because the powers of $T$ commute.

**Theorem 7.4.9.** $R_L$ *is closed.*

*Proof. Let $X_1$ be a complement of $N_L$ in $H$. One exists by example 6 in section 6.4. We can choose $X_1 = N_L^{\perp}$, but we are not making this election because the rest of the proof below works well with any complement of $N_L$. We first prove the following fact: There exists a constant $\delta > 0$ such that $\|Lu\| \geq \delta \|u\|$ for every $u \in X_1$. Suppose not. Then there exists a sequence $(u_n)$ in $X_1$ such that $\|u_n\| = 1, \|Lu_n\| < 1/n$. Clearly, $Lu_n \to 0$ as $n \to \infty$. Since $T$ is compact, $Tu_n$ contains a convergent subsequence, $Tw_n$. Thus $w_n = \frac{1}{\lambda} Tw_n - \frac{1}{\lambda} Lw_n$ is convergent. Let $w = \lim_n w_n$. Since $X_1$ is closed, $w \in X_1$. Now $w = \lim_n w_n = \frac{1}{\lambda} \lim_n (Tw_n - Lw_n) = \frac{1}{\lambda} Tw$. Thus $Tw = \lambda w$; hence $w \in N_L \cap X_1 = \{0\}$. This contradicts the fact that $\|w\| = \lim_n \|w_n\| = 1$ and establishes the fact. We now prove that $R_L$ is closed.*

*Suppose $Ly_n$ is a convergent sequence in $R_L$. We need to show that $\lim_n Ly_n \in R_L$. Write $y_n = u_n + w_n$, where $u_n \in X_1$, and $w_n \in N_L$. Note that $Ly_n = Lu_n$, so $Lu_n$ is convergent. By the above fact, $\|u_n - u_m\| \leq \frac{1}{\delta}\|Lu_n - Lu_m\| \to 0$ as $m, n \to \infty$. Thus $u_n$ is a Cauchy sequence, so $u = \lim u_n$ exists. Finally, $\lim_n Lu_n = Lu$.* ∎

**Remarks.** (a) An immediate consequence of theorem 7.4.9 is that $H = N_{L^*} \oplus R_L$, because, by theorem 7.3.12, $\overline{R_L} = N_{L^*}^{\perp}$. Since $R_L$ is closed, $R_L = N_{L^*}^{\perp}$, which is the result we seek.

(b) Since $L^* = T^* - \overline{\lambda}I$, and $T^*$ is compact, the above theorem implies that $N_{L^*}$ is finite dimensional and that $R_{L^*}$ is closed. As in remark a, $H = N_L \oplus R_{L^*}$.

(c) By the above remarks, $codim(R_L) = dim(N_{L^*})$, and $codim(R_{L^*}) = dim(N_L)$. It is also true that $dim(N_L) = dim(N_{L^*})$. It follows that the numbers $dim(N_L)$, $dim(N_{L^*})$, $codim(R_L)$, and $codim(R_{L^*})$ are all finite and equal. The proof that $dim(N_{L^*}) = dim(N_L)$ appears at the end of this subsection. See theorem 7.4.15.

**Lemma 7.4.10.** *Let $N_{L^n}$ denote $Ker(L^n)$. Then $N_{L^n}$ is finite dimensional, and $N_{L^n} \subseteq N_{L^{n+1}}$. Moreover, there exists an integer $n$ such that, for every $k \geq n$, $N_{L^k} = N_{L^n}$.*

*Proof.* Observe that

$$L^n = (T - \lambda I)^n = \sum_{i=0}^{n} \binom{n}{i} T^{n-i}(-\lambda I)^i$$

$$= (T^n - n\lambda T^{n-1} + \dots + n(-\lambda)^{n-1}T) - [(-1)^{n+1}\lambda^n]I.$$

*The operator $K = T^n - n\lambda T^{n-1} + \dots + n(-\lambda)^{n-1}T$ is compact by theorems 7.4.2 and 7.4.3. Since $(-1)^{n+1}\lambda^n \neq 0$, the kernel $N_{L^n}$ of $L^n$ is finite dimensional by theorem 7.4.8. The fact that $N_{L^n} \subseteq N_{L^{n+1}}$ is obvious. Now suppose, for a contradiction, that $N_{L^n} \neq N_{L^{n+1}}$ for all $n \in \mathbb{N}$. By Riesz's lemma, choose a unit vector $u_n \in N_{L^{n+1}}$ such that $dist(u_n, N_{L^n}) \geq 1/2$. We claim that $Tu_n$ contains no convergent subsequence, contradicting the compactness of $T$, and concluding the proof. For $n > m$,*

$$Tu_n - Tu_m = \lambda u_n - (Tu_m - Lu_n) \tag{4}$$

*Now $L^n(Tu_m -, Lu_n) = T(L^n u_m) - L^{n+1}u_n = 0 - 0 = 0$. Thus $Tu_m - Lu_n \in N_{L^n}$, and, by (4), $\|Tu_n - Tu_m\| = |\lambda|\|u_n - \frac{1}{\lambda}(Tu_m - Lu_n)\| \geq |\lambda|/2$, which is the contradiction we seek. In the above computation, we used the fact that $T$ and $L^n$ commute.* ∎

**Lemma 7.4.11.** *Let $R_{L^n} = \mathfrak{R}(T - \lambda I)^n$. Then each $R_{L^n}$ is closed, $R_{L^n} \supseteq R_{L^{n+1}}$, and there exists a positive integer $n$ such that $R_{L^k} = R_{L^n}$ for all $k \geq n$.*

*Proof.* As in the proof of the previous lemma, $L^n = K - [(-1)^{n+1}\lambda^n]I$, where $K$ is a compact operator; hence, by theorem 7.4.9, $R_{L^n}$ is closed. The inclusions $R_{L^n} \supseteq R_{L^{n+1}}$ are obvious. If $R_{L^n} \neq R_{L^{n+1}}$ for all $n$, then, by Riesz's lemma, choose a unit vector $u_n \in R_{L^n}$ such that $\text{dist}(u_n, R_{L^{n+1}}) \geq 1/2$. We claim that $Tu_n$ contains no convergent subsequence, contradicting the compactness of $T$, and concluding the proof. For $n > m$,

$$Tu_m - Tu_n = \lambda u_m - (Tu_n - Lu_m). \tag{5}$$

Now $Tu_n - Lu_m \in R_{L^{m+1}}$; hence, by (5),

$$\|Tu_m - Tu_n\| = |\lambda| \|u_m - \frac{1}{\lambda}(Tu_n - Lu_m)\| \geq |\lambda|/2. \blacksquare$$

**Proposition 7.4.12 (the Fredholm alternative theorem).** *The operator $L$ is surjective if and only if it in injective. Symbolically, $R_L = H$ if and only if $N_L = \{0\}$.*

*Proof.* Suppose $R_L = H$. If $N_L \neq \{0\}$, then there exists a vector $u_0 \neq 0$ such that $Lu_0 = 0$. Since $L$ is onto, there is a vector $u_1$ such that $Lu_1 = u_0$, and, by induction there exists a sequence of nonzero vectors $u_1, u_2, \ldots$ such that $Lu_i = u_{i-1}$. Now, for all $n$, $L^n u_n = u_0 \neq 0$, but $L^{n+1}u_n = 0$. Thus $u_n \in N_{L^{n+1}} - N_{L^n}$. This contradicts lemma 7.4.10.

Conversely, suppose $N_L = \{0\}$. Note that $N_{L^n} = \{0\}$, that is, $L^n$ is one-to-one. If $R_L \neq H$, then there is an element $x \notin R_L$. In this case, for all $y \in H$, $L^n x - L^{n+1} y = L^n(x - Ly) \neq 0$, because $x \neq Ly$ and $L^n$ is injective. Hence $L^n x \neq L^{n+1}y$ for all $y \in H$. This means that $R_{L^n}$ strictly contains $R_{L^{n+1}}$ for all $n$, thus contradicting lemma 7.4.11. $\blacksquare$

**Theorem 7.4.13.** *Let $T$ be compact. If $\lambda \neq 0$, and $\lambda$ is not an eigenvalue, then $T - \lambda I$ is invertible. In other words, all the nonzero elements of the spectrum of a compact operator are eigenvalues.*

*Proof.* Suppose $\lambda \neq 0$ and that $\lambda$ is not an eigenvalue, that is, $N_L = \{0\}$. By proposition 7.4.12, $R_L = H$. Hence $L = T - \lambda I$ is one-to-one and onto and hence is invertible by theorem 6.3.7. $\blacksquare$

**Remark.** It follows from theorem 7.4.8 and the previous theorem that the spectrum of a compact operator on an infinite-dimensional separable Hilbert space is $\{0, \lambda_1, \lambda_2, \ldots\}$.

The following result is an immediate consequence of proposition 7.4.12 and theorem 7.4.13.

**Theorem 7.4.14 (the Fredholm alternative theorem).** *Let T be a compact operator, and let $\lambda \neq 0$. Then exactly one of the following holds:*

(a) *$T - \lambda I$ is invertible.*
(b) *$\lambda$ is an eigenvalue of T.* ■

We conclude this subsection by furnishing the proof of a result we mentioned earlier.

**Theorem 7.4.15.** *Let T be a compact operator on a separable Hilbert space H, and, for a complex number $\lambda \neq 0$, let $L = T - \lambda I, L^* = T^* - \bar{\lambda} I$. Then $N_L$ and $N_{L^*}$ have the same (finite) dimension.*

*Proof.* Suppose, for a contradiction, that $dim(N_L) = m < n = dim(N_{L^*})$, and let $\{u_1, \ldots, u_m\}$ and $\{v_1, \ldots, v_n\}$ be orthonormal bases for $N_L$ and $N_{L^*}$ respectively. Define a finite rank operator on H by

$$F(x) = \sum_{i=1}^{m} \langle x, u_i \rangle v_i.$$

Notice that $Fu_i = v_i$ for $1 \leq i \leq m$ and that the restriction of F to $N_L$ is one-to-one. The operator $K = T + F$ is compact. We claim that $K - \lambda I$ is one-to-one. If $(K - \lambda I)(x) = 0$, then $(T - \lambda I)(x) = -Fx \in R_L \cap N_{L^*} = \{0\}$. Thus $(T - \lambda I)(x) = 0 = Fx$. In particular, $x \in N_L$. Because $F|_{N_L}$ is one-to-one, $x = 0$, and this proves our claim. By the Fredholm alternative theorem, $K - \lambda I$ is onto, which is a contradiction because $\mathfrak{R}(K - \lambda I) \subseteq R_L \oplus Span\{v_1, \ldots, v_m\}$ and $v_{m+1} \notin R_L \oplus Span\{v_1, \ldots, v_m\}$. This contradiction shows that $n \leq m$. By the preceding part of the proof and the fact that $L^{**} = L$, $m \leq n$, and the proof is complete. ■

## The Spectral Theorem

The discussion so far shows that compact operators, like self-adjoint operators, share some properties with operators on finite-dimensional spaces. When we limit our attention to compact, self-adjoint operators, we obtain results that directly extend those of the finite-dimensional case.

**Lemma 7.4.16.** *If T is a nonzero compact, self-adjoint operator on a separable Hilbert space, then $\|T\|$ or $-\|T\|$ is an eigenvalue of T. In particular, every nonzero compact, self-adjoint operator has a nonzero eigenvalue.*

*Proof.* By the proof of theorem 7.3.8, there exists a real number $\lambda$ and a sequence of unit vectors $(y_n)$ such that $|\lambda| = \|T\|$ and $\lim_n Ty_n - \lambda y_n = 0$. Since $T$ is compact, $(y_n)$ contains a subsequence $(u_n)$ such that $(Tu_n)$ is convergent. It follows that $(u_n)$ is convergent (it is the difference between the two convergent sequences $\frac{1}{\lambda} Tu_n$ and $\frac{1}{\lambda}[Tu_n - \lambda u_n]$). Let $u = \lim_n u_n$. Now $\lim_n Tu_n - \lambda u_n = 0$; hence $Tu - \lambda u = 0$. Since $u \neq 0$, $\lambda$ is an eigenvalue of $T$. ∎

In light of theorems 7.3.8 and 7.4.13, $r(T) = \|T\| = |\lambda_1|$ (the largest eigenvalue of $T$). Thus the previous lemma is, in fact, redundant. However, we decided to include it here in order to make this subsection self-contained and independent of the Fredholm theory.

**Theorem 7.4.17 (the Hilbert-Schmidt theorem).** *Let $T$ be a compact self-adjoint operator on a separable Hilbert space $H$. Then $H$ possesses an orthonormal basis of eigenvectors of $T$.*

*Proof.* Let $\lambda_1, \lambda_2, \ldots$ be the nonzero eigenvalues of $T$, and, for each $n$, let $B_n$ be an orthonormal basis for the (finite-dimensional) eigenspace, $V_n$, that corresponds to $\lambda_n$. The reader should keep in mind that the set of eigenvalues may be finite. Since the eigenspaces are mutually orthogonal, the set $B = \cup_n B_n$ is an orthonormal set. Let $M$ be the closure of the span of $B$, and let $N = M^{\perp}$. Since each $V_n$ is $T$-invariant, so is $M$. It follows that $N$ is also $T$-invariant (see problem 13 on section 7.3). If $N = \{0\}$, then $M = H$ and $B$ is the desired orthonormal basis for $H$.

If $N \neq \{0\}$, the restriction of $T$ to $N$ is compact and self-adjoint. If $T|_N$ is not the zero operator, then, by lemma 7.4.16, $T|_N$ has a nonzero eigenvalue $\lambda$, which is also an eigenvalue of $T$. Since the set $\{\lambda_1, \lambda_2, \ldots\}$ contains all the nonzero eigenvalues of $T$, $\lambda = \lambda_n$ for some $n$. This is a contradiction because then an eigenvector $v$ that corresponds to $\lambda$ would be in $M \cap N = \{0\}$. This shows that $N \subseteq Ker(T)$. In particular, $Ker(T) \neq \{0\}$; hence $\lambda = 0$ is an eigenvalue of $T$. Now we show that $N = Ker(T)$. If $x \in Ker(T)$, then by the orthogonality of eigenvectors belonging to distinct eigenvalues, $x \perp u$ for every $u \in B$. Thus $x \in M^{\perp} = N$, and $N = Ker(T)$. Now choose an orthonormal basis $B_0$ of $N$. The set $B \cup B_0$ is an orthonormal basis of $M \oplus N = H$ consisting entirely of eigenvectors of $T$. ∎

We now arrive at the spectral theorem for compact self-adjoint operators.

**Theorem 7.4.18 (the spectral theorem).** *Let $T$ be a compact self-adjoint operator on a separable Hilbert space $H$, and let $\{u_n\}$ be an orthonormal basis of* *eigenvectors of T corresponding to the eigenvalues $\{\lambda_n\}$. Then, for every $x \in H$,*

$$Tx = \sum_{n=1}^{\infty} \lambda_n \langle x, u_n \rangle u_n.$$

*Proof.* Write $x = \sum_{n=1}^{\infty} \langle x, u_n \rangle u_n$. Then

$$Tx = T\left( \sum_{n=1}^{\infty} \langle x, u_n \rangle u_n \right) = \sum_{n=1}^{\infty} \langle x, u_n \rangle Tu_n = \sum_{n=1}^{\infty} \lambda_n \langle x, u_n \rangle u_n. \quad \blacksquare$$

The spectral theorem is the exact analog of the finite-dimensional case for a Hermitian matrix. If we define $P_n$ to be the projection on the one-dimensional subspace spanned by $u_n$, then $P_n$ is a rank-1 operator, and $T = \sum_{n=1}^{\infty} \lambda_n P_n$. Notice that the series $\sum_{n=1}^{\infty} \lambda_n P_n$ converges in the operator norm by theorem 7.3.11.

**Remarks.**  1. If $\lambda = 0$ is an eigenvalue of $T$, then $\lambda = 0$ contributes nothing to the sum in the statement of theorem 7.4.18. Consequently, if $y \in \mathcal{R}(T)$, then $y = \sum_{n=1}^{\infty} \langle y, u_n \rangle u_n$, where the series involves only the eigenvectors that correspond to the nonzero eigenvalues.
2. The proof of theorem 7.4.17 and remark 1 reveal that $T$ projects $H$ onto the orthogonal complement of $\mathcal{N}(T)$, which is nothing other than the closure of the span of the eigenvectors that belong to the nonzero eigenvalues of $T$.

**Example 10.** Let $T$ be a compact self-adjoint operator, let $\{\lambda_n\}$ be the nonzero eigenvalues of $T$, and let $u_n$ be the corresponding eigenvectors. For a fixed $g \in H$, consider the equation $Tf - \lambda f = g$. We work out two cases:

(a) Suppose $\lambda \neq 0$ is not an eigenvalue. In this case, $T - \lambda I$ is invertible, and the equation has a unique solution $f$. To find $f$, observe that the equation can be written as $Tf = \lambda f + g$; hence $\lambda f + g \in \mathcal{R}(T)$. Remark 1 implies that

$$\lambda f + g = \sum_{n=1}^{\infty} \langle \lambda f + g, u_n \rangle u_n = \sum_{n=1}^{\infty} \left[ \lambda \hat{f}_n + \hat{g}_n \right] u_n. \tag{6}$$

By theorem 7.4.18,

$$Tf = \sum_{n=1}^{\infty} \lambda_n \hat{f}_n u_n. \tag{7}$$

Equating the Fourier coefficients of the two series in (6) and (7), we obtain $\lambda_n \hat{f}_n = \lambda \hat{f}_n + \hat{g}_n$, which gives $\hat{f}_n = \frac{\hat{g}_n}{\lambda_n - \lambda}$, and the unique solution of the equation is

$$f = \frac{-g}{\lambda} + \frac{1}{\lambda} Tf = \frac{-g}{\lambda} + \sum_{n=1}^{\infty} \frac{\lambda_n \langle g, u_n \rangle}{\lambda (\lambda_n - \lambda)} u_n.$$

(b) If $\lambda = \lambda_n$, and $\lambda_n$ is a simple eigenvalue with eigenvector $v$, then the equation has a solution if and only if $\langle g,v\rangle = 0$ and, in this case, the solutions are of the form

$$f = -\frac{g}{\lambda_n} + av + \sum\{\frac{\lambda_k\langle g,u_n\rangle}{\lambda_n(\lambda_k-\lambda_n)}u_k : k \neq n\},\text{ where } a \text{ is an arbitrary scalar.}$$

To see this, one duplicates case (a) to obtain the equations

$$\lambda_k\hat{f}_k = \lambda_n\hat{f}_k + \hat{g}_k \text{ for all } k \in \mathbb{N} \tag{8}$$

when $k = n$, the equation $\lambda_n\hat{f}_n = \lambda_n\hat{f}_n + g_n$ is satisfied if and only if $\hat{g}_n = 0$. In this case, $\hat{f}_n$ is arbitrary and, for $k \neq n$, $\hat{f}_k$ is uniquely determined by equation (8), so we have arrived at the stated solution. ◆

See problem 8 at the end of the section for the continuation of this example.

We are now ready to prove the spectral theorem for compact normal operators.

**Theorem 7.4.19.** *Let $T$ be a compact normal operator on a separable Hilbert space $H$. Then $H$ possesses an orthonormal basis of eigenvectors of $T$.*

*Proof. Consider the self-adjoint operator $U = TT^* = T^*T$. Observe that $T$ and $U$ commute: $TU = TT^*T = T^*TT = UT$.*

*We show that if $\lambda_0 = 0$ is an eigenvalue of $U$, then $\lambda_0$ is an eigenvalue of $T$ and $Ker(T) = Ker(U)$. If $U(x) = 0$ then*

$$\|T(x)\|^2 = \langle Tx, Tx\rangle = \langle x, T^*Tx\rangle = \langle x, U(x)\rangle = \langle x,0\rangle = 0.$$

*Conversely, if $Tx = 0$, then $U(x) = T^*(T(x)) = 0$. Now let $\lambda_1,\lambda_2,...$ be the distinct nonzero eigenvalues of $U$, and let $V_1, V_2,...$ be the corresponding finite-dimensional, mutually orthogonal eigenspaces. We show that each $V_n$ is $T$-invariant. If $x \in V_n$, then $(U-\lambda_nI)(Tx) = (UT - \lambda_nT)(x) = (TU - \lambda_nT)(x) = T(U-\lambda_nI)(x) = T(0) = 0$. Thus $Tx \in V_n$. Now $T|_{V_n}$ is a normal operator on the finite-dimensional space $V_n$. By theorem 3.7.15, $V_n$ has a basis $B_n$ of eigenvectors of $T$. Choose an orthonormal basis $B_0$ for $V_0 = Ker(T) = Ker(U)$. By theorem 7.4.17, $H = \overline{Span\{\cup_{n=0}^{\infty}V_n\}}$. Since $V_n = Span(B_n)$, $\cup_{n=0}^{\infty}B_n$ is an orthonormal basis for $H$.* ∎

**Example 11.** Let $T : l^2 \to l^2$ be the finite-rank operator

$$T(x_1,x_2,...) = (ix_1, -ix_2, (1+i)x_3, (1-i)x_4, 0,0,...).$$

It is easy to verify that $T^*(x_1, x_2, ...) = (-ix_1, ix_2, (1-i)x_3, (1+i)x_4, 0, 0, ...)$ and that $T$ is a normal operator. The self-adjoint operator $U = T^*T$ is given by

$$U(x_1, x_2, ...) = (x_1, x_2, 2x_3, 2x_4, 0, 0, ...).$$

The three eigenvalues of $U$ are $\lambda_0 = 0$, $\lambda_1 = 1$, and $\lambda_2 = 2$ with eigenspaces $Ker(U) = \{(0,0,0,0,x_5,x_6,...)\}$, $V_1 = Span\{e_1, e_2\}$ and $V_2 = Span\{e_3, e_4\}$, respectively.[3]

The nonzero eigenvalues of $T$ are $\pm i$ and $1 \pm i$, and the corresponding eigenvectors are $e_1$, $e_2$, $e_3$, and $e_4$, respectively. Also, $\lambda_0 = 0$ is an eigenvalues of $T$ and $Ker(T) = Ker(U)$. In the notation of the previous theorem, $B_0 = \{e_5, e_6, ...\}$, $B_1 = \{e_1\}$, $B_2 = \{e_2\}$, $B_3 = \{e_3\}$, and $B_4 = \{e_4\}$. ◆

## Excursion: Integral Equations

The theory of compact operators has deep roots in the study of integral equations, and this section would not be complete without a brief mention of integral equations.

Consider the **Fredholm integral equation**

$$Tu - \lambda u = f,$$

where $T$ is the integral operator generated by the function $K(x, \xi)$,

$$Tu(x) = \int_a^b K(x, \xi) u(\xi) d\xi.$$

The complex function $K(x, \xi)$ on the square $[a, b] \times [a, b]$ is called the kernel of the operator, and we limit ourselves to Hilbert-Schmidt kernels since these, as it turns out, define compact integral operators on $\mathfrak{L}^2 = \mathfrak{L}^2[a, b]$.

**Definition.** The function $K(x, \xi)$ is said to be a **Hilbert-Schmidt kernel** if $\int_a^b \int_a^b |K(x, \xi)|^2 dx d\xi < \infty$.

We now prove that a Hilbert-Schmidt kernel generates a compact integral operator on $\mathfrak{L}^2$, and we achieve this in a number of steps.

**Theorem 7.4.20.** *If $K(x, \xi)$ is continuous on the closed square $[a, b] \times [a, b]$ and $u \in \mathfrak{L}^2$, then $Tu$ is continuous on $[a, b]$.*

---

[3] The set $\{e_n\}$ is the canonical basis for $\mathbb{K}(\mathbb{N})$.

*Proof.* Let $\epsilon > 0$. *By the uniform continuity of $K$ on $[a,b] \times [a,b]$, there exists a number $\delta > 0$ such that if $|x_1 - x_2| < \delta$, then $|K(x_1, \xi) - K(x_2, \xi)| < \epsilon$. Now using the Cauchy-Schwarz inequality,*

$$|Tu(x_1) - Tu(x_2)| = \left| \int_a^b (K(x_1, \xi) - K(x_2, \xi))u(\xi)d\xi \right|$$

$$\leq \int_a^b |(K(x_1, \xi) - K(x_2, \xi))u(\xi)|d\xi$$

$$\leq \left( \int_a^b |K(x_1, \xi) - K(x_2, \xi)|^2 d\xi \right)^{1/2} \left( \int_a^b |u(\xi)|^2 d\xi \right)^{1/2}$$

$$\leq \left( \int_a^b \epsilon^2 d\xi \right)^{1/2} \|u\|_2 = \epsilon(b-a)^{1/2}\|u\|_2.$$

*This proves the continuity of $Tu$.* ∎

**Corollary 7.4.21.** *If $K$ is continuous on $[a,b] \times [a,b]$, and $\mathfrak{F}$ is a bounded subset of $\mathfrak{L}^2$, then $T(\mathfrak{F})$ is equicontinuous.*

*Proof.* *This is obvious from the proof of the previous theorem since if $\|u\|_2 \leq C$ for all $u \in \mathfrak{F}$, then $|Tu(x_1) - Tu(x_2)| \leq C(b-a)^{1/2}\epsilon$ for all $x_1, x_2 \in [a,b]$ with $|x_1 - x_2| < \delta$.* ∎

**Theorem 7.4.22.** *If $K(x, \xi)$ is continuous on $[a,b] \times [a,b]$, then the integral operator it generates is a compact operator on $\mathfrak{L}^2$.*

*Proof.* *This result follows from the previous corollary and Ascoli's theorem. If $\{u_n\}$ is a bounded sequence in $\mathfrak{L}^2$, then $T(u_n)$ is equicontinuous and bounded in $\mathcal{C}[a,b]$; hence it contains a subsequence $Tu_{n_k}$ that converges uniformly in $\mathcal{C}[a,b]$. Since, for any function $u \in \mathcal{C}[a,b]$, $\|u\|_2 \leq (b-a)^{1/2}\|u\|_\infty$, the subsequence $Tu_{n_k}$ is convergent in $\mathfrak{L}^2$.* ∎

We now prove the result we seek.

**Theorem 7.4.23.** *If $K(x, \xi)$ is a Hilbert-Schmidt kernel, then the integral operator it generates is compact.*

*Proof.* *We utilize the fact that $\mathcal{C}([a,b] \times [a,b])$ is dense in $\mathfrak{L}^2([a,b] \times [a,b])$. Let $K_n(x, \xi)$ be a sequence of continuous functions on $[a,b] \times [a,b]$ such that $\lim_n \|K_n - K\|_2 = 0$. It suffices to show that if $T_n$ is the compact integral operator* *generated by $K_n$, then $\lim_n \|T_n - T\| = 0$ in $\mathcal{L}(\mathcal{Q}^2)$. Now*

$$\|T_n u - Tu\|_2^2 = \int_a^b |\int_a^b (K_n(x,\xi) - K(x,\xi))u(\xi)d\xi|^2 dx$$

$$\leq \int_a^b |K_n(x,\xi) - K(x,\xi)|^2 d\xi\, dx \int_a^b |u(\xi)|^2 d\xi = \|K_n - K\|_2^2 \|u\|_2^2.$$

*Thus $\|T_n - T\| \leq \|K_n - K\|_2 \to 0$ as $n \to \infty$.* ∎

**Example 12.** Consider the Hilbert-Schmidt kernel $K(x,\xi) = \cos x \cos \xi, 0 \leq x, \xi \leq \pi$, and let $T$ be the corresponding integral operator.

If $\lambda \neq 0$ is an eigenvalue of $T$ and $u$ is the corresponding eigenvector, then $(Tu)(x) = \cos x \int_0^\pi \cos \xi u(\xi)d\xi = \lambda u = \lambda u(x)$. It follows that $u$ is a multiple of $\cos x$, the only nonzero eigenvalue is $\lambda_1 = \pi/2$, and the normalized eigenvector is $u_1(x) = \sqrt{\frac{2}{\pi}} \cos x$. In this case, 0 is an eigenvalue of infinite multiplicity, and the null-space of $T$ is the orthogonal complement of the one-vector set $\{\cos x\}$. Using the result of example 10, we find the solutions of the equation $Tf - \lambda f = g$ in two cases:

(a) $0 \neq \lambda \neq \pi/2$. The unique solution of the equation is $f(x) = \frac{-g(x)}{\lambda} + \frac{c \cos x}{\lambda(\frac{\pi}{2} - \lambda)}$,

where $c = \int_0^\pi g(\xi)\cos \xi d\xi$.
(b) The equation $Tf - \frac{\pi}{2}f = g$ has a solution if and only if $\langle g, \cos x \rangle = 0$, and, in this case, $f = \frac{-2g(x)}{\pi} + k \cos x$, where $k$ is an arbitrary constant. ◆

## Exercises

1. Prove that $\mathcal{K}$ is a subspace of $\mathcal{L}(H)$.
2. The following characterization of compact operators is sometimes handy. Let $T \in \mathcal{L}(H)$. Prove that $T$ is compact if and only if $x_n \to^w x$, implies that $Tx_n$ converges in the norm to $Tx$. Hint: If $x_n \to^w x$ then $Tx_n \to^w Tx$. Now see problem 10 on section 6.7.
3. Let $\{\lambda_n\}$ be a bounded sequence of complex numbers, and let $\{u_n\}$ be an orthonormal basis for $H$. Prove that the function $T : H \to H$ defined by $Tx = \sum_{n=1}^\infty \lambda_n \langle x, u_n \rangle u_n$ is a bounded operator. Also show that $T^*x = \sum_{n=1}^\infty \bar{\lambda}_n \langle x, u_n \rangle u_n$ and hence $T$ is self-adjoint if and only if each $\lambda_n$ is real. Finally, show that $\{\lambda_n\}$ are all the eigenvalues of $T$.
4. This is a continuation of the previous exercise. Show that every compact subset $C$ of the complex plane is the spectrum of a bounded operator. Hint:

Let $\{\lambda_n\}$ be a dense subset of $C$, and define $T$ as in the previous exercise. Since $\{\lambda_n\} \subseteq \sigma(T)$, $C \subseteq \sigma(T)$. To show that $\sigma(T) \subseteq C$, let $\lambda \notin C$, and show that $(T - \lambda I)^{-1} x = \sum_{n=1}^{\infty} (\lambda_n - \lambda)^{-1} \langle x, u_n \rangle u_n$.

5. In this exercise, we construct a compact operator which has no eigenvalues. Define a bounded operator $T : l^2 \to l^2$ by $T(x) = (0, x_1, x_2/2, x_3/3, ...)$. Show that $T$ is compact and that $\sigma(T) = \{0\}$. Hint: $T = RoD$, where $R$ is the right shift operator and $D(x) = (x_1, x_2/2, ...)$. $D$ is compact by example 4. Show directly that $T$ has no eigenvalues.

6. Give a direct proof of the following theorem without using the subsection on the Fredholm theory. Let $T$ be a compact self-adjoint operator on a separable Hilbert space. Prove that the nonzero points of the spectrum of $T$ are eigenvalues. Hint: Use problem 15 on section 7.3, and examine the proof of theorem 7.4.16.

7. Let $T$ be a compact self-adjoint operator on a separable Hilbert space $H$. Prove that there exists a set of orthonormal vectors $\{u_n\}$ corresponding to nonzero eigenvalues $\{\lambda_n\}$ such that every element $x \in H$ can be written uniquely as $x = \sum_{n=1}^{\infty} \langle x, u_n \rangle u_n + v$, where $v \in \mathcal{N}(T)$. Some books refer to this result as the **Hilbert-Schmidt theorem**. It is clearly equivalent to theorem 7.4.17.

8. Let $T$ be a compact self-adjoint operator, let $\{\lambda_n\}$ be the nonzero eigenvalues of $T$, and let $u_n$ be the corresponding eigenvectors. For a fixed $g \in H$, consider the equation $Tf - \lambda f = g$.
   (a) Prove that if $\lambda = \lambda_n$, and $\lambda_n$ has multiplicity $m$, with eigenvectors $v_1, ..., v_m$, then the equation has a solution if and only if $\langle g, v_i \rangle = 0$ for all $1 \le i \le m$ and, in this case, the solutions are of the form

   $$f = -\frac{g}{\lambda_n} + \sum_{i=1}^{m} a_i v_i + \sum \{\frac{\lambda_k \langle g, u_k \rangle}{\lambda_n(\lambda_k - \lambda_n)} u_k : k \ne n\}.$$

   Here $a_1, ..., a_m$ are arbitrary scalars.
   (b) Prove that if $\lambda = 0$ is not an eigenvalue, and $\sum_{n=1}^{\infty} |\frac{\langle g, u_n \rangle}{\lambda_n}|^2 < \infty$, then $f = \sum_{n=1}^{\infty} \frac{\langle g, u_n \rangle}{\lambda_k} u_n$ is the unique solution of the equation.
   (c) What can you say about the case when $\lambda = 0$ is an eigenvalue of $T$?

9. Let $T$ be a self-adjoint operator. Show that if $T^k$ is compact for some integer $k \ge 2$, then $T$ is compact. Hint: It is enough to show that $T^{k-1}$ is compact.

10. Let $K(x, \xi) = \sin x \cos \xi, 0 \le x, \xi \le \pi$. Show that $\lambda = 0$ is the only eigenvalue of the corresponding integral operator.

11. Let $K(x, \xi)$ be a Hilbert-Schmidt kernel, and let $T$ be the corresponding integral operator.

(a) Show that if $K(x, \xi) = \overline{K(\xi, x)}$, then $T$ is self-adjoint.

(b) Show that $\|T\| \leq \|K\|_2 = \left\{ \int_a^b \int_a^b |K(x, \xi)|^2 dx d\xi \right\}^{1/2}$

12. For a fixed $0 < k < 1$, let $K(x, \xi) = \frac{x\xi}{1 - x^2 \xi^2}$, $0 \leq x, \xi \leq k$, and let $T$ be the corresponding integral operator. Show that $\|T\| \leq \frac{1}{4} \log(\frac{1+k}{1-k}) - \frac{1}{2} \tan^{-1} k$.

13. Let $K(x, \xi) = 1 + \sin x \sin \xi$, $0 \leq x, \xi \leq 2\pi$, and let $T$ be the corresponding integral operator. Find all the eigenvalues of $T$. Then find the solution of the integral equation $\int_0^{2\pi} (1 + \sin x \sin \xi) u(\xi) d\xi = \lambda u(x) + x$, where $\lambda$ is not an eigenvalue of $T$. Hint: If $\lambda \neq 0$ is an eigenvalue of $T$, then the corresponding eigenfunction must be of the form $u(x) = A + B \sin x$.

14. Let $K(x, \xi) = \cos(x - \xi)$, $0 \leq x, \xi \leq 2\pi$, and let $T$ be the corresponding integral operator. Find all the eigenvalues of $T$, and describe the corresponding eigenspaces.

15. Let $K(x, \xi)$ be a Hilbert-Schmidt kernel, and let $T$ be the corresponding integral operator. Show that $T^2 u(x) = \int_a^b K_2(x, \xi) u(\xi) d\xi$, where $K_2(x, \xi) = \int_a^b K(x, t) K(t, \xi) dt$. In general, show that if $K_n(x, \xi) = \int_a^b K_{n-1}(x, t) K(t, \xi) dt$, then $T^n u(x) = \int_a^b K_n(x, \xi) u(\xi) d\xi$.

16. Let $K(x, \xi)$ be a Hilbert-Schmidt kernel, and let $T$ be the corresponding integral operator. Show that if $|\lambda| > \|T\|$, then the function $F : \mathfrak{L}^2 \to \mathfrak{L}^2$ defined by $F(u) = \frac{1}{\lambda}[Tu - f]$ is a contraction on $\mathfrak{L}^2$. In this case, show that the solution of the equation $Tu - \lambda u = f$ is $\frac{-1}{\lambda} \sum_{n=0}^{\infty} \frac{T^n f}{\lambda^n}$.

17. Let $K(x, \xi) = x\xi$, $0 \leq x, \xi \leq 1$. Show that $K_n(x, \xi) = x\xi/3^{n-1}$. Also show that if $|\lambda| > 1/3$, then the solution of the integral equation $Tu - \lambda u = f$ is $u(x) = \frac{-f}{\lambda} - \frac{3x}{\lambda(3\lambda - 1)} \int_0^1 \xi f(\xi) d\xi$.

## 7.5  Compact Operators on Banach Spaces

The reader may have observed that the definition of a compact operator makes perfectly good sense for an operator on a Banach space. We state the definition again. A linear operator on a Banach space $X$ is **compact** if it maps bounded subsets of $X$ into relatively compact subsets of $X$. All the results in theorems 7.4.1 through 7.4.15 are valid for compact operators on Banach spaces. All the proofs we presented for theorems 7.4.1 through 7.4.15, are valid without alteration for compact operators on Banach spaces, with the exception of theorems 7.4.5, 7.4.6, and 7.4.15. The proofs of theorems 7.4.1 through 7.4.15 (with the exceptions noted above) were deliberately made more general than is needed for Hilbert spaces. For example, we used Riesz's theorem at several places when a simpler alternative was available. As an illustration, in the proof of lemma 7.4.10, we could simply choose a unit vector $u_n \in N_{L^{n+1}}$ such that $u_n \perp N_{L^n}$. Another place where the proof could be simplified is theorem 7.4.9, where we could have used the orthogonal complement $N_L^\perp$ instead of a complement $X_1$ of $N_L$. We now furnish the proofs of theorems 7.4.5, 7.4.6, and 7.4.15 for compact operators on Banach spaces.

**Lemma 7.5.1.** *Let $T_n$ be a sequence of bounded operators on a Banach space $X$, and let $T$ be a bounded operator on $X$ such that $\lim_n T_n(x) = T(x)$ for every $x \in X$. Then, for every compact subset $K$ of $X$, $T_n$ converges uniformly to $T$ on $K$.*

*Proof.* By the Banach-Steinhaus theorem, $\sup_n\|T_n\| < \infty$. Choose a constant $M > 0$ such that $M > \|T\|$ and $M > \sup_n\|T_n\|$. Suppose, for a contradiction, that there exists a compact subset $K$ of $X$ on which $(T_n)$ does not converge uniformly to $T$. Then there exists a sequence $(x_n)$ of $K$, a subsequence $(S_n)$ of $(T_n)$, and a positive number $\epsilon$ such that $\|S_n(x_n) - T(x_n)\| > \epsilon$ for every $n \in \mathbb{N}$. By the compactness of $K$, $(x_n)$ contains a convergent subsequence $(y_n)$. Let $y = \lim_n y_n$. Now $\|S_n(y_n) - T(y_n)\| \leq \|S_n(y) - T(y)\| + \|(S_n - T)(y_n - y)\| \leq \|S_n(y) - T(y)\| + 2M\|y_n - y\| \to 0$. This contradicts $\|S_n(y_n) - T(y_n)\| \geq \epsilon$ and concludes the proof. ∎

The following is a partial generalization of theorem 7.4.5.

**Theorem 7.5.2.** *If a Banach space $X$ has a Schauder basis, then every compact operator $T$ on $X$ is the limit of a sequence of finite-rank operators.*

*Proof.* Let $\{u_n\}$ be a Schauder basis for $X$, and let $P_n$ be the canonical projection of $X$ onto $\text{Span}\{u_1, \ldots, u_n\}$ (see the definition before problem 25 on section 6.2). We prove that the sequence $T_n = P_n T$ of finite-rank operators converges in $\mathcal{L}(X)$ to $T$. For every $x \in X$, $\lim_n(P_n - I)(x) = 0$. By the previous lemma, sequence $(P_n - I)$ converges uniformly to 0 on compact subsets of $X$. In particular, $P_n - I$ converges uniformly to 0 on $\overline{T(B)}$. Now $\|T_n - T\| = \sup_{x \in B}\|T_n(x) - T(x)\| = \sup_{x \in B}\|(P_n - I)(Tx)\| \to 0$. ∎

**Theorem 7.5.3.** *A bounded operator $T$ on a Banach space $X$ is compact if and only if $T^*$ is compact.*

*Proof.* Suppose $T$ is compact. The proof that $T^*$ is compact is a slight modification of the proof of theorem 7.4.6. Let $B$ and $B^*$ denote the closed unit balls in $X$ and $X^*$, respectively. We need to show that $T^*(B^*)$ is relatively compact in $X^*$. Let $(\lambda_n)$ be a sequence of functionals in $B^*$. For $x, x' \in X$, $|\lambda_n(x) - \lambda_n(x')| \leq \|x - x'\|$. It follows that the sequence $(\lambda_n)$ is equicontinuous on $X$ and, in particular, on $\overline{T(B)}$, which is compact by assumption. Ascoli's theorem guarantees a subsequence $(\lambda_{n_k})$ of $(\lambda_n)$ that converges uniformly on $\overline{T(B)}$. Now

$$\|T^*\lambda_{n_i} - T^*\lambda_{n_j}\| = sup_{x \in B}|\langle x, T^*(\lambda_{n_i} - \lambda_{n_j})\rangle| = sup_{x \in B}|\langle Tx, \lambda_{n_i} - \lambda_{n_j}\rangle|$$

$$= sup_{x \in B}|\langle Tx, \lambda_{n_i}\rangle - \langle Tx, \lambda_{n_j}\rangle|$$

$$= sup_{x \in B}|\lambda_{n_i}(Tx) - \lambda_{n_j}(Tx)|.$$

*The uniform convergence of $\lambda_{n_k}$ on $\overline{T(B)}$ guarantees that the last quantity can be made less than $\epsilon$ for sufficiently large integers i and j. Thus $T^*\lambda_{n_k}$ is Cauchy and hence convergent.*

*We now prove that if $T^*$ is compact, then T is compact. If $T^*$ is compact, then, by the first part of the theorem, $T^{**}$ is compact. Let $\hat{B}$ be the image of B under the natural embedding of X into $X^{**}$. By the compactness of $T^{**}$, $\overline{T^{**}(B^{**})}$ is compact, and hence its subset $\overline{T^{**}(\hat{B})}$ is also compact. By theorem 6.6.3, $T^{**}(\hat{B}) = \widehat{(T(B))}$. Therefore $\overline{\widehat{(T(B))}}$ is compact, and hence $\overline{T(B)}$ is compact since it is isometric to $\overline{\widehat{(T(B))}}$. This proves the compactness of T.* ∎

Using annihilators is not as simple as using orthogonal complements, for the simple reason that the annihilator of a subspace M of a Banach space X resides in a different space, $X^*$. Thus the fact that $H = R_L \oplus N_{L^*}$ makes no sense if H is replaced with a Banach space X. However, we will generalize the fact that the dimensions of the spaces $N_L, N_{L^*}, X/R_L$, and $X^*/R_{L^*}$ are all finite and equal (theorem 7.4.15). We adopt the standing assumption that T is a compact operator on X, and use the notation that $\lambda$ is a nonzero complex number, $L = T - \lambda I$, $L^* = T^* - \lambda I$, $N_L = Ker(L), R_L = \mathfrak{R}(L), N_{L^*} = Ker(L^*)$, and $R_{L^*} = \mathfrak{R}(L^*)$.

Recall that we have already established (theorem 7.4.8) that $N_L$ and $N_{L^*}$ are finite dimensional and that $R_L$ and $R_{L^*}$ are closed (theorem 7.4.9).

**Lemma 7.5.4.** $dim(X/R_L) \le dim(N_{L^*})$.

*Proof.* Let $x_1, ..., x_n \in X$ be such that $\tilde{x}_i = x_i + R_L$ are linearly independent in $X/R_L$. Then, for each i, $x_i \notin R_L + Span\{x_1, ..., x_{i-1}, x_{i+1}, ..., x_n\} = M_i$. Because the spaces $M_i$ are closed (see problem 18 on section 6.6), there exist bounded linear functionals $\lambda_1, ..., \lambda_n \in X^*$ such that $\lambda_i(x_i) = 1, \lambda_i(M_i) = 0$. Clearly, $\{\lambda_1, ..., \lambda_n\}$ are independent in $X^*$ (reason: $\lambda_i(x_j) = \delta_{ij}$), and since $\lambda_i(R_L) = 0, \lambda_i \in R_L^{\perp} = N_{L^*}$. Thus $dim(X/R_L) \le dim(N_{L^*})$. ∎

**Theorem 7.5.5.** $dim(X/R_L) = dim(N_{L^*})$.

*Proof.* Since $X/R_L$ is finite dimensional by the previous lemma, $dim(X/R_L) = dim(X/R_L)^*$. Applying theorem 6.6.7 with $M = R_L, (X/R_L)^*$ is isometrically isomorphic to $R_L^\perp = N_{L*}$. Thus $dim(X/R_L)^* = dim(N_{L*})$, and $dim(X/R_L) = dim(X/R_L)^* = dim(N_{L*})$. ∎

**Theorem 7.5.6.** $dim(X/R_L) = dim(N_L)$.

*Proof.* Let $y_1, \dots, y_n$ be such that $\tilde{y}_i = y_i + R_L$ form a basis for $X/R_L$, let $x_1, \dots, x_m$ be a basis for $N_L$, and let $X_1$ be a closed complement of $N_L$. We will show that $m = n$. Suppose that $m < n$. Define a finite-rank operator $F \in \mathcal{L}(X)$ by $F|_{X_1} = 0, Fx_i = y_i$ for $1 \le i \le m$. The operator $K = T + F$ is compact, and we claim that $K - \lambda I$ is one-to-one. If $(K - \lambda I)(x) = 0$, then $(T - \lambda I)(x) = -Fx \in R_L \cap Span\{y_1, \dots, y_n\} = \{0\}$. Thus $(T - \lambda I)x = 0 = Fx$, and hence $x \in N_L$. The restriction of $F$ to $N_L$ is clearly one-to-one; hence $Fx = 0$ implies that $x = 0$, and we have proved the claim. By the Fredholm alternative, $K - \lambda I$ is onto, which contradicts the fact that $y_{m+1}$ is not in the range of $K - \lambda I$. (Note that $\mathfrak{R}(K - \lambda I) \subseteq R_L \oplus Span\{y_1, \dots, y_m\}$).

We have proved that $m \ge n$. If $m > n$, define a finite rank operator $F$ by $F|_{X_1} = 0$, $Fx_i = y_i$ for $1 \le i \le n$, and $Fx_i = y_n$ for $n \le i \le m$. In this case, $K - \lambda I$ is onto (note that $\mathfrak{R}(K - \lambda I) = R_L \oplus Span\{y_1, \dots, y_n\} = X$) and hence one-to-one by the Fredholm alternative theorem. But this contradicts the fact that $(K - \lambda I)(x_n) = Fx_n = Fx_{n+1} = (K - \lambda I)(x_{n+1})$. Therefore $m = n$. ∎

The following result follows immediately.

**Theorem 7.5.7.** *The following numbers are finite and equal:*

$$dim(N_L), dim(X/R_L), dim(N_{L*}), and \ dim(X^*/R_{L*}).$$ ∎

## Exercises

1. Find an example of an unbounded, finite-rank linear operator on a Banach space.
2. Verify the details of the proof of theorem 7.5.6.
3. Let $X = \mathcal{C}[0,1]$, and define $(Tu)(x) = \int_0^x u(t)dt$. Prove that $T$ is compact.
4. Let $X = \mathcal{C}[0,1]$, and define $(Tu)(x) = \int_0^1 e^{xt}u(t)dt$. Prove that $T$ is compact.
5. Let $X = \mathcal{C}[-1,1]$, and define $(Tu)(x) = \frac{1}{\pi}\int_{-1}^1 \frac{u(\xi)}{1+(x-\xi)^2}d\xi$. Prove that $T$ is compact, and estimate $\|T\|$.
6. Let $X = \mathcal{C}[-1,1]$, and define $(Tu)(x) = x\int_{-1}^1 \xi u(\xi)d\xi$. It is clear that $T$ is compact. Show that $\lambda = 0$, and $\lambda = 2/3$ are the only eigenvalues of $T$ and that if $0 \ne \lambda \ne 2/3$, $(T - \lambda I)^{-1}f = \frac{-1}{\lambda}\left[\frac{3x}{3\lambda - 2}\int_{-1}^1 \xi f(\xi)d\xi + f(x)\right]$. Is it true that $X$ is the direct sum of the two eigenspaces?

# 8

# Integration Theory

*The only teaching that a professor can give, in my opinion, is that of thinking in front of his students.*

Henri Lebesgue

Henri Lebesgue. 1875–1941

Lebesgue entered the École Normale Supérieure in Paris in 1894 and was awarded his teaching diploma in mathematics in 1897. He studied Baire's papers on discontinuous functions and realized that much more could be achieved in this area. Building on the work of others, including that of Émile Borel and Camille Jordan, Lebesgue formulated measure theory, which he published in 1901. He generalized the definition of the Riemann integral by extending the concept of the area (or measure), and his definition allowed the integrability of a much wider class of functions, including many discontinuous functions. This generalization of the Riemann integral revolutionized integral calculus. Up to the end of the nineteenth century, mathematical analysis was limited to continuous functions, based largely on the Riemann method of integration.

*Fundamentals of Mathematical Analysis.* Adel N. Boules, Oxford University Press (2021). © Adel N. Boules.
DOI: 10.1093/oso/9780198868781.003.0008

Hawkins writes,[1]

> In Lebesgue's work ... the generalized definition of the integral was simply the starting point of his contributions to integration theory. What made the new definition important was that Lebesgue was able to recognize in it an analytic tool capable of dealing with—and to a large extent overcoming— the numerous theoretical difficulties that had arisen in connection with Riemann's theory of integration. In fact, the problems posed by these difficulties motivated all of Lebesgue's major results.

After he received his doctorate in 1902, Lebesgue held appointments in regional colleges. In 1910 he was appointed to the Sorbonne, where he was promoted to Professor of the Application of Geometry to Analysis in 1918. In 1921 he was named as Professor of Mathematics at the Collège de France, a position he held until his death in 1941. He also taught at the École Supérieure de Physique et de Chimie Industrielles de la Ville de Paris between 1927 and 1937 and at the École Normale Supérieure in Sèvres.

Lebesgue did not concentrate throughout his career on the field which he started. He also made major contributions in other areas of mathematics, including topology, potential theory, the Dirichlet problem, the calculus of variations, set theory, the theory of surface area, and dimension theory.

## 8.1 The Riemann Integral

In this section, we treat the definition and the fundamental properties of the Riemann integral of a bounded function on a compact box. The main reason for the inclusion of this section is that our definition of Lebesgue measure is, loosely stated, based on the notion that the Riemann integral of a continuous function $f$ on a compact box measures the volume of the region below the graph of $f$. The presentation in this section is standard and reflects almost exactly the standard approach to the Riemann integral on a compact interval found in undergraduate real analysis textbooks.

Let $I = [a, b]$ be a compact interval. A **grid** in $I$ is a sequence of points $x_0 = a < x_1 < x_2 < ... < x_k = b$.

Each grid in $I$ defines a partition of $I$ into a finite set of closed intervals $\mathcal{P} = \{[x_0, x_1], [x_1, x_2], ..., [x_{k-1}, x_k]\}$. We make no distinction between a grid in

---

[1] T. Hawkins, "Lebesgue, Henri Léon" in C. C. Gillispie, F. L. Holmes, and N. Koertge (eds.), *Complete Dictionary of Scientific Biography* (Detroit: Charles Scribner's Sons, 2008), 110–12.

$I$ and the partition it generates. We also denote a partition of $I$ by the sequence that defines it, $\{x_0, \dots, x_k\}$. We say that a partition $\mathcal{P}' = \{y_0, \dots, y_m\}$ is a refinement of a partition $\mathcal{P} = \{x_0, \dots, x_k\}$ if $\{x_0, \dots, x_k\} \subseteq \{y_0, \dots, y_m\}$. This simply means that $\mathcal{P}'$ is obtained from $\mathcal{P}$ by inserting additional grid points between some (or all) consecutive points $x_i$ and $x_{i+1}$. Note that if $\mathcal{P}'$ is a refinement of $\mathcal{P}$, then every interval in $\mathcal{P}$ is the union of intervals in $\mathcal{P}'$. If $\mathcal{P}$ and $\mathcal{P}'$ are partitions of $[a, b]$, then $\mathcal{P}$ and $\mathcal{P}'$ have a common refinement, namely, the partition generated by the grid $\{x_0, \dots, x_k\} \cup \{y_0, \dots, y_m\}$.

Let $I_1, \dots, I_n$ be compact intervals. The closed box in $\mathbb{R}^n$ determined by $I_1, \dots, I_n$ is $Q = I_1 \times \dots \times I_n$. Thus if $I_i = [a_i, b_i]$, then $Q = \{x = (x_1, \dots, x_n) : a_i \leq x_i \leq b_i\}$. By definition, the volume of the box $Q$ is $vol(Q) = \prod_{i=1}^{n}(b_i - a_i)$. It is easy to show that $diam(Q) = \left(\sum_{i=1}^{n}(b_i - a_i)^2\right)^{1/2}$. Now if, for each $1 \leq i \leq n$, $\mathcal{P}_i$ is a partition of $I_i$, then the corresponding partition of $Q$ is $\Delta = \mathcal{P}_1 \times \dots \times \mathcal{P}_n$. We often use the notation $\sigma$ to denote a typical sub-box in $\Delta$. Thus we use the following notation to denote the partition of $Q$ generated by $\mathcal{P}_1, \dots, \mathcal{P}_n$:

$$\Delta = \{\sigma = J_1 \times J_2 \times \dots \times J_n : J_i \in \mathcal{P}_i\}.$$

By a refinement $\Delta'$ of $\Delta$, we mean a sequence of refinements $\mathcal{P}_1', \dots, \mathcal{P}_n'$ of $\mathcal{P}_1, \dots, \mathcal{P}_n$, respectively, and

$$\Delta' = \{\sigma' = J_1' \times \dots \times J_n' : J_i' \in \mathcal{P}_i'\}.$$

Again, if $\Delta'$ is a refinement of $\Delta$, then every sub-box $\sigma$ in $\Delta$ is the union of sub-boxes $\{\sigma_1', \dots, \sigma_s'\}$ in $\Delta'$ such that $vol(\sigma) = \sum_{j=1}^{s} vol(\sigma_j')$.

Now let $f$ be a bounded real-valued function on $Q$, and let $\Delta$ be a partition of $Q$. Let the sub-boxes in $\Delta$ be enumerated as $\sigma_1, \dots, \sigma_K$. We use the notation

$$f^{\sigma_i} = \sup_{x \in \sigma_i} f(x), \text{ and } f_{\sigma_i} = \inf_{x \in \sigma_i} f(x).$$

Both numbers are finite because $f$ is assumed to be bounded. We define the **upper** and **lower Riemann sums**, respectively, of $f$ corresponding to the partition $\Delta$ on $Q$ by

$$S^{\Delta}(f) = \sum_{i=1}^{K} f^{\sigma_i} vol(\sigma_i), \text{ and } s_{\Delta}(f) = \sum_{i=1}^{K} f_{\sigma_i} vol(\sigma_i).$$

Clearly, $s_{\Delta}(f) \leq S^{\Delta}(f)$. Since $f$ is bounded, there exist real numbers $m$ and $M$ such that, for every $x \in Q, m \leq f(x) \leq M$. For an arbitrary partition $\Delta$ of $Q$, $f^{\sigma_i} \geq m$, so $S^{\Delta}(f) \geq m \sum_{i=1}^{K} vol(\sigma_i) = m \, vol(Q)$. Thus the set $\{S^{\Delta}(f)\}$ of upper Riemann sums is bounded below, and hence the number $\beta = inf_\Delta S^\Delta(f)$ is finite, where *inf* is taken over all partitions $\Delta$ of Q.

Similarly, $\alpha = sup_\Delta s_\Delta(f)$ is a finite number. The numbers $\alpha$ and $\beta$ are called, respectively, the **lower** and **upper Riemann integrals** of $f$ over Q.

**Definition.** A bounded function $f$ on a box Q is **Riemann integrable** over Q if $\alpha = \beta$. In this case, we use the notation $\int_Q f(x)dx$ to denote the common value of $\alpha$ and $\beta$, and we call this number the **Riemann integral** of $f$ over Q.

An important property of refinements is the following: if $\Delta'$ is a refinement of $\Delta$, then

$$S^{\Delta'}(f) \le S^\Delta(f), \text{ and } s_{\Delta'}(f) \ge s_\Delta(f).$$

The reason is as follows: Consider the contribution $f^{\sigma_i} vol(\sigma_i)$ of one sub-box $\sigma_i$ to the upper Riemann sum of $f$ corresponding to the partition $\Delta$. Since $\sigma_i$ is the union of sub-boxes $\sigma_1', ..., \sigma_s'$ in $\Delta'$, $f^{\sigma_j'} = sup_{x \in \sigma_j'} f(x) \le sup_{x \in \sigma_i} f(x) = f^{\sigma_i}$. Therefore the sum of the contributions of $\sigma_1', ..., \sigma_s'$ to the upper Riemann sum corresponding to $\Delta'$ is $\sum_{j=1}^{s} sup_{x \in \sigma_j'} f^{\sigma_j'} vol(\sigma_j') \le f^{\sigma_i} \sum_{j=1}^{s} vol(\sigma_j') = f^{\sigma_i} vol(\sigma_i)$. This shows that $S^{\Delta'}(f) \le S^\Delta(f)$. The fact that $s_{\Delta'}(f) \ge s_\Delta(f)$ is justified using a similar estimate. The reader can now check that, for any two partitions $\Delta_1$ and $\Delta_2$ of Q,

$$s_{\Delta_1}(f) \le S^{\Delta_2}(f).$$

See problem 2 at the end of this section. Therefore, $\alpha \le \beta$.

**Theorem 8.1.1.** *A bounded function $f$ on a box Q is Riemann integrable if and only if, for every $\epsilon > 0$, there exists a partition $\Delta$ of Q such that $S^\Delta(f) - s_\Delta(f) < \epsilon$.*

*Proof.* Let $\epsilon > 0$, and let $\Delta$ be such that $S^\Delta(f) - s_\Delta(f) < \epsilon$. Now $s_\Delta(f) \le \alpha \le \beta \le S^\Delta(f)$. Therefore $\beta \le \alpha + \epsilon$. Since $\epsilon$ is arbitrary, $\beta \le \alpha$, and hence $\alpha = \beta$. Conversely, if $\alpha = \beta$, and $\epsilon > 0$, then there exist partitions $\Delta_1$ and $\Delta_2$ of Q such that $S^{\Delta_1} - \alpha < \epsilon/2$, and $\alpha - s_{\Delta_2} < \epsilon/2$. Let $\Delta$ be a common refinement of $\Delta_1$ and $\Delta_2$. Then $S^\Delta(f) \le S^{\Delta_1}(f)$, and $s_\Delta(f) \ge s_{\Delta_2}(f)$. Therefore, $S^\Delta(f) - s_\Delta(f) = S^\Delta(f) - \alpha + \alpha - s_\Delta(f) < \epsilon/2 + \epsilon/2 = \epsilon$. ∎

**Example 1.** If $f: [a, b] \to \mathbb{R}$ is integrable, then $|f|$ is integrable.

Let $\Delta$ be a partition of $[a, b]$, and let $\sigma_i$ be one of the subintervals in $\Delta$. It is easy to see that, for $x, y \in \sigma_i$, $||f(x)| - |f(y)|| \le f^{\sigma_i} - f_{\sigma_i}$. It follows that $|f|^{\sigma_i} - |f|_{\sigma_i} \le f^{\sigma_i} - f_{\sigma_i}$; hence $S^\Delta(|f|) - s_\Delta(|f|) \le S^\Delta(f) - s_\Delta(f)$. Since $f$ is integrable, there is a partition $\Delta$ such that $S^\Delta(f) - s_\Delta(f) < \epsilon$. The result now follows from theorem 8.1.1. ◆

**Example 2.** Under the assumptions of example 1, $f^2$ is integrable.

Since $f^2 = |f|^2$, we may assume that $f$ is a nonnegative function; hence

$$(f^2)^{\sigma_i} = (f^{\sigma_i})^2, \text{ and } (f^2)_{\sigma_i} = (f_{\sigma_i})^2.$$

Now

$$(f^2)^{\sigma_i} - (f^2)_{\sigma_i} = (f^{\sigma_i} + f_{\sigma_i})(f^{\sigma_i} - f_{\sigma_i}) \le 2M(f^{\sigma_i} - f_{\sigma_i}),$$

where $M$ is an upper bound of $f$ on $I$. The result now follows from theorem 8.1.1. ◆

**Example 3.** If $f$ and $g$ are integrable on an interval $[a, b]$, then so is $fg$.

By problem 1 at the end of this section, the functions $f + g$ and $f - g$ are integrable. By example 2, $fg$ is integrable since $fg = \frac{1}{4}[(f + g)^2 - (f - g)^2]$. ◆

**Example 4.** The converse of the result in example 2 is false.

For example, the function

$$f(x) = \begin{cases} 1 & \text{if } x \in \mathbb{Q}, \\ -1 & \text{if } x \notin \mathbb{Q} \end{cases}$$

is not integrable on $[0, 1]$, but $f^2$ is. ∎

Now we consider a special sequence of partitions of $Q$ that is very useful in proving results, especially when $f$ is continuous. As before, $Q = I_1 \times ... \times I_n$. For each $k \in \mathbb{N}$, and for $1 \le i \le n$, let $\mathcal{P}_i$ be the partition of $I_i$ into $2^k$ subintervals of equal length $\frac{b_i - a_i}{2^k}$, and let $\Delta_k$ be the corresponding partition of $Q$. This is the construction described earlier except that each of the intervals $I_1, ..., I_n$ is divided into the same number of congruent subintervals, which is a power of 2. It follows that each $\Delta_{k+1}$ is a refinement of $\Delta_k$. Thus $\Delta_k$ consists of $2^{nk}$ congruent sub-boxes, and each sub-box $\sigma$ has dimensions $\frac{b_1 - a_1}{2^k}, ..., \frac{b_n - a_n}{2^k}$, $vol(\sigma) = \frac{vol(Q)}{2^{nk}}$, and $diam(\sigma) = \frac{1}{2^k}(\sum_{i=1}^{n}(b_i - a_i)^2)^{1/2}$. Denote the sub-boxes in the partition $\Delta_k$ by $\sigma_1, ..., \sigma_{2^{nk}}$. As before, we form the upper and lower Riemann sums of $f$ corresponding to the partition $\Delta_k$ and write

$$S_k(f) = \sum_{i=1}^{2^{nk}} f^{\sigma_i} vol(\sigma_i), \text{ and } s_k(f) = \sum_{i=1}^{2^{nk}} f_{\sigma_i} vol(\sigma_i).$$

Since $\Delta_{k+1}$ is a refinement of $\Delta_k$,

$$S_1(f) \ge S_2(f) \ge ..., \text{ and } s_1(f) \le s_2(f) \le ....$$

As we discussed, the above sequences are bounded; hence

$$\alpha_0 = \lim_k s_k(f) \text{ and } \beta_0 = \lim_k S_k(f)$$

are finite and $\alpha_0 \le \beta_0$. Also $\alpha_0 \le \alpha \le \beta \le \beta_0$.

For the rest of this section, we assume that $f$ is continuous on $Q$.

**Theorem 8.1.2.** *If $f$ is a continuous real-valued function on $Q$, then $f$ is Riemann integrable, and*

$$\int_Q f(x)dx = \lim_k s_k(f) = \lim_k S_k(f).$$

*Proof.* In the notation of the previous paragraph, we prove that $\alpha_0 = \beta_0$. This will establish all the assertions of the theorem. Let $\epsilon > 0$, and let $k$ be a positive integer such that $|S_k(f) - \beta_0| < \epsilon/3$, and $|s_k(f) - \alpha_0| < \epsilon/3$. Since $f$ is uniformly continuous on $Q$, there exists $\delta > 0$ such that $|f(x) - f(y)| < \dfrac{\epsilon}{3 \, vol(Q)}$ whenever $\|x - y\| < \delta$. We may assume, without loss of generality, that the integer $k$ is such that the diameter of each sub-box in $\Delta_k$ is less than $\delta$. Since $f$ assumes its maximum and minimum values on $\sigma_i$ in $\sigma_i$, $|f^{\sigma_i} - f_{\sigma_i}| < \dfrac{\epsilon}{3 \, vol(Q)}$, for each $1 \le i \le 2^{nk}$. Now

$$|S_k(f) - s_k(f)| \le \sum_{i=1}^{2^{nk}} |f^{\sigma_i} - f_{\sigma_i}| vol(\sigma_i) \le \frac{\epsilon}{3 \, vol(Q)} \sum_{i=1}^{2^{nk}} vol(\sigma_i) = \epsilon/3. \qquad Finally,$$

$|\alpha_0 - \beta_0| \le |\alpha_0 - s_k(f)| + |s_k(f) - S_k(f)| + |S_k(f) - \beta_0| < \epsilon/3 + \epsilon/3 + \epsilon/3 = \epsilon$.
Since $\epsilon$ is arbitrary, $\alpha_0 = \beta_0$. ∎

**Theorem 8.1.3.** *If $f$ and $g$ are continuous on $Q$, then*

$$\int_Q (f+g)dx = \int_Q f dx + \int_Q g dx.$$

*Proof.* $S_k(f+g) = \sum_{i=1}^{2^{nk}} (f+g)^{\sigma_i} \, vol(\sigma_i)$. Now $(f+g)^{\sigma_i} = max_{x \in \sigma_i}(f(x) + g(x)) \le max_{x \in \sigma_i} f(x) + max_{x \in \sigma_i} g(x) = f^{\sigma_i} + g^{\sigma_i}$. Therefore, $S_k(f+g) \le S_k(f) + S_k(g)$. Taking the limit of both sides as $k \to \infty$, $\int_Q (f+g)dx \le \int_Q f + \int_Q g dx$. Similarly, $s_k(f+g) \ge s_k(f) + s_k(g)$; hence $\int_Q (f+g)dx \ge \int_Q f + \int_Q g dx$. ∎

**Example 5.** Let $f : \mathbb{R}^n \to \mathbb{C}$ be continuous. If $\int_Q |f(x)|dx = 0$ for every cube $Q$, then $f = 0$.

Suppose, contrary to our assertion, there is a point $x_0 \in \mathbb{R}^n$ such that $m = |f(x_0)| > 0$. By the continuity of $f$, there is a cube $Q$ centered at $x_0$ such that, for $x \in Q$, $|f(x) - f(x_0)| < m/2$. Now, for $x \in Q$, $m - |f(x)| = |f(x_0)| - |f(x)| \le |f(x_0) - f(x)| < m/2$. Thus $|f(x)| > m/2$ for all $x \in Q$. Consequently, $\int_Q f(x)dx \ge \dfrac{m \, vol(Q)}{2}$. This contradiction proves the result. ◆

**Lemma 8.1.4.** *Let f be continuous on Q. Then $\int_Q(-f)dx = -\int_Q dx$.*

*Proof.* Since $(-f)^{\sigma_i} = max_{x \in \sigma_i}(-f(x)) = -min_{x \in \sigma_i} f(x) = -f_{\sigma_i}$,

$$\int_Q(-f)dx = \lim_k S_k(-f) = \lim_k \sum_{i=1}^{2^{nk}}(-f)^{\sigma_i} vol(\sigma_i)$$

$$= -\lim_k \sum_{i=1}^{2^{nk}} f_{\sigma_i} vol(\sigma_i) = -\lim_k s_k(f) = -\int_Q f dx. \blacksquare$$

**Theorem 8.1.5.** *If f is a continuous real-valued function on Q, and $a \in \mathbb{R}$, then $\int_Q(af)dx = a\int_Q f dx$.*

*Proof.* If $a \geq 0$, the proof is simple. If $a < 0$, then

$$\int_Q(af)dx = \int_Q |a|(-f)dx = |a| \int_Q(-f)dx = -|a| \int_Q f dx = a \int_Q f dx. \blacksquare$$

It is now easy to verify the linearity of the integral: if $f$ and $g$ are continuous on $Q$, and $a, b \in \mathbb{R}$, then $\int_Q(af + bg)dx = a\int_Q f dx + b\int_Q g dx$.

**Theorem 8.1.6.** *Let f and g be continuous real-valued functions on Q. Then*

(a) *if $f \geq 0$, then $\int_Q f \geq 0$; and*
(b) *if $f \leq g$ on Q, then $\int_Q f dx \leq \int_Q g dx$.*

*Proof.* Part (a) follows from the definition.
To prove (b), let $h = g - f$. Then $h \geq 0$; hence, by (a),
$\int_Q h dx \geq 0$, so $\int_Q g dx - \int_Q f dx = \int_Q(g - f)dx = \int_Q h dx \geq 0. \blacksquare$

**Definition.** Let $f$ be a continuous, complex-valued function on $Q$, and write $f = f_1 + if_2$, where $f_1$ and $f_2$ are continuous real-valued functions. Define

$$\int_Q f dx = \int_Q f_1 + i\int_Q f_2 dx.$$

**Theorem 8.1.7.** *For continuous complex-valued functions f and g, and all $a, b \in \mathbb{C}$,*

$$\int_Q(af + bg)dx = a\int_Q f dx + b\int_Q g dx.$$

*The proof is purely computational and is left as an exercise.* $\blacksquare$

Theorems 8.1.6(a) and 8.1.7 are often summarized by the terminology that the Riemann integral is a positive linear functional on the space $\mathcal{C}(Q)$ of continuous complex-valued functions on $Q$. The positivity of the integral means that, for $f \geq 0$, $\int_Q f dx \geq 0$.

## Exercises

In all the exercises below, we assume that $f$ is a bounded function on a box $Q$.

1. Prove that the sum (difference) of two integrable functions $f$ and $g$ is integrable, and $\int_Q (f \pm g) dx = \int_Q f dx \pm \int_Q g dx$.
2. Prove that, for any two partitions $\Delta_1$ and $\Delta_2$ of $Q$, $s_{\Delta_1} \leq S^{\Delta_2}$. Hint: Consider a common refinement, $\Delta$, of $\Delta_1$ and $\Delta_2$.
3. Let $f$ and $g$ be integrable on $[a, b]$, and let $f \leq h \leq g$. Give an example to show that $h$ need not be integrable.
4. Suppose $f$ is integrable, and $f(x) \geq m > 0$ for some constant $m$ and all $x \in [a, b]$. Prove that $1/f$ is integrable on $[a, b]$.
5. Let $a = x_0 < x_1 < ... < x_n = b$ be a partition of the interval $[a, b]$. For $1 \leq k \leq n-1$, let $E_k = [x_{k-1}, x_k)$, and let $E_n = [x_{n-1}, x_n]$. For constants $a_1, ..., a_n$, define $s = \sum_{k=1}^{n} a_k \chi_{E_k}$. Such a function is called a **step function**. Prove that $\int_a^b s(x) dx = \sum_{k=1}^{n} a_k (x_k - x_{k-1})$.
6. Let $f : [a, b] \to [0, \infty)$ be an integrable function. Prove that $\int_a^b f(x) dx = sup\{\int_a^b s(x) dx\}$, where the supremum is taken over all step functions $s$ such that $s \leq f$.

   In all of the remaining exercises, we assume that $f$ is continuous on $Q$.
7. (a) Prove theorem 8.1.7.
   (b) Prove that $|\int_Q f dx| \leq \int_Q |f| dx$. Prove the statement for complex-valued functions $f$.
8. Let $f \geq 0$ be such that $\int_Q f dx = 0$. Prove that $f = 0$.
9. Suppose that the sequence $f_k$ converges to $f$ in $\mathcal{C}(Q)$, that is, in the uniform norm. Prove that $\lim_k \int_Q f_k dx = \int_Q f dx$
10. Define the average value of the function $f$ on $Q$ by $f_{av} = \frac{1}{vol(Q)} \int_Q f dx$. Prove that there exists a point $\xi \in Q$ such that $f(\xi) = f_{av}$.
11. Let $Q = I_1 \times ... \times I_n$, let $J_i$ be a closed subinterval of $I_i$ for $1 \leq i \leq n$, and let $Q_1 = J_1 \times ... \times J_n$. Prove that if $f \geq 0$ on $Q$, then $\int_{Q_1} f dx \leq \int_Q f dx$.
12. Let $I_1, ..., I_n$ be compact intervals, and let $c$ be an interior point in $I_1 = [a, b]$. Suppose $Q_1 = [a, c] \times I_2 \times ... \times I_n$ and that $Q_2 = [c, b] \times I_2 \times ... \times I_n$. Show that if $f$ is continuous on $Q_1 \cup Q_2$, then $\int_{Q_1 \cup Q_2} f dx = \int_{Q_1} f dx + \int_{Q_2} f dx$.
13. **Fubini's theorem.** Let $\{I_i = [a_i, b_i], 1 \leq i \leq n\}$ be compact intervals, let $Q = I_1 \times ... \times I_n$, and let $Q' = I_2 \times ... \times I_n$. For a point $x = (x_1, x_2, ..., x_n) \in Q$,

we write $x = (x_1, x')$, where $x' = (x_2, \ldots, x_n) \in Q'$. Prove that $\int_Q f(x) dx = \int_{a_1}^{b_1} \int_{Q'} f(x_1, x') dx' dx_1$. It follows that $\int_Q f dx$ can be computed by evaluating the iterated integral $\int_{a_1}^{b_1} \int_{a_2}^{b_2} \ldots \int_{a_n}^{b_n} f dx_n \ldots dx_1$.

14. **The fundamental theorem of calculus.** Let $f \in \mathcal{C}[a, b]$.
    (a) For $x \in [a, b]$, define $F(x) = \int_a^x f(t) dt$. Show that $F$ is differentiable and that $F'(x) = f(x)$.
    (b) Show that if $F$ is differentiable in an open interval containing $[a, b]$ such that $F' = f$ on $[a, b]$, then $\int_a^b f(x) dx = F(b) - F(a)$.

## 8.2 Measure Spaces

Let us consider the problem of measuring the volume of objects (sets) in $\mathbb{R}^3$. Strictly speaking, volume is a function that assigns a nonnegative number to a subset of $\mathbb{R}^3$. A natural question is whether it is possible to measure the volume of an arbitrary subset of $\mathbb{R}^3$. For the most natural measure on $\mathbb{R}^3$, namely, the Lebesgue measure, the answer to the question is no. In other words, there are subsets of $\mathbb{R}^3$ to which a volume cannot be assigned. The question then becomes that of finding a large enough collection of $\mathbb{R}^3$ for which a volume can be assigned. Such sets are called measurable. It is clearly desirable for the finite union of measurable sets to be measurable. It was a paradigm shift when it was realized that a successful formulation of a measure theory necessitates that we allow the countable union of measurable sets to be measurable, and this leads to the definition of a $\sigma$-algebra. The definition of a measure as a set function on a $\sigma$-algebra is quite intuitive. This section develops the basics of abstract measure theory and measurable functions. The picture continues to evolve and culminates in section 8.4 with the construction of the Lebesgue measure.

For the remainder of this chapter, we use the notation $E'$ for the complement $X - E$ of a subset $E$ of a set $X$.

**Definitions.** A collection $\mathfrak{M}$ of subsets of a nonempty set $X$ is said to be an **algebra of sets** in $X$ if the following two conditions are met:

(a) if $E \in \mathfrak{M}$, then $E' \in \mathfrak{M}$; and
(b) if $E_1, E_2 \in \mathfrak{M}$, then $E_1 \cup E_2 \in \mathfrak{M}$.

An algebra $\mathfrak{M}$ is called a $\sigma$**-algebra** if it satisfies the additional condition

(c) if $(E_n)$ is a sequence in $\mathfrak{M}$, then $\cup_{n=1}^{\infty} E_n \in \mathfrak{M}$.

**Example 1.** For an arbitrary set $X$, the power set $\mathcal{P}(X)$ is a $\sigma$-algebra in $X$. ◆

**Example 2.** Let $X$ be an uncountable set. A subset $E$ of $X$ is called co-countable if $E'$ is countable. The collection of countable and co-countable subsets of $X$ is a $\sigma$-algebra. ◆

**Theorem 8.2.1.**
- (a) If $\mathfrak{M}$ is an algebra, then $\varnothing, X \in \mathfrak{M}$.
- (b) If $\mathfrak{M}$ is an algebra and $E_1, E_2 \in \mathfrak{M}$, then $E_1 \cap E_2 \in \mathfrak{M}$, and $E_1 - E_2 \in \mathfrak{M}$. It follows by induction that an algebra is closed under the formation of finite unions and intersections.
- (c) If $\mathfrak{M}$ is a $\sigma$-algebra, and $E_n \in \mathfrak{M}$, then $\cap_{n=1}^{\infty} E_n \in \mathfrak{M}$.

*Proof.*
- (a) Let $E \in \mathfrak{M}$. Then $E' \in \mathfrak{M}$; hence $X = E \cup E' \in \mathfrak{M}$, and $\varnothing = X' \in \mathfrak{M}$.
- (b) Using De Morgan's laws, if $E_1, E_2 \in \mathfrak{M}$, then $E_1 \cap E_2 = (E_1' \cup E_2')' \in \mathfrak{M}$. Also $E_1 - E_2 = E_1 \cap E_2' \in \mathfrak{M}$.
- (c) This follows from De Morgan's law, since $\cap_{n=1}^{\infty} E_n = (\cup_{n=1}^{\infty} E_n')'$. ∎

**Theorem 8.2.2.** *Let $\mathfrak{C}$ be an arbitrary collection of subsets of a set $X$. Then there exits a (unique) smallest $\sigma$-algebra $\mathfrak{M}$ that contains $\mathfrak{C}$.*

*Proof.* It is clear that the intersection of a family of $\sigma$-algebras is a $\sigma$-algebra. The collection of $\sigma$-algebras on $X$ containing $\mathfrak{C}$ is not empty since $\mathcal{P}(X)$ is such an algebra. Now take $\mathfrak{M}$ to be the intersection of all the $\sigma$-algebras in $X$ that contain $\mathfrak{C}$. ∎

**Definition.** The smallest $\sigma$-algebra that contains a collection of sets $\mathfrak{C}$ is called the **$\sigma$-algebra generated** by $\mathfrak{C}$.

**Definition.** Let $X$ be a metric (topological) space. The smallest $\sigma$-algebra in $X$ containing the collection of open subsets of $X$ is called the **Borel algebra** in $X$, and its members are called the Borel subsets of $X$. The collection of **Borel sets** of $X$ is denoted by $\mathcal{B}(X)$. In particular, the $\sigma$-algebras $\mathcal{B}(\mathbb{R}^n)$ are of central importance.

**Example 3.** The collection $\mathfrak{C} = \{(a, b) : a, b \in \mathbb{R}, a < b\}$ generates $\mathcal{B}(\mathbb{R})$.
Since every member of $\mathfrak{C}$ is an open set, the $\sigma$-algebra generated by $\mathfrak{C}$ is contained in $\mathcal{B}(\mathbb{R})$. Now $\mathfrak{C}$ generates $\mathcal{B}(\mathbb{R})$ because every open subset of $\mathbb{R}$ is a countable union of members of $\mathfrak{C}$. ◆

**Definition.** Let $X$ be a metric (topological) space. The intersection of a countable collection of open subsets of $X$ is known as a $G_\delta$ **set**. The countable union of closed subsets of $X$ is called an $F_\sigma$ **set**.

It follows from theorem 8.2.1 that $\mathcal{B}(X)$ contains all open sets, closed sets, $F_\sigma$ sets, and $G_\delta$ sets.

**Definitions.** Let $\mathfrak{M}$ be a $\sigma$-algebra of subsets in $X$. A **positive measure** on $\mathfrak{M}$ is a set function $\mu : \mathfrak{M} \to [0, \infty]$ such that

(a) $\mu \not\equiv \infty$, in the sense that $\mu(E) < \infty$ for at least one $E \in \mathfrak{M}$; and
(b) if $\{E_n\}$ is a countable collection of mutually disjoint members of $\mathfrak{M}$, then

$$\mu(\cup_{n=1}^{\infty} E_n) = \sum_{n=1}^{\infty} \mu(E_n).$$

The pair $(X, \mathfrak{M})$ is called a **measurable space**, the members of $\mathfrak{M}$ are called **measurable sets**, and $(X, \mathfrak{M}, \mu)$ is called a **measure space**. If $\mathfrak{M}$ and $\mu$ are understood, we loosely say that $X$ is a measure space.

Property (b) is known as the **countable additivity of positive measures**. If $\mu(X) < \infty$, we say that $\mu$ is a **finite positive measure**.

**Example 4.** the (**counting measure**). Let $X$ be a nonempty set, and let $\mathfrak{M} = \mathcal{P}(X)$. Define $\mu : \mathfrak{M} \to \mathbb{R}$ as follows: $\mu(E) = Card(E)$ if $E$ is finite, and $\mu(E) = \infty$ otherwise. Then $\mu$ is a measure on $\mathcal{P}(X)$. ◆

**Example 5.** the (**Dirac measure**). Let $X$ be a nonempty set, and let $\mathfrak{M} = \mathcal{P}(X)$. Fix an element $x_0 \in X$, and define $\mu : \mathfrak{M} \to \mathbb{R}$ as follows: $\mu(E) = 1$ if $x_0 \in E$, and $\mu(E) = 0$ otherwise. Then $\mu$ is a measure on $\mathcal{P}(X)$. ◆

**Example 6.** Let $X = \mathbb{N}$. A subset $E$ of $X$ is at most countable, so we can write $E = \{n_1, n_2, ...\}$. Define $\mu(E) = \sum_k 2^{-n_k}$. It is easy to see that $\mu$ is a measure on $\mathcal{P}(X)$. Observe that $\mu(X) = 1$. ◆

**Theorem 8.2.3.** *If $X$ is a measure space, then*

(a) *The **monotonicity of positive measures**: if $E, F \in \mathfrak{M}$ and $E \subseteq F$, then*

$$\mu(E) \leq \mu(F).$$

(b) The **countable subadditivity of positive measures**: if $(E_n)$ is a sequence in $\mathfrak{M}$, then

$$\mu(\cup_{n=1}^{\infty} E_n) \leq \sum_{n=1}^{\infty} \mu(E_n).$$

(c) If $E_1 \subseteq E_2 \subseteq \ldots$ is an ascending sequence of subsets in $\mathfrak{M}$, then

$$\mu(\cup_{n=1}^{\infty} E_n) = \lim_n \mu(E_n).$$

(d) If $E_1 \supseteq E_2 \supseteq \ldots$ is a descending sequence of subsets in $\mathfrak{M}$ and $\mu(E_1) < \infty$, then

$$\mu(\cap_{n=1}^{\infty} E_n) = \lim_n \mu(E_n).$$

Proof. (a) Since $F = E \cup (F - E)$, $\mu(F) = \mu(E) + \mu(F - E) \geq \mu(E)$.

(b) Let $B_1 = E_1$, and, for $n \geq 2$, let $B_n = E_n - \cup_{i=1}^{n-1} E_i$. The sequence $\{B_n\}$ is pairwise disjoint, and $\cup_{n=1}^{\infty} E_n = \cup_{n=1}^{\infty} B_n$; hence $\mu(\cup_{n=1}^{\infty} E_n) = \mu(\cup_{n=1}^{\infty} B_n) = \sum_{n=1}^{\infty} \mu(B_n) \leq \sum_{n=1}^{\infty} \mu(E_n)$.

(c) Let $B_1 = E_1$, and, for $n \geq 2$, let $B_n = E_n - E_{n-1}$. The sequence $\{B_n\}$ is pairwise disjoint, and $\cup_{i=1}^{n} B_i = E_n$. Now $\mu(\cup_{n=1}^{\infty} E_n) = \mu(\cup_{n=1}^{\infty} B_n) = \sum_{n=1}^{\infty} \mu(B_n) = \lim_n \sum_{i=1}^{n} \mu(B_i) = \lim_n \mu(\cup_{i=1}^{n} B_i) = \lim_n \mu(E_n)$.

(d) The sequence $E_1 - E_n$ is ascending, and $E_1 - \cap_{n=1}^{\infty} E_n = \cup_{n=1}^{\infty}(E_1 - E_n)$. By part (c), $\mu(E_1) - \mu(\cap_{n=1}^{\infty} E_n) = \mu(\cup_{n=1}^{\infty}(E_1 - E_n)) = \lim_n \mu(E_1 - E_n) = \mu(E_1) - \lim_n \mu(E_n)$. Hence $\mu(\cap_{n=1}^{\infty} E_n) = \lim_n \mu(E_n)$. ∎

**Example 7.** The condition $\mu(E_1) < \infty$ in part (d) of the previous theorem cannot be omitted. For example, if $\mu$ is the counting measure on $\mathbb{N}$, and $E_n = [n, \infty) \cap \mathbb{N}$, then $\lim_n \mu(E_n) = \infty$, while $\mu(\cap_{n=1}^{\infty} E_n) = \mu(\varnothing) = 0$. ◆

## Outer Measures

We now discuss an important general construction which we need in section 8.4 for the construction of the Lebesgue measure on $\mathbb{R}^n$.

**Definition.** Let $X$ be a nonempty set. A set function $m^* : \mathcal{P}(X) \to [0, \infty]$ is called an **outer measure** on $X$ if the following conditions are satisfied:

(a) if $E \subseteq F \subseteq X$, then $m^*(E) \leq m^*(F)$; and

(b)  for a countable sequence $(E_n)$ of subsets of $X$,

$$m^*(\cup_{n=1}^{\infty} E_n) \le \sum_{n=1}^{\infty} m^*(E_n).$$

Thus an outer measure is a nonnegative set function on $\mathcal{P}(X)$ that is **monotone** and **countably subadditive**. Outer measures have little intrinsic importance. However, an outer measure can be restricted to a positive measure on a certain $\sigma$-algebra of sets in $X$, as we detail below.

**Definition.**  Let $m^*$ be an outer measure on $X$. A subset $E$ of $X$ is said to be $m^*$-measurable (or simply measurable, in this discussion) if

$$m^*(A) = m^*(A \cap E) + m^*(A \cap E')$$

for all subsets $A$ of $X$.

   The above condition is known as the **Carathéodory condition**. Let $\mathfrak{M}$ denote the set of all $m^*$-measurable subsets of $X$.

The Carathéodory condition is not a very intuitive idea. However, it immediately guarantees the finite additivity of $m^*$ on $\mathfrak{M}$. Indeed, if $E_1$ and $E_2$ are disjoint subsets of $X$, and $E_1$ is measurable, then applying the Carathéodory condition with $A = E_1 \cup E_2$, we obtain

$$m^*(E_1 \cup E_2) = m^*((E_1 \cup E_2) \cap E_1) + m^*((E_1 \cup E_2) \cap E_1') = m^*(E_1) + m^*(E_2).$$

The Carathéodory condition also implies without too much difficulty that $\mathfrak{M}$ is an algebra (see lemma 8.2.4). In fact, it turns out that $\mathfrak{M}$ is a $\sigma$-algebra and that the restriction of $m^*$ to $\mathfrak{M}$ is a positive measure. We prove this in three steps.

**Lemma 8.2.4.**  $\mathfrak{M}$ *is an algebra.*

*Proof.  If $E \in \mathfrak{M}$, then $E' \in \mathfrak{M}$. This follows from the symmetry of the definition of a measurable set. Now let $E_1, E_2 \in \mathfrak{M}$. We need to prove that $E_1 \cup E_2$ is measurable. Because $m^*$ is subadditive, it is sufficient to show that*

$$m^*(A \cap (E_1 \cup E_2)) + m^*(A \cap (E_1 \cup E_2)') \le m^*(A)$$

*for all subsets $A$ of $X$.*

   *Using the identity $A \cap (E_1 \cup E_2) = (A \cap E_1) \cup (A \cap E_2 \cap E_1')$ and the measurability of $E_1$ and $E_2$,*

$$m^*((A \cap E_1) \cup (A \cap E_2 \cap E_1')) + m^*(A \cap E_1' \cap E_2')$$
$$\leq m^*(A \cap E_1) + m^*(A \cap E_2 \cap E_1') + m^*(A \cap E_1' \cap E_2')$$
$$= m^*(A \cap E_1) + m^*(A \cap E_1') = m^*(A). \blacksquare$$

**Lemma 8.2.5.** *If $(E_n)$ is a disjoint sequence of measurable sets and $A \subseteq X$, then*

(a) $m^*(A \cap \cup_{i=1}^n E_i) = \sum_{i=1}^n m^*(A \cap E_i)$,
(b) $m^*(A \cap \cup_{i=1}^\infty E_i) = \sum_{i=1}^\infty m^*(A \cap E_i)$, *and*
(c) $m^*(\cup_{i=1}^\infty E_i) = \sum_{i=1}^\infty m^*(E_i)$.

*Proof.* Using the fact that $E_1$ is measurable, we have

$$m^*(A \cap (E_1 \cup E_2)) = m^*(A \cap (E_1 \cup E_2) \cap E_1) + m^*(A \cap (E_1 \cup E_2) \cap E_1')$$
$$= m^*(A \cap E_1) + m^*(A \cap E_2).$$

To complete the proof of part (a), we use induction coupled with the fact we just established ($n = 2$) and the fact that $\mathfrak{M}$ is an algebra.

To prove (b),

$$\sum_{i=1}^n m^*(A \cap E_i) = m^*(A \cap \cup_{i=1}^n E_i) \leq m^*(A \cap \cup_{i=1}^\infty E_i)$$

$$= m^*(\cup_{i=1}^\infty (A \cap E_i)) \leq \sum_{i=1}^\infty m^*(A \cap E_i).$$

Taking the limit as $n \to \infty$, we obtain (b). Part (c) follows from (b) by taking $A = \cup_{i=1}^\infty E_i$. $\blacksquare$

**Theorem 8.2.6 (Carathéodory's theorem).** $\mathfrak{M}$ *is a $\sigma$-algebra, and the restriction of $m^*$ to $\mathfrak{M}$ is a positive measure.*

*Proof.* The fact that $m^*$ is countably additive on $\mathfrak{M}$ is part (c) of the previous theorem. We need to show that $\mathfrak{M}$ is closed under the formation of countable unions. Let $E_n \in \mathfrak{M}$, and write $E = \cup_{n=1}^\infty E_n$. Define $B_1 = E_1$, and, for $n \geq 2$, $B_n = E_n - \cup_{i=1}^{n-1} E_i$. Since $\mathfrak{M}$ is an algebra, each $B_n \in \mathfrak{M}$. Notice that the sets $B_n$ are mutually disjoint, and $\cup_{n=1}^\infty B_n = \cup_{n=1}^\infty E_n$. Therefore, without loss of generality, we may assume the sets $E_n$ are mutually disjoint. We need to show that, for $A \subseteq X$, $m^*(A) \geq m^*(A \cap E) + m^*(A \cap E')$. Using the facts that $\cup_{i=1}^n E_i \in \mathfrak{M}$, $A \cap (\cup_{i=1}^n E_i)' \supseteq A \cap E'$, and lemma 8.2.5, we obtain

$$m^*(A) = m^*(A \cap (\cup_{i=1}^{n} E_i)) + m^*(A \cap (\cup_{i=1}^{n} E_i)')$$

$$\geq m^*(A \cap (\cup_{i=1}^{n} E_i)) + m^*(A \cap E') = \sum_{i=1}^{n} m^*(A \cap E_i) + m^*(A \cap E').$$

*Taking the limit as $n \to \infty$ in the above string, then applying part (b) of the previous theorem, we obtain*

$$m^*(A) \geq \sum_{i=1}^{\infty} m^*(A \cap E_i) + m^*(A \cap E') = m^*(A \cap E) + m^*(A \cap E'). \blacksquare$$

**Definition.** Let $(X, \mathfrak{M}, \mu)$ be a measure space. We say that $\mu$ is a **complete measure** if whenever $E \in \mathfrak{M}$ is such that $\mu(E) = 0$, then any subset of $E$ is in $\mathfrak{M}$. Thus $\mathfrak{M}$ contains all subsets of sets of measure 0.

We have now reached the culmination of this construction.

**Theorem 8.2.7.** *Let $m^*$ be an outer measure on a set $X$, and let $\mathfrak{M}$ be the $\sigma$-algebra of measurable subsets of $X$. Then the restriction of $m^*$ to $\mathfrak{M}$ is a complete measure.*

*Proof.* We have already established the fact that $m^*$ is a measure on $\mathfrak{M}$. It remains to show the completeness of $m^*$. We first show that if $Z \subseteq X$ and $m^*(Z) = 0$, then $Z \in \mathfrak{M}$. Let $A \subseteq X$. Then $0 \leq m^*(A \cap Z) \leq m^*(Z) = 0$. Thus $m^*(A) \leq m^*(A \cap Z) + m^*(A \cap Z') = m^*(A \cap Z') \leq m^*(A)$. This proves that $Z$ is measurable. Now if $E \subseteq Z$, then $0 \leq m^*(E) \leq m^*(Z) = 0$; hence $m^*(E) = 0$. By what we have already established, $E \in \mathfrak{M}$. $\blacksquare$

A word about complete measures is very much in order here. It is an inconvenient fact that incomplete measures can occur quite naturally. For example, the product of Lebesgue measures, which are complete, is not a complete measure (see section 8.8). It is desirable to know whether an incomplete measure space can be *completed*. The answer is yes, and the completion of a measure turns out to be a rather simple construction. See problems 3 and 4 at the end of this section.

## Measurable Functions

For the remainder of this section, $(X, \mathfrak{M})$ is a measurable space. We allow real-valued functions on $X$ to take infinite values. This is essential because, for example, the limit of a sequence of functions $f_n(x)$ may well diverge to $\pm\infty$, or it may not even exist for some $x \in X$. It will turn out that this is largely a technicality because, in practice, the exceptional set of points where a reasonable measurable function $f$ takes infinite values has measure 0 (see, e.g., example 1 in section 8.3). In this section, we have to contend with the nuisance that functions can assume infinite values.

**Definition.** An extended real-valued function $f : X \to \overline{\mathbb{R}}$ is said to be **measurable** if $f^{-1}((a, \infty])$ is measurable for every $a \in \mathbb{R}$.

**Proposition 8.2.8.** *For a function $f : X \to \overline{\mathbb{R}}$, and $a \in \mathbb{R}$, the following are equivalent:*

*(a) $f$ is measurable.*
*(b) $f^{-1}([a, \infty]) \in \mathfrak{M}$.*
*(c) $f^{-1}([-\infty, a)) \in \mathfrak{M}$.*
*(d) $f^{-1}([-\infty, a]) \in \mathfrak{M}$.*

*Proof.* (a) *implies* (b): $f^{-1}([a, \infty]) = \cap_{n=1}^{\infty} f^{-1}((a - 1/n, \infty])$.
    (b) *implies* (c): $f^{-1}([-\infty, a)) = X - f^{-1}([a, \infty])$.
    (c) *implies* (d): $f^{-1}([-\infty, a]) = \cap_{n=1}^{\infty} f^{-1}([-\infty, a + 1/n))$.
    (d) *implies* (a): $f^{-1}((a, \infty]) = X - f^{-1}([-\infty, a])$. ∎

**Theorem 8.2.9.** *(a) A constant function is measurable.*
    *(b) If $A \subseteq X$, then $\chi_A$ is measurable if and only if $A$ is measurable.*
    *(c) If $f : X \to \mathbb{R}$ is measurable and $c \in \mathbb{R}$, then $f + c$ and $cf$ are measurable.*

*The proof is left as an exercise.* ∎

**Lemma 8.2.10.** *Let $f$ and $g$ be measurable, extended real-valued functions. Then the following subsets of $X$ are measurable:*

$A = \{x \in X : f(x) > g(x)\}$.
$B = \{x \in X : f(x) \geq g(x)\}$.
$C = \{x \in X : f(x) = g(x)\}$.

*Proof. The set*

$$A = \cup_{r \in \mathbb{Q}} \left[ \{x \in X : f(x) > r\} \cap \{x \in X : g(x) < r\} \right]$$

*is measurable because $\mathbb{Q}$ is countable.*
    *The set $B$ is measurable because it is the complement of the set*

$$\{x \in X : g(x) > f(x)\},$$

*which is measurable by part (a).*
    *Finally,*

$$C = \{x \in X : f(x) \geq g(x)\} \cap \{x \in X : g(x) \geq f(x)\}$$

*is measurable by part (b).* ∎

**Proposition 8.2.11.** *An extended real-valued function f is measurable if and only if the following two conditions hold:*

*(a)* $f^{-1}(-\infty)$ *and* $f^{-1}(\infty)$ *are measurable subsets of X, and*
*(b)* $f^{-1}(V)$ *is measurable for every open subset V of* $\mathbb{R}$.

*Proof.* Suppose f is measurable. Because $f^{-1}(\infty) = \cap_{n=1}^{\infty} f^{-1}((n,\infty])$, and $f^{-1}(-\infty) = \cap_{n=1}^{\infty} f^{-1}([-\infty,-n))$, $f^{-1}(\infty)$ and $f^{-1}(-\infty)$ are measurable. Since an open subset of $\mathbb{R}$ is a countable union of open bounded intervals, it is enough to show that the inverse image under f of an open bounded interval is in $\mathfrak{M}$. But, for a bounded interval $(a,b)$, $f^{-1}((a,b)) = f^{-1}((a,\infty]) \cap f^{-1}([-\infty,b))$, which is in $\mathfrak{M}$.
    Conversely, since $(a,\infty)$ is open and $f^{-1}((a,\infty]) = f^{-1}((a,\infty)) \cup f^{-1}(\infty)$, $f^{-1}((a,\infty])$ is in $\mathfrak{M}$. ∎

**Lemma 8.2.12.** *Let* $f : X \to \overline{\mathbb{R}}$ *be a measurable function, and let* $\varphi : \mathbb{R} \to \mathbb{R}$ *be continuous. Then the function* $h : X \to \mathbb{R}$ *defined below is measurable:*

$$h(x) = \begin{cases} \varphi(f(x)) & \text{if } f(x) \in \mathbb{R}, \\ 0 & \text{if } |f(x)| = \infty. \end{cases}$$

*Proof.* We use proposition 8.2.11. By construction, h takes only finite values, so $h^{-1}(\infty) = \varnothing = h^{-1}(-\infty)$. By the continuity of $\varphi$, if V is an open subset of $\mathbb{R}$, then $U = \varphi^{-1}(V)$ is open in $\mathbb{R}$. By proposition 8.2.11, $f^{-1}(U)$ is measurable. Now

$$h^{-1}(V) = \begin{cases} f^{-1}(U) & \text{if } 0 \notin V, \\ f^{-1}(U) \cup f^{-1}(\infty) \cup f^{-1}(-\infty) & \text{if } 0 \in V. \end{cases}$$

*In either case,* $h^{-1}(V)$ *is measurable. Again by proposition 8.2.11, h is a measurable function.* ∎

This lemma can be applied to infer the measurability of a wide class of functions. The following is a sample.

**Theorem 8.2.13.** *If* $f : X \to \overline{\mathbb{R}}$ *is measurable, then so are* $\max\{f, 0\}$, $\min\{f, 0\}$, $|f|^p$ *for all positive p, and* $f^m$ *for* $m \in \mathbb{N}$.

*Proof.* This follows from lemma 8.2.12 applied with $\varphi(t) = \max\{t, 0\}$, $\varphi(t) = \min\{t, 0\}$, $\varphi(t) = |t|^p$, and $\varphi(t) = t^m$, respectively. Here it is assumed, in accordance with the lemma, that when $|f(x)| = \infty$, $(\varphi \circ f)(x)$ is defined to be 0. ∎

**Lemma 8.2.14.** *Let $A$ be a measurable subset of $X$, and let $f : A \to \overline{\mathbb{R}}$ be such that $\{x \in A : f(x) > a\} \in \mathfrak{M}$ for every $a \in \mathbb{R}$. Define $h : X \to \overline{\mathbb{R}}$ as follows: $h|_A = f$, and $h(X - A) = 0$. Then $h$ is measurable.*

*Proof.* *Let $a \in \mathbb{R}$. If $a \geq 0$, then*

$$h^{-1}((a, \infty]) = \{x \in A : h(x) > a\} \cup \{x \in X - A : h(x) > a\} = \{x \in A : f(x) > a\},$$

*which is measurable.*
   *If $a < 0$,*

$$h^{-1}((a, \infty]) = \{x \in A : f(x) > a\} \cup (X - A),$$

*which is also measurable.* ∎

A function satisfying the conditions of lemma 8.2.14 is said to be **measurable on** $A$. Loosely speaking, lemma 8.2.14 says that a measurable function on a measurable subset of $X$ can be extended to a measurable function on $X$. Another way to look at it is that altering the values of a measurable function on a measurable subset produces a measurable function. Assigning the value 0 to $h(X - A)$ is arbitrary, and any (extended) real number can be used instead of 0.

**Theorem 8.2.15.** *Let $f$ and $g$ be measurable, extended real-valued functions on a measurable space $X$, and let*

$$A = \{x \in X : f(x) \in \mathbb{R}\} \cap \{x \in X : g(x) \in \mathbb{R}\}.$$

*Then the following functions are measurable:*

$$h(x) = \begin{cases} f(x) + g(x) & \text{if } x \in A, \\ 0 & \text{if } x \notin A, \end{cases}$$

$$k(x) = \begin{cases} f(x)g(x) & \text{if } x \in A, \\ 0 & \text{if } x \notin A. \end{cases}$$

*Proof.* *By proposition 8.2.11, the set $A$ is measurable. By lemma 8.2.14, it is enough to check that $h$ and $k$ are measurable on $A$. Now if $a \in \mathbb{R}$, $\{x \in A : f(x) + g(x) > a\} = A \cap \{x \in X : f(x) > a - g(x)\}$, which is measurable by lemma 8.2.10. Thus $h$ is measurable on $A$. Now that $f + g$ and $f - g$ are measurable on $A$, $(f + g)^2$ and $(f - g)^2$ are measurable on $A$ by theorem 8.2.13. It follows that $k$ is measurable on $A$ because $fg = [(f + g)^2 - (f - g)^2]/4$.* ∎

**Theorem 8.2.16.** *Let $f_n$ be a sequence of measurable functions. Then the following functions are measurable:*

(a) $sup_n f_n$,
(b) $inf_n f_n$,
(c) $\limsup_n f_n$, and
(d) $\liminf_n f_n$.

*Also, the set $\{x \in X : \lim_n f_n(x) \text{ exists}\}$ is measurable.*

*Proof.* Parts (a) and (b) are true because

$$\{x \in X : sup_n f_n(x) > a\} = \cup_{n=1}^{\infty}\{x \in X : f_n(x) > a\},$$

and

$$\{x \in X : inf_n f_n(x) < a\} = \cup_{n=1}^{\infty}\{x \in X : f_n(x) < a\},$$

respectively. Now parts (c) and (d) follow from parts (a) and (b) because

$$\limsup_n f_n = inf_n\{sup_{k \geq n} f_k\}, \text{ and } \liminf_n f_n = sup_n\{inf_{k \geq n} f_k\}.$$

The last assertion follows from parts (c) and (d) and from lemma 8.2.10, because the set in question is

$$\{x \in X : \limsup_n f_n(x) = \liminf_n f_n(x)\}. \blacksquare$$

**Definition.** A complex function $f : X \to \mathbb{C}$ is said to be **measurable** if its real and imaginary parts are measurable.

**Theorem 8.2.17.** *A complex function $f : X \to \mathbb{C}$ is measurable if and only if $f^{-1}(V) \in \mathfrak{M}$ for every open subset $V$ of the complex plane.*

*Proof.* Write $f = f_1 + if_2$, and suppose $f$ is measurable. An open set $V$ in $\mathbb{C}$ is a countable union of open bounded rectangles, so it is enough to show that, for the rectangle $R = (a, b) \times (c, d)$, $f^{-1}(R)$ is measurable. But this is obvious since $f^{-1}(R) = f_1^{-1}((a, b)) \cap f_2^{-1}((c, d))$. To prove the converse, let $a \in \mathbb{R}$, and consider the open set $V = \{\xi + i\eta \in \mathbb{C} : \xi > a\}$. By assumption, $f^{-1}(V)$ is in $\mathfrak{M}$. But $f^{-1}(V) = f_1^{-1}((a, \infty))$. One shows that $f_2$ is measurable by considering the open set $V = \{\xi + i\eta \in \mathbb{C} : \eta > a\}$. $\blacksquare$

## Excursion: The Hopf Extension Theorem[2]

The motivation for the Hopf extension included below is not entirely precise, but we hope it will help the reader gain some insight into the construction of important measures such as the Lebesgue measure on $\mathbb{R}^2$. The plane contains a collection of subsets for which a *natural measure* exists, namely, the collection of rectangles.[3] The measure (area) of a rectangle ought to be the product of its dimensions. The collection $\mathfrak{C}$ of finite disjoint unions of rectangles in the plane is known to be an algebra in $\mathbb{R}^2$, and the measure of a member of $\mathfrak{C}$ is defined in the obvious way: it is the (finite) sum of the measures of the rectangles in the union. The immediate question is whether the *natural measure* we just described can be extended to the $\sigma$-algebra $\mathfrak{M}$ generated by $\mathfrak{C}$.

The Hopf extension abstracts the above motivation and provides an affirmative answer (theorem 8.2.19). Theorem 8.2.20 gives a sufficient condition for the uniqueness of such an extension.

We will construct measures on product spaces (section 8.8) using a different approach, and this excursion can be bypassed without affecting the continuity of the rest of this chapter.

We will adopt the following standing assumptions throughout this excursion:

1. $\mathfrak{C}$ is an algebra of subsets in $X$. (Thus $X \in \mathfrak{C}$, and $\varnothing \in \mathfrak{C}$.)
2. $\mu : \mathfrak{C} \to [0, \infty]$ is a set function such that $\mu(\varnothing) = 0$.
3. $\mu$ is a countably additive on $\mathfrak{C}$ in the sense that if $\{C_n\}$ is a disjoint sequence in $\mathfrak{C}$, and $\cup_{n=1}^{\infty} C_n \in \mathfrak{C}$, then $\mu(\cup_{n=1}^{\infty} C_n) = \sum_{n=1}^{\infty} \mu(C_n)$. Observe that such a function is monotone.

**Lemma 8.2.18.** *Under the standing assumptions, define a set function* $n^* : \mathcal{P}(X) \to [0, \infty]$ *by*

$$n^*(E) = \inf\left\{ \sum_{n=1}^{\infty} \mu(C_n) : C_n \in \mathfrak{C}, E \subseteq \cup_{n=1}^{\infty} C_n \right\}.$$

(a) $n^*$ *is an outer measure on* $X$.
(b) *For every* $E \in \mathfrak{C}$, $n^*(E) = \mu(E)$.
(c) *Every* $E \in \mathfrak{C}$ *is* $n^*$-*measurable*.

---

[2] This theorem is also attributed to Lebesgue.
[3] Intuitively, a rectangle is the product of two intervals. More precisely, a rectangle is the product of two Lebesgue measurable subsets of $\mathbb{R}$.

*Proof.* The definition of $n^*$ is meaningful because $X \in \mathfrak{C}$, and the monotonicity of $n^*$ is obvious. To prove that $n^*$ is subadditive, let $\{E_n\}$ be a countable collection of subsets of $X$, and, without loss of generality, assume that $\sum_{n=1}^{\infty} n^*(E_n) < \infty$. Let $\epsilon > 0$ and, for each $n \in \mathbb{N}$, let $\{C_{n,j}\} \subseteq \mathfrak{C}$ be such that $E_n \subseteq \cup_{j=1}^{\infty} C_{n,j}$ and $\sum_{j=1}^{\infty} \mu(C_{n,j}) < n^*(E_n) + \epsilon/2^n$. Now $\cup_{n=1}^{\infty} E_n \subseteq \cup_{n,j=1}^{\infty} C_{n,j}$; hence $n^*(\cup_{n=1}^{\infty} E_n) \leq \sum_{n=1}^{\infty} \sum_{j=1}^{\infty} \mu(C_{n,j}) \leq \sum_{n=1}^{\infty} n^*(E_n) + \epsilon/2^n = \sum_{n=1}^{\infty} n^*(E_n) + \epsilon$. Since $\epsilon$ is arbitrary, the proof of part (a) is complete.

(b) If $E \in \mathfrak{C}$, then $E \subseteq \cup_{n=1}^{\infty} C_n$, where $C_1 = E$, and $C_n = \emptyset$ for $n \geq 2$. Thus $n^*(E) \leq \mu(E)$. Suppose $E \subseteq \cup_{n=1}^{\infty} C_n$, and define $D_1 = C_1$ and, for $n \geq 2$, $D_n = C_n - \cup_{i=1}^{n-1} C_i$. Because $\mathfrak{C}$ is an algebra, each $D_n \in \mathfrak{C}$. Clearly, $\cup_{n=1}^{\infty} D_n = \cup_{n=1}^{\infty} C_n$; hence $E = \cup_{n=1}^{\infty} (E \cap D_n)$. By the additivity of $\mu$ on $\mathfrak{C}$, $\mu(E) = \mu(\cup_{n=1}^{\infty}(E \cap D_n)) = \sum_{n=1}^{\infty} \mu(E \cap C_n) \leq \sum_{n=1}^{\infty} \mu(C_n)$. By the very definition of $n^*$, $\mu(E) \leq n^*(E)$.

(c) Let $E \in \mathfrak{C}$, $A \subseteq X$, and, without loss of generality, assume that $n^*(A) < \infty$. For every $\epsilon > 0$, there exists a sequence $\{C_n\}$ in $\mathfrak{C}$ such that $A \subseteq \cup_{n=1}^{\infty} C_n$ and $\sum_{n=1}^{\infty} \mu(C_n) \leq n^*(A) + \epsilon$. By the additivity of $\mu$ on $\mathfrak{C}$, $n^*(A) + \epsilon \geq \sum_{n=1}^{\infty} \mu(C_n) = \sum_{n=1}^{\infty} \mu(C_n \cap E) + \sum_{n=1}^{\infty} \mu(C_n \cap E') \geq n^*(A \cap E) + \nu^*(A \cap E')$. Since $\epsilon$ is arbitrary, the result follows. ∎

**Theorem 8.2.19 (the Hopf extension theorem).** *Under the standing assumptions, the set function $\mu$ has an extension to a positive measure on the $\sigma$-algebra $\mathfrak{M}$ generated by $\mathfrak{C}$.*

*Proof.* By theorem 8.2.6 (Carathéodory's theorem), the collection $\mathfrak{N}$ of $n^*$-measurable subsets of $X$ is a $\sigma$-algebra, and the restriction, $\nu$, of $n^*$ to $\mathfrak{N}$ is a positive measure. Since every member of $\mathfrak{C}$ is $n^*$-measurable, $\mathfrak{M} \subseteq \mathfrak{N}$. The measure space we seek is $(X, \mathfrak{M}, \nu)$. ∎

The next corollary establishes a sufficient condition for the uniqueness of the Hopf extension.

**Theorem 8.2.20.** *Suppose, in addition to the standing assumptions, that the following assumption is satisfied:*

$\sigma$-**finiteness:** *there exists a sequence $(X_n)$ in $\mathfrak{C}$ such that $X = \cup_{n=1}^{\infty} X_n$, and, for every $n \in \mathbb{N}$, $\mu(X_n) < \infty$*

*Then the extension $\nu$ provided by the previous theorem is unique.*

*Proof.* The following two facts are consequences of the $\sigma$-finiteness assumption.

(a) *The sequence $(X_n)$ may be assumed to be mutually disjoint, because we can replace it with the sequence $Y_1 = X_1$, and, for $n \geq 2$, $Y_n = X_n - \cup_{i=1}^{n-1} X_i$. Clearly, $Y_n \in \mathfrak{C}$, $\mu(Y_n) \leq \mu(X_n) < \infty$, and $\cup_{n=1}^{\infty} Y_n = X$.*

(b) *An arbitrary member $E \in \mathfrak{M}$ can be written as $E = \cup_{n=1}^{\infty} E_n$, where $(E_n)$ is a disjoint sequence in $\mathfrak{M}$ such that $\nu(E_n) < \infty$. We simply set $E_n = E \cap Y_n$. Then $\nu(E_n) \leq \nu(Y_n) = \mu(Y_n) < \infty$.*

*We now prove the result. Suppose there is another measure that extends $\mu$ from $\mathfrak{C}$ to $\mathfrak{M}$. We continue to use the symbol $\mu$ to denote this extension. Thus we assume that $\mu(C) = \nu(C)$ for every $C \in \mathfrak{C}$ and prove that $\mu(E) = \nu(E)$ for every $E \in \mathfrak{M}$.*
*Observe the following facts:*

(c) *If $\{C_n\} \subseteq \mathfrak{C}$, and $C = \cup_{n=1}^{\infty} C_n$, then $\nu(C) = \lim_n \nu(\cup_{i=1}^{n} C_i) = \lim_n \mu(\cup_{i=1}^{n} C_i) = \mu(C)$.*

(d) *For every $E \in \mathfrak{M}$, $\mu(E) \leq \nu(E)$. If $E \subseteq \cup_{n=1}^{\infty} C_n$, where $\{C_n\} \subseteq \mathfrak{C}$, then $\mu(E) \leq \mu(\cup_{n=1}^{\infty} C_n) \leq \sum_{n=1}^{\infty} \mu(C_n)$. By the definition of $\nu$, $\mu(E) \leq \nu(E)$.*

(e) *If $E \in \mathfrak{M}$ and $\nu(E) < \infty$, then $\mu(E) = \nu(E)$. Let $\epsilon > 0$. There exists a sequence $(C_n)$ in $\mathfrak{C}$ such that $E \subseteq C = \cup_{n=1}^{\infty} C_n$ and $\sum_{n=1}^{\infty} \mu(C_n) < \nu(E) + \epsilon$. Now $\nu(C) \leq \sum_{n=1}^{\infty} \nu(C_n) = \sum_{n=1}^{\infty} \mu(C_n) < \nu(E) + \epsilon$. In particular, $\nu(C - E) < \epsilon$. Using fact (c), we have $\nu(E) \leq \nu(C) = \mu(C) = \mu(E) + \mu(C - E) \leq \mu(E) + \nu(C - E) < \mu(E) + \epsilon$. Since $\epsilon$ is arbitrary, $\nu(E) \leq \mu(E)$. Now $\mu(E) = \nu(E)$ by fact (d).*

*Finally, for an arbitrary set $E \in \mathfrak{M}$, we use fact (b) to write $E = \cup_{n=1}^{\infty} E_n$, where $(E_n)$ is a disjoint sequence in $\mathfrak{M}$, such that $\nu(E_n) < \infty$. Using fact (e), $\nu(E) = \sum_{n=1}^{\infty} \nu(E_n) = \sum_{n=1}^{\infty} \mu(E_n) = \mu(E)$.* ∎

## Exercises

1. Let $\mathcal{A} = \{A_1, ..., A_n\}$ be distinct subsets of a nonempty set $X$. Show that the $\sigma$-algebra generated by $\mathcal{A}$ contains at most $2^{2^n}$ members.

2. Show that if a $\sigma$-algebra $\mathfrak{M}$ is infinite, then it is uncountable.

3. **Completion of an incomplete measure.** Let $(X, \mathfrak{M}, \mu)$ be an incomplete measure space, and let $\mathfrak{Z}$ be the collection of subsets of sets of $\mu$-measure 0. Let $\overline{\mathfrak{M}}$ be the smallest $\sigma$-algebra in $X$ that contains $\mathfrak{M} \cup \mathfrak{Z}$. Prove that every member of $\overline{\mathfrak{M}}$ has the form $E \cup Z$, where $E \in \mathfrak{M}$ and $Z \in \mathfrak{Z}$. Extend the definition of $\mu$ to $\overline{\mathfrak{M}}$ as follows: for $E \in \mathfrak{M}$ and $Z \in \mathfrak{Z}$, $\overline{\mu}(E \cup Z) = \mu(E)$.

Show that $\overline{\mu}$ is well defined and that $\overline{\mu}$ is a complete measure. Hint: Show that the set $\mathfrak{M}_1 = \{E \cup Z : E \in \mathfrak{M}, Z \in \mathfrak{Z}\}$ is a $\sigma$-algebra.

4. This exercise provides a useful alternative characterization of the completion of a measure space. In the notation of the previous exercise, prove that, for a subset $E$ of $X$, $E \in \overline{\mathfrak{M}}$ if and only if there exists two sets $A$ and $B$ in $\mathfrak{M}$ such that $A \subseteq E \subseteq B$ and $\mu(B - A) = 0$.

5. Prove that each of the following collections of sets generates $\mathcal{B}(\mathbb{R})$:
   (a) $\{(a, \infty) : a \in \mathbb{R}\}$
   (b) $\{(-\infty, b) : b \in \mathbb{R}\}$

6. Prove that the collection of open boxes $\{\prod_{i=1}^{n}(a_i, b_i) : a_i, b_i \in \mathbb{Q}\}$ generates $\mathcal{B}(\mathbb{R}^n)$.

7. Suppose $\mathfrak{M}$ is a $\sigma$-algebra generated by a collection $\mathfrak{C}$ of subsets of a nonempty set $X$. Prove that $\mathfrak{M}$ is the union of the $\sigma$-algebras generated by $\mathfrak{F}$ where $\mathfrak{F}$ ranges over all the countable subsets of $\mathfrak{C}$. Hint: The latter union is a $\sigma$-algebra.

8. Prove that if $E$ and $F$ are measurable sets such that $\mu(E \Delta F) = 0$, then $\mu(E) = \mu(F) = \mu(E \cup F) = \mu(E \cap F)$.

9. Let $E_n$ be a sequence of measurable sets such that $\sum_{n=1}^{\infty} \mu(E_n) < \infty$. Prove that the set $\bigcap_{n=1}^{\infty} \bigcup_{k \geq n} E_k$ has measure 0. Conclude that, except for a set of measure 0, every $x \in X$ belongs to finitely many of the sets $E_n$.

10. Let $E_1, \ldots, E_n$ be measurable sets and, for $1 \leq j \leq n$, let $F_j$ to be the set of points in $X$ that belong to exactly $j$ of the sets $E_1, \ldots, E_n$. Prove that $\mu(\bigcup_{i=1}^{n} E_i) = \sum_{j=1}^{n} \mu(F_j)$, and $\sum_{i=1}^{n} \mu(E_i) = \sum_{j=1}^{n} j\mu(F_j)$. Hint: $F_j = \{x \in X : \sum_{i=1}^{n} \chi_{E_i}(x) = j\}$.

11. Prove theorem 8.2.9.

12. Show that if $f$ is measurable and $a \in \overline{\mathbb{R}}$, then $f^{-1}(a)$ is measurable.

13. Let $(X, \mathfrak{M})$ be a measurable space such that $\mathfrak{M} \neq \mathcal{P}(X)$. Prove that there is a function $f$ such that $|f|$ is measurable but $f$ is not.

14. Suppose that $(X, \mathfrak{M})$ is a measurable space and that $Y$ is a nonempty set. Show that if $f : X \to Y$, then the collection $\mathfrak{N} = \{E \subseteq Y : f^{-1}(E) \in \mathfrak{M}\}$ is a $\sigma$-algebra.

15. Let $(X, \mathfrak{M})$ be a measurable space, and let $f : X \to \mathbb{R}$ be a measurable function. Show that $f^{-1}(B)$ is measurable for every Borel subset $B$ of $\mathbb{R}$. Hint: The collection $\Omega = \{E \subseteq \mathbb{R} : f^{-1}(E) \in \mathfrak{M}\}$ contains all open subsets of $\mathbb{R}$.

16. Let $X$ be a topological space, and let $f : X \to \mathbb{R}$ be a continuous function. Show that $f^{-1}(B)$ is a Borel subset of $X$ for every Borel subset $B$ of $\mathbb{R}$.

17. Show that if $E \in \mathcal{B}(\mathbb{R}^s)$ and $F \in \mathcal{B}(\mathbb{R}^r)$, then $E \times F \in \mathcal{B}(\mathbb{R}^{r+s})$. Hint: For an open subset $E$ of $\mathbb{R}^r$, consider the collection $\Omega = \{F \subseteq \mathbb{R}^s : E \times F \in \mathcal{B}(\mathbb{R}^{r+s})\}$. Show that $\mathcal{B}(\mathbb{R}^s) \subseteq \Omega$. Then, for a Borel subset $F$ of $\mathbb{R}^s$, consider the collection $\{E \subseteq \mathbb{R}^r : E \times F \in \mathcal{B}(\mathbb{R}^{r+s})\}$.

18. Let $C$ be the Cantor set, and define a function $f : [0,1] \to C$ as follows: $f(0) = 0$, and, for $x \in (0,1]$, write $x = \sum_{i=1}^{\infty} \frac{a_i}{2^i}$ and set $f(x) = \sum_{i=1}^{\infty} \frac{2a_i}{3^i}$.[4] Show that $f$ is Borel measurable. Hint: For a fixed $i \in \mathbb{N}$, define $f_i(x) = a_i$. It is enough to show that $f_i$ is measurable. To this end, show that $f_i = \sum \{ \chi_{E_{i,k}} : k = 1, 3, 5, \ldots, 2^i - 1 \}$, where $E_{i,k} = (\frac{k}{2^i}, \frac{k+1}{2^i}]$.

## 8.3 Abstract Integration

In this section, we examine Lebesgue's revolutionary approach to the definition of the integral. The motivation below is imprecise and does not rigorously develop any particular set of ideas. For the sake of simplicity, we assume that $f$ is a positive continuous function on a compact interval.

The Riemann integral is based on the simple geometrical idea of dividing the region below the graph of $f$ into thin vertical strips, where the area below the graph is approximated by the integral of a step function (the Riemann sum). Lebesgue's idea was to divide the range of $f$ by points $y_0, \ldots, y_n$, and, for $k = 1, \ldots, n$, we consider the sets $E_k = f^{-1}([y_k, y_{k+1}))$. Even for an uncomplicated function, the set $E_k$ may come in several fragments, as shown in figure 8.1, where $E_k$ has three fragments. When $y_{k+1} - y_k$ is small, the approximate combined area of the three shaded strips is the approximate common height, $y_k$, times the sum of the lengths of the three fragments that comprise the set $E_k$ or, more precisely, the measure of $E_k$. Thus

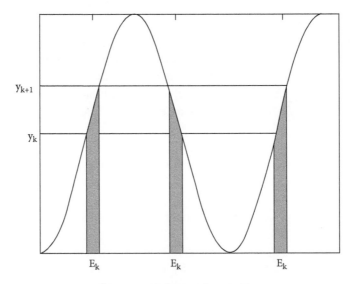

**Figure 8.1** Lebesgue integration

---

[4] We use the series representation of $x$ if $x$ has a terminating binary expansion.

the approximate area below the graph is $\sum_{k=1}^{n} y_k \mu(E_k)$, which, by definition, is the integral of a simple function. Needless to say, as the partition of the range of $f$ gets finer, we expect the integrals of the simple functions to converge to the integral of $f$. This is the overarching idea in Lebesgue integration. As it turns out, we can integrate far more functions under the Lebesgue definition than under the Riemann definition. For example, the integral of any positive measurable function is defined, although it may not be finite. Additionally, the definition of the integral extends seamlessly to abstract measure spaces. The section results capture the above ideas. First we define the integral of a positive measurable function $f$, then we show that $f$ is the limit of simple functions, $s_n$, and then we show that $\int_X f d\mu = \lim_n \int_X s_n d\mu$. Extending the definition of the integral to complex functions follows without difficulty. The section concludes with three important convergence theorems.

**Definition.** Let $(X, \mathfrak{M})$ be a measurable space. A **simple function** on $X$ is a function $s : X \to \mathbb{C}$ of finite range. If $a_1, \dots, a_m$ are the distinct values of $s$, then $s = \sum_{i=1}^{m} a_i \chi_{E_i}$, where $E_i = s^{-1}(a_i)$. Clearly, $E_i \cap E_j = \varnothing$ if $i \neq j$.

**Remarks.** (a) It is clear that a simple function is measurable if and only if each $E_i$ is a measurable set. Also, a simple function need not have bounded support. For example, $s = \chi_{(-\infty,0)} + \chi_{(1,2)}$ is not supported on a bounded set.

(b) Our definition of a simple function is sometimes referred to as the **standard form of a simple function**. It is important to understand that any finite linear combination of characteristic functions of disjoint sets is a simple function. For example, if $s = \sum_{i=1}^{m} a_i \chi_{E_i}$, where $a_1, \dots, a_m$ are not all distinct, we can rewrite $s$ in standard form as follows: Let $b_1, \dots, b_n$ be the distinct values of $s$ and, for $1 \leq i \leq n$, let $T_i = \{j \in \mathbb{N}_m : a_j = b_i\}$. Note that $\{T_1, \dots, T_n\}$ is a partition of $\mathbb{N}_m$. Set $B_i = \cup_{j \in T_i} E_j$. The sets $B_i$ are disjoint since the sets $E_i$ are. Clearly, $s = \sum_{i=1}^{n} b_i \chi_{B_i}$. It is, in fact, true that a finite linear combination of characteristic functions of subsets of $X$ is a simple function, even when the sets $E_i$ overlap. An inductive proof is possible. We invite the reader to try it.

**Definition. The integral of a simple function.** Let $(X, \mathfrak{M}, \mu)$ be a measure space, and let $s = \sum_{i=1}^{m} a_i \chi_{E_i}$ be a measurable simple function in standard form. The integral of $s$ with respect to $\mu$ is

$$\int_X s d\mu = \sum_{i=1}^{m} a_i \mu(E_i).$$

The above formula is robust in the sense that it is valid even when $s$ is not in standard form. This follows from remark (b) above. If $a_1, \dots, a_m$ are not all distinct, write $s$ in standard form using remark (b): $s = \sum_{i=1}^{n} b_i \chi_{B_i}$. Then $\sum_{i=1}^{n} b_i \mu(B_i) = \sum_{i=1}^{n} b_i \sum_{j \in T_i} \mu(E_j) = \sum_{j=1}^{m} a_j \mu(E_j)$.

**Theorem 8.3.1.** *If $s$ and $t$ are simple functions and $c \in \mathbb{K}$, then*

$$\int_X (s+t)d\mu = \int_X s\,d\mu + \int_X t\,d\mu, \text{ and } \int_X cs\,d\mu = c\int_X s\,d\mu.$$

*Proof.* The second identity follows trivially from the definition.

Let $s = \sum_{i=1}^{m} a_i \chi_{E_i}$ and $t = \sum_{j=1}^{n} b_j \chi_{F_j}$ be simple functions in standard form, and let $B_{ij} = E_i \cap F_j, 1 \leq i \leq m, 1 \leq j \leq n$. The collection $\{B_{ij}\}$ is disjoint, $E_i = \cup_{j=1}^{n} B_{ij}$, and $F_j = \cup_{i=1}^{m} B_{ij}$. Now $s = \sum_{i=1}^{m} \sum_{j=1}^{n} a_i \chi_{B_{ij}}$, $t = \sum_{j=1}^{n} \sum_{i=1}^{m} b_j \chi_{B_{ij}}$, and $s + t = \sum_{i=1}^{m} \sum_{j=1}^{n} (a_i + b_j) \chi_{B_{ij}}$. By definition,

$$\int_X (s+t)d\mu = \sum_{i=1}^{m}\sum_{j=1}^{n}(a_i + b_j)\mu(B_{ij})$$

$$= \sum_{i=1}^{m} a_i \sum_{j=1}^{n} \mu(B_{ij}) + \sum_{j=1}^{n} b_j \sum_{i=1}^{m} \mu(B_{ij})$$

$$= \sum_{i=1}^{m} a_i \mu(E_i) + \sum_{j=1}^{n} b_j \mu(F_j) = \int_X s\,d\mu + \int_X t\,d\mu. \blacksquare$$

**Remark.** This proof includes a proof of the fact that the sum of two simple functions is a simple function.

**Definition. The integral of a positive function.** Let $f : X \to [0, \infty]$ be a measurable function. We define

$$\int_X f\,d\mu = \sup\left\{ \int_X s\,d\mu : 0 \leq s \leq f, \text{s simple}\right\}.$$

Observe that this definition is reminiscent of the fact that the Riemann integral of a function is the supremum of the lower Riemann sums of the function and that a lower Riemann sum of a function is the integral of a step function dominated by $f$.

The following facts are immediately obvious:

(a) $\int_X cf\,d\mu = c\int_X f\,d\mu$ for $c \geq 0$, and
(b) If $g : X \to [0, \infty]$ is measurable and $0 \leq f \leq g$, then $\int_X f\,d\mu \leq \int_X g\,d\mu$.

The fact that, for positive functions $f$ and $g$, $\int_X (f+g)d\mu = \int_X f d\mu + \int_X g d\mu$ requires the development of some machinery. First we show that a positive measurable function $f$ is the limit of a sequence, $s_n$, of simple functions, then we show that $\lim_n \int_X s_n d\mu = \int_X f d\mu$. The details appear below.

**Definition.** The **positive and negative parts of a measurable real-valued function** $f$ are, respectively,

$$f^+(x) = \max\{f(x), 0\}, \text{ and } f^-(x) = -\min\{f(x), 0\}.$$

Observe that $f^+$ and $f^-$ are positive, measurable functions,

$$f = f^+ - f^-, \text{ and } |f| = f^+ + f^-.$$

**Theorem 8.3.2.** *(a) Let $f : X \to [0, \infty]$ be a measurable function. Then there exists an increasing sequence of simple functions $s_1, s_2, \ldots$ such that $\lim_n s_n(x) = f(x)$ for every $x \in X$.*

*(b) Let $f : X \to \mathbb{C}$ be a measurable function. Then there exists a sequence of simple functions $u_1, u_2, \ldots$ such that $\lim_n u_n(x) = f(x)$ for every $x \in X$ and $|u_1| \leq |u_2| \leq \ldots \leq |f|$.*

*Proof.* For each $n \in \mathbb{N}$, define $E_{n,k} = \{x \in X : \frac{k-1}{2^n} \leq f(x) < \frac{k}{2^n}\}$, $k = 1, 2, \ldots, n2^n$, and $F_n = \{x \in X : f(x) \geq n\}$. Let $s_n = \sum_{k=1}^{n2^n} \frac{k-1}{2^n} \chi_{E_{n,k}} + n\chi_{F_n}$.

The fact that $s_n(x) \leq f(x)$ is clear. Now every $x \in X$ belongs to exactly one of the sets $E_{n,k}$ or to $F_n$. We show that $s_n$ is an increasing sequence of functions. If $\frac{2(k-1)}{2^{n+1}} \leq f(x) < \frac{2k-1}{2^{n+1}}$, then $s_n(x) = \frac{2(k-1)}{2^{n+1}} = s_{n+1}(x)$. If $\frac{2k-1}{2^{n+1}} \leq f(x) < \frac{2k}{2^{n+1}}$, then $s_n(x) = \frac{2(k-1)}{2^{n+1}} < \frac{2k-1}{2^{n+1}} = s_{n+1}(x)$. If $f(x) \geq n, n = s_n(x) \leq s_{n+1}(x)$. Now we show that $\lim_n s_n(x) = f(x)$. If $f(x) < \infty$, $0 \leq f(x) - s_n(x) \leq 1/2^n$. If $f(x) = \infty, s_n(x) = n$ for all $n \in \mathbb{N}$. In either case, $\lim_n s_n(x) = f(x)$.

To prove (b), write $f = (f_1^+ - f_1^-) + i(f_2^+ - f_2^-)$. By part (a), there are sequences of positive, increasing, measurable simple functions $s_n^+, s_n^-, t_n^+$, and $t_n^-$ such that $\lim_n s_n^+ = f_1^+$, $\lim_n s_n^- = f_1^-$, $\lim_n t_n^+ = f_2^+$, and $\lim_n t_n^- = f_2^-$. The sequence of simple functions $u_n = (s_n^+ - s_n^-) + i(t_n^+ - t_n^-)$ satisfies the requirements of part (b). ∎

**Remark.** This proof shows that if $f$ is a bounded, positive, measurable function, then $s_n$ converges uniformly to $f$ because $\|s_n - f\|_\infty \leq 1/2^n$.

**Lemma 8.3.3.** *Let $f_n : X \to [0, \infty]$ be an increasing sequence of measurable functions such that $\lim_n f_n(x) = f(x)$ for all $x \in X$. If $s$ is a simple measurable function such that $0 \leq s \leq f$, then $\lim_n \int_X f_n d\mu \geq \int_X s d\mu$.*

*Proof.* Let $s = \sum_{j=1}^m a_j \chi_{E_j}$. Fix $0 < \alpha < 1$, and define $B_n = \{x \in X : f_n(x) > \alpha s(x)\}$. It is easy to see that $B_n \subseteq B_{n+1}$ and that $\cup_{n=1}^{\infty} B_n = X$. Notice that $(E_j \cap B_n)$ is an ascending sequence of sets, and $E_j = \cup_{n=1}^{\infty}(E_j \cap B_n)$; hence $\mu(E_j) = \lim_n \mu(E_j \cap B_n)$. Now $\int_X f_n d\mu \geq \int_X f_n \chi_{B_n} d\mu \geq \alpha \int_X s \chi_{B_n} d\mu = \alpha \sum_{j=1}^m a_j \mu(E_j \cap B_n)$. Taking the limit as $n \to \infty$, we obtain $\lim_n \int_X f_n d\mu \geq \alpha \sum_{j=1}^m a_j \mu(E_j) = \alpha \int_X s d\mu$. The result we need follows by letting $\alpha \to 1$. ∎

**Theorem 8.3.4.** *Let $f \geq 0$ be a measurable function, and let $0 \leq s_n \leq s_{n+1}$ be simple functions such that $s_n \leq f$, and $\lim_n s_n(x) = f(x)$. Then $\int_X f d\mu = \lim_n \int_X s_n d\mu$.*

*Proof.* Since $0 \leq s_n \leq f$, $\int_X s_n \leq \int_X f d\mu$, and $\lim_n \int_X s_n d\mu \leq \int_X f d\mu$. Now if $t$ is a simple function such that $0 \leq t \leq f$, then, by lemma 8.3.3, $\int_X t d\mu \leq \lim_n \int_X s_n d\mu$. Therefore, $\int_X f d\mu = \sup\{\int_X t d\mu : 0 \leq t \leq f, t \text{ simple }\} \leq \lim_n \int_X s_n d\mu$. ∎

**Theorem 8.3.5.** *If $f \geq 0$ and $g \geq 0$ are measurable functions and $a, b \geq 0$, then $\int_X (af + bg) d\mu = a \int_X f d\mu + b \int_X g d\mu$.*

*Proof.* By theorem 8.3.2, there exist sequences of simple functions $s_1 \leq s_2 \leq ...$, and $t_1 \leq t_2 \leq ...$ such that $\lim_n s_n(x) = f(x)$, and $\lim_n t_n(x) = g(x)$. By theorem 8.3.1, $\int_X (as_n + bt_n) d\mu = a \int_X s_n d\mu + b \int_X t_n d\mu$. Now the sequence of simple functions $as_n + bt_n$ is increasing, and $\lim_n (as_n(x) + bt_n(x)) = af(x) + bg(x)$. Thus, by theorem 8.3.4, $\int_X (af + bg) d\mu = \lim_n \int_X (as_n + bt_n) d\mu = a \lim_n \int_X s_n d\mu + b \lim_n \int_X t_n d\mu = a \int_X f d\mu + b \int_X g d\mu$. ∎

**Definition. The integral of a real function.** Let $f : X \to [-\infty, \infty]$ be a measurable function, and write $f = f^+ - f^-$. By definition,

$$\int_X f d\mu = \int_X f^+ - \int_X f^- d\mu,$$

provided that at least one of the integrals on the right-hand side of the definition is finite. We say $f$ is **integrable** if both $\int_X f^+ d\mu$ and $\int_X f^- d\mu$ are finite, which is equivalent to the condition that $\int_X |f| d\mu < \infty$. This is because $|f| = f^+ + f^-$, $f^+ \leq |f|$, and $f^- \leq |f|$.

**Theorem 8.3.6.** *If $f$ and $g$ are real and integrable, then $\int_X (f + g) d\mu = \int_X f d\mu + \int_X g d\mu$. Also, $\int_X af d\mu = a \int_X f d\mu$ for every real number $a$.*

*Proof.* Write $f=f^+-f^-, g=g^+-g^-$, and let $h=f+g$. Writing $h=h^+-h^-$ yields $h^++f^-+g^-=h^-+f^++g^+$. Integrating both sides of the last identity and using the previous theorem, we obtain $\int_X h^+ d\mu + \int_X f^- d\mu + \int_X g^- d\mu = \int_X h^- d\mu + \int_X f^+ d\mu + \int_X g^+ d\mu$. The result we seek is obtained by rearranging the last identity. To complete the proof of the theorem, we only need to show that $\int_X(-f)d\mu = -\int_X f d\mu$. This is a simple calculation if we use the identities $(-f)^+=f^-$, and $(-f)^-=f^+$. ∎

**Definition. The integral of a complex function.** If $f: X \to \mathbb{C}$, and $f=f_1+if_2$, define $\int_X f d\mu = \int_X f_1 d\mu + i\int_X f_2 d\mu$. We say that $f$ is **integrable** if $f_1$ and $f_2$ are integrable.

Notice that a complex function is integrable if and only if $\int_X |f| d\mu < \infty$. This is because $|f_1| \le |f|, |f_2| \le |f|$, and $|f| \le |f_1|+|f_2|$.

**Theorem 8.3.7.** *If $f$ and $g$ are integrable complex functions, and $a,b \in \mathbb{C}$, then $af+bg$ is integrable and*

$$\int_X (af+bg)d\mu = a\int_X f d\mu + b\int_X g d\mu.$$

*Proof.* $\int_X |af+bg| d\mu \le \int_X |a||f| + |b||g| d\mu = |a|\int_X |f| d\mu + |b|\int_X |g| d\mu < \infty$. Thus $af+bg$ is integrable. The verification that $\int_X (f+g)d\mu = \int_X f d\mu + \int_X g d\mu$ is a routine calculation, as is the fact that $\int_X cf d\mu = c\int_X f d\mu$ when $c$ is a real constant. It now suffices to show that $\int_X if d\mu = i\int_X f d\mu$. Indeed, $\int_X if d\mu = \int_X i(f_1+if_2)d\mu = \int_X(-f_2+if_1)d\mu = \int_X -f_2 d\mu + i\int_X f_1 d\mu = i\int_X f d\mu$. ∎

It is easy to see that the set of complex integrable functions is a vector space. We denote it by $\mathfrak{L}^1(\mu)$. In fact, if a norm is defined on $\mathfrak{L}^1(\mu)$ by $\|f\|_1 = \int_X |f| d\mu$, then $\mathfrak{L}^1(\mu)$ is a normed linear space, as the reader can easily verify.

**Definition.** Let $(X, \mathfrak{M}, \mu)$ be a measure space, and let $P(x)$ be a property that may or may not be satisfied by a point $x \in X$. For example, for a given extended real-valued function $f$, $P(x)$ may be the property that $f(x)$ is finite. Another example is the property that $f(x)=g(x)$ for two measurable functions $f$ and $g$. We say that property $P$ holds for almost every $x$ in a measurable set $E$, or that $P$ holds **almost everywhere** in $E$, if $\mu(\{x \in E : P(x) \text{ is false}\}) = 0$. In this situation, we often write "$P$ holds for a.e. $x \in E$." The examples below are good illustrations of the concept.

**Example 1.** If $f \in \mathfrak{L}^1(\mu)$, then $f$ is finite almost everywhere.

Let $E_n = \{x \in X : |f(x)| \geq n\}$. It is clear that $E_1 \supseteq E_2 \supseteq E_3 \supseteq ...$, and that, for each $n \in \mathbb{N}, \int_X |f| d\mu \geq \int_X |f| \chi_{E_n} d\mu \geq \int_X n \chi_{E_n} d\mu = n\mu(E_n)$. Thus $\mu(E_n) \leq \frac{1}{n} \int_X |f| d\mu \to 0$ as $n \to \infty$. Therefore $\mu(\cap_{n=1}^{\infty} E_n) = \lim_n \mu(E_n) = 0$. But $\cap_{n=1}^{\infty} E_n = \{x \in X : |f(x)| = \infty\}$. This shows that $f$ is finite almost everywhere. ◆

**Example 2.** If $f \in \mathfrak{L}^1(\mu)$, then $|\int_X f d\mu| \leq \int_X |f| d\mu$.

Assume that $z = \int_X f d\mu \neq 0$ because, otherwise, there is nothing to prove. Let $\alpha = \frac{|z|}{z}$. Then $|\alpha| = 1$, and $\alpha z = |z|$. Now $|\int_X f d\mu| = |z| = \alpha z = \alpha \int_X f d\mu = \int_X (\alpha f) d\mu$. It follows that $\int_X \alpha f d\mu$ is real and positive; therefore $\int_X \alpha f d\mu = \int_X u d\mu$, where $u = \text{Re}(\alpha f)$. Now $u \leq |\alpha f| = |f|$, and $\int_X u d\mu \leq \int_X |\alpha f| d\mu = \int_X |f| d\mu$. ◆

**Definition.** If $f : X \to [0, \infty]$ is measurable and $E \in \mathfrak{M}$, we define

$$\int_E f d\mu = \int_X f \chi_E d\mu.$$

Note that if $E_1 \subseteq E_2$, then $\int_{E_1} f d\mu \leq \int_{E_2} f d\mu$. Also, if $0 \leq f \leq g$, then $\int_E f d\mu \leq \int_E g d\mu$.

When $s = \sum_{j=1}^{m} a_j \chi_{E_j}$ is a simple function, then $s \chi_E = \sum_{j=1}^{m} a_j \chi_{E_j \cap E}$ is also a simple function and

$$\int_E s d\mu = \sum_{j=1}^{m} a_j \mu(E_j \cap E).$$

This equation can very well be used to define $\int_E s d\mu$. One can then take the alternative approach of defining

$$\int_E f d\mu = \sup \left\{ \int_E s d\mu : 0 \leq s \leq f, s \text{ simple} \right\}.$$

The two methods of defining $\int_E f d\mu$ are clearly equivalent, and the interested reader is encouraged to work out the details of reconciling the two definitions.

Another detail must be mentioned here. If $E \in \mathfrak{M}$, one can restrict $\mathfrak{M}$ and $\mu$ to $E$ in the obvious way: Define $\mathfrak{M}_E$ to be the members of $\mathfrak{M}$ contained in $E$, and define $\mu_E$ to be the restriction of $\mu$ to $\mathfrak{M}_E$. This clearly turns $(E, \mathfrak{M}_E, \mu_E)$ into a measure space, and it makes sense to define $\int_E f d\mu$ to be the integral of $f|_E$ with respect to $(E, \mathfrak{M}_E, \mu_E)$. Again, this definition is consistent with the above definitions of $\int_E f d\mu$, and, again, we leave the details to the interested reader.

**Example 3.** Suppose $f : X \to [0, \infty]$ is a measurable function. If $\int_E f d\mu = 0$ for some measurable set $E$, then $f = 0$ a.e. on $E$. In particular, if $\int_X f d\mu = 0$, then $f = 0$ a.e. on $X$.

Let $E_n = \{x \in E : f(x) > \frac{1}{n}\}$. Then $\frac{1}{n}\mu(E_n) \le \int_{E_n} f d\mu \le \int_E f d\mu = 0$. Thus $\mu(E_n) = 0$. The result now follows from the fact that $\{x \in E : f(x) > 0\} = \cup_{n=1}^{\infty} E_n$, and $\mu(\cup_{n=1}^{\infty} E_n) \le \sum_{n=1}^{\infty} \mu(E_n) = 0$. ◆

**Example 4.** If $f$ is a measurable function and $\int_E f d\mu = 0$, for every measurable set $E$, then $f = 0$ a.e.

Without loss of generality, assume $f$ is real. Let $E = \{x \in X : f(x) \ge 0\}$. By assumption, $\int_E f d\mu = 0$. But $\int_E f d\mu = \int_X f^+ d\mu$. By example 3, $f^+ = 0$ a.e. on $X$. Similarly, $f^- = 0$ a.e. on $X$. ◆

## Convergence Theorems

**Theorem 8.3.8 (Fatou's theorem).** *Let $f_n : X \to [0, \infty]$ be a sequence of measurable functions. Then*

$$\int_X \liminf_n f_n d\mu \le \liminf_n \int_X f_n d\mu.$$

*Proof.* Let $g_n = \inf_{k \ge n} f_k$. Then $0 \le g_1 \le g_2 \le ...$, and let $f(x) = \lim_n g_n(x)$. Note that $f(x) = \liminf_n f_n(x)$. If $s$ is a simple function such that $0 \le s \le f$, then, by lemma 8.3.3, $\int_X s d\mu \le \lim_n \int_X g_n d\mu$. Hence $\int_X f d\mu = \sup\{\int_X s d\mu : s \le f\} \le \lim_n \int_X g_n d\mu$. Since $g_n \le f_n$, $\int_X g_n d\mu \le \int_X f_n d\mu$, and $\lim_n \int_X g_n d\mu \le \liminf_n \int_X f_n d\mu$. ∎

**Example 5.** Let $(f_n)$ be a convergent sequence in $\mathfrak{L}^1(\mu)$, and let $f$ be its $\mathfrak{L}^1$-limit. Then $(f_n)$ contains a subsequence that converges to $f$ for almost every $x \in X$.

Choose a subsequence $(f_{n_i})$ of $(f_n)$ such that, for $i \in \mathbb{N}$, $\|f_{n_i} - f\|_1 < 2^{-i}$. Let $g_k = \sum_{i=1}^{k} |f_{n_i} - f|$. The functions $g_k$ are in $\mathfrak{L}^1$ and, by construction, $0 \le g_1 \le g_2 \le ...$, and $\|g_k\|_1 \le 1$. Let $g(x) = \lim_k g_k(x)$. By Fatou's theorem, $\int_X g d\mu \le \liminf_n \|g_k\|_1 \le 1$. This shows that $g \in \mathfrak{L}^{1}$.[5] Since $g(x) = \sum_{i=1}^{\infty} |f_{n_i}(x) - f(x)|$, it follows that the series $\sum_{i=1}^{\infty} |f_{n_i}(x) - f(x)|$ is convergent for a.e. $x \in X$ (by example 1). In particular, $\lim_{i \to \infty} |f_{n_i}(x) - f(x)| = 0$ for a.e. $x \in X$. ◆

**Theorem 8.3.9 (the monotone convergence theorem).** *If $f_n : X \to [0, \infty]$ is an increasing sequence of measurable functions such that $f(x) = \lim_n f_n(x)$ exists for every $x \in X$, then*

$$\int_X f d\mu = \lim_n \int_X f_n d\mu.$$

[5] One can also use the monotone convergence theorem to show that $g \in \mathfrak{L}^1$.

*Proof.* Since $f_n$ is increasing, $\liminf_n f_n = f$, and since $\int_X f_n d\mu$ is increasing, $\liminf_n \int_X f_n = \lim_n \int_X f_n d\mu$. By Fatou's theorem, $\int_X f d\mu \le \lim_n \int_X f_n d\mu$. Since $f_n \le f, \int_X f_n d\mu \le \int_X f d\mu$, and $\lim_n \int_X f_n d\mu \le \int_X f d\mu$, and the proof is complete. ∎

**Example 6.** Let $f \in \mathfrak{L}^1$, and suppose that $X = \cup_{n=1}^\infty E_n$, where $E_n$ is an ascending sequence of measurable sets. The $\lim_n \int_{E_n} |f| d\mu = \int_X |f| d\mu$.

For $n \in \mathbb{N}$ define $f_n = |f| \chi_{E_n}$. It is clear that $f_n$ is increasing and that $\lim_n f_n(x) = |f|(x)$. By the monotone convergence theorem, $\lim_n \int_{E_n} |f| d\mu = \lim_n \int_X f_n d\mu = \int_X |f| d\mu$. ♦

**Theorem 8.3.10 (the dominated convergence theorem).** *Let $f_n$ be a sequence of complex measurable functions, and let $g \in \mathfrak{L}^1(\mu)$ be such that $|f_n(x)| \le |g(x)|$. If $f(x) = \lim_n f_n(x)$ exists for every $x \in X$, then*

$$f \in \mathfrak{L}^1(\mu) \text{ and } \lim_n \int_X |f_n - f| d\mu = 0, \text{ that is }, f_n \to f \text{ in } \mathfrak{L}^1(\mu).$$

*Proof.* Notice that $|f_n(x)| \le |g(x)|$ implies that $|f(x)| \le |g(x)|$. Hence $f_n \in \mathfrak{L}^1(\mu)$, and $f \in \mathfrak{L}^1(\mu)$. Since $|f_n - f| \le 2g$, we can apply Fatou's theorem to the sequence $2g - |f_n - f|$ to obtain $\int_X 2g d\mu \le \liminf_n \int_X 2g - |f_n - f| d\mu = \int_X 2g d\mu - \limsup_n \int_X |f_n - f| d\mu$. Hence $\limsup_n \int_X |f_n - f| d\mu \le 0$, so $\limsup_n \int_X |f_n - f| d\mu = 0$. Since $\int_X |f_n - f| d\mu$ is a nonnegative sequence, $\lim_n \int_X |f_n - f| d\mu = 0$, as desired. ∎

**Example 7.** Let $f \in \mathfrak{L}^1(\mu)$. Then, for every $\epsilon > 0$, there exists $\delta > 0$ such that whenever $\mu(E) < \delta, \int_E |f| d\mu < \epsilon$.

Suppose there exists a number $\epsilon > 0$ such that, for every $n \in \mathbb{N}$, there is a measurable set $E_n$ such that $\mu(E_n) < 2^{-n}$, and $\int_{E_n} |f| d\mu \ge \epsilon$. Let $F_k = \cup_{n \ge k} E_n$, and let $F = \cap_{k=1}^\infty F_k$. On the one hand, $\mu(F_k) \le \sum_{n=k}^\infty 2^{-n} = 2^{-k+1}$; hence $\mu(F) = \lim_k \mu(F_k) = 0$, and $\int_F |f| d\mu = 0$. On the other hand, by the dominated convergence theorem, $\int_F |f| d\mu = \lim_k \int_{F_k} |f| d\mu \ge \liminf_k \int_{E_k} |f| d\mu \ge \epsilon$. This contradiction establishes the result. ♦

## Exercises

In the problems below, $(X, \mathfrak{M}, \mu)$ is a measure space.

1. Let $f$ be a measurable function, and let $g$ be a function such that $f(x) = g(x)$ for a.e. $x \in X$. Prove that $g$ is measurable.
2. Define a relation on the collection of measurable functions as follows: $f \equiv g$ if $f(x) = g(x)$ for a.e. $x \in X$. Prove that $\equiv$ is an equivalence relation.

3. Let $f$ be an integrable function, and let $g$ be such that $f(x) = g(x)$ a.e. Prove that $g$ is integrable and that $\int_X f d\mu = \int_X g d\mu$.

4. Let $f \in \mathfrak{L}^1(\mu)$, and let $E = \{x \in X : |f(x)| > c\}$, where $c > 0$. Prove the inequality (Tchebychev) $\mu(E) \le \frac{1}{c} \int_E |f| d\mu$. More generally, if $f$ is measurable and $|f|^p \in \mathfrak{L}^1(\mu)$, then $\mu(E) \le \frac{1}{c^p} \int_E |f|^p d\mu$. Here $1 \le p < \infty$.

5. Let $f \in \mathfrak{L}^1(\mu)$. Show that the set $E = \{x \in X : f(x) \ne 0\}$ is a countable union of sets of finite measure.

6. Let $f$ be a positive measurable function. Show that if $E$ and $F$ are measurable sets such that $\mu(E \Delta F) = 0$, then $\int_E f d\mu = \int_F f d\mu$.

7. Let $f_n$ be a sequence of measurable functions such that $\sum_{n=1}^\infty \int_X |f_n| d\mu < \infty$. Show that the series $\sum_{n=1}^\infty |f_n(x)|$ converges a.e. in $X$.

8. Show that if $\mu$ is a finite measure and $(f_n)$ is a sequence of bounded measurable functions such that $f_n$ converges uniformly to $f$, then $\lim_n \int_X |f_n - f| d\mu = 0$.

9. Let $f \in \mathfrak{L}^1(\mu)$. Prove that for every $\epsilon > 0$, there exists a set $E$ of finite measure such that $\int_E |f| d\mu > \|f\|_1 - \epsilon$.

10. Let $(f_n)$ be a decreasing sequence of nonnegative measurable functions, and let $f = \lim_n f_n$. Show that if $f_1$ is integrable, then $\int_X f d\mu = \lim_n \int_X f_n d\mu$.

## 8.4 Lebesgue Measure on $\mathbb{R}^n$

This section is the centerpiece of the chapter. The motivation for the definition of the Lebesgue measure, as well as an extensive development of its properties, appear later in the section. We must furnish some needed background. The four leading results in this section are valid for locally compact Hausdorff spaces, and this is made abundantly clear in the excursion on Radon measures. We chose to limit the bulk of the section to the Lebesgue measure because we do not wish to base this section too heavily on chapter 5.

### Preliminaries

**Lemma 8.4.1 (Urysohn's lemma).** *Let $E$ and $F$ be disjoint closed subsets of $\mathbb{R}^n$. Then there exists a continuous function $f : \mathbb{R}^n \to [0, 1]$ such that $f(E) = 1$, and $f(F) = 0$.*

*Proof.* *The functions $g(x) = dist(x, F)$ and $h(x) = dist(x, E)$ are continuous and are never simultaneously zero since $E$ and $F$ are closed and disjoint. Furthermore, $g(x) > 0$ for every $x \in E$, and $h(x) > 0$ for every $x \in F$.*

*The function $f(x) = \frac{g(x)}{g(x) + h(x)}$ has the stated properties.* ∎

**Lemma 8.4.2.** *Let $K$ be a compact subset of an open subset $V$ of $\mathbb{R}^n$. Then there exists an open set $U$ such that $\overline{U}$ is compact and $K \subseteq U \subseteq \overline{U} \subseteq V$.*

*Proof. For every $x \in K$, there exists a ball $B(x, \delta_x)$ such that $\overline{B(x, \delta_x)} \subseteq V$. Since $K$ is compact, and $K \subseteq \cup_{x \in K} B(x, \delta_x)$, there exists a finite number of points $x_1, \dots, x_m \in K$ such that $K \subseteq \cup_{i=1}^m B(x_i, \delta_{x_i})$. The set $U = \cup_{i=1}^m B(x_i, \delta_{x_i})$ satisfies the requirements.* ∎

**Definition.** Let $f : \mathbb{R}^n \to \mathbb{C}$ be a continuous function. The **support** of $f$, written $supp(f)$, is the closure of the set $\{x \in \mathbb{R}^n : f(x) \neq 0\}$. A continuous function $f : \mathbb{R}^n \to \mathbb{C}$ is said to be of **compact support** if $supp(f)$ is compact.

We use the notation $\mathcal{C}_c(\mathbb{R}^n)$ to denote the set of continuous, complex-valued functions of compact support on $\mathbb{R}^n$. Clearly, $\mathcal{C}_c(\mathbb{R}^n)$ is a vector space. We also use $\mathcal{C}_c^r(\mathbb{R}^n)$ to denote the set of continuous, real-valued functions of compact support on $\mathbb{R}^n$.

**Notation.** Let $K$ be a compact subset of $\mathbb{R}^n$, and let $V$ be an open subset of $\mathbb{R}^n$. For a function $f \in \mathcal{C}_c^r(\mathbb{R}^n)$, we write $f \prec V$ to mean that $0 \leq f \leq 1$ and $supp(f) \subseteq V$. We use the notation $K \prec f$ to mean that $0 \leq f \leq 1$ and $f(x) = 1$ for all $x \in K$. Many books refer to the following result as Urysohn's lemma.

**Lemma 8.4.3 (Urysohn's lemma).** *Let $K$ be a compact subset of an open subset $V$ of $\mathbb{R}^n$. Then there exists a function $f \in \mathcal{C}_c^r(\mathbb{R}^n)$ such that $K \prec f \prec V$.*

*Proof. By lemma 8.4.2, there exists an open set $U$ such that $\overline{U}$ is compact and $K \subseteq U \subseteq \overline{U} \subseteq V$. Applying lemma 8.4.1 with $E = K$ and $F = \mathbb{R}^n - U$, we find the function we seek.* ∎

**Lemma 8.4.4.** *Suppose $K \subseteq \mathbb{R}^n$ is compact, and let $\{V_1, \dots, V_m\}$ be an open cover of $K$. Then there exist continuous, compactly supported functions $h_1, \dots, h_m$ such that $h_i \prec V_i$ and $(h_1 + \dots + h_m)(x) = 1$ for all $x \in K$.*

*Proof. First we show that there exists an open cover $\{U_1, \dots, U_m\}$ of $K$ such that each $\overline{U_i}$ is compact and $\overline{U_i} \subseteq V_i$. The proof is by induction on $m$. When $m = 2$, let $K_1 = K - V_2$. Then $K_1$ is compact and contained in $V_1$. By lemma 8.4.2, there exists an open set $U_1$ with compact closure such that $K_1 \subseteq U_1 \subseteq \overline{U_1} \subseteq V_1$. Clearly, $\{U_1, V_2\}$ is an open cover of $K$. Now let $K_2 = K - U_1$, and repeat the above argument to find an open set $U_2$ with compact support such that $K_2 \subseteq U_2 \subseteq \overline{U_2} \subseteq V_2$. Clearly, $\{U_1, U_2\}$ is an open cover of $K$. This proves the base case when $m = 2$. We outline the inductive step. Let $\{V_1, \dots, V_m\}$ be an open cover of $K$, and write $W = V_2 \cup \dots \cup V_m$. Then $\{V_1, W\}$ is an open cover of $K$. By what* *we already established, there are open sets $U_1$ and $W_1$ with compact closures such that $K \subseteq U_1 \cup W_1$, and $\overline{U_1} \subseteq V_1$ and $\overline{W_1} \subseteq W = V_2 \cup \ldots \cup V_m$. Now apply the inductive hypothesis to the compact set $\overline{W_1}$ and its open cover $\{V_2, \ldots, V_m\}$.*

*By lemma 8.4.3, there exist functions $g_i \in \mathcal{C}_c^r(\mathbb{R}^n)$ such that $\overline{U_i} \prec g_i \prec V_i$. Define*

$$h_1 = g_1, h_2 = (1 - g_1)g_2, \ldots, h_m = (1 - g_1) \ldots (1 - g_{m-1})g_m.$$

*The fact that $h_i \prec V_i$ is obvious. Simple induction shows that, for $2 \leq i \leq m$, $h_1 + \ldots + h_i = 1 - (1 - g_1) \ldots (1 - g_i)$. Now define $h = h_1 + \ldots + h_m$. Thus $h = 1 - (1 - g_1) \ldots (1 - g_m)$. If $x \in K$, then $x \in U_i$ for some $i$, so $g_i(x) = 1$, and $h(x) = 1$.* ∎

The functions $h_1, \ldots, h_m$ in the above lemma are called a **partition of unity** on $K$ subordinate to the open cover $\{V_i\}_{i=1}^m$.

## Dicing $\mathbb{R}^n$

For a fixed natural number $k$, consider the following partition of $\mathbb{R}$:

$$[\frac{\nu}{2^k}, \frac{\nu+1}{2^k}), \nu \in \mathbb{Z}.$$

This partitions each interval $[m, m+1)(m \in \mathbb{Z})$ into $2^k$ congruent half-open intervals, each of length $\frac{1}{2^k}$.

The above partition of $\mathbb{R}$ can be employed to partition $\mathbb{R}^n$ into a collection of half-open cubes:

$$\mathcal{S}_k = \{\sigma = [\frac{\nu_1}{2^k}, \frac{\nu_1+1}{2^k}) \times \ldots \times [\frac{\nu_n}{2^k}, \frac{\nu_n+1}{2^k}) : (\nu_1, \ldots, \nu_n) \in \mathbb{Z}^n\}.$$

Note that, for $\sigma \in \mathcal{S}_k$, $diam(\sigma) = \sqrt{n}2^{-k}$ and that if $\sigma$ and $\sigma'$ are distinct cubes in $\mathcal{S}_k$, then $\sigma \cap \sigma' = \emptyset$.

Observe that the half-open unit cube $[0, 1) \times \ldots \times [0, 1)$ is the union of $2^{nk}$ cubes in $\mathcal{S}_k$, that $\mathcal{S}_{k+1}$ is a refinement of $\mathcal{S}_k$, and that each cube in $\mathcal{S}_k$ is the union of $2^n$ cubes in $\mathcal{S}_{k+1}$.

Now, given an open set $V \subseteq \mathbb{R}^n$, let

$$\mathcal{S}_k(V) = \{\sigma \in \mathcal{S}_k : \overline{\sigma} \subseteq V\}, \text{ and } G_k = \cup\{\sigma : \sigma \in \mathcal{S}_k(V)\}.$$

Note that $G_1 \subseteq G_2 \subseteq \ldots$. We claim that $V = \cup_{k=1}^{\infty} G_k$. Clearly, $V \supseteq \cup_{k=1}^{\infty} G_k$. Conversely, if $x \in V$, there exists $\delta > 0$ such that $B(x, 2\delta) \subseteq V$. Choose $k \in \mathbb{N}$ such that $\sqrt{n}2^{-k} < \delta$. (Reminder: $\sqrt{n}2^{-k}$ is the diameter of a cube in $\mathcal{S}_k$.) Since $\mathbb{R}^n = \cup\{\sigma : \sigma \in \mathcal{S}_k\}$, $x \in \sigma$ for some $\sigma \in \mathcal{S}_k$. Since $diam(\sigma) < \delta, \overline{\sigma} \subseteq B(x, 2\delta) \subseteq V$. This proves that $x \in G_k$ and that $V = \cup_{k=1}^{\infty} G_k$.

(a)                                              (b)

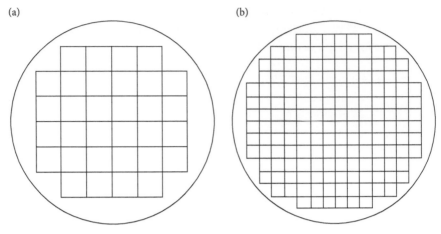

**Figure 8.2** (a) $S_3(U)$ (b)$S_4(U)$

This construction should be geometrically obvious. The set $S_k(V)$ is the largest set of cubes in $S_k$ that fits inside $V$. It is also clear that $S_{k+1}(V)$ is a refinement of $S_k(V)$ that also contains all the additional cubes in $S_{k+1}$ that fit in $V$. Figure 8.2 illustrates the construction: figure 8.2(a) depicts all the squares of length $1/8$ that fit in the unit disk $U$, and figure 8.2(b) shows all the squares of length $1/16$ that fit in the disk. The union of the squares are $G_3(U)$ and $G_4(U)$, respectively.

**Lemma 8.4.5.** *Let $V$ be an open subset of $\mathbb{R}^n$. Then $V$ is the countable union of disjoint cubes $\sigma$ of the type discussed in the previous paragraph. More specifically, $V = \cup_{i=1}^{\infty}\sigma_i$, where $\sigma_i \in \cup_{k=1}^{\infty}S_k$, and $\sigma_i \cap \sigma_j = \varnothing$ if $i \neq j$.*

*Proof.* *Let $B_1 = G_1$, and, for $k \geq 1$, let $B_{k+1} = G_{k+1} - G_k$. The family $\{B_k\}$ is mutually disjoint, and $\cup_{k=1}^{\infty}B_k = V$. Each $B_k$ is the union of cubes in $S_k$. The collection of all such cubes is countable, and their union (over $k \in \mathbb{N}$) is $V$. Renumbering those cubes as $\sigma_1, \sigma_2, ...$, we obtain $V = \cup_{k=1}^{\infty}\sigma_k$. Finally, consider two distinct cubes, $\sigma_i$ and $\sigma_j$. If $\sigma_i \subseteq B_r$, where $\sigma_j \subseteq B_s$ and $r \neq s$, then $\sigma_i \cap \sigma_j = \varnothing$ because $B_r \cap B_s = \varnothing$ if $r \neq s$. If $\sigma_i$ and $\sigma_j$ are subsets of $B_r$, for some integer $r$, then $\sigma_i \cap \sigma_j = \varnothing$ because the cubes in $S_r$ are disjoint.* ∎

As an illustration of the above construction, figure 8.3 depicts the set

$$B_4(U) = G_4(U) - G_3(U)$$

for the unit disk $U$.

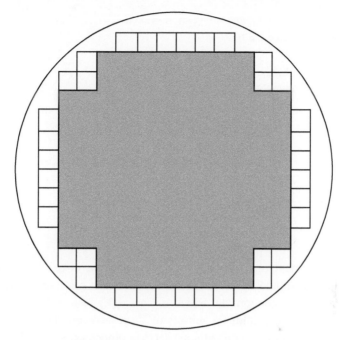

**Figure 8.3** $B_4(U)$ shown as the union of the unshaded squares

## Lebesgue measure: Motivation and Overview

This subsection is included for the sole purpose of building the reader's intuition. It is not meant to be a rigorous development of any particular set of ideas.

It must be emphasized at the outset that the Lebesgue measure is not an artificial construct but rather a very natural kind of measure, as the reader will see below. The broad goals are intuitively clear; we wish to find a large enough $\sigma$-algebra $\mathcal{L}^n$ in $\mathbb{R}^n$ and a positive measure $\lambda$ on $\mathcal{L}^n$ that extends and is consistent with our common geometric perceptions about length, area, and volume. It is therefore entirely reasonable to expect (indeed, require) that every closed box $Q$ must be in $\mathcal{L}^n$ and that the Lebesgue measure of such a box must be the product of its dimensions, consistent with our definition of the volume of a closed box in section 8.1. Surprisingly, those two simple requirements allow us to achieve most of our broad goals. Because every open subset of $\mathbb{R}^n$ is a countable union of closed boxes, every open subset of $\mathbb{R}^n$ must be in $\mathcal{L}^n$; hence $\mathcal{L}^n$ contains all Borel subsets of $\mathbb{R}^n$. The requirement that $\lambda(Q) = vol(Q)$ uniquely extends the Lebesgue measure to all open sets, as we explain below.

In theorem 8.4.5, if we define $K_i = \cup_{j=1}^{i} \overline{\sigma}_j$, then $\lambda(K_i) = \sum_{j=1}^{i} vol(\overline{\sigma}_j)$, and it follows directly from theorem 8.2.3 that

$$\lambda(V) = \lim_{i} \lambda(K_i) \tag{1}$$

This discussion strongly suggests equation (1) as a possible definition of the Lebesgue measure of an open set.[6] However, we will take a different path.

Another way to view equation (1) is as follows: Since, for any compact subset $K$ of $V$, $\lambda(K) \le \lambda(V)$ and since there is a sequence $K_i$ of compact subsets of $V$ such that $\lambda(V) = \lim_{i} \lambda(K_i)$, it must be true that, for an open set $V \subseteq \mathbb{R}^n$,

$$\lambda(V) = sup\{\lambda(K) : K \text{compact}, K \subseteq V\}. \tag{2}$$

We will use a variant of equation (2) as the definition of the Lebesgue outer measure of an open subset $V$ of $\mathbb{R}^n$. However, this raises a serious question: why would we abandon equation (1), which defines $\lambda(V)$ in terms of the measure of a sequence of simple compact subsets of $V$, in favor of equation (2), which involves the measure of general compact sets? In other words, how do we define the measure of an arbitrary compact subset $K$ of $\mathbb{R}^n$? The answer is, we do not! We use the Riemann integral as an instrument for the approximation of $\lambda(K)$ for a compact subset $K$ of $V$, and this is why Urysohn's lemma is crucially important for our development of the Lebesgue measure. Figures 8.4 and 8.5 illustrate the idea. In figure 8.4, the outer disk depicts the open set $V$, and the inner disk depicts a compact subset $K$ of $V$. If $f$ is a continuous function such that $K \prec f \prec V$, then the Riemann integral $\int_{\mathbb{R}^n} f(x)dx$ can be regarded as an approximation of both $\lambda(K)$ and $\lambda(V)$. Figure 8.5 further illustrates the point. In that figure, the measure of $K$ is the volume of the cylinder above $K$, which differs from $\int_{\mathbb{R}^n} f(x)dx$ by the volume of the thin shell between the cylinder and the wall of the graph of $f$. Since we can construct a compact subset $K$ that fills as much of $V$ as we wish (the compact sets $K_i$ in equation (1)), $\int_{\mathbb{R}^n} f(x)dx$ can be used to simultaneously approximate $\lambda(K)$ and $\lambda(V)$ with arbitrary precision. We hope that the preceding discussion motivates the definition below of the outer measure of an open subset of $\mathbb{R}^n$.

---

[6] Equation (1) can be written more explicitly as $\lambda(V) = \lim_{k} \frac{Card(\mathcal{S}_k(V))}{2^{nk}}$. This is a perfectly viable approach, and some recent books have adopted this as the definition of the measure of an open subset of $\mathbb{R}^n$. Observe that this definition accepts as a axiom the fact that the measure of the half-open cube is the product of its dimensions; hence the quantity $\frac{1}{2^{nk}}$. Another implied assumption is that all the cubes in $\mathcal{S}_k(V)$ have the same measure. This is the seed of the translation invariance of the Lebesgue measure. See the proof of theorem 8.4.14.

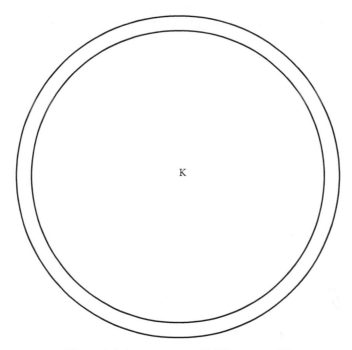

**Figure 8.4** A compact set $K$ filling *most of V*

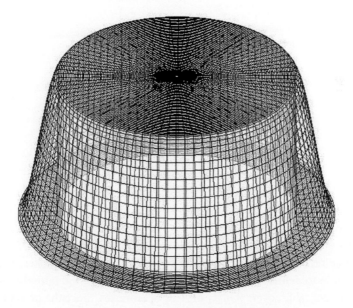

**Figure 8.5** A function $f$ such that $K \prec f \prec V$

In figure 8.4 we depict $V$ as a bounded open set. However, the discussion points are valid even for open sets of infinite measure. Specifically, this means that if $V$ is an open set of infinite measure, then $V$ contains compact subsets of arbitrarily large measure.

## Lebesgue Measure

As explained in the above motivation, the Riemann integral will play a pivotal role in our development of the Lebesgue measure. For a function $f \in \mathcal{C}_c(\mathbb{R}^n)$, let $Q$ be a closed box that contains $supp(f)$, and define

$$\int_{\mathbb{R}^n} f(x)dx = \int_Q f(x)dx.$$

The Riemann integral is clearly a positive linear functional on $\mathcal{C}_c(\mathbb{R}^n)$. For the remainder of this section, we will use the following notation: for a function $f \in \mathcal{C}_c(\mathbb{R}^n)$,

$$\text{write } I(f) = \int_{\mathbb{R}^n} f(x)dx.$$

**Definition. The Lebesgue outer measure** is the set function

$$m^* : \mathcal{P}(\mathbb{R}^n) \to [0, \infty],$$

which is as follows: for an open set $V \subseteq \mathbb{R}^n$,

$$m^*(V) = sup\{I(f) : f < V\},$$

and for an arbitrary set $A \subseteq \mathbb{R}^n$,

$$m^*(A) = inf\{m^*(V) : A \subseteq V, V \text{ open}\}.$$

The definition of $m^*(A)$ requires some justification. It is a well-known fact that an open subset of $\mathbb{R}$ is the disjoint union of a countable collection $\{(a_i, b_i)\}_{i=1}^{\infty}$ of open intervals. Therefore $m^*(V) = \sum_{i=1}^{\infty}(b_i - a_i)$. It follows, trivially, that $m^*(V) = inf\{\sum_{i=1}^{\infty}(b_i - a_i) : V \subseteq \cup_{i=1}^{\infty}(a_i, b_i)\}$. While an arbitrary subset $A$ of $\mathbb{R}^n$ is not the countable union of open boxes, it can be covered by such a set of boxes. Therefore it makes sense to define $m^*(A) = inf\{\sum_{i=1}^{\infty} vol(Q_i)\}$, where the infimum is taken over all the countable covers $\{Q_i\}_{i=1}^{\infty}$ of $A$ by open boxes. Since the union of open boxes is an open set, and since every open set is the countable union of open boxes, the definition of $m^*(A)$ is well justified.

**Proposition 8.4.6.** *The set function $m^*$ is an outer measure on $\mathbb{R}^n$.*

*Proof. The monotonicity of $m^*$ is obvious. First we show that, for open sets $V_1,\dots,V_m$, $m^*(\cup_{i=1}^m V_i) \leq \sum_{i=1}^m m^*(V_i)$. Let $f < \cup_{i=1}^m V_i$. By lemma 8.4.4, there exist functions $h_i < V_i, 1 \leq i \leq m$, such that that $h_1(x) + \dots + h_m(x) = 1$ for every $x \in \operatorname{supp}(f)$. Now $I(f) = I(\sum_{i=1}^m h_i f) = \sum_{i=1}^m I(h_i f) \leq \sum_{i=1}^m m^*(V_i)$. This shows that $m^*(\cup_{i=1}^n V_i) \leq \sum_{i=1}^n m^*(V_i)$.*

*We now show that $m^*$ is countably subadditive. Let $(E_i)$ be a sequence of subsets of $\mathbb{R}^n$. We must prove that $m^*(\cup_{i=1}^\infty E_i) \leq \sum_{i=1}^\infty m^*(E_i)$. If $\sum_{i=1}^\infty m^*(E_i) = \infty$, there is nothing to prove, so assume that $\sum_{i=1}^\infty m^*(E_i) < \infty$. Let $\epsilon > 0$, and choose open sets $V_i$ such that $E_i \subseteq V_i$ and $m^*(V_i) < m^*(E_i) + \epsilon/2^i$. Let $V = \cup_{i=1}^\infty V_i$, and let $f < V$, that is, $K = \operatorname{supp}(f) \subseteq \cup_{i=1}^\infty V_i$. The compactness of $K$ produces a finite subcover $V_1,\dots,V_m$ of $K$. Now $I(f) \leq m^*(V_1 \cup \dots \cup V_m) \leq \sum_{i=1}^m m^*(V_i) \leq \sum_{i=1}^\infty m^*(V_i) \leq \sum_{i=1}^\infty [m^*(E_i) + \epsilon/2^i] = \sum_{i=1}^\infty m^*(E_i) + \epsilon$. Since the last inequality is true for an arbitrary $f < V$, $m^*(V) \leq \sum_{i=1}^\infty m^*(E_i) + \epsilon$. Since $\cup_{i=1}^\infty E_i \subseteq V$ and $m^*$ is monotone, $m^*(\cup_{i=1}^\infty E_i) \leq m^*(V) \leq \sum_{i=1}^\infty m^*(E_i) + \epsilon$. Because $\epsilon$ is arbitrary, $m^*(\cup_{i=1}^\infty E_i) \leq \sum_{i=1}^\infty m^*(E_i)$.* ∎

**Example 1.** The outer measure of an open interval $V = (a,b)$ is $b - a$.
For any function $f$ such that $f < V$, it is clear that $\int_a^b f(x)dx \leq b - a$. Thus $m^*(V) \leq b - a$. Let $g$ be the continuous, piecewise linear function whose graph contains the points $(a,0),(a+\epsilon,0),(a+2\epsilon,1),(b-2\epsilon,1)$, and $(b-\epsilon,0), (b,0)$. The function $g$ is supported in $V$, and $\int_a^b g(x)dx = b - a - 3\epsilon$. Thus $m^*(V) = b - a$. ♦

**Example 2.** The outer measure of a point set $\{x\}$ in $\mathbb{R}$ is zero.
For any $\epsilon > 0$, the interval $V_\epsilon = (x - \epsilon, x + \epsilon)$ contains $\{x\}$, and $m^*(V_\epsilon) = 2\epsilon$. Since $\epsilon$ is arbitrary, $m^*(\{x\}) = 0$. ♦

**Example 3.** The following facts follow directly from example 2. The outer measure of a countable subset of $\mathbb{R}$ is zero. The outer measure of the closed interval $[a,b]$ is $b - a$. ♦

**Example 4.** The outer measure of the Cantor set is zero.
In the notation of section 4.2, the Cantor set, $C$, is contained in the set $C_n$ for each $n \in \mathbb{N}$. By the previous example and the subadditivity of $m^*$, $m^*(C_n) \leq 2^n/3^n$. Since $m^*(C) \leq m^*(C_n)$ and since $n$ is arbitrary, $m^*(C) = 0$. ♦

**Example 5.** The two-dimensional outer measure of the set $E = \{(x,0) : x \in \mathbb{R}\}$ (the x-axis) is zero.
In a manner quite similar to that in example 1, one can show that the two-dimensional outer measure of an open rectangle is the product of its

382 FUNDAMENTALS OF MATHEMATICAL ANALYSIS

dimensions. Let $R_n$ be the open rectangle $(-n,n) \times (-\epsilon/n^3, \epsilon/n^3)$. Now $m^*(R_n) = 4\epsilon/n^2$, and $E \subseteq \cup_{n=1}^{\infty} R_n$. By the subadditivity of $m^*$, $m^*(E) \le m^*(\cup_{n=1}^{\infty} R_n) \le \sum_{n=1}^{\infty} \frac{4\epsilon}{n^2}$. Since $\epsilon$ is arbitrary, $m^*(E) = 0$. ◆

**Definition.** A subset $E$ of $\mathbb{R}^n$ is **Lebesgue measurable** if it satisfies the Carathéodory condition, for every $A \subseteq \mathbb{R}^n$,

$$m^*(A) = m^*(A \cap E) + m^*(A \cap E').$$

By theorem 8.2.7, the set $\mathcal{L}(\mathbb{R}^n)$ of **Lebesgue measurable sets** is a $\sigma$-algebra, and the restriction of $m^*$ to $\mathcal{L}(\mathbb{R}^n)$ is a complete positive measure: the **Lebesgue measure** on $\mathcal{L}(\mathbb{R}^n)$. We will reserve the notation $\lambda(E)$ exclusively to denote the Lebesgue measure of a set $E \in \mathcal{L}(\mathbb{R}^n)$. We continue to write $m^*(E)$ for the outer measure of a set $E$ whose Lebesgue measurability has not been established. We will frequently write $\mathcal{L}^n$ as an abbreviation of $\mathcal{L}(\mathbb{R}^n)$.

The immediate task now is to show that every open subset of $\mathbb{R}^n$ is Lebesgue measurable (theorem 8.4.9). We first need to establish the finite additivity of $m^*$ for compact and open sets.

**Theorem 8.4.7.**

(a) *If $K$ is compact, then*

$$m^*(K) = \inf\{I(f) : K \prec f\}.$$

*In particular, compact subsets of $\mathbb{R}^n$ have finite Lebesgue outer measures.*

(b) *If $K_1$ and $K_2$ are disjoint compact subsets of $\mathbb{R}^n$, then*

$$m^*(K_1 \cup K_2) = m^*(K_1) + m^*(K_2).$$

*Proof.* Let $K \prec f$. If $0 < \alpha < 1$, then the set $V_\alpha = \{x \in \mathbb{R}^n : f(x) > \alpha\}$ is open and contains $K$. Now if $g \prec V_\alpha$, then $\alpha g < f$, and $m^*(K) \le m^*(V_\alpha) = \sup\{I(g) : g \prec V_\alpha\} \le \frac{1}{\alpha} I(f)$. Letting $\alpha \to 1$, we obtain $m^*(K) \le I(f)$. Let $\epsilon > 0$. There exists an open set $V$ containing $K$ such that $m^*(V) < m^*(K) + \epsilon$. Choose a function $f \in C_c^r(\mathbb{R}^n)$ such that $K \prec f \prec V$. Then $I(f) \le m^*(V) < m^*(K) + \epsilon$. This establishes part (a).

To prove part (b), let $\epsilon > 0$. By part (a), there exists a function $g \in C_c^r(\mathbb{R}^n)$ such that $K_1 \cup K_2 \prec g$, and $I(g) < m^*(K_1 \cup K_2) + \epsilon$. By lemma 8.4.2, there exists an open subset $W$ with compact closure such that $K_1 \subseteq W \subseteq \overline{W} \subseteq \mathbb{R}^n - K_2$. By theorem 8.4.3, there exists a function $f \in C_c^r(\mathbb{R}^n)$ such that $K_1 \prec f \prec W$. In particular, $f(K_1) = 1$, and $f(K_2) = 0$. Note that $K_1 \prec fg$ and that $K_2 \prec (1 - f)g$.

Now $m^*(K_1) + m^*(K_2) \leq I(fg) + I(g - fg) = I(g) < m^*(K_1 \cup K_2) + \epsilon$. Since $\epsilon$ is arbitrary, $m^*(K_1) + m^*(K_2) \leq m^*(K_1 \cup K_2)$. Now the subadditivity of $m^*$ delivers the result. ∎

**Theorem 8.4.8.** *For an open set $V$,*

(a) $m^*(V) = \sup\{m^*(K) : K compact, K \subseteq V\}$,
(b) $m^*(V) = \sup\{m^*(U) : U open, \overline{U} compact, \overline{U} \subseteq V\}$, and
(c) *if $V_1, V_2$ are disjoint open sets, then $m^*(V_1 \cup V_2) = m^*(V_1) + m^*(V_2)$.*

*Proof.* Let $\alpha < m^*(V)$. By the definition of $m^*(V)$, there exists a function $f \in C_c^r(\mathbb{R}^n)$ such that $f \prec V$ and $I(f) > \alpha$. Let $K = \text{supp}(f)$. If $K \subseteq W$ for some open set $W$, then $f \prec W$, so $I(f) \leq m^*(W)$. This shows that

$$m^*(K) = \inf\{m^*(W) : W open, K \subseteq W\} \geq I(f) > \alpha.$$

Thus $\alpha < m^*(K) \leq m^*(V)$. This proves part (a). Observe that this proof is valid even when $m^*(V) = \infty$.

Part (b) follows from part (a) and lemma 8.4.2.

To prove (c), we may, without loss of generality, assume that $V_1$ and $V_2$ have finite outer measure. Let $\epsilon > 0$. By part (a), there exist compact sets $K_1$ and $K_2$ such that $m^*(V_i) < m^*(K_i) + \epsilon/2, i = 1, 2$. The set $K = K_1 \cup K_2$ is compact, and $K_1 \cup K_2 \subseteq V_1 \cup V_2$. Now $m^*(V_1) + m^*(V_2) \leq m^*(K_1) + m^*(K_2) + \epsilon = m^*(K_1 \cup K_2) + \epsilon \leq m^*(V_1 \cup V_2) + \epsilon$. Since $\epsilon$ is arbitrary, and $m^*$ is subadditive, $m^*(V_1) + m^*(V_2) = m^*(V_1 \cup V_2)$. ∎

**Theorem 8.4.9.** *Every open subset of $\mathbb{R}^n$ is Lebesgue measurable. Consequently, every Borel subset of $\mathbb{R}^n$ is in $\mathcal{L}^n$.*

*Proof.* Let $U$ be an open subset of $\mathbb{R}^n$, and let $A$ be an arbitrary subset of $\mathbb{R}^n$. Since $m^*(A) \leq m^*(A \cap U) + m^*(A \cap U')$, we may assume that $m^*(A) < \infty$. Without loss of generality, assume that $A \cap U \neq \emptyset \neq A \cap U'$. Let $\epsilon > 0$. There exists an open set $V$ containing $A$ such that $m^*(V) < m^*(A) + \epsilon/2$. By theorem 8.4.8, there exists an open set $W$ such that $\overline{W}$ is compact, $\overline{W} \subseteq V \cap U$, and $m^*(W) + \epsilon/2 > m^*(V \cap U)$.

Let $W_0 = V \cap (\overline{W})'$. Notice that $W_0 \cap W = \emptyset$, that $W_0 \cup W \subseteq V$, and that $W_0$ has finite outer measure. Now $V \cap U' \subseteq W_0$, so $m^*(W_0) \geq m^*(V \cap U')$. Using this information and theorem 8.4.8(c),

$$m^*(A) + \epsilon > m^*(V) + \epsilon/2 \geq m^*(W \cup W_0) + \epsilon/2 = m^*(W) + m^*(W_0) + \epsilon/2$$
$$\geq m^*(V \cap U) + m^*(V \cap U') \geq m^*(A \cap U) + m^*(A \cap U').$$

*Since $\epsilon$ is arbitrary, $m^*(A) \geq m^*(A \cap U) + m^*(A \cap U')$.* ■

**Theorem 8.4.10.** *Let $E \in \mathcal{L}^n$.*

(a) *For $\epsilon > 0$, there exists a closed subset $F$ and an open subset $V$ such that $F \subseteq E \subseteq V$ and $\lambda(V - F) < \epsilon$.*
(b) *$\lambda(E) = \sup\{\lambda(K) : K \text{compact}, K \subseteq E\}$.*
(c) *There exists an $F_\sigma$ set $A$ and a $G_\delta$ set $B$ such that $A \subseteq E \subseteq B$ and $\lambda(B - A) = 0$.*

*Proof.* $\mathbb{R}^n$ *is the countable union of the nest of compact balls $K_i = \overline{B(0,i)}, i \in \mathbb{N}$. For each $i \in \mathbb{N}, \lambda(K_i \cap E) < \infty$. Thus there exists open sets $V_i \supseteq K_i \cap E$ such that $\lambda(V_i - (K_i \cap E_i)) < \epsilon/2^{i+1}$. Let $V = \cup_{i=1}^\infty V_i$. Then $E \subseteq V$, $V - E \subseteq \cup_{i=1}^\infty (V_i - (K_i \cap E))$, and $\lambda(V - E) < \epsilon/2$. Applying this result to $E'$, we can find an open set $W$ containing $E'$ such that $\lambda(W - E') < \epsilon/2$. Let $F = W'$. Then $F \subseteq E, E - F = W - E'$ and $\lambda(E - F) = \lambda(W - E') < \epsilon/2$. Thus $\lambda(V - F) = \lambda(V - E) + \lambda(E - F) < \epsilon/2 + \epsilon/2 = \epsilon$. This proves part (a).*

*If $F$ is closed, $F = \cup_{i=1}^\infty (K_i \cap F)$. Each $K_i \cap F$ is compact and $\lim_i \lambda(K_i \cap F) = \lambda(F)$. Thus (b) holds for closed subsets of $\mathbb{R}^n$. If $E \in \mathcal{L}^n$ and $\epsilon > 0$, by part (a) we can choose a closed set $F \subseteq E$ such that $\lambda(E - F) < \epsilon/2$. If $\lambda(F) = \infty$, then $\sup\{\lambda(K) : K \text{compact}, K \subseteq E\} \geq \sup\{\lambda(K) : K \text{compact}, K \subseteq F\} = \infty$. If $\lambda(F) < \infty$, there exists a compact subset $K$ of $F$ such that $\lambda(F - K) < \epsilon/2$. Now $\lambda(E) = \lambda(K) + \lambda(E - F) + \lambda(F - K) < \lambda(K) + \epsilon$.*

*To prove part (c), find open sets $V_i$ and closed sets $F_i$ such that $F_i \subseteq E \subseteq V_i$ and $\lambda(V_i - F_i) < 1/i$. Set $A = \cup_{i=1}^\infty F_i$, and $B = \cap_{i=1}^\infty V_i$. Then $\lambda(B - A) < 1/i$ for every $i \in \mathbb{N}$; hence $\lambda(B - A) = 0$. Observe that these results are valid even when $\lambda(E) = \infty$.* ■

Observation. Let $\mathcal{B}^n$ be the $\sigma$-algebra of Borel subsets of $\mathbb{R}^n$, and consider the measure space $(\mathbb{R}^n, \mathcal{B}^n, \lambda)$. It is known that the restriction of $\lambda$ to $\mathcal{B}^n$ is not a complete measure (see problem 4 at the end of this section). Theorem 8.4.10(c) implies that $(\mathbb{R}^n, \mathcal{L}^n, \lambda)$ is the completion of $(\mathbb{R}^n, \mathcal{B}^n, \lambda)$. See problems 3 and 4 on section 8.2. The following corollary is an affirmation of the same fact.

**Corollary 8.4.11.** $\mathcal{L}^n$ *is the smallest $\sigma$-algebra that contains $\mathcal{B}^n$ and all sets of Lebesgue (outer) measure 0.*

*Proof.* We have already seen that $\mathcal{B}^n \subseteq \mathcal{L}^n$ and that all subsets of Lebesgue outer measure 0 are Lebesgue measurable. We show that if $\mathfrak{M}$ is a $\sigma$-algebra containing $\mathcal{B}^n$ and all subsets of Lebesgue measure 0, then $\mathcal{L}^n \subseteq \mathfrak{M}$. If $E \in \mathcal{L}^n$, then, by theorem 8.4.10, there exists an $F_\sigma$ set $A$ such that $A \subseteq E$ and $\lambda(E - A) = 0$. Thus $E = A \cup (E - A)$, where $A \in \mathcal{B}^n$ and $\lambda(E - A) = 0$. ∎

Recall from section 8.1 that the volume of a closed box

$$Q = \{x \in \mathbb{R}^n : a_i \leq x_i \leq b_i\}$$

is, by definition, $vol(Q) = \prod_{i=1}^n (b_i - a_i)$.

**Lemma 8.4.12.** *For an open box $Q = (a_1, b_1) \times \ldots \times (a_n, b_n)$,*

$$\lambda(Q) = vol(\overline{Q}) = \prod_{i=1}^n (b_i - a_i).$$

*Proof.* Let $f \prec Q$. Then $I(f) = \int_{\overline{Q}} f(x)dx \leq \int_{\overline{Q}} 1 dx = vol(\overline{Q})$. Thus $\lambda(Q) \leq vol(\overline{Q})$. For small-enough positive constants $\epsilon$, define $Q_\epsilon = [a_1 + \epsilon, b_1 - \epsilon] \times \ldots \times [a_n + \epsilon, b_n - \epsilon]$. There exists a function $f \in C_c^r(\mathbb{R}^n)$ such that $Q_\epsilon \prec f \prec Q$. Therefore $\lambda(Q) \geq I(f) = \int_{\overline{Q}} f(x)dx \geq \int_{Q_\epsilon} f(x)dx = \int_{Q_\epsilon} 1 dx = vol(Q_\epsilon) = \prod_{i=1}^n (b_i - a_i - 2\epsilon)$. Since $\epsilon$ is arbitrary, $\lambda(Q) \geq \prod_{i=1}^n (b_i - a_i) = vol(\overline{Q})$. ∎

**Example 6.** Consider an open box $Q = (a_1, b_1) \times \ldots \times (a_n, b_n)$, and let $\overline{Q}$ be the closed box $[a_1, b_1] \times \ldots \times [a_n, b_n]$. For every $k \in \mathbb{N}$, let $Q_k$ be the open box $(a_1 - \frac{1}{k}, b_1 + \frac{1}{k}) \times \ldots \times (a_n - \frac{1}{k}, b_n + \frac{1}{k})$. Since $\{Q_k\}$ is a descending sequence and $\overline{Q} = \cap_{k=1}^\infty Q_k$, $\lambda(\overline{Q}) = \lim_k \lambda(Q_k) = \lim_k \prod_{i=1}^n (b_i - a_i + \frac{2}{k}) = \prod_{i=1}^n (b_i - a_i) = vol(\overline{Q}) = \lambda(Q)$. Therefore $\lambda(Q) = \lambda(\overline{Q})$, and the boundary of any box has Lebesgue measure zero. Therefore the Lebesgue measure of any box (open, closed, or half-open) is the product of its dimensions. We will continue to refer to the Lebesgue measure of a box as its volume. ♦

**Example 7.** The Lebesgue measure of an open ball of radius $r$ in $\mathbb{R}^n$ is $c_n r^n$, where $c_n$ is the measure of the unit open ball in $\mathbb{R}^n$.

Let $B_r$ be the open ball of radius $r$ centered at the origin, and let $B$ be the open unit ball. By lemma 8.4.5, $B = \cup_{i=1}^\infty \sigma_i$, where $\{\sigma_i\}$ is a sequence of disjoint half-open cubes. Because $B_r = rB = \cup_{i=1}^{\infty} r\sigma_i$, $\lambda(B_r) = \sum_{i=1}^{\infty} \lambda(r\sigma_i) = \sum_{i=1}^{\infty} r^n \lambda(\sigma_i) = r^n \lambda(B) = c_n r^n$. ♦

We are now ready to prove that the Riemann integral of a function of compact support is the same as its integral with respect to Lebesgue measure.

**Theorem 8.4.13.** *For a function $f \in C_c(\mathbb{R}^n)$, $\int_{\mathbb{R}^n} f d\lambda = \int_{\mathbb{R}^n} f(x) dx$.*

*Proof. It is sufficient to prove the result for a positive function $f$. Let $Q$ be a half-open cube containing $\text{supp}(f)$ in its interior. Consider the partition of $Q$ into $2^{nk}$ congruent, half-open sub-cubes $\{\sigma_1, \ldots, \sigma_{2^{nk}}\}$, and let $s_k(f) = \sum_{i=1}^{2^{nk}} f_{\sigma_i} \, \text{vol}(\sigma_i)$ be the lower Riemann sums of $f$. Here $f_{\sigma_i} = \min\{f(x) : x \in \overline{\sigma_i}\}$. By theorem 8.1.2, $\lim_{k \to \infty} s_k(f) = \int_{\mathbb{R}^n} f(x) dx$. On the other hand, the simple functions $f_k(x) = \sum_{i=1}^{2^{nk}} f_{\sigma_i} \chi_{\sigma_i}(x)$ satisfy $0 \le f_1 \le f_2 \le \ldots$, and $\lim_k f_k(x) = f(x)$ for every $x \in \mathbb{R}^n$. By the monotone convergence theorem, $\lim_k \int_{\mathbb{R}^n} f_k d\lambda = \int_{\mathbb{R}^n} f d\lambda$. But $\int_{\mathbb{R}^n} f_k d\lambda = \sum_{i=1}^{2^{nk}} f_{\sigma_i} \lambda(\sigma_i) = \sum_{i=1}^{2^{nk}} f_{\sigma_i} \, \text{vol}(\sigma_i) = s_k(f)$. Therefore, $\lim_k s_k(f) = \int_{\mathbb{R}^n} f d\lambda$.* ∎

The previous theorem is commonly cast in the following language: the Lebesgue integral extends the Riemann integral from $C_c(\mathbb{R}^n)$ to $\mathfrak{L}^1(\mathbb{R}^n, \mathcal{L}^n, \lambda)$. We also say that the Lebesgue measure $\lambda$ represents the positive linear functional $I(f) = \int_{\mathbb{R}^n} f(x) dx$.

**Theorem 8.4.14.** *The Lebesgue measure is translation invariant. Thus if $E \in \mathcal{L}^n$, and $x \in \mathbb{R}^n$, then $\lambda(E + x) = \lambda(E)$.*

*Proof. It is easy to see that $\text{vol}(Q + x) = \text{vol}(Q)$ for every box $Q$. Now let $V$ be an open subset of $\mathbb{R}^n$. By lemma 8.4.5, we can write $V$ as a disjoint union of half-open cubes, $V = \cup_{i=1}^{\infty} \sigma_i$. Thus $V + x = \cup_{i=1}^{\infty} (\sigma_i + x)$, and $\lambda(V + x) = \lambda(\cup_{i=1}^{\infty} (\sigma_i + x)) = \sum_{i=1}^{\infty} \lambda(\sigma_i + x) = \sum_{i=1}^{\infty} \lambda(\sigma_i) = \lambda(V)$. Thus the result holds for open subsets of $\mathbb{R}^n$. The general result for an arbitrary measurable set $E$ follows from the special case we just established and the fact that $\lambda(E) = \inf\{\lambda(V) : E \subseteq V, V \text{open}\}$. See the definition of the Lebesgue outer measure.* ∎

We now summarize the properties of Lebesgue measure.

**Theorem 8.4.15.** *Lebesgue measure is a complete, translation-invariant measure on $\mathcal{L}^n$, and*

*(a) every Borel subset of $\mathbb{R}^n$ is Lebesgue measurable;*
*(b) for every open set $V$, $\lambda(V) = \sup\{\int_{\mathbb{R}^n} f(x) dx : f \prec V\}$;*
*(c) for every $E \in \mathcal{L}^n$, $\lambda(E) = \inf\{\lambda(V) : E \subseteq V, V \text{open}\}$;*

(d) for every compact set K, $\lambda(K) = \inf\{\int_{\mathbb{R}^n} f(x)dx : K \prec f\}$;

(e) for every $E \in \mathcal{L}^n$, $\lambda(E) = \sup\{\lambda(K) : K \subseteq E, K\,compact\}$, and

(f) $\lambda$ extends (represents) the Riemann integral in the sense that

$$\int_{\mathbb{R}^n} f(x)dx = \int_{\mathbb{R}^n} fd\lambda$$

for every $f \in \mathcal{C}_c(\mathbb{R}^n)$.

Property (a) is by theorem 8.4.9.

Properties (b) and (c) are the definitions of Lebesgue outer measure.

Property (d) is theorem 8.4.7(a).

Property (e) is theorem 8.4.10(b).

Property (f) is theorem 8.4.13.

Finally, the completeness of $\lambda$ is by theorem 8.2.7, and the translation invariance of $\lambda$ is by theorem 8.4.14. ∎

**Definitions.** Let $X$ be a locally compact metric (topological) space, and let $\mathfrak{M}$ be a $\sigma$-algebra of subsets of $X$ that contains all Borel subsets of $X$.

A positive measure $\mu$ on $\mathfrak{M}$ is said to be

(a) **outer regular** if, for every $E \in \mathfrak{M}, \mu(E) = \inf\{\mu(V) : E \subseteq V, V open\}$, and

(b) **inner regular** if, for every$E \in \mathfrak{M}, \mu(E) = \sup\{\mu(K) : K \subseteq E, K compact\}$.

We say that $\mu$ is **regular** if it is both inner and outer regular.

Lebesgue measure is outer regular by the very definition of the Lebesgue outer measure, $m^*$. Theorem 8.4.10(b) states that Lebesgue measure is inner regular.

We conclude this section with two uniqueness results that characterize Lebesgue measure.

**Theorem 8.4.16.** Let $\mu$ be a regular measure on $\mathcal{L}^n$ such that $\mu(K) < \infty$ for every compact subset $K$ of $\mathbb{R}^n$, and $\int_{\mathbb{R}^n} fd\mu = \int_{\mathbb{R}^n} f(x)dx$ for every $f \in \mathcal{C}_c(\mathbb{R}^n)$. Then $\mu = \lambda$.

*Proof.* It is sufficient to prove that $\mu(K) = \lambda(K)$ for every compact set K. The result then follows from the regularity of $\lambda$ and $\mu$.

Let $\epsilon > 0$. There exists an open subset V such that $K \subseteq V$ and $\mu(V) < \mu(K) + \epsilon$. Let $f \in \mathcal{C}_c^r(\mathbb{R}^n)$ be such that $K \prec f \prec V$. Then $\lambda(K) = \int_{\mathbb{R}^n} \chi_K d\lambda \leq \int_{\mathbb{R}^n} fd\lambda = \int_{\mathbb{R}^n} f(x)dx = \int_{\mathbb{R}^n} fd\mu \leq \int_{\mathbb{R}^n} \chi_V d\mu = \mu(V) < \mu(K) + \epsilon$. Since $\epsilon$ is arbitrary, $\lambda(K) \leq \mu(K)$. Switching the roles of $\lambda$ and $\mu$, we obtain $\mu(K) \leq \lambda(K)$. ∎

**Theorem 8.4.17.** *Let $\mu$ be a translation invariant measure on $\mathcal{L}^n$ such that $c = \mu([0,1)^n) > 0$. Then $\mu = c\lambda$, that is, $\mu(E) = c\lambda(E)$ for every $E \in \mathcal{L}^n$.*

*Proof. Let $Q$ be the half-open unit cube $[0,1)^n$. For a fixed $k \in \mathbb{N}$, partition $Q$ into $2^{nk}$ congruent half-open sub-cubes $\sigma_1, \ldots, \sigma_{2^{nk}}$ in $\mathcal{S}_k$. Each cube in $\mathcal{S}_k$ is a translation of any other cube in $\mathcal{S}_k$, therefore, by assumption, $\mu(\sigma_i) = \mu(\sigma_1)$ for $i = 1, 2, \ldots, 2^{nk}$. Since $\sigma_1, \ldots, \sigma_{2^{nk}}$ are disjoint, $c = \mu(Q) = 2^{nk}\mu(\sigma_1)$. Also $c = c.1 = c\lambda(Q) = c\sum_{i=1}^{2^{nk}} \lambda(\sigma_i) = c2^{nk}\lambda(\sigma_1)$. It follows that $\mu(\sigma_1) = c\lambda(\sigma_1)$; hence $\mu(\sigma) = c\lambda(\sigma)$ for every cube $\sigma$ in $\mathcal{S}_k$. Since $k$ is arbitrary, $\mu(\sigma) = c\lambda(\sigma)$ for any cube $\sigma \in \cup_{k=1}^{\infty}\mathcal{S}_k$. By lemma 8.4.5, an arbitrary open set $V$ is a countable union of disjoint cubes in $\cup_{k=1}^{\infty}\mathcal{S}_k$. The countable additivity of $\lambda$ and $\mu$ produces $\mu(V) = c\lambda(V)$ for every open subset $V$.*

*Now let $E \in \mathcal{L}^n$. For an open set $V \supseteq E$, $\mu(E) \leq \mu(V) = c\lambda(V)$. Hence $\mu(E) \leq c \inf\{\lambda(V) : V \supseteq E, V \text{ open}\} = c\lambda(E)$. To show that $\mu(E) \geq c\lambda(E)$, we first assume that $E$ is bounded. Choose a large enough open box $\Omega$ that contains $E$. Then $\mu(\Omega) - \mu(E) = \mu(\Omega - E) \leq c\lambda(\Omega - E) = c\lambda(\Omega) - c\lambda(E) = \mu(\Omega) - c\lambda(E)$. Thus $\mu(E) \geq c\lambda(E)$.*

*If $E$ is unbounded, let $B_i$ be the open ball of radius $i$ and centered at the origin. Then $E = \cup_{i=1}^{\infty}(E \cap B_i)$, and $\mu(E) = \lim_i \mu(E \cap B_i) = c \lim_i \lambda(E \cap B_i) = c\lambda(E)$.* ∎

## Excursion: Radon Measures

A close examination of the constructions and the results of this section so far reveals that most of the theory we developed can be extended to locally compact Hausdorff spaces. Specifically, if $\mathbb{R}^n$ is replaced with a locally compact Hausdorff space $X$ and the Riemann integral is replaced with a positive linear functional $I$ on $\mathcal{C}_c(X)$, then we can construct a measure $\mu$ that represents $I$ and has most (but not all) of the regularity properties we derived for Lebesgue measure.[7]

The following results can be established by replicating the proofs of the corresponding results for the Lebesgue integral. Theorem 5.9.2 must be used instead of theorem 8.4.2, and lemma 5.11.6 instead of theorem 8.4.3. The proof we included for lemma 8.4.4 is valid for any locally compact Hausdorff; hence we state the lemma below for the sake of completeness. We urge the reader to scrutinize our claim that the proofs of the theorems below for Radon measures are identical to those provided for the Lebesgue measure. The exercise is illuminating.

Throughout this subsection, $X$ is a locally compact Hausdorff space, and $I$ is a positive linear functional on $\mathcal{C}_c(X)$. Explicitly, for $f, g \in \mathcal{C}_c(X)$, and $\alpha, \beta \in \mathbb{K}$, $I(\alpha f + \beta g) = \alpha I(f) + \beta I(g)$, and if $f \geq 0$, then $I(f) \geq 0$. Observe that such a functional is monotone in the sense that if $f \leq g$, then $I(f) \leq I(g)$.

We continue to use the notation $K \prec f \prec V$ to indicate that $f \in \mathcal{C}_c^r(X)$, $0 \leq f \leq 1$, $f(K) = 1$, and $supp(f) \subseteq V$.

---

[7] In the sense that $\int_X f d\mu = I(f)$ for all $f \in \mathcal{C}_c(X)$.

**Lemma 8.4.18.** *Suppose $K \subseteq X$ is compact and that $V_1, \ldots, V_m$ are open subsets of $X$ such that $K \subseteq \bigcup_{i=1}^m V_i$. Then there exist functions $h_1, \ldots, h_m$ such that $h_i \prec V_i$ and $(h_1 + \ldots + h_m)(x) = 1$ for all $x \in K$.* ∎

The **Radon outer measure induced by the positive linear functional** $I$ is the set function
$$m^* : \mathcal{P}(X) \to [0, \infty],$$
defined as follows: for an open set $V \subseteq X$,
$$m^*(V) = \sup\{I(f) : f \prec V\},$$
and for an arbitrary set $A \subseteq X$,
$$m^*(A) = \inf\{m^*(V) : A \subseteq V,\ V \text{ open}\}.$$

**Proposition 8.4.19.** *The set function $m^*$ is an outer measure on $X$.* ∎

**Definition.** A subset $E$ of $X$ is **Radon measurable**, or simply measurable, if it satisfies the Carathéodory condition: for every $A \subseteq X$,
$$m^*(A) = m^*(A \cap E) + m^*(A \cap E').$$

By theorem 8.2.7, the set $\mathfrak{M}$ of **measurable sets** is a $\sigma$-algebra, and the restriction of $m^*$ to $\mathfrak{M}$ is a complete positive measure: the **Radon measure on $\mathfrak{M}$ induced by** $I$. We will reserve the notation $\mu(E)$ exclusively to denote the $\mu$-measure of a set $E \in \mathfrak{M}$. We continue to write $m^*(E)$ for the outer measure of a set $E$ whose Radon measurability has not been established.

**Theorem 8.4.20.**

(a) *If $K$ is compact, then*
$$m^*(K) = \inf\{I(f) : K \prec f\}.$$

*In particular, compact subsets of $X$ have finite outer measure.*

(b) *If $K_1$ and $K_2$ are disjoint compact subsets of $X$, then $m^*(K_1 \cup K_2) = m^*(K_1) + m^*(K_2)$.* ∎

**Theorem 8.4.21.** *For an open set $V$,*

(a) $m^*(V) = \sup\{m^*(K) : K \text{ compact}, K \subseteq V\}$,
(b) $m^*(V) = \sup\{m^*(U) : U \text{ open}, \overline{U} \text{ compact}, \overline{U} \subseteq V\}$, *and*
(c) *if $V_1, V_2$ are disjoint open sets, then $m^*(V_1 \cup V_2) = m^*(V_1) + m^*(V_2)$.* ∎

**Theorem 8.4.22.** *Every open subset of X is Radon measurable. Consequently, every Borel subset of X is in $\mathfrak{M}$.* ∎

We now arrive at the main distinction between the Lebesgue measure and general Radon measures. Part (a) of theorem 8.4.21 states that $\mu$ is inner regular on open subsets of $X$. Inner regularity does not extend to all Radon measurable sets, however. But we do have the following result.

**Theorem 8.4.23.** *If $E \in \mathfrak{M}$ and $\mu(E) < \infty$, then $\mu(E) = sup\{\mu(K) : K \subseteq E, K compact\}$. Thus $\mu$ is inner regular on sets of finite measure.*

*Proof.* Let $\epsilon > 0$, and choose an open set $U \supseteq E$ such that $\mu(U) < \mu(E) + \epsilon$. Since $\mu(U - E) = \mu(U) - \mu(E) < \epsilon$, there exists an open set $V \supseteq U - E$ such that $\mu(V) < \epsilon$. By theorem 8.4.21, $U$ contains a compact subset $H$ such that $\mu(U) < \mu(H) + \epsilon$. The set $K = H - V$ is clearly compact, and $K \subseteq U - V \subseteq E$. Now

$$\mu(E) < \mu(U) = \mu(U) - \mu(H) + \mu(H)$$
$$= \mu(U - H) + \mu(H - V) + \mu(H \cap V) < \epsilon + \mu(K) + \mu(V) < \mu(K) + 2\epsilon. \blacksquare$$

We now prove the generalization of theorem 8.4.13.

**Theorem 8.4.24.** *For a function $f \in \mathcal{C}_c(X)$, $\int_X f d\mu = I(f)$.*

*Proof.* It is enough to prove that

$$I(f) \le \int_X f d\mu \text{ for every } f \in \mathcal{C}_c^r(X) \tag{3}$$

because we then have $-I(f) = I(-f) \le \int_X -f d\mu = -\int_X f d\mu$.

It is further sufficient to establish (3) when $0 \le f \le 1$. Let $K = supp(f)$, and let $n$ be an arbitrary positive integer. For notational convenience, set $\epsilon = 1/n$. For $0 \le i \le n$, let $y_i = i/n$, and define $E_1 = f^{-1}([0, y_1]) \cap K$ and $E_i = f^{-1}(y_{i-1}, y_i]$ for $2 \le i \le n$. Clearly, the sets $E_i$ are disjoint and $\cup_{i=1}^n E_i = K$. Since $f$ is continuous, $f^{-1}(B) \in \mathcal{B}(X)$ for every Borel subset $B \subseteq \mathbb{R}$; see problem 16 on section 8.2. In particular, the sets $E_i$ are in $\mathfrak{M}$. Because $y_i - \epsilon = y_{i-1} \le f(x)$ for all $x \in E_i$, the simple function $s = \sum_{i=1}^n (y_i - \epsilon)\chi_{E_i}$ satisfies $0 \le s \le f$. Therefore

$$\sum_{i=1}^n (y_i - \epsilon)\mu(E_i) \le \int_X f d\mu. \tag{4}$$

For $1 \le i \le n$, choose open subsets $V_i \supseteq E_i$ such that $\mu(V_i) < \mu(E_i) + \epsilon/n$ and $f(x) < y_i + \epsilon$ for all $x \in V_i$, and let $\{h_i\}$ be a partition of unity of $K$ subordinate to $\{V_i\}$. Since $h_i \prec V_i$,

$$I(h_i) \le \mu(V_i) < \mu(E_i) + \epsilon/n. \tag{5}$$

Since $f = \sum_{i=1}^{n} h_i f$ and since $h_i f \le (y_i + \epsilon)h_i$, we have (using inequalities (4) and (5)),

$$I(f) = \sum_{i=1}^{n} I(h_i f) \le \sum_{i=1}^{n} (y_i + \epsilon)I(h_i) \le \sum_{i=1}^{n} (y_i + \epsilon)(\mu(E_i) + \epsilon/n)$$

$$= \sum_{i=1}^{n} (y_i - \epsilon)\mu(E_i) + 2\epsilon\mu(K) + \epsilon/n \sum_{i=1}^{n} (y_i + \epsilon)$$

$$\le \int_X f d\mu + 2\epsilon\mu(K) + \epsilon/n \sum_{i=1}^{n} (1 + \epsilon) = \int_X f d\mu + 2\epsilon\mu(K) + \epsilon(1 + \epsilon).$$

This establishes inequality (3) because n is arbitrary. ■

The following theorem summarizes the properties of Radon measures.

**Theorem 8.4.25.** Suppose $X$ is a locally compact Hausdorff space, and let $I$ be a positive linear functional on $\mathcal{C}_c(X)$. Then the Radon measure induced by $I$ is a complete measure on $\mathfrak{M}$, and

(a) every Borel subset of $X$ is Radon measurable;
(b) for every open subset $V$ $\mu(V) = \sup\{I(f) : f \prec V\}$;
(c) $\mu$ is outer regular;
(d) $\mu$ is inner regular on open sets and sets of finite $\mu$-measure;
(e) for every compact set $K$, $\mu(K) = \inf\{I(f) : K \prec f\}$, and
(f) $\lambda$ extends (represents) $I$ in the sense that $\int_X f d\mu = I(f)$ for every $f \in \mathcal{C}_c(X)$.

Additionally, $\mu$ is unique, subject to these properties. ■

To prove the uniqueness part of this theorem, mimic the proof of theorem 8.4.16. Observe that the proof of theorem 8.4.16 is based only on the outer regularity of the measures in question and their inner regularity for open sets.

The following theorem provides a sufficient condition for the inner regularity of Radon measures (for all $E \in \mathfrak{M}$). The proof is identical to that of theorem 8.4.10.

**Theorem 8.4.26.** Suppose $X$ is an $\sigma$-compact, locally compact Hausdorff space, and let $E \in \mathfrak{M}$.

(a) For $\epsilon > 0$, there exists a closed subset $F$ and an open subset $V$ such that $F \subseteq E \subseteq V$ and $\mu(V - F) < \epsilon$.

(b) $\mu(E) = \sup\{\mu(K) : K \text{compact}, K \subseteq E\}$.

(c) There exists an $F_\sigma$ set $A$ and a $G_\delta$ set $B$ such that $A \subseteq E \subseteq B$ and $\mu(B - A) = 0$. ∎

## Exercises

1. Let $f \in \mathcal{C}_c(\mathbb{R}^n)$. Prove that $f$ is uniformly continuous.

2. Prove that every countable subset of $\mathbb{R}^n$ has measure 0, and find an example of a subset $E$ of $\mathbb{R}^n$ such that $\lambda(E) = 0$ but $\lambda(\partial E) = \infty$.

3. The goal of this exercise is to show the existence of non-Lebesgue measurable subsets of $\mathbb{R}$. Complete the following sketch of the proof. Define an equivalence relation $\approx$ on $\mathbb{R}$ by $x \approx y$ if $x - y \in \mathbb{Q}$. Each equivalence class of $\approx$ intersects the interval $[0, 1/2]$. Let $P$ be a subset of $[0, 1/2]$ containing exactly one element from each of the equivalence classes of $\approx$. Enumerate the rational numbers in $[-1/2, 1/2]$ as $\{r_n : n \in \mathbb{N}\}$, and let $P_n = r_n + P$. Show that the family $\{P_n : n \in \mathbb{N}\}$ is disjoint and that the union $A = \cup_{n=1}^\infty P_n$ satisfies $[0, 1/2] \subseteq A \subseteq [-1/2, 1]$. If $P$ were measurable, then $1/2 \leq \lambda(A) \leq 3/2$. But $\lambda(A) = \sum_{n=1}^\infty \lambda(P_n)$.

4. Show that not every Lebesgue measurable set is a Borel set. Construction: Let $C$ be the Cantor set, and define a function $f : [0, 1] \to C$ as follows: $f(0) = 0$, and, for $x \in (0, 1]$, write $x = \sum_{i=1}^\infty \frac{a_i}{2^i}$, and set $f(x) = \sum_{i=1}^\infty \frac{2a_i}{3^i}$.[8] The function $f$ is measurable by problem 18 on section 8.2, and is one-to-one by theorem 4.2.18. Choose a subset $P$ of $[0, 1]$ which is not Lebesgue measurable. The set $A = f(P)$ is the set you need. Recall that the Cantor set has measure 0; see example 4.

5. Use the fact that the Cantor set has measure 0 to show that $Card(\mathcal{L}(\mathbb{R})) = 2^c$. It can be shown that $Card(\mathcal{B}(\mathbb{R})) = c$. Thus there are *many more* Lebesgue measurable subsets than Borel subsets of $\mathbb{R}$.

6. Let $V = \{(x_1, ..., x_n) \in \mathbb{R}^n : x_n = 0\}$. Prove that $\lambda(V) = 0$.

7. Compute the Lebesgue measure of the set $\{x \in (0, \frac{1}{\pi}) : sin(1/x) > 0.\}$

8. Let $E$ be a subset of $\mathbb{R}^n$.

   (a) Prove that $E$ is Lebesgue measurable if and only if, for every $\epsilon > 0$, there exists an open set $V$ containing $E$ such that $m^*(V - E) < \epsilon$. Hint: The necessity of the condition is by theorem 8.4.10. To prove the sufficiency, use the identity $A \cap E' = (A \cap V') \cup (A \cap (V - E))$.

---

[8] We use the series representation of $x$ if $x$ has a terminating binary expansion.

(b) Prove that $E$ is Lebesgue measurable if and only if, for every $\epsilon > 0$, $E$ contains a closed subset $F$ such that $m^*(E - F) < \epsilon$.

The importance of this problem and the next is that they provide more intuitive characterizations of Lebesgue measurability than the Carathédory condition does. Intuitively, a subset $E$ of $\mathbb{R}^n$ is Lebesgue measurable if it can be approximated from the outside by an open set or from the inside by a closed set.

9. Let $E$ be a subset of $\mathbb{R}^n$.
   (a) Prove that $E$ is Lebesgue measurable if and only if there exists a $G_\delta$ set $G$ containing $E$ such that $m^*(G - E) = 0$.
   (b) Prove that $E$ is Lebesgue measurable if and only if $E$ contains an $F_\sigma$ set $F$ such that $m^*(E - F) = 0$.

10. In this exercise, we use $\lambda_k$ to denote the Lebesgue measure on $\mathbb{R}^k$. Let $r$ and $s$ be positive integers, and let $n = r + s$. Prove that if $U \subseteq \mathbb{R}^r$ and $V \subseteq \mathbb{R}^s$ are open sets, then $\lambda_n(U \times V) = \lambda_r(U)\lambda_s(V)$. Hint: $\mathcal{S}_k(U \times V) = \mathcal{S}_k(U) \times \mathcal{S}_k(V)$.

11. Let $r_n$ be an enumeration of $\mathbb{Q}$, and let $G = \cup_{n=1}^\infty (r_n - \frac{1}{n^2}, r_n + \frac{1}{n^2})$. Prove that $\lambda(G \Delta F) > 0$ for every closed subset $F$ of $\mathbb{R}$. Hint: Show that if $\lambda(G - F) = 0$, then $F = \mathbb{R}$.

12. Let $f$ be a continuous function in $\mathfrak{L}^1(\mathbb{R}^n)$. Show that if $\lim_{\|x\| \to \infty} f(x)$ exists, then $\lim_{\|x\| \to \infty} f(x) = 0$. Also give an example to show that a continuous positive integrable function need not be bounded.

13. Let $f \in \mathfrak{L}^1(\mathbb{R}^n)$, and let $a \in \mathbb{R}^n$ be fixed. Define $(\tau_a f)(x) = f(x - a)$. Show that $\int_{\mathbb{R}^n} f d\lambda = \int_{\mathbb{R}^n} (\tau_a f) d\lambda$. This is a familiar linear change of variables formula. Using more conventional notation, $\int_{\mathbb{R}^n} f(x) d\lambda(x) = \int_{\mathbb{R}^n} f(x - a) d\lambda(x)$.

14. For a subset $E$ of $\mathbb{R}^n$, let $-E = \{-x : x \in E\}$. Prove that $E$ is measurable if and only if $-E$ is measurable and, in this case, $\lambda(-E) = \lambda(E)$.

15. For $r > 0$ and $E \subseteq \mathbb{R}^n$, define $rE = \{rx : x \in E\}$. Prove that $E$ is measurable if and only if $rE$ is measurable and that $\lambda(rE) = r^n \lambda(E)$.

16. Let $f \in \mathfrak{L}^1(\mathbb{R}^n)$. For $r > 0$, define $f_r(y) = f(ry)$. Show that $\int_{\mathbb{R}^n} f d\lambda = r^n \int_{\mathbb{R}^n} f_r d\lambda$. Using more familiar notation, if $x = ry$, then $d\lambda(x) = r^n d\lambda(y)$.

17. Let $X$ be an infinite-dimensional normed linear space. Prove that there does not exist a translation-invariant measure on $\mathcal{B}(X)$ that assigns finite measure to bounded sets in $\mathcal{B}(X)$. Hint: Use Riesz's theorem to find a sequence $\{u_n\}$ of unit vectors such that $\|u_i - u_j\| \geq 1/2$.

## 8.5 Complex Measures

Complex measures do not really *measure* anything in the strict geometric sense of the word, but they do share the defining property of a positive measure, namely, countable additivity. Although they are rather abstract, real and complex measures have applications in differentiation and probability theories, among many other applications. We study the notion of differentiating one measure with respect to another measure, and our main result is the Radon-Nikodym theorem, which we apply in section 8.6 to study duals of $\mathfrak{L}^p$ spaces. Although the section results are limited to the basics, example 2 and the section exercises significantly expand the scope of the section, where we introduce such topics as the total variation of real and complex measures, uniform integrability, and measurable dissections. The properties of the Radon-Nikodym derivatives are also explored in the section exercises.

**Definition.** Let $(X, \mathfrak{M})$ be a measurable space. A **real measure** on $\mathfrak{M}$ is a countably additive function $\nu : \mathfrak{M} \to \mathbb{R}$. Thus if $(E_n)$ is a disjoint sequence in $\mathfrak{M}$, and $E = \cup_{n=1}^{\infty} E_n$, then $\nu(E) = \sum_{n=1}^{\infty} \nu(E_n)$. By definition, $\nu(\varnothing) = 0$. Observe that finite positive measures are real measures.

In this definition, it is tacitly assumed that the series is absolutely convergent. We call the reader's attention to the fact that, according to the above definition, $\nu$ takes finite values, that is, $\nu(E) = \infty$ and $\nu(E) = -\infty$ are specifically not permitted.[9]

**Theorem 8.5.1.** *Let $(X, \mathfrak{M}, \nu)$ be a real measure space.*

(a) *If $(E_n)$ is a ascending sequence in $\mathfrak{M}$, then $\nu(\cup_{n=1}^{\infty} E_n) = \lim_n \nu(E_n)$.*
(b) *If $(E_n)$ is a descending sequence in $\mathfrak{M}$, then $\nu(\cap_{n=1}^{\infty} E_n) = \lim_n \nu(E_n)$.*

*Proof. The proofs parallel those of theorem 8.2.3 and are therefore omitted.* ∎

**Definition.** Let $(X, \mathfrak{M}, \nu)$ be a real measure space. A measurable set $E$ is said to be a $\nu$-**positive set** (or simply positive) if $\nu(F) \geq 0$ for every measurable subset $F$ of $E$. A measurable set $E$ is said to be a $\nu$-**negative set** (or simply negative) if $\nu(F) \leq 0$ for every measurable subset $F$ of $E$. A measurable set $E$ is said to be $\nu$-null if $\nu(F) = 0$ for every measurable subset $F$ of $E$.

Clearly, if a measurable set $E$ is both negative and positive, then $E$ is $\nu$-null.

Warning. Monotonicity does not hold for real measures. It is possible for a set of positive measure to contain a subset of negative measure, and conversely. Monotonicity does hold, however, for positive and negative sets: if $F$ is a measurable subset of a positive set $E$, then $\nu(F) \leq \nu(E)$.

---

[9] This definition is not standard. Most books allow a real measure to take extended real values, $\infty$ or $-\infty$, but not both. The standard term used in this case is *signed measure*.

**Proposition 8.5.2.** *A measurable subset of a positive set is positive, and the count-able union of positive sets is positive. The corresponding statements are true for negative sets.*

*Proof.* The first assertion follows from the definition. To prove the second assertion, let $(E_n)$ be a sequence of positive measurable subsets of $X$. Define $A_1 = E_1$, and, for $n \geq 2$, let $A_n = E_n - \cup_{i=1}^{n-1} E_i$. The sequence $A_n$ is disjoint, and each $A_n$ is a positive set. Now let $E \subseteq \cup_{n=1}^{\infty} E_n = \cup_{n=1}^{\infty} A_n$. Since $\nu(E \cap A_n) \geq 0$, $\nu(E) = \sum_{n=1}^{\infty} \nu(E \cap A_n) \geq 0$. ∎

**Lemma 8.5.3.** *Every set of positive measure contains a positive set of positive measure.*

*Proof.* Suppose, for a contradiction, that $S$ is a set of positive measure that contains no positive sets of positive measure. We first establish the following:

If $A \subseteq S$ and $\nu(A) > 0$, then there is a subset of $B$ of $A$ such that $\nu(B) > \nu(A)$.
$$(*)$$

Since $A$ is not a positive set, $A$ contains a subset $C$ such that $\nu(C) < 0$. Set $B = A - C$. Then $\nu(B) = \nu(A) - \nu(C) > \nu(A)$. This proves $(*)$.

Set $A_1 = S$, and let $n_1$ be the least natural number for which $\nu(A_1) > \frac{1}{n_1}$. By $(*)$, $A_1$ contains a set $B$ such that $\nu(B) > \nu(A_1)$. Let $n_2$ be the least natural number for which $A_1$ contains a set $B$ such that $\nu(B) > \nu(A_1) + \frac{1}{n_2}$, and let $A_2$ be such a set. Continue inductively to construct a sequence of natural numbers $n_1, n_2, \ldots$, and measurable sets $A_1 \supseteq A_2 \supseteq \ldots$ such that $n_j$ is the least positive integer for which $A_{j-1}$ contains a set $B$ with $\nu(B) > \nu(A_{j-1}) + \frac{1}{n_j}$, and $A_j$ is such a set. Now $\nu(A_3) > \nu(A_2) + \frac{1}{n_3} > \nu(A_1) + \frac{1}{n_2} + \frac{1}{n_3} > \frac{1}{n_1} + \frac{1}{n_2} + \frac{1}{n_3}$. Inductively, $\nu(A_j) > \sum_{i=1}^{j} \frac{1}{n_i}$. Define $A = \cap_{j=1}^{\infty} A_j$. Then $\infty > \nu(A) = \lim_j \nu(A_j) \geq \sum_{j=1}^{\infty} \frac{1}{n_j}$. In particular, $\sum_{j=1}^{\infty} \frac{1}{n_j}$ is convergent, and $\lim_j n_j = \infty$. Again by $(*)$, $A$ contains a subset $B$ such that $\nu(B) > \nu(A)$, and there is a natural number $n$ such that $\nu(B) > \nu(A) + \frac{1}{n}$. But there is an integer $j$ such that $n_j > n$. Thus $\nu(B) > \nu(A) + \frac{1}{n} > \nu(A_{j-1}) + \frac{1}{n}$. This contradicts the definition of $n_j$ because $B \subseteq A_{j-1}$. ∎

**Theorem 8.5.4 (the Hahn decomposition theorem).** *If $(X, \mathfrak{M}, \nu)$ is a real measure space, then there exist a positive set $P$ and a negative set $N$ such that $X = P \cup N$ and $P \cap N = \emptyset$. The sets $P$ and $N$ are essentially unique in the sense that if $(Q, M)$ is another pair of subsets of $X$ satisfying the conclusion of the theorem, then $P \Delta Q$ and $N \Delta M$ are $\nu$-null sets. Here $P \Delta Q = (P - Q) \cup (Q - P)$.*

*Proof.* Let $K = \sup\{\nu(E) : E \in \mathfrak{M}, E \text{ positive}\}$, let $P_n$ be a sequence of positive measurable subsets such that $\lim_n \nu(P_n) = K$, and let $P = \cup_{n=1}^{\infty} P_n$. By proposition
8.5.2, $P$ is positive; hence $\nu(P) \leq K$. Now $\nu(P_n) \leq \nu(P) \leq K$ implies that $K =
\lim_n \nu(P_n) \leq \nu(P) \leq K$. Therefore $\nu(P) = K$. Notice that this proves that $K < \infty$.
Let $N = X - P$. We show that $N$ is a negative set, and this will prove the existence
part of the theorem. If $N$ is not negative, then it contains a subset $S$ of positive
measure. By lemma 8.5.3, $S$ contains a positive subset $G$ of positive measure. This
contradicts the definition of $K$, since $P \cup G$ would be a positive set and $\nu(P \cup G) =
\nu(P) + \nu(G) > \nu(P) = K$. If the pair $(Q, M)$ also satisfies the conclusion of the
theorem, then $P - Q = P \cap M$ is both positive and negative and hence $P - Q$ is
a $\nu$-null set. Similarly, $Q - P$ is $\nu$-null and so is $P \Delta Q$. One shows that $N \Delta M$ is
$\nu$-null using an argument identical to this one. ∎

**Corollary 8.5.5.** *A real measure is bounded.*

*Proof.* We use the notation of the proof of the previous theorem. Let $k = \nu(N)$. For a
measurable set $E$, $\nu(E) = \nu(E \cap P) + \nu(E \cap N)$. Since $0 \leq \nu(E \cap P) \leq K$, and $k \leq
\nu(E \cap N) \leq 0$, $k \leq \nu(E) \leq K$. ∎

**Definition.** Two positive measures $\nu$ and $\mu$ on a $\sigma$-algebra $\mathfrak{M}$ are called **mutually
singular** if there exist disjoint measurable subsets $Q$ and $M$ such that $Q \cup M = X$
and $\mu(Q) = 0 = \nu(M)$.

**Theorem 8.5.6 (the Jordan decomposition theorem).** *If $\nu$ is a real measure, then
there exist unique, finite, positive, mutually singular measures $\nu^+$ and $\nu^-$ such
that, for every $E \in \mathfrak{M}$, $\nu(E) = \nu^+(E) - \nu^-(E)$.*

*Proof.* Let $(P, N)$ be a Hahn decomposition of $\nu$, and define $\nu^+(E) = \nu(E \cap P)$, and
$\nu^-(E) = -\nu(E \cap N)$. The pair $\nu^+$ and $\nu^-$ has the desired properties since $\nu^+(N) =
0 = \nu^-(P)$. If $\mu^+$ and $\mu^-$ is another pair satisfying the stated properties with
$\mu^+(M) = 0 = \mu^-(Q)$, where $Q \cap M = \emptyset, Q \cup M = X$, then the pair $(Q, M)$ is a
Hahn decomposition of $\nu$ and hence $P \Delta Q$ is $\nu$-null. Therefore, for $E \in \mathfrak{M}$,

$$\mu^+(E) = \mu^+(E \cap Q) + \mu^+(E \cap M) = \mu^+(E \cap Q)$$
$$= \mu^+(E \cap Q) - \mu^-(E \cap Q) = \nu(E \cap Q) = \nu(E \cap P) = \nu^+(E).$$

*Thus $\mu^+ = \nu^+$; hence $\mu^- = \nu^-$.* ∎

**Definitions.** The finite positive measures $\nu^+$ and $\nu^-$ are called the **positive** and
**negative variations** of $\nu$, respectively. The finite positive measure $|\nu| = \nu^+ + \nu^-$
is called the **total variation** of $\nu$. Notice that, for every $E \in \mathfrak{M}$, $|\nu(E)| \leq |\nu|(E)$.
Define $\|\nu\| = |\nu|(X)$.

**Example 1.** Let $\lambda$ be Lebesgue measure on $\mathbb{R}$, and let $\mathcal{L}$ be the set of Lebesgue measurable subsets of $\mathbb{R}$. Let $f(x) = xe^{-|x|}$, and define a set function $\nu : \mathcal{L} \to \mathbb{R}$ by $\nu(E) = \int_E f d\lambda$ $(E \in \mathcal{L})$. One can easily check that $\nu$ is a real measure (also see theorem 8.5.7). The Hahn decomposition of $\nu$ consists of the sets $P = [0, \infty)$, and $N = (-\infty, 0)$. Let $A = (-1, 2)$, and $B = (-1, 0)$, and let $C = (0, 2)$. Notice that while $B \subseteq A$, $A$ has positive measure and $B$ has negative measure. Also $C \subseteq A$, but $\nu(C) = \int_0^2 xe^{-x} dx > \nu(A) = \int_0^2 xe^{-x} + \int_{-1}^0 xe^x dx$. One can see that $\nu^+(E) = \int_{E \cap (0,\infty)} f d\lambda$, $\nu^-(E) = -\int_{E \cap (-\infty, 0)} f d\lambda$ and that $|\nu|(E) = \int_E |f| d\lambda$. In particular, $\nu(\mathbb{R}) = 0$, while $\|\nu\| = |\nu|(\mathbb{R}) = 2$. ♦

**Definition.** Let $(X, \mathfrak{M})$ be a measurable space. A **complex measure** on $\mathfrak{M}$ is a countably additive complex-valued function on $\mathfrak{M}$.

Now let $\nu$ be a complex measure, and let $\nu_r$ and $\nu_i$ be the real and imaginary parts of $\nu$, that is for $E \in \mathfrak{M}$, $\nu(E) = \nu_r(E) + i\nu_i(E)$. Clearly, $\nu_r$ and $\nu_i$ are real measures; hence $\|\nu_r\| < \infty$, $\|\nu_i\| < \infty$, and $|\nu(E)| \le |\nu_r(E)| + |\nu_i(E)| \le \|\nu_r\| + \|\nu_i\| < \infty$. Therefore, complex measures, like real measures, are bounded. Notice that the set of complex measures contains the set of real measures and, in particular, the set of finite positive measures. The set of complex measures on a $\sigma$-algebra $\mathfrak{M}$ is a vector space under the obvious operations: for complex measures $\nu$ and $\mu$ and for a complex scalar $\alpha$, $(\nu + \mu)(E) = \nu(E) + \mu(E)$, and $(\alpha\nu)(E) = \alpha\nu(E)(E \in \mathfrak{M})$.

The following theorem generalizes example 1 and provides a rich source of real and complex measures.

**Theorem 8.5.7.** *Let $(X, \mathfrak{M})$ be a measurable space, and let $\mu$ be a positive (not necessarily finite) measure on $\mathfrak{M}$. If $h \in \mathfrak{L}^1(\mu)$, then the following set function defines a complex measure on $\mathfrak{M}$:*

$$\nu(E) = \int_E h d\mu.$$

*Proof.* Let $\{E_n\}$ be a disjoint family of members of $\mathfrak{M}$, and let $E = \cup_{n=1}^\infty E_n$; $h\chi_E = \sum_{n=1}^\infty h\chi_{E_n} = \lim_n \sum_{i=1}^n h\chi_{E_i}$. Since $|\sum_{i=1}^n h\chi_{E_i}| \le \sum_{i=1}^n |h|\chi_{E_i} \le |h| \in \mathfrak{L}^1(\mu)$, the dominated convergence theorem implies that

$$\nu(E) = \int_X h\chi_E d\mu = \int_X \sum_{n=1}^\infty h\chi_{E_n} d\mu$$

$$= \lim_n \int_X \sum_{i=1}^n h\chi_{E_i} d\mu = \lim_n \sum_{i=1}^n \nu(E_i) = \sum_{n=1}^\infty \nu(E_n). \blacksquare$$

**Definition.** Let $\mu$ and $\nu$ be as in theorem 8.5.7. The function $h$ is called the **Radon-Nikodym derivative** of $\nu$ with respect to $\mu$, and we symbolically write $h = \frac{d\nu}{d\mu}$, or $d\nu = hd\mu$, to indicate that $\nu(E) = \int_E hd\mu$. The following theorem justifies the definition and the notation.

**Theorem 8.5.8.** *Let $\mu$ be a positive measure, let $h \in \mathfrak{L}^1(\mu)$ be a positive function, and let $d\nu = hd\mu$. Then, for a positive measurable function $f$,*

$$\int_X fd\nu = \int_X fhd\mu.$$

*In particular, if $f \in \mathfrak{L}^1(\nu)$, then $fh \in \mathfrak{L}^1(\mu)$ and $\|f\|_{1,\nu} = \|fh\|_{1,\mu}$.*

*Proof.* For a measurable set $E$, $\int_X \chi_E d\nu = \nu(E) = \int_E hd\mu = \int_X \chi_E hd\mu$. Linearity guarantees that $\int_X sd\nu = \int_X shd\mu$ for every simple function $s$. Now, for a positive function $f$, let $s_n$ be an increasing sequence of simple functions such that $\lim_n s_n = f$. Then $s_n h$ increases to $fh$, and, by the monotone convergence theorem, $\int_X fd\nu = \lim_n \int_X s_n d\nu = \lim_n \int_X s_n hd\mu = \int_X fhd\mu$. The remaining parts of the theorem are obvious. ∎

Theorem 8.5.8 justifies the definition of the Radon-Nikodym derivative and the notation $h = \frac{d\nu}{d\mu}$. Indeed, this theorem can be stated using the notation $\int_X fd\nu = \int_X f\frac{d\nu}{d\mu}d\mu$. Observe that the last formula is reminiscent of the change of variables formula. Problem 8 at the end of this section is what one might call the chain rule for Radon-Nikodym derivatives.

**Definition.** Let $\mu$ be a positive (not necessarily finite) measure on a $\sigma$-algebra $\mathfrak{M}$, and let $\nu$ be an arbitrary complex measure on $\mathfrak{M}$. We say that $\nu$ is **absolutely continuous** with respect to $\mu$ if, for every $E \in \mathfrak{M}$, $\mu(E) = 0$ implies $\nu(E) = 0$. In this situation, we write $\nu \ll \mu$.

Notice that if $\mathfrak{M}, \mu, h$, and $\nu$ are as in theorem 8.5.7, then $\nu \ll \mu$. The Radon-Nikodym theorem, in effect, is the converse of theorem 8.5.7.

If $\nu$ is a real measure and $\nu = \nu^+ - \nu^-$ is the Jordan decomposition of $\nu$, then $\nu \ll \mu$ if and only if $\nu^+ \ll \mu$ and $\nu^- \ll \mu$. Also if $\nu = \nu_r + i\nu_i$ is a complex measure, then $\nu \ll \mu$ if and only if $\nu_r$ and $\nu_i$ are absolutely continuous with respect to $\mu$. We leave the details to the reader to verify.

**Lemma 8.5.9.** *Let $\nu$ and $\mu$ be finite positive measures on a $\sigma$-algebra $\mathfrak{M}$, and suppose that $\nu << \mu$. Then there exists a positive number $\epsilon$ and a set $P \in \mathfrak{M}$ such that*

(a) $\mu(P) > 0$, and

(b) *$P$ is positive for the measure $\nu - \epsilon \mu$, that is, $\nu(E \cap P) - \epsilon\mu(E \cap P) \geq 0$ for every $E \in \mathfrak{M}$.*

*Proof.* Since $0 < \nu(X) < \infty$ and $0 < \mu(X) < \infty$, there exists a positive number $\epsilon$ such that $\nu(X) - \epsilon\mu(X) > 0$. Let $(P,N)$ be the Hahn decomposition of the real measure $\nu - \epsilon\mu$. Then $P$ is positive for $\nu - \epsilon\mu$. If $\mu(P) = 0$, then $\nu(P) = 0$; hence $\nu(X) - \epsilon\mu(X) = \nu(N) - \epsilon\mu(N) \leq 0$, since $N$ is negative for $\nu - \epsilon\mu$. This contradicts $\nu(X) - \epsilon\mu(X) > 0$ and proves that $\mu(P) > 0$. ∎

**Theorem 8.5.10 (the Radon-Nikodym theorem—the real version).** *If $\nu$ and $\mu$ are finite positive measures and $\nu << \mu$, then there exists a unique positive function $f \in \mathfrak{L}^1(\mu)$ such that $d\nu = f d\mu$, that is,*

$$\nu(E) = \int_E f d\mu \text{ for every } E \in \mathfrak{M}.$$

*Proof. The uniqueness of $f$ follows from example 4 in section 8.3.[10] Let $\mathfrak{F}$ be the following set of measurable functions:*

$$\mathfrak{F} = \{f \geq 0 : \int_E f d\mu \leq \nu(E) \, \forall \, E \in \mathfrak{M}\}.$$

*Since $f = 0 \in \mathfrak{F}, \mathfrak{F} \neq \emptyset$. Every $f \in \mathfrak{F}$ is $\mu$-integrable since $\int_X f d\mu \leq \nu(X) < \infty$. It follows that $\alpha = \sup\{\int_X f d\mu : f \in \mathfrak{F}\}$ is finite. We first prove the fact that if $f, g \in \mathfrak{F}$, then $h = \max\{f,g\} \in \mathfrak{F}$. Let $A = \{x \in X : f(x) \geq g(x)\}$, and $B = \{x \in X : f(x) < g(x)\}$. For every $E \in \mathfrak{M}$, $\int_E h d\mu = \int_{A \cap E} h d\mu + \int_{B \cap E} h d\mu = \int_{A \cap E} f d\mu + \int_{B \cap E} g d\mu \leq \nu(A \cap E) + \nu(B \cap E) = \nu(E)$. Hence $h \in \mathfrak{F}$. By the definition of $\alpha$, there exists a sequence $g_1, g_2, \ldots \in \mathfrak{F}$ such that $\lim_n \int_X g_n d\mu = \alpha$. Let $f_1 = g_1, f_2 = \max\{g_1, g_2\}, \ldots, f_n = \max\{g_1, g_2, \ldots, g_n\}$. By the above fact, $f_n \in \mathfrak{F}, 0 \leq f_1 \leq f_2 \leq \ldots$, and $\lim_n \int_X f_n d\mu = \alpha$. Set $f(x) = \lim_n f_n(x)$. By the monotone convergence theorem, $\int_E f d\mu = \lim_n \int_E f_n d\mu \leq \nu(E)$; hence $f \in \mathfrak{F}$. Also, $\int_X f d\mu = \alpha$. We claim that $\nu(E) = \int_E f d\mu$ for every $E \in \mathfrak{M}$. To this end, it is enough to show that the measure $\zeta(E) = \nu(E) - \int_E f d\mu$ is identically equal to zero. If not, then $\zeta$ would be a positive measure, and $\zeta << \mu$. By lemma 8.5.9, there exists a positive number $\epsilon$ and a set $P$ with $\mu(P) > 0$ such that $P$ is positive for $\zeta - \epsilon\mu$. Thus, for every*

---

[10] More precisely, if $f_1 d\mu = f_2 d\mu$ for $f_i \in \mathfrak{L}^1(\mu)$, then $f_1 = f_2, \mu$-a.e.

$E \in \mathfrak{M}, \zeta(E) \geq \zeta(E \cap P) \geq \epsilon \mu(E \cap P) = \epsilon \int_E \chi_P d\mu$, and $\nu(E) = \int_E f d\mu + \zeta(E) \geq \int_E (f + \epsilon \chi_P) d\mu$. Thus $f + \epsilon \chi_P \in \mathfrak{F}$. But this leads to the following violation of the definition of $\alpha$: $\int_X (f + \epsilon \chi_P) d\mu = \int_X f d\mu + \epsilon \mu(P) > \int_X f d\mu = \alpha$. ∎

**Theorem 8.5.11 (the Radon-Nikodym theorem).** *If $\mu$ is a finite positive measure on $\mathfrak{M}$ and $\nu$ is a complex measure on $\mathfrak{M}$ such that $\nu << \mu$, then there exists a unique complex-valued function $f \in \mathfrak{L}^1(\mu)$ such that, for every $E \in \mathfrak{M}$, $\nu(E) = \int_E f d\mu$.*

*Proof. If $\nu$ is real and $\nu << \mu$, then $\nu^+ << \mu$ and $\nu^- << \mu$. By theorem 8.5.10, we find positive $\mu$-integrable functions $f^+$ and $f^-$ such that, for every $E \in \mathfrak{M}$, $\nu^+(E) = \int_E f^+ d\mu$, and $\nu^-(E) = \int_E f^- d\mu$. Thus $\nu(E) = \int_E f d\mu$, where $f = f^+ - f^- \in \mathfrak{L}^1(\mu)$. If $\nu$ is a complex measure, apply the result we just established to the real and imaginary parts of $\nu$, since each part is absolutely continuous with respect to $\mu$.* ∎

As an application of the Radon-Nikodym theorem, we develop the definition of the **total variation of a complex measure**.

**Example 2.** Let $\nu$ be a complex measure, and let $\mu$ be a finite positive measure such that $\nu << \mu$. Such $\mu$ exists; one can take, for example, $\mu = |\nu_r| + |\nu_i|$. By the Radon-Nikodym theorem, there exists a function $f \in \mathfrak{L}^1(\mu)$ such that $d\nu = f d\mu$. We define the total variation of $\nu$ to be the finite positive measure given by $d|\nu| = |f| d\mu$. Notice that this definition is consistent with the result of problem 6(c) for real measures. We need to prove that $|\nu|$ is well defined in the sense that it does not depend on the particular choice of $\mu$. Suppose that, for finite positive measures $\mu_1$ and $\mu_2$, and for functions $f_i \in \mathfrak{L}^1(\mu_i)$, $f_1 d\mu_1 = f_2 d\mu_2$. Let $\xi = \mu_1 + \mu_2$. Then $\xi$ is a finite positive measure and $\mu_i << \xi$. By problem 8 in the section exercises, $f_1 \frac{d\mu_1}{d\xi} d\xi = f_2 \frac{d\mu_2}{d\xi} d\xi$. By the uniqueness of the Radon-Nikodym derivative, $f_1 \frac{d\mu_1}{d\xi} = f_2 \frac{d\mu_2}{d\xi}$, $\xi$-a.e. Since $\frac{d\mu_i}{d\xi} \geq 0$, $|f_1| \frac{d\mu_1}{d\xi} = |f_1 \frac{d\mu_1}{d\xi}| = |f_2 \frac{d\mu_2}{d\xi}| = |f_2| \frac{d\mu_2}{d\xi}$, $\xi$-a.e. Now, for a measurable set $E$,

$$\int_E |f_1| d\mu_1 = \int_E |f_1| \frac{d\mu_1}{d\xi} d\xi = \int_E |f_2| \frac{d\mu_2}{d\xi} d\xi = \int_E |f_2| d\mu_2.$$

## Exercises

In the following exercises, $(X, \mathfrak{M})$ is a measurable space and $\mu$ is a positive measure on $\mathfrak{M}$.

1. Prove that if $P$ and $Q$ are positive sets for a real measure $\nu$ such that $P \triangle Q$ is $\nu$-null, then $\nu(E \cap P) = \nu(E \cap Q) = \nu(E \cap P \cap Q)$ for every measurable set $E$.

2. Show that, for a real measure $\nu$, $\nu^+ = \frac{1}{2}(|\nu| + \nu)$ and $\nu^- = \frac{1}{2}(|\nu| - \nu)$.

3. Let $\nu$ be a real measure. Prove that if $\xi$ and $\eta$ are finite positive measures such that $\nu = \xi - \eta$, then $\xi \geq \nu^+$ and $\eta \geq \nu^-$.

4. Define the following function on the space of real measures on $\mathfrak{M}$: $\|\nu\| = |\nu|(X)$. Prove that $\|.\|$ is a norm.

5. Show that if $\nu$ is a real measure on $\mathfrak{M}$, then $\nu \ll \mu$ if and only if $\nu^+ \ll \mu$, and $\nu^- \ll \mu$ if and only if $|\nu| \ll \mu$.

6. Let $f \in \mathfrak{L}^1(\mu)$ be a real-valued function, and define $\nu(E) = \int_E f d\mu$, $(E \in \mathfrak{M})$. Prove that
   (a) the pair $(P, N)$ is a Hahn decomposition of $\nu$, where $P = \{x \in X : f(x) \geq 0\}$, and $N = \{x \in X : f(x) < 0\}$;
   (b) $\nu^+(E) = \int_E f^+ d\mu$, and $\nu^-(E) = -\int_E f^- d\mu$; and
   (c) $|\nu|(E) = \int_E |f| d\mu$; using our notation for Radon-Nikodym derivatives, this result can be written as $\frac{d|\nu|}{d\mu} = |\frac{d\nu}{d\mu}|$.

**Definition.** A subset $\mathfrak{F}$ of $\mathfrak{L}^1(\mu)$ is said to be uniformly integrable if, for every $\epsilon > 0$, there exists $\delta > 0$ such that $\int_E |f| d\mu < \epsilon$ for every $f \in \mathfrak{F}$ and for every measurable set $E$ with $\mu(E) < \delta$.

7. Prove that if $(f_n)$ is a convergent sequence in $\mathfrak{L}^1(\mu)$, then $\{f_n\}$ is uniformly integrable. Hint: See example 7 in section 8.3.

8. Let $\zeta, \nu$, and $\mu$ be finite positive measures such that $\zeta \ll \nu \ll \mu$. Show that $\frac{d\zeta}{d\mu} = \frac{d\zeta}{d\nu}\frac{d\nu}{d\mu}$. Conclude that if $\nu \ll \mu \ll \nu$, then $\frac{d\nu}{d\mu} = (\frac{d\mu}{d\nu})^{-1}$ ($\mu$-or $\nu$-a.e.)

9. Prove that if $\nu$ is a complex measure, then, for every measurable set $E$, $|\nu(E)| \leq |\nu|(E)$.

10. For a complex measure $\nu$, define $\|\nu\| = |\nu|(X)$. Prove that $\|.\|$ is a norm on the space of complex measures on $\mathfrak{M}$.

11. Let $\nu$ be a complex measure on $\mathfrak{M}$. Show that $\nu \ll \mu$ if and only if, for every $\epsilon > 0$, there exists $\delta > 0$ such that, for every $E \in \mathfrak{M}$, $\mu(E) < \delta$ implies that $|\nu(E)| < \epsilon$. Hint: See example 7 in section 8.3.

**Definition.** Let $\nu$ be a real measure on $\mathfrak{M}$. A **measurable dissection** of $E$ is a disjoint collection $\{E_1, \dots, E_n\}$ of measurable subsets such that $E = \cup_{i=1}^n E_i$.

12. Prove that, for every $E \in \mathfrak{M}$, $|\nu|(E) = \sup \sum_{i=1}^n |\nu(E_i)|$, where the supremum is taken over all measurable dissections of $E$. The result also holds

when $\{E_i\}$ is a countable dissection of $E$. Hint: If $(P, N)$ is a Hahn decomposition of $\nu$, then $E_1 = E \cap P$ and $E_2 = E \cap N$ is a measurable dissection of $E$.

**Definition.** A positive measure $\mu$ on $(X, \mathfrak{M})$ is said to be $\sigma$-**finite** if $X = \cup_{n=1}^{\infty} X_n$, where $X_n \in \mathfrak{M}$ and $\mu(X_n) < \infty$. We may, and often do, choose $(X_n)$ to be a disjoint sequence.

13. Prove that theorem 8.5.11 is valid when $\mu$ is a $\sigma$-finite positive measure and $\nu$ is a complex measure such that $\nu << \mu$.

    Here is a proof outline. It is sufficient to prove the result when $\nu$ is a finite positive measure. For $n \in \mathbb{N}$, define two finite positive measures on $\mathfrak{M}$ as follows: $\mu_n(E) = \mu(E \cap X_n)$ and $\nu_n(E) = \nu(E \cap X_n)$. Show that $\nu_n << \mu_n$. By theorem 8.5.10, there exist positive functions $h_n \in \mathfrak{L}^1(\mu_n)$ such that $d\nu_n = h_n d\mu_n$. Without loss of generality, $h_n$ vanishes outside $X_n$. Set $h = \sum_{n=1}^{\infty} h_n$. Argue that $h \in \mathfrak{L}^1(\mu)$ and that $d\nu = h d\mu$.

## 8.6  $\mathfrak{L}^p$ Spaces

In addition to the function spaces $\mathcal{B}(X)$, $\mathcal{C}(X)$, and $\mathcal{BC}(X)$, the $\mathfrak{L}^p$ spaces are prototypical examples of Banach spaces and play a prominent role in modern analysis. By far, the most important of the $\mathfrak{L}^p$ spaces is the Hilbert space $\mathfrak{L}^2(X, \mathfrak{M}, \mu)$, where $(X, \mathfrak{M}, \mu)$ is a positive measure space, such as a Lebesgue measure restricted to a subset $X$ of $\mathbb{R}^n$. The section results parallel those for the sequence spaces $l^p$. We prove the completeness of $\mathfrak{L}^p$ and derive the representation theorem that, for $1 < p < \infty$, $\mathfrak{L}^q$ is the dual of $\mathfrak{L}^p$. In fact, the sequence spaces $l^p$ are special cases of the $\mathfrak{L}^p$ spaces. See problem 1 at the end of this section. The next section is a continuation of this one.

Throughout this section, $p \geq 1$ and $q \geq 1$ denote conjugate Hölder exponents; thus $\frac{1}{p} + \frac{1}{q} = 1$. It is understood that $p = 1$, and $q = \infty$ are conjugate exponents.

**Lemma 8.6.1.** *For all $x, y \in \mathbb{C}$,*

*(a)* $|xy| \leq \frac{|x|^p}{p} + \frac{|y|^q}{q}, 1 < p, q < \infty$.
*(b)* $|x + y|^p \leq 2^p(|x|^p + |y|^p), 1 \leq p < \infty$.

*Proof. Part (a) was established in lemma 3.6.1.*

*For part (b), $|x + y| \leq |x| + |y| \leq 2 \max\{|x|, |y|\}$. Thus $|x + y|^p \leq 2^p(\max\{|x|, |y|\})^p = 2^p \max\{|x|^p, |y|^p\} \leq 2^p(|x|^p + |y|^p)$.* ∎

**Definition.** Let $(X, \mathfrak{M})$ be a measurable space, and let $\mu$ be a positive measure on $\mathfrak{M}$. For $1 \leq p < \infty$, we define $\mathfrak{L}^p(\mu)$ to be the set of all measurable functions $f : X \to \mathbb{C}$ such that $\int_X |f|^p d\mu < \infty$. If the measure $\mu$ is understood, we sometimes write $\mathfrak{L}^p$ for $\mathfrak{L}^p(\mu)$. In anticipation of the fact that $\mathfrak{L}^p$ is a normed linear space, we write

$$\|f\|_p = \left( \int_X |f|^p d\mu \right)^{1/p} \text{ for } f \in \mathfrak{L}^p(\mu).$$

**Definition.** We define $\mathfrak{L}^\infty(\mu)$ as follows. A measurable function $f$ is said to be **essentially bounded** if there exists a positive constant $M$ such that $|f(x)| \leq M$ for almost every $x \in X$. Thus $f$ is bounded by $M$ a.e. Such a constant $M$ is called an **essential upper bound** of $f$. The space $\mathfrak{L}^\infty(\mu)$ is the set of all **essentially bounded functions** on $X$. For $f \in \mathfrak{L}^\infty(\mu)$, we define

$$\|f\|_\infty = \inf\{M > 0 : M \text{ is an essential upper bound of } f\}.$$

We leave it to the reader to prove that $\|f\|_\infty$ is an essential upper bound of $f$. Thus $\|f\|_\infty$ is the least nonnegative constant such that $|f(x)| \leq \|f\|_\infty$ a.e.

Observe that if $0 < \epsilon < \|f\|_\infty$, then the set $\{x \in X : |f(x)| > \|f\|_\infty - \epsilon\}$ has a positive measure.

**Theorem 8.6.2 (Hölder's inequality).** *If $f \in \mathfrak{L}^p(\mu)$ and $g \in \mathfrak{L}^q(\mu)$, then $fg \in \mathfrak{L}^1(\mu)$, and $\|fg\|_1 \leq \|f\|_p \|g\|_q$.*

*Proof. By lemma 8.6.1, $\dfrac{|f(x)|}{\|f\|_p} \dfrac{|g(x)|}{\|g\|_q} \leq \dfrac{1}{p} \dfrac{|f(x)|^p}{\|f\|_p^p} + \dfrac{1}{q} \dfrac{|g(x)|^q}{\|g\|_q^q}$. Integrating both sides, we obtain $\dfrac{1}{\|f\|_p \|g\|_q} \int_X |fg| d\mu \leq \dfrac{1}{p} \dfrac{\|f\|_p^p}{\|f\|_p^p} + \dfrac{1}{q} \dfrac{\|g\|_q^q}{\|g\|_q^q} = \dfrac{1}{p} + \dfrac{1}{q} = 1$. Therefore $\int_X |fg| d\mu \leq \|f\|_p \|g\|_q$.*

*If $f \in \mathfrak{L}^1$ and $g \in \mathfrak{L}^\infty$, then $\int_X |fg| d\mu \leq \int_X |f| \|g\|_\infty d\mu = \|f\|_1 \|g\|_\infty$.* ∎

When $p = 2 = q$, Hölder's inequality is the familiar Cauchy-Schwarz inequality.

**Example 1.** If $f, g \in \mathfrak{L}^1(\mu)$, then $\sqrt{|fg|} \in \mathfrak{L}^1$.

By assumption, the functions $\sqrt{|f|}$ and $\sqrt{|g|}$ are in $\mathfrak{L}^2$. By the Cauchy-Schwarz inequality, $\sqrt{|fg|} \in \mathfrak{L}^1$. ◆

**Example 2.** We show that $\int_1^\infty \dfrac{e^{-x}}{x} dx \leq \dfrac{1}{e\sqrt{2}}$.

Let $f(x) = 1/x$, and $g(x) = e^{-x}$. Then $f$ and $g$ are in $\mathfrak{L}^2((1, \infty))$, and $\|f\|_2 = 1$, $\|g\|_2 = \dfrac{1}{e\sqrt{2}}$. The desired inequality now follows from the Cauchy-Schwartz inequality. ◆

**Example 3.** If $\mu(X) < \infty$, then, for $p \in (1, \infty)$, $\mathfrak{L}^p(\mu) \subseteq \mathfrak{L}^1(\mu)$, and, for $f \in \mathfrak{L}^p(\mu)$, $\|f\|_1 \leq \|f\|_p (\mu(X))^{1/q}$. In particular, if $\mu(X) = 1$, then $\|f\|_1 \leq \|f\|_p$.

Because $\mu(X) < \infty$, the constant function $g(x) = 1 \in \mathfrak{L}^q$. Using Hölder's inequality, we have $\int_X |f| d\mu \leq \|f\|_p \|g\|_q = \|f\|_p (\mu(X))^{1/q}$. ◆

**Example 4.** Suppose that $\omega$ is a positive integrable function on $\mathbb{R}^n$ and that $\int_{\mathbb{R}^n} \omega(x) dx = 1$. If $f$ is a measurable function such that $|f|^p \omega$ is Lebesgue integrable, so is $|f| \omega$, and

$$\int_{\mathbb{R}^n} |f(x)| \omega(x) dx \leq \left( \int_{\mathbb{R}^n} |f(x)|^p \omega(x) dx \right)^{1/p}.$$

Here $p \in [1, \infty)$.

Define a finite measure on $\mathbb{R}^n$ by $d\mu = \omega d\lambda$, where $\lambda$ is a Lebesgue measure. The measure $\mu$ and the function $f$ satisfy the conditions of example 1; hence the result. ◆

**Theorem 8.6.3 (Minkowsi's inequality).** *For $f, g \in \mathfrak{L}^p$, $f + g \in \mathfrak{L}^p$ and $\|f + g\|_p \leq \|f\|_p + \|g\|_q$.*

*Proof.* We leave the cases $p = 1$ and $p = \infty$ to the reader. Assume $1 < p < \infty$. By lemma 8.6.1, $\int_X |f + g|^p d\mu \leq \int_X 2^p (|f|^p + |g|^p) d\mu = 2^p (\|f\|_p^p + \|g\|_p^p) < \infty$. This shows that $f + g \in \mathfrak{L}^p$. To prove Minkowski's inequality when $1 < p < \infty$, notice that if $h \in \mathfrak{L}^p$, then $|h|^{p-1} \in \mathfrak{L}^q$ because $(p-1)q = p$. Now

$$\|f + g\|_p^p = \int_X |f + g|^p d\mu$$

$$= \int_X |f + g|^{p-1} |f + g| d\mu \leq \int_X |f + g|^{p-1} |f| + |f + g|^{p-1} |g| d\mu$$

$$\leq \left( \int_X |f|^p \right)^{1/p} \left( \int_X |f + g|^{(p-1)q} \right)^{1/q} + \left( \int_X |g|^p \right)^{1/p} \left( \int_X |f + g|^{(p-1)q} \right)^{1/q}$$

$$= \|f + g\|_p^{p/q} (\|f\|_p + \|g\|_p);$$

*hence* $\|f + g\|_p \leq \|f\|_p + \|g\|_p$. ∎

It is now easy to verify that $\mathfrak{L}^p$ is a normed linear space for $1 \leq p \leq \infty$.

**Theorem 8.6.4 (completeness of $\mathfrak{L}^p(\mu)$).** *For an arbitrary positive measure $\mu$, $\mathfrak{L}^p(\mu)$ is a Banach space for $1 \leq p \leq \infty$.*

*Proof.* We do two cases.

*Case 1.* $1 \leq p < \infty$. We use the result of problem 10 on section 6.1. Let $(f_k)$ be a sequence in $\mathfrak{L}^p$ such that $K = \sum_{k=1}^{\infty} \|f_k\|_p < \infty$. We show that the series $\sum_{k=1}^{\infty} f_k$ converges in $\mathfrak{L}^p$. Define $g_n = \sum_{k=1}^{n} |f_k|$, and let $g = \sum_{k=1}^{\infty} |f_k|$. Then $g_n \in \mathfrak{L}^p$ and $\|g_n\|_p \leq \sum_{k=1}^{n} \|f_k\|_p \leq K$. By the monotone convergence theorem, $\int_X g^p d\mu = \lim_n \int_X g_n^p d\mu \leq K^p$. Thus $g \in \mathfrak{L}^p$. In particular, the series $\sum_{k=1}^{\infty} f_k(x)$ converges for a.e. $x \in X$. Define $f(x) = \sum_{k=1}^{\infty} f_k(x)$. Since $|f| \leq g$, $f \in \mathfrak{L}^p$. Finally, we show that the sequence $\sum_{k=1}^{n} f_k$ converges to $f$ in $\mathfrak{L}^p$. Now $|f - \sum_{k=1}^{n} f_k|^p \leq (2g)^p$, and the dominated convergence theorem implies that $\lim_n \|f - \sum_{k=1}^{n} f_k\|_p = 0$.

*Case 2.* $p = \infty$. Suppose for $m, n > N$, $\|f_n - f_m\|_\infty < \epsilon$. Let $E_n = \{x \in X : |f_n(x)| \geq \|f_n\|_\infty\}$, and $E_{n,m} = \{x \in X : |f_n(x) - f_m(x)| \geq \|f_n - f_m\|_\infty\}$. By definition of $\|\cdot\|_\infty$, the set $E = \cup_{n=1}^{\infty} E_n \cup \cup_{n,m=1}^{\infty} E_{m,n}$ has measure 0.

For $m, n > N$, $\sup\{|f_n(x) - f_m(x)| : x \in X - E\} < \epsilon$. Thus $\{f_n\}$ is a Cauchy sequence in the space $\mathcal{B}(X - E)$. Therefore, by 4.8.1, $f_n$ converges uniformly to some function $f \in \mathcal{B}(X - E)$. Extend $f$ to $X$ by defining $f(x) = 0$ for $x \in E$. Clearly, $\|f_n - f\|_\infty \to 0$ as $n \to \infty$. ∎

## Representation of Bounded Linear Functionals on $\mathfrak{L}^p$

**Definition.** If $g$ is a measurable function, the sign of $g$ is the function

$$(sgn(g))(x) = \begin{cases} \overline{\frac{g(x)}{|g(x)|}} & \text{if } g(x) \neq 0, \\ 1 & \text{otherwise}. \end{cases}$$

Notice that $g.sgn(g) = |g|$ and that $|sgn(g)| = 1$.

**Theorem 8.6.5.** Let $1 < p \leq \infty$, and let $g \in \mathfrak{L}^q(\mu)$. Then the functional

$$\Phi_g(f) = \int_X fg d\mu$$

is a bounded linear functional on $\mathfrak{L}^p(\mu)$, and $\|\Phi_g\| = \|g\|_q$. The same is true for $p = 1$ under the additional assumption that $\mu$ is $\sigma$-finite.

*Proof.* By Hölder's inequality, $|\Phi_g(f)| \leq \int_X |fg| d\mu \leq \|f\|_p \|g\|_q$. Since the linearity of $\Phi_g$ is obvious, this inequality shows that $\Phi_g$ is bounded and that $\|\Phi_g\| \leq \|g\|_q$.

It remains to show that $\|\Phi_g\| = \|g\|_q$.

If $1 < p < \infty$, let $f = |g|^{q-1} sgn(g)$. Then $f \in \mathfrak{L}^p(\mu)$, and $\|f\|_p^p = \|g\|_q^q$. Now $\Phi_g(f) = \int_X fg d\mu = \int_X |g|^q d\mu = \|g\|_q^q = \|g\|_q^{q-1} \|g\|_q = \|f\|_p \|g\|_q$. This concludes the proof of the case $1 < p < \infty$.

*If $p = \infty$, set $f = sgn(g)$. Then $\|f\|_\infty = 1$, and $\int_X fg d\mu = \int_X |g| d\mu = \|g\|_1$.*

*Now suppose that $p = 1$ and that $\mu$ is $\sigma$-finite. Let $0 < \epsilon < \|g\|_\infty$, and let $E = \{x \in X : |g(x)| > \|g\|_\infty - \epsilon\}$. By definition of $\|g\|_\infty$, $\mu(E) > 0$. Since $\mu$ is $\sigma$-finite, $X = \cup_{n=1}^\infty X_n$, where each $X_n$ has finite measure. Since $E = \cup_{n=1}^\infty (E \cap X_n)$ and since $0 < \mu(E) \le \sum_{n=1}^\infty \mu(E \cap X_n)$, $\mu(E \cap X_n) > 0$ for some integer $n$. Let $A = E \cap X_n$, and let $f = sgn(g)\chi_A/\mu(A)$. Then $f \in \mathfrak{L}^1(\mu), \|f\|_1 = 1$, and $|\Phi_g(f)| = \frac{1}{\mu(A)} \int_A |g| d\mu \ge \|g\|_\infty - \epsilon$.* ∎

Theorem 8.6.5 establishes the fact that $\Phi : g \mapsto \Phi_g$ is an isometry from $\mathfrak{L}^q(\mu)$ into $(\mathfrak{L}^p(\mu))^*$. Theorem 8.6.7 establishes sufficient conditions for $\Phi$ to be an isometric isomorphism. First we need a technical result.

**Lemma 8.6.6.** *Suppose $\mu(X) < \infty$. If $g \in \mathfrak{L}^1(\mu)$ and there exists a constant $M$ such that $|\int_X sg d\mu| \le M\|s\|_p$ for every simple function $s$, then $g \in \mathfrak{L}^q$.*

*Proof. Because $\mu(X) < \infty$, all measurable simple functions are in $\mathfrak{L}^p(\mu)$, and $\mathfrak{L}^\infty(\mu) \subseteq \mathfrak{L}^p(\mu)$ for all $p \ge 1$. We work out two separate cases.*

*Case 1. $1 < p < \infty$. First we show that $|\int_X fg d\mu| \le M\|f\|_p$ for every function $f \in \mathfrak{L}^\infty$. To see that, let $(s_n)$ be a sequence of simple functions that converges to $f$ in $\mathfrak{L}^\infty$ (see theorem 8.3.2.) In this case, $s_n$ converges to $f$ in $\mathfrak{L}^p$ for every $p \ge 1$. Now $|\int_X fg - s_n g d\mu| \le \|s_n - f\|_\infty \|g\|_1 \to 0$ as $n \to \infty$. Thus $|\int_X fg d\mu| = \lim_n |\int_X s_n g d\mu| \le \lim_n M\|s_n\|_p = M\|f\|_p$.*

*We show that $g \in \mathfrak{L}^q$. Let $E_n = \{x \in X : |g(x)| \le n\}$, and let $f = |g|^{q-1} sgn(g)\chi_{E_n}$. Then $f \in \mathfrak{L}^\infty$, $fg = |g|^q \chi_{E_n}$, and $|f|^p = |g|^q \chi_{E_n}$. Hence $\int_{E_n} |g|^q d\mu = \int_X fg d\mu \le M(\int_X |f|^p d\mu)^{1/p} = M(\int_{E_n} |g|^q d\mu)^{1/p}$. Thus $(\int_{E_n} |g|^q d\mu)^{1/q} \le M$. Taking the limit of the left side of the last inequality, the monotone convergence theorem yields $\|g\|_q \le M < \infty$.*

*Case 2. $p = 1$. For every measurable set $E$, $|\int_E g d\mu| \le M\|\chi_E\|_1 = M\mu(E)$. It follows that, for every measurable set $E$ of positive measure, $\frac{1}{\mu(E)} \int_E g d\mu$ is in the closed disk $D$ of radius $M$ and centered at the origin of the complex plane. We claim that $|g(x)| \le M$ for a.e. $x \in X$, that is, $\|g\|_\infty \le M$. We show that if $B = B(z_0, r)$ is an open disk of radius $r$ in $\mathbb{C} - D$, then $\mu(g^{-1}(B)) = 0$. This will establish the claim, since $\mathbb{C} - D$ is a countable union of open disks. Let $E = g^{-1}(B)$. If $\mu(E) > 0$, then $|\frac{1}{\mu(E)} \int_E g d\mu - z_0| = |\frac{1}{\mu(E)} \int_E (g - z_0) d\mu| \le \frac{1}{\mu(E)} \int_E |g - z_0| d\mu \le r$. Thus $\frac{1}{\mu(E)} \int_E g d\mu \in B \cap D = \emptyset$. This contradiction completes the proof.* ∎

**Theorem 8.6.7.** *If $\mu$ is $\sigma$-finite, then the function $\Phi$ in theorem 8.6.5 is onto for $1 \leq p < \infty$.*

*Proof.* Let $\varphi \in (\mathfrak{L}^p(\mu))^*$. We need to prove the existence of a function $g \in \mathfrak{L}^q$ such that $\varphi = \Phi_g$. Equivalently, for all $f \in \mathfrak{L}^p(\mu)$,

$$\varphi(f) = \int_X fg d\mu. \tag{6}$$

We first prove the result in the special case when $\mu(X) < \infty$. For a measurable set $E$, define $\nu(E) = \varphi(\chi_E)$. Since $\varphi$ is linear and since, for disjoint measurable sets $E_1$ and $E_2$, $\chi_{E_1 \cup E_2} = \chi_{E_1} + \chi_{E_2}$, $\nu(E_1 \cup E_2) = \nu(E_1) + \nu(E_2)$.) Thus $\nu$ is finitely additive. We show that $\nu$ is countably additive, and this will establish the fact that $\nu$ is a complex measure. Let $(E_n)$ be a disjoint sequence in $\mathfrak{M}$, and let $E = \cup_{n=1}^{\infty} E_n$. Let $A_k = \cup_{i=1}^{k} E_i$. Then $\lim_k \mu(A_k) = \mu(E)$; hence $\|\chi_E - \chi_{A_k}\|_p^p = \int_X |\chi_E - \chi_{A_k}|^p d\mu = \mu(E - A_k) \to 0$ as $k \to \infty$. Thus $\chi_{A_k}$ converges to $\chi_E$ in $\mathfrak{L}^p(\mu)$. By the continuity of $\varphi$, $\lim_k \varphi(\chi_{A_k}) = \varphi(\chi_E)$. that is, $\sum_{n=1}^{\infty} \nu(E_n) = \nu(E)$.

If $\mu(E) = 0$, then $\chi_E = 0$ $\mu$-a.e.; thus $\nu(E) = 0$.

The summary of the proof so far is that $\nu$ is a complex measure and $\nu << \mu$. By the Radon-Nikodym theorem, there exists a function $g \in \mathfrak{L}^1(\mu)$ such that $\varphi(\chi_E) = \int_E g d\mu$ for every measurable set $E$. The linearity of the functionals on the two sides of the last identity implies that $\varphi(s) = \int_X sg d\mu$ for every simple measurable function $s$. Now, for a simple function $s$, $|\int_X sg d\mu| = |\varphi(s)| \leq \|\varphi\| \|s\|_p$. By the previous lemma, $g \in \mathfrak{L}^q$. The functional on the left-hand side of identity (6) is continuous on $\mathfrak{L}^p$ by assumption, and the functional on the right side of (6) is continuous by theorem 8.6.5. Since the two functionals agree on a dense subset of $\mathfrak{L}^p$, namely, the set of simple functions,[11] identity (6) holds for all $f \in \mathfrak{L}^p$. This completes the proof of the theorem when $\mu(X) < \infty$.

Now suppose that $\mu(X) = \infty$ and that $X$ is the disjoint union of a countable sequence $(E_n)$ of sets of finite measure. Define $h(x) = \sum_{n=1}^{\infty} \frac{\chi_{E_n}}{2^n \mu(E_n)}$. By the monotone convergence theorem, $\int_X h d\mu = \sum_{n=1}^{\infty} \int_{E_n} \frac{\chi_{E_n} d\mu}{2^n \mu(E_n)} = \sum_{n=1}^{\infty} 1/2^n = 1$. Thus $h \in \mathfrak{L}^1(\mu)$. Let $\nu$ be the finite positive measure such that $d\nu = h d\mu$. For $1 \leq p < \infty$, the correspondence $F \mapsto h^{1/p} F$ defines a linear isometry from $\mathfrak{L}^p(\nu)$ onto $\mathfrak{L}^p(\mu)$ (theorem 8.5.8 is relevant here and for the rest of the proof). In particular, $\psi(F) = \varphi(h^{1/p}F)$ defines a bounded linear functional on $\mathfrak{L}^p(\nu)$. By the first part of the proof, there exists a function $G \in \mathfrak{L}^q(\nu)$ such that $\psi(F) = \int_X FG d\nu$, for every $F \in \mathfrak{L}^p(\nu)$.

---

[11] see theorem 8.7.3.

If $1 < p < \infty$, define $g = h^{1/q}G$. By theorem 8.5.8, $\int_X |g|^q d\mu = \int_X |G|^q d\nu < \infty$; hence $g \in \mathfrak{L}^q(\mu)$. For $f \in \mathfrak{L}^p(\mu)$, $\varphi(f) = \psi(h^{-1/p}f) = \int_X h^{-1/p}fG d\nu = \int_X h^{-1/p}fGh d\mu = \int_X fh^{1/q}G d\mu = \int_X fg d\mu$, as desired.

If $p = 1$, define $g = G$. Because $\|G\|_{\mu,\infty} = \|G\|_{\nu,\infty}$, $g \in \mathfrak{L}^\infty(\mu)$. If $f \in \mathfrak{L}^1(\mu)$, then $\varphi(f) = \psi(h^{-1}f) = \int_X h^{-1}fG d\nu = \int_X h^{-1}fgh d\mu = \int_X fg d\mu$. ∎

## Exercises

1. Let $\mu$ be the counting measure on $\mathbb{N}$, and let $f : \mathbb{N} \to \mathbb{K}$. Show that $f \in \mathfrak{L}^p(\mu)$ if and only if $f \in l^p$ as defined in section 3.6.

2. Show that, for an essentially bounded function $f$, $\|f\|_\infty$ is an essential upper bound of $f$.

3. Prove that if $\mu(X) < \infty$ and $1 \le p < q \le \infty$, then $\mathfrak{L}^q(\mu) \subseteq \mathfrak{L}^p(\mu)$.

4. Let $f_n$ be a convergent sequence in $\mathfrak{L}^1(\mu)$, and let $f = \lim_n f_n$. For $\epsilon > 0$, define $E_{n,\epsilon} = \{x \in X : |f_n(x) - f(x)| \ge \epsilon\}$. Show that $\lim_n \mu(E_{n,\epsilon}) = 0$.

5. Let $f \in \mathfrak{L}^\infty(\mu)$, and suppose that $\mu(X) < \infty$. Prove that $\lim_n \|f\|_n = \|f\|_\infty$.

6. Let $f \in \mathfrak{L}^\infty(\mu)$, and suppose that $\mu(X) < \infty$. Show that $\lim_n \frac{\|f^{n+1}\|_1}{\|f^n\|_1} = \|f\|_\infty$.

7. Show that if $p_1, \ldots, p_m > 1$ are such that $\frac{1}{p_1} + \ldots + \frac{1}{p_m} = 1$, and $f_i \in \mathfrak{L}^{p_i}$, then $f_1 \ldots f_m \in \mathfrak{L}^1$, and $\|f_1 \ldots f_m\|_1 = \|f_1\|_{p_1} \cdots \|f_m\|_{p_m}$.

8. Show that if $f \in \mathfrak{L}^{p_1}$ and $g \in \mathfrak{L}^{p_2}$, then $fg \in \mathfrak{L}^p$ for some $p$.

9. Let $f : X \to [0, \infty)$ be in $\mathfrak{L}^p$, and let $f_m = min\{f, m\}$. Prove that $f_m$ converges to $f$ in $\mathfrak{L}^p$.

10. Show that if $f_n \to f$ in $\mathfrak{L}^p$ and $g_n \to f$ in $\mathfrak{L}^q$, then $f_n g_n \to fg$ in $\mathfrak{L}^1$. Here $p$ and $q$ are conjugate Hölder exponents.

11. Let $\mu$ and $\nu$ be finite positive measures such that $\nu << \mu << \nu$. Prove that $\mathfrak{L}^\infty(\mu) = \mathfrak{L}^\infty(\nu)$.

## 8.7 Approximation

In this section, we prove a large collection of approximation theorems. The highlights include approximating $\mathfrak{L}^p$ functions by simple functions and continuous functions of compact support. We prove that trigonometric polynomials are dense in $\mathfrak{L}^2(-\pi, \pi)$, which is the last piece of information we need to settle the question of the convergence of Fourier series of functions in $\mathfrak{L}^2(-\pi, \pi)$. The important operation of convoluting functions makes its first debut in this section. Finally, we study approximations by $C^\infty$ functions, prove the $C^\infty$ version of Urysohn's lemma, and prove that $\mathcal{C}_c^\infty(\mathbb{R}^n)$ is dense in $\mathfrak{L}^p(\mathbb{R}^n)$.

**Lemma 8.7.1 (the Tietze extension theorem).** *Let K be a compact subset of $\mathbb{R}^n$ and let $f : K \to [0,1]$ be continuous. Then f can be extended to a continuous function $g \in \mathcal{C}_c(\mathbb{R}^n)$ such that $0 \le g \le 1$. If K is contained in an open set U, then g can be constructed in such a way that $supp(g) \subseteq U$.*

*Proof.* Let W be an open set such that $\overline{W}$ is compact and $K \subseteq W \subseteq \overline{W} \subseteq U$. Let $K_1 = f^{-1}([0,1/3])$, and $K_2 = f^{-1}([2/3,1])$. Applying lemma 8.4.1 to the closed sets $E = K_1 \cup (\mathbb{R}^n - W)$, and $F = K_2$, produces a continuous function $g_1 : \mathbb{R}^n \to [0,1/3]$ such that $g_1(E) = 0$, and $g_1(F) = 1/3$. By construction, $supp(g_1) \subseteq \overline{W}$, and $0 \le f - g_1 \le 2/3$ on K. Applying the same construction to the function $f - g_1$, we can find a function $g_2 : \mathbb{R}^n \to [0, \frac{1}{3} \cdot \frac{2}{3}]$ such that $supp(g_2) \subseteq \overline{W}$, and $0 \le f - g_1 - g_2 \le (\frac{2}{3})^2$ on the set K. Continuing this construction yields a sequence of continuous functions $g_i$ on $\mathbb{R}^n$ such that $supp(g_i) \subseteq \overline{W}$, $0 \le g_i \le \frac{2^{i-1}}{3^i}$, and $0 \le f - g_1 - g_2 \cdots - g_i \le (\frac{2}{3})^i$ on K. The sequence $G_i = g_1 + \ldots + g_i$ is supported in $\overline{W}$ and is Cauchy in the uniform norm on the compact set $\overline{W}$. Therefore $G_i$ converges uniformly to a continuous function g. Since $0 \le f - G_i \le (\frac{2}{3})^i$ on K, $g = f$ on K. Because each $G_i$ is supported in $\overline{W} \subseteq U$, so is g. ∎

**Remark.** The Tietze extension theorem is valid for locally compact Hausdorff spaces. See problem 1 at the end of this section.

**Proposition 8.7.2 (Egoroff's theorem).** *Let $(X, \mathfrak{M}, \mu)$ be a finite measure space. Suppose the functions f and $(f_n)_{n=1}^{\infty}$ are measurable and finite a.e. and that $\lim_n f_n(x) = f(x)$ for a.e. $x \in X$. Then, for every positive number $\delta$, there exists a measurable set E such that $\mu(E) < \delta$ and $f_n$ converges to f uniformly on $X - E$.*

*Proof.* First we show that, for every pair of positive real numbers $\epsilon$ and $\delta$, there exists a measurable set A and an integer $N \ge 1$ such that $\mu(A) < \delta$ and $\sup\{|f_k(x) - f(x)| : x \in X - A\} < \epsilon$ for every $k \ge N$. Define $C_k = \{x \in X : |f_k(x) - f(x)| < \epsilon\}$, and let $D_n = \cap_{k=n}^{\infty} C_k = \{x \in X : |f_k(x) - f(x)| < \epsilon$ for every $k \ge n\}$. Clearly, $D_1 \subseteq D_2 \subseteq \ldots$. The set $X - \cup_{n=1}^{\infty} D_n$ is contained in the set $\{x \in X : \lim_n f_n(x) \neq f(x)\}$, which, by assumption, has measure 0. It follows that $\lim_n \mu(D_n) = \mu(X)$. Therefore there exists a positive integer N such that $\mu(X - D_N) < \delta$. Set $A = X - D_N$. This proves our assertion because if $x \notin A$, then $x \in C_k$ for every $k \ge N$, and $|f_k(x) - f(x)| < \epsilon$ for every $k \ge N$.

For a fixed $\delta > 0$, and each $k \in \mathbb{N}$, let $\delta_k = \delta/2^k$. Applying the above construction to the pair $\epsilon_k = 1/k$, and $\delta_k$, we find a measurable set $A_k$ such that $\mu(A_k) < \delta/2^k$ and a positive integer $n_k$ such that $\sup\{|f_n(x) - f(x)| : x \in X - A_k\} < 1/k$ for $n > n_k$. Define $E = \cup_{k=1}^{\infty} A_k$. Then $\mu(E) \le \sum_{k=1}^{\infty} \mu(A_k) \le \sum_{k=1}^{\infty} \delta/2^k = \delta$.

Now let $\epsilon > 0$, and choose a positive integer $k$ such that $1/k < \epsilon$.
    Now, for $m > n_k = N$,

$$sup\{|f_m(x) - f(x)| : x \in X - E\} \leq sup\{|f_m(x) - f(x)| : x \in X - A_k\} < 1/k < \epsilon.$$

Thus $f_n$ converges uniformly to $f$ on $X - E$. ∎

Before we proceed to the next theorem, we call the reader's attention to the fact that $\mathfrak{L}^\infty$ contains all simple functions, while a simple function $s = \sum_{i=1}^m a_i \chi_{E_i}$ belongs to $\mathfrak{L}^p$ if and only if the support of $s$ has finite measure, that is, if $\mu(E_i) < \infty$ for all $1 \leq i \leq m$.

**Theorem 8.7.3.** *Let $(X, \mathfrak{M}, \mu)$ be a measure space. For $1 \leq p \leq \infty$, the simple functions that belong to $\mathfrak{L}^p(\mu)$ are dense in $\mathfrak{L}^p(\mu)$.*

*Proof.* We will show that if $f \in \mathfrak{L}^p$, there is a sequence of simple functions $s_n$ such that $\lim_n \|f - s_n\|_p = 0$. By theorem 8.3.2, there exists a sequence $s_1, s_2, \ldots$ of simple functions such that $|s_1| \leq |s_2| \leq \ldots \leq |f|$ and $\lim_n s_n(x) = f(x)$.
Clearly, $|f - s_n| \leq 2|f|$; hence, for $1 \leq p < \infty$, $|f - s_n|^p \leq 2^p |f|^p \in \mathfrak{L}^1$. By the dominated convergence theorem, $\lim_n \|f - s_n\|_p = 0$.
    For $p = \infty$, notice that if $n > \|f\|_\infty$, then, for a.e. $x \in X$, $0 \leq f(x) - s_n(x) \leq \frac{1}{2^n}$ (theorem 8.3.2). Clearly, $\lim_n \|f - s_n\|_\infty = 0$. ∎

**Lemma 8.7.4.** *Let $s = \sum_{i=1}^m a_i \chi_{E_i}$ be a simple function on $\mathbb{R}^n$, and let $E = \cup_{i=1}^m E_i$. It is assumed that $E_1, \ldots, E_m$ are pairwise disjoint. If $\lambda(E) < \infty$, then, for every $\epsilon > 0$, there exists a function $g \in \mathcal{C}_c(\mathbb{R}^n)$ such that $\lambda(\{x \in \mathbb{R}^n : s(x) \neq g(x)\}) < \epsilon$, and $\|g\|_\infty \leq \|s\|_\infty$.*

*Proof.* Let $U$ be an open set containing $E$ such that $\lambda(U - E) < \epsilon/2$. For each $1 \leq i \leq m$, let $K_i$ be a compact subset of $E_i$ such that $\lambda(E_i - K_i) < \frac{\epsilon}{2m}$, and set $H = \cup_{i=1}^m K_i$. Notice that $\lambda(E - H) < \epsilon/2$. For $1 \leq i \leq n$, define $V_i = U - \cup_{j \neq i} K_j$. By theorem 8.4.3, there exist functions $g_i \in \mathcal{C}_c(\mathbb{R}^n)$ such that $K_i \prec g_i \prec V_i$. Now define $g = \sum_{i=1}^m a_i g_i$. Clearly, $g \in \mathcal{C}_c(\mathbb{R}^n)$, $g|_H = s|_H$, and $g$ vanishes outside $U$. The set $\{x \in \mathbb{R}^n : s(x) \neq g(x)\}$ is contained in the union of $U - E$ and $E - H$, and the Lebesgue measure of each of the two sets is less than $\epsilon/2$. If $\|g\|_\infty > \|s\|_\infty$, we modify $g$ as follows to satisfy the last requirement of the theorem. Let $S = \{x \in \mathbb{C} : |z| \leq \|s\|_\infty\}$, and $T = \{z \in \mathbb{C} : |z| \leq \|g\|_\infty\}$. Define $\varphi : T \to S$ by

$$\varphi(z) = \begin{cases} z & \text{if } z \in S, \\ \frac{z\|s\|_\infty}{|z|} & \text{if } z \in T - S. \end{cases}$$

$\varphi$ is continuous,[12] and $|\varphi(z)| = ||s||_\infty$ for every $z \in T - S$. Now let $h = \varphi \circ g$. Clearly, $h(x) = 0$ when $g(x) = 0$, and hence $h \in \mathcal{C}_c(\mathbb{R}^n)$, and $||h||_\infty = ||s||_\infty$. ∎

Lemma 8.7.4 is a very special case of the following well-known theorem, which says, loosely speaking, that a measurable function on a set of finite Lebesgue measure is *not too far from being continuous*.

**Theorem 8.7.5 (Luzin's theorem).** *Let $f : \mathbb{R}^n \to \mathbb{C}$ be a measurable function, and suppose that there exists a set $E$ of finite Lebesgue measure such that $f(x) = 0$ for every $x \in \mathbb{R}^n - E$. Then, for every $\epsilon > 0$, there exists a function $g \in \mathcal{C}_c(\mathbb{R}^n)$ such that the set $\{x \in E : f(x) \neq g(x)\}$ has a Lebesgue measure less than $\epsilon$. Moreover, if $f$ is bounded, $g$ can be chosen in such a way that $||g||_\infty \leq ||f||_\infty$.*

*Proof.* Let $U$ be an open set such that $E \subseteq U$ and $\lambda(U - E) < \epsilon/3$. Let $s_i$ be a sequence of simple measurable functions such that $|s_1| \leq |s_2| \leq ... \leq |f|$ and $\lim_i s_i(x) = f(x)$. Since $f$ is supported in $E$, each $s_i$ is supported in $E$. By Egoroff's theorem, there exists a subset $A$ of $E$ such that $\lambda(A) < \epsilon/3$, and the sequence $s_i$ converges uniformly to $f$ on $E - A$. By the proof of lemma 8.7.4, there exist compact sets $H_i \subseteq E - A$ such that $\lambda((E - A) - H_i) < \frac{\epsilon/3}{2^i}$ and functions $g_i \in \mathcal{C}_c(\mathbb{R}^n)$ such that and $g_i|_{H_i} = s_i|_{H_i}$, and each $g_i$ vanishes outside $U$. Now let $K = \cap_{i=1}^\infty H_i$. Clearly, $K$ is compact, and $\lambda((E - A) - K) < \epsilon/3$. The sequence of continuous functions $g_i$ converges uniformly to $f$ on $K$. Thus $f|_K$ is continuous. By the Tietze extension theorem, there exists a function $g \in \mathcal{C}_c(\mathbb{R}^n)$ that extends $f|_K$ and $g(x) = 0$ for every $x \notin U$. The set $\{x \in \mathbb{R}^n : f(x) \neq g(x)\}$ is contained in the union of $U - E$, $A$, and $(E - A) - K$, and each of the three sets has Lebesgue measure less than $\epsilon/3$.

If $||f||_\infty < \infty$, and $||g||_\infty > ||f||_\infty$, we modify $g$ as in the proof of lemma 8.7.4. to satisfy the requirement $||g||_\infty \leq ||f||_\infty$. ∎

**Theorem 8.7.6.** *For $1 \leq p < \infty$, $\mathcal{C}_c(\mathbb{R}^n)$ is dense in $\mathfrak{L}^p(\mathbb{R}^n)$ for all $1 \leq p < \infty$.*

*Proof.* Let $f \in \mathfrak{L}^p(\mathbb{R}^n)$, and let $\epsilon > 0$. By lemma 8.7.3, we may assume that $f = s$, a simple function with $\lambda(\text{supp}(s)) < \infty$. Lemma 6.7.4 produces a set $A$ of measure less than $\epsilon$ and a function $g \in \mathcal{C}_c(\mathbb{R}^n)$ such that $s(x) = g(x)$ for $x \notin A$, and $||g||_\infty \leq ||s||_\infty$. Thus $|g - s| \leq |g| + |s| \leq 2||s||_\infty$. Hence $||g - s||_p^p = \int_A |g - s|^p d\mu \leq 2^p ||s||_\infty^p \lambda(A) < 2^p ||s||_\infty^p \epsilon$. ∎

**Remark.** Lemma 8.7.4 and theorems 8.7.5 and 8.7.6 are valid for Radon measures on locally compact Hausdorff spaces without any alterations to the proofs

---

[12] Observe that $\varphi$ simply fixes $S$ and retracts the annulus between the disks $T$ and $S$ radially onto the boundary of $S$.

included for the last three results. For example, in the proof of lemma 8.7.4, we only used the inner regularity of measurable sets of finite measure.

**Example 1.** Let $[a, b]$ be a compact interval in $\mathbb{R}$. For $1 \leq p < \infty$, $\mathcal{C}[a, b]$ is dense in $\mathfrak{L}^p(a, b)$.

Let $f \in \mathfrak{L}^p(a, b)$, and extend $f$ to a function $\tilde{f} \in \mathfrak{L}^p(\mathbb{R})$ by defining $\tilde{f}$ to be 0 outside $(a, b)$. By theorem 8.7.6, there exists a function $g \in \mathcal{C}_c(\mathbb{R})$ such that $\|\tilde{f} - g\|_p < \epsilon$. The restriction of $g$ to $[a, b]$ is in $\mathcal{C}[a, b]$, and $\int_a^b |f(x) - g(x)|^p \, dx \leq \int_{\mathbb{R}} |\tilde{f} - g|^p \, dx < \epsilon^p$. ◆

**Example 2.** For a function $f \in \mathcal{C}[-\pi, \pi]$ (not necessarily periodic), and for every $\epsilon > 0$, there exists a $2\pi$-periodic function $g$ such that $\|f - g\|_p < \epsilon$. Here $1 \leq p < \infty$.

The function $g$ is be obtained by modifying $f$ near $\pm\pi$ in the exact same manner as in the proof of lemma 4.10.2. ◆

**Example 3.** The space $\mathcal{C}(\mathcal{S}^1)$ is dense in $\mathfrak{L}^p(-\pi, \pi)$. Also, trigonometric polynomials are dense in $\mathfrak{L}^p(-\pi, \pi)$ for $p \in [1, \infty)$.

This result follows immediately from the last two examples and theorem 4.10.1. ◆

We conclude this subsection by proving the following separability result.

**Theorem 8.7.7.** *For $1 \leq p < \infty$, $\mathfrak{L}^p(\mathbb{R}^n)$ is separable.*

*Proof.* Let $\mathfrak{C}$ be the collection of half-open boxes of the form $\sigma = [a_1, b_1) \times \ldots \times [a_n, b_n)$, where $a_i, b_i \in \mathbb{Q}$. Define $\mathfrak{D}$ to be the collection of linear combinations of characteristic functions of members of $\mathfrak{C}$ with rational coefficients. Thus a member of $\mathfrak{D}$ is a simple function of the form $s = \sum_{i=1}^m c_i \chi_{\sigma_i}$, where $m \in \mathbb{N}$, the coefficients $c_i$ are rational numbers, and $\sigma_i \in \mathfrak{C}$. It is clear that $\mathfrak{D}$ is countable. We prove that it is dense in $\mathfrak{L}^p(\mathbb{R}^n)$. In light of theorem 8.7.6, it suffices to show that if $f \in \mathcal{C}_c^r(\mathbb{R}^n)$ and $\epsilon > 0$, then there is a function $s \in \mathfrak{D}$ such that $\|f - s\|_p < c\epsilon$ for some constant $c$, which is independent of $\epsilon$.

By the uniform continuity of $f$, there exists a number $\delta > 0$ such that $|f(x) - f(y)| < \epsilon$ whenever $\|x - y\| < \delta$. Let $Q$ be a box in $\mathfrak{C}$ that contains $\text{supp}(f)$ in its interior. Partition $Q$ into disjoint sub-boxes $\sigma_1, \ldots, \sigma_m$, where each $\sigma_i \in \mathfrak{C}$, and $\text{diam}(\sigma_i) < \delta$. For each $1 \leq i \leq m$, choose a rational number $c_i$ such that $\min_{x \in \bar{\sigma}_i} f(x) \leq c_i \leq \max_{x \in \bar{\sigma}_i} f(x)$. Finally, define $s = \sum_{i=1}^m c_i \chi_{\sigma_i}$. By construction, $\|f - s\|_\infty < \epsilon$.

Now $\|f - s\|_p^p = \int_Q |f - s|^p \, d\lambda \leq \|f - s\|_\infty^p \, \text{vol}(Q) < \epsilon^p \, \text{vol}(Q)$. ∎

## Approximation by $\mathcal{C}^\infty$ Functions

**Definition.** For Lebesgue measurable functions $f$ and $g$ on $\mathbb{R}^n$, the **convolution** of $f$ and $g$ is the function

$$(f*g)(x) = \int_{\mathbb{R}^n} f(x-y)g(y)dy.$$

It is clear that if $(f*g)(x)$ is finite, then $(f*g)(x) = (g*f)(x)$. Thus

$$(f*g)(x) = \int_{\mathbb{R}^n} f(x-y)g(y)dy = \int_{\mathbb{R}^n} f(y)g(x-y)dy.$$

A variety of conditions can be imposed on $f$ and $g$ to guarantee the finiteness of the integral, at least for a.e. $x \in \mathbb{R}^n$. We take for granted the measurability of the function $f(x-y)g(y)$.

In this subsection, we will limit the functions $f$ and $g$ to be continuous functions of compact support. The reader can look at the section exercises for a slightly expanded discussion of the properties of convolutions.

**Lemma 8.7.8.** Let $f, g \in \mathcal{C}_c(\mathbb{R}^n)$. Then

(a) $(f*g)(x)$ exists for all $x \in \mathbb{R}^n$, and
(b) $f*g \in \mathcal{C}_c(\mathbb{R}^n)$.

*Proof.* (a) Let $K = supp(g)$. Then

$$|(f*g)(x)| \le \int_{\mathbb{R}^n} |f(x-y)g(y)|dy \le \|f\|_\infty \int_{\mathbb{R}^n} |g(y)|dy \le \|f\|_\infty \|g\|_\infty \lambda(K) < \infty.$$

(b) Let $F$ be the closure of the (bounded) set $\{x+y : x \in supp(f), y \in supp(g)\}$. We claim that $f*g$ is supported inside $F$. If $x \notin F$, then, for every $y \in supp(g)$, $x-y \notin supp(f)$. Thus $f(x-y)g(y) = 0$ for all $y \in \mathbb{R}^n$; hence $(f*g)(x) = 0$.

Let $\epsilon > 0$. Since $f$ is uniformly continuous, there exists a number $\delta > 0$ such that, for $\xi, \eta \in \mathbb{R}^n$, $|f(\xi) - f(\eta)| < \epsilon$ whenever $\|\xi - \eta\| < \delta$. Now, for such $\xi$ and $\eta$,

$$|(f*g)(\xi) - (f*g)(\eta)| \le \int_{\mathbb{R}^n} |f(\xi-y) - f(\eta-y)||g(y)|dy \le \epsilon \int_K |g(y)|dy$$

$$\le \epsilon \|g\|_\infty \lambda(K). \ \blacksquare$$

**Lemma 8.7.9.** Let $f \in \mathcal{L}^p(\mathbb{R}^n)$, where $1 \le p < \infty$. Then $\lim_{a\to 0} \|\tau_a f - f\|_p = 0$.

*Proof.* Recall that $(\tau_a f)(x) = f(x - a)$. First we prove the result for a function $g \in \mathcal{C}_c(\mathbb{R}^n)$. Without loss of generality, assume that $\|a\| < 1$. Thus the functions $\tau_a g$ have a common support, say, $K$. Let $\epsilon > 0$. By the uniform continuity of $g$, there exists a number $\delta > 0$ such that whenever $\|a\| < \delta$, then $|(\tau_a g)(x) - g(x)| = |g(x - a) - g(x)| < \epsilon$. Now $\|\tau_a g - g\|_p^p = \int_K |g(x - a) - g(x)|^p\, dx \leq \epsilon^p \lambda(K)$.

Now let $f \in \mathfrak{L}^p(\mathbb{R}^n)$, and let $\epsilon > 0$. By theorem 8.7.6, there is a function $g \in \mathcal{C}_c(\mathbb{R}^n)$ such that $\|f - g\|_p < \epsilon/3$. By the first part of the proof, there is $\delta > 0$ such that for $\|a\| < \delta$, $\|\tau_a g - g\|_p < \epsilon/3$. Now if $\|a\| < \delta$, then

$$\|f - \tau_a f\|_p \leq \|f - g\|_p + \|g - \tau_a g\|_p + \|\tau_a g - \tau_a f\|_p$$
$$= \|f - g\|_p + \|g - \tau_a g\|_p + \|g - f\|_p < \epsilon. \quad \blacksquare$$

**Definition.** A multi-index $\alpha$ is a sequence $\alpha = (\alpha_1, \ldots, \alpha_n)$, where each $\alpha_i$ is a nonnegative integer. The length of $\alpha$ is the integer $|\alpha| = \sum_{i=1}^{n} \alpha_i$.

**Notation.** Let $f$ be a scalar-valued function on $\mathbb{R}^n$, and let $\alpha = (\alpha_1, \ldots, \alpha_n)$ be a multi-index. The notation $D^\alpha f$ stands for the derivative $\frac{\partial^{|\alpha|} f}{\partial x_1^{\alpha_1} \partial x_2^{\alpha_2} \ldots \partial x_n^{\alpha_n}}$, if it exists. For example, if $n = 5$, and $\alpha = (1, 0, 2, 0, 1)$, then $D^\alpha f = \frac{\partial^4 f}{\partial x_1 \partial x_3^2 \partial x_5}$.

**Definition.** A function $f$ is said to be infinitely differentiable if $D^\alpha f$ exists for every multi-index $\alpha$. The space of infinitely differentiable functions is denoted by $\mathcal{C}^\infty(\mathbb{R}^n)$, and the space of infinitely differentiable functions of compact support is given the symbol $\mathcal{C}_c^\infty(\mathbb{R}^n)$. We will shortly see that there is an abundance of such functions.

**Example 4.** Consider the function

$$f(x) = \begin{cases} exp\{-1/x\} & \text{if } x > 0, \\ 0 & \text{if } x \leq 0. \end{cases}$$

It is easily seen that, for $x > 0$, and $k \in \mathbb{N}$, $f^{(k)}(x) = p(1/x)\, exp\{-1/x\}$, where $p$ is a polynomial of degree $2k$. Therefore $\lim_{x \downarrow 0} f^{(k)}(x) = 0$. Hence $f^{(k)}(0) = 0$, and $f$ is infinitely differentiable at $x = 0$. Since the differentiability of $f$ at $x \neq 0$ is obvious, $f \in \mathcal{C}^\infty(\mathbb{R})$. ♦

**Example 5** (*the* bump function). For a fixed $h > 0$, consider the function

$$\varphi(x) = \begin{cases} exp\{\frac{-h^2}{h^2 - |x|^2}\} & \text{if } |x| < h, \\ 0 & \text{if } |x| \geq h. \end{cases}$$

As $|x| \uparrow h$, $h^2 - |x|^2 \downarrow 0$, so, by example 1, $\varphi \in \mathcal{C}_c^\infty(\mathbb{R})$. ♦

**Example 6** (*the* bump kernel). We can use the function $\varphi$ in example 5 to construct a continuously parameterized family of functions as follows. For a fixed $h > 0$,

$$\delta_h(x) = \begin{cases} A_n h^{-n} exp\{\frac{-h^2}{h^2 - \|x\|^2}\} & \text{if } \|x\| < h, \\ 0 & \text{if } \|x\| \geq h, \end{cases}$$

where $A_n^{-1} = \int_{\|y\|<1} exp\{\frac{-1}{1-\|y\|^2}\}dy$. By the above examples, $\delta_h \in C_c^\infty(\mathbb{R}^n)$. ♦

Observe that $max_{x \in \mathbb{R}^n} \delta_h(x) = \delta_h(0) = A_n h^{-n}/e$, and $\int_{\mathbb{R}^n} \delta_h(x)dx = 1$.

The first assertion is obvious. For the second assertion, use the change of variable $x = hy$. By problem 14 on section 8.4, $dx = h^n dy$, and

$$\int_{\mathbb{R}^n} \delta_h(x)dx = \int_{\|x\|<h} A_n h^{-n} exp\left\{\frac{-h^2}{h^2 - \|x\|^2}\right\} dx$$

$$= A_n h^{-n} \int_{\|y\|<1} exp\left\{\frac{-1}{1 - \|y\|^2}\right\} h^n dy = 1.$$

We call the family $\{\delta_h : h > 0\}$ the **bump kernel**.

**Lemma 8.7.10.** *If $f \in C_c^\infty(\mathbb{R}^n)$, and $g \in C_c(\mathbb{R}^n)$, then $f * g \in C_c^\infty(\mathbb{R}^n)$, and, for every multi-index $\alpha$, $D^\alpha(f * g) = D^\alpha f * g$.*

*Proof. The proof is by induction on $|\alpha|$. It is sufficient to prove the result when $|\alpha| = 1$. Thus we need to show that $\frac{\partial}{\partial x_i}(f * g) = \frac{\partial f}{\partial x_i} * g$. For simplicity of notation, fix $x_1, \ldots, x_{i-1}, x_{i+1}, \ldots, x_n$, and consider $f$ as a function of the single variable $x_i$, which we rename $x$. Thus we need to prove that $\frac{d}{dx}(f * g) = f' * g$.*

*We will show that $\lim_{t \to 0} \frac{(f*g)(x+t) - (f*g)(x)}{t} - (f' * g)(x) = 0$.*

*Let $\epsilon > 0$. By the uniform continuity of $f'$, there is $\delta > 0$ such that $|f'(\xi) - f'(\eta)| < \epsilon$ whenever $|\xi - \eta| < \delta$. Now $|\frac{(f*g)(x+t) - (f*g)(x)}{t} - (f' * g)(x)| = |\int_{\mathbb{R}} \{\frac{f(x+t-y) - f(x-y)}{t} - f'(x-y)\}g(y)dy| = |\int_{\mathbb{R}} \{f'(x+\theta t - y) - f'(x-y)\}g(y)dy|$, where $0 < \theta < 1$. Now if $|t| < \delta$, then $|f'(x + \theta t - y) - f'(x - y)| < \epsilon$ and $|\int_{\mathbb{R}} \{f'(x + \theta t - y) - f'(x - y)\}g(y)dy| \leq \epsilon \int_{\mathbb{R}} |g(y)|dy \leq \epsilon \|g\|_\infty \lambda(K)$, where $K = supp(g)$.* ∎

As a corollary of the last result, for every $f \in C_c(\mathbb{R}^n)$, $f * \delta_h \in C_c^\infty(\mathbb{R}^n)$.

The following is the $C^\infty$ version of Urysohn's lemma.

**Corollary 8.7.11.** *Let $K$ be a compact subset of an open subset $V$ of $\mathbb{R}^n$. Then there exists a function $f \in \mathcal{C}_c^\infty(\mathbb{R}^n)$ such that $K < f < V$.*

*Proof.* Let $\delta = \text{dist}(K, \mathbb{R}^n - V)$. Since $K$ is compact, $\delta$ is positive. Define $K_1 = \{x \in \mathbb{R}^n : \text{dist}(x, K) \leq \delta/4\}$, and $V_1 = \{x \in \mathbb{R}^n : \text{dist}(x, K) < \delta/2\}$. Since $K_1$ is compact, $V_1$ is open, and $K_1 \subseteq V_1$, theorem 8.4.3 produces a function $g$ such that $K_1 < g < V_1$. Now choose a number $h < \delta/4$, and define $f = g * \delta_h$. By lemma 8.7.8, $\text{supp}(f) \subseteq \{x \in \mathbb{R}^n : \text{dist}(x, K) \leq 3\delta/4\} \subseteq V$.

Clearly, $f(x) \geq 0$. Since $0 \leq g(x) \leq 1$, $f(x) = \int_{\mathbb{R}^n} g(x-y)\delta_h(y)dy \leq \int_{\mathbb{R}^n} \delta_h(y)dy = 1$.

It remain to show that $K < f$. If $x \in K$, then $B(x, h) \subseteq B(x, \delta/4) \subseteq K_1$; hence $f(x) = \int_{\|x-y\|<h} g(y)\delta_h(x-y)dy = \int_{\|x-y\|<h} \delta_h(x-y)dy = 1$. ∎

**Proposition 8.7.12.** *If $f \in \mathcal{C}_c(\mathbb{R}^n)$, then, for $1 \leq p < \infty$, $f * \delta_h \to f$ in $\mathfrak{L}^p(\mathbb{R}^n)$.*

*Proof.* We will make use of the fact that $\int_{\mathbb{R}^n} \delta_h(y)dy = 1$ and example 4 on section 8.6:

$$|(f * \delta_h)(x) - f(x)| = \left| \int_{\mathbb{R}^n} (f(x-y) - f(x))\delta_h(y)dy \right|$$

$$\leq \int_{\mathbb{R}^n} |f(x-y) - f(x)|\delta_h(y)dy$$

$$\leq \left( \int_{\mathbb{R}^n} |f(x-y) - f(x)|^p \delta_h(y)dy \right)^{1/p}.$$

Integrating the $p^{th}$ power of the extreme sides of the above string, we have

$$\|f * \delta_h - f\|_p^p \leq \int_{\mathbb{R}^n} \int_{\mathbb{R}^n} |f(x-y) - f(x)|^p \delta_h(y)dydx$$

$$= \int_{\mathbb{R}^n} \delta_h(y) \int_{\mathbb{R}^n} |f(x-y) - f(x)|^p dxdy$$

$$= \int_{\mathbb{R}^n} \|\tau_y f - f\|_p^p \delta_h(y)dy = \int_{\|y\|\leq h} \|\tau_y f - f\|_p^p \delta_h(y)dy.^{13}$$

By lemma 8.7.9, there exists a number $h_0 > 0$ such that, for $\|y\| < h_0$, $\|\tau_y f - f\|_p < \epsilon$. Hence, for $h < h_0$, $\int_{\|y\|<h} \|\tau_y f - f\|_p^p \delta_h(y)dy \leq \epsilon^p \int_{\|y\|\leq h} \delta_h(y)dy = \epsilon^p$. ∎

Part (a) of the following result is a vast generalization of theorem 8.7.6.

**Theorem 8.7.13.** *(a) For $1 \leq p < \infty$, $C_c^\infty(\mathbb{R}^n)$ is dense in $\mathfrak{L}^p(\mathbb{R}^n)$.*
*(b) $\mathcal{C}_c^\infty(\mathbb{R}^n)$ is dense in $\mathcal{C}_0(\mathbb{R}^n)$.*

---

[13] Fubini's theorem is used below to switch the order of integration.

*Proof.* (a) Let $f \in \mathfrak{L}^p(\mathbb{R}^n)$, and let $\epsilon > 0$. By theorem 8.7.6, there is a function $g \in \mathcal{C}_c(\mathbb{R}^n)$ such that $\|f - g\|_p < \epsilon/2$. By the previous proposition, we can choose $h > 0$ small enough so that $\|g - g * \delta_h\|_p < \epsilon/2$. The function $g * \delta_h$ is in $\mathcal{C}_c^\infty(\mathbb{R}^n)$, and $\|f - g * \delta_h\|_p < \epsilon$.

(b) Let $f \in \mathcal{C}_0(\mathbb{R}^n)$, and let $\epsilon > 0$. By theorem 5.11.8, there exits a function $g \in \mathcal{C}_c(\mathbb{R}^n)$ such that $\|f - g\|_\infty < \epsilon$. By the uniform continuity of $g$, there is a number $\delta > 0$ such that $|g(x) - g(y)| < \epsilon$ whenever $\|x - y\| < \delta$. Choose a positive number $h < \delta$. The proof will be complete if we show that $\|g * \delta_h - g\|_\infty < \epsilon$. Since $\int_{\mathbb{R}^n} \delta_h(x - y) dy = 1$,

$$\begin{aligned} |g * \delta_h(x) - g(x)| &= \left| \int_{\mathbb{R}^n} \{g(y) - g(x)\} \delta_h(x - y) dy \right| \\ &\leq \int_{\|x-y\|<\delta} |g(x) - g(y)| \delta_h(x - y) dy \\ &< \epsilon \int_{\|x-y\|<h} \delta_h(x - y) dy = \epsilon. \blacksquare \end{aligned}$$

## Exercises

1. Let $K$ be a compact subset of a locally compact Hausdorff space $X$, and let $f : K \to [0,1]$ be continuous. Show $f$ can be extended to a continuous function $g \in \mathcal{C}_c(X)$ such that $0 \leq g \leq 1$. If $K$ is contained in an open set $U$, then $g$ can be constructed in such a way that $supp(g) \subseteq U$. Hint: Mimic the proof of lemma 8.7.1, and use theorem 5.11.5.

2. Let $F$ be a closed subset of normal space $X$, and let $f : F \to [0,1]$ be continuous. Show $f$ can be extended to a continuous function $g \in \mathcal{C}(X)$ such that $0 \leq g \leq 1$. If $F$ is contained in an open set $U$, then $g$ can be constructed in such a way that $supp(g) \subseteq U$. Hint: Modify the proof of lemma 8.7.1, and use theorem 5.11.2. In this case, convergence of the functions $G_i$ takes place in the space $\mathcal{BC}(X)$.

3. Let $\mu$ be a $\sigma$-finite measure on $X$, let $(f_n)$ be a sequence of measurable functions, and suppose that $\lim_n f_n(x) = f(x)$. Prove that there exists a sequence $(E_j)$ of measurable sets such that $f_n$ converges uniformly to $f$ on each $E_j$ and $\mu(X - \cup_{j=1}^\infty E_j) = 0$.

4. Let $f \in \mathfrak{L}^p(\mathbb{R}^n)$, where $1 \leq p < \infty$. Show that the mapping $\mathbb{R}^n \to \mathfrak{L}^p(\mathbb{R}^n)$ defined by $a \to \tau_a f$ is uniformly continuous. Observe that, in lemma 8.7.9, we established the continuity at $a = 0$.

5. Assuming that $(f * g)(x)$ exists for every $x \in \mathbb{R}^n$, prove that
   (a) $f * g = g * f$, and
   (b) $\tau_a(f * g) = (\tau_a f) * g = f * (\tau_a g)$.

6. Let $f \in \mathfrak{L}^p(\mathbb{R}^n)$ and $g \in \mathfrak{L}^q(\mathbb{R}^n)$, where $p$ and $q$ are conjugate exponents. Prove that $f * g \in \mathfrak{L}^\infty(\mathbb{R}^n)$ and that $\|f * g\|_\infty \le \|f\|_p \|g\|_q$.

7. This problem is a continuation of the previous exercise. Prove that if $1 \le p < \infty$, then $f * g$ is uniformly continuous.

8. This problem is a continuation of exercise 6. Show that if $1 < p < \infty$, then $f * g \in \mathcal{C}_0(\mathbb{R}^n)$.

9. Prove that $\dim \mathcal{C}_c^\infty(\mathbb{R}^n) = \infty$.

## 8.8 Product Measures

Throughout this section, $(X, \mathfrak{M}, \mu)$ and $(Y, \mathfrak{N}, \nu)$ denote a pair of measure spaces. The objective of this section is to find a reasonable definition of the product measure on $X \times Y$. Fubini's theorem is one of the section's main results. We also settle questions about the products of Lebesgue measures in this section.

The basic definitions are motivated by the ideas found in standard calculus textbooks. Let us look at the simplest case, which is the product of two copies of the real line with Lebesgue measure, $\lambda$. The problem of computing the area of a plane region contains all the motivations for the ideas behind the definitions in this section. Figure 8.6 depicts a (bounded) plane region $E$ in $\mathbb{R}^2$. To compute the area of $E$, we take a vertical cross section $S_x$ in $E$, and the area of $E$ is obtained by integrating the length (the Lebesgue measure) of the cross section. The same can be achieved by taking a horizontal cross section $S^y$ in $E$. Thus the area (two-dimensional measure) of $E$, denoted $\rho(E)$, is given by

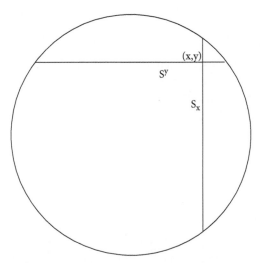

**Figure 8.6** Computing the area of a plane region

$$\rho(E) = \int_{\mathbb{R}} \lambda(S_x)dx = \int_{\mathbb{R}} \lambda(S^y)dy.$$

We also wish the two-dimensional measure $\rho$ to preserve the property that the area of a rectangle is the product of its dimensions. More generally, if $A$ and $B$ are measurable subsets of $\mathbb{R}$, then it should be the case that

$$\rho(A \times B) = \lambda(A)\lambda(B).$$

Now see theorem 8.8.9, where the definition of the product measure appears.

Before we can achieve any of the above goals, we need to define a reasonable $\sigma$-algebra in $X \times Y$ where our expectations can materialize. Geometry dictates that the product of two intervals (or, more generally, measurable subsets) $A$ and $B$ in $\mathbb{R}$ ought to be measurable in the product space. This immediately suggests that we look at the smallest $\sigma$-algebra that contains all rectangles, and this provides the motivation of the definitions below of the product of measurable spaces.

## Products of Measurable Spaces

**Definition.** A subset of $X \times Y$ of the form $A \times B$, where $A \in \mathfrak{M}, B \in \mathfrak{N}$ is called a **measurable rectangle** in $X \times Y$.

**Definition.** The **product of the measurable spaces** $(X, \mathfrak{M})$ and $(Y, \mathfrak{N})$ is the measurable space $(X \times Y, \mathfrak{M} \otimes \mathfrak{N})$, where $\mathfrak{M} \otimes \mathfrak{N}$ is the $\sigma$-algebra generated by the collection of measurable rectangles.

**Definition.** For a subset $E \subseteq X \times Y$, and for a fixed element $x \in X$, we define the $x$-**section** of $E$ to be the set $E_x = \{y \in Y : (x,y) \in E\}$. Similarly, for $y \in Y$, the $y$-**section** of $E$ is the set $E^y = \{x \in X : (x,y) \in E\}$.

The following lemma will be used without explicit reference. Its proof is simple.

**Lemma 8.8.1.** *If* $(E_n)$ *is a sequence of subsets of* $X \times Y$, *then*

$$(\cup_n E_n)_x = \cup_n (E_n)_x, \text{ and } (\cap_n E_n)_x = \cap_n (E_n)_x.$$

*The corresponding statements for y-sections are true.* ∎

**Proposition 8.8.2.** *If* $E \in \mathfrak{M} \otimes \mathfrak{N}$, *then, for every* $x \in X$, $E_x \in \mathfrak{N}$. *Likewise, for every* $y \in Y$, $E^y \in \mathfrak{M}$.

*Proof.* Let $\Omega = \{E \subseteq X \times Y : E_x \in \mathfrak{N} \; \forall x \in X\}$. *Clearly,* $\Omega$ *contains all elementary rectangles because if* $A \in \mathfrak{M}$ *and* $B \in \mathfrak{N}$, *then*

$$(A \times B)_x = \begin{cases} B & \text{if } x \in A, \\ \varnothing & \text{if } x \notin A. \end{cases}$$

*If* $E \in \Omega$, $(E')_x = (E_x)' \in \mathfrak{N}$; *hence* $E' \in \Omega$. *Here* $E'$ *denotes the complement of* $E$ *in* $X \times Y$.

*Finally, if for every* $n \in \mathbb{N}$, $E_n \in \Omega$, $(\cup_{n=1}^{\infty} E_n)_x = \cup_{n=1}^{\infty} (E_n)_x \in \mathfrak{N}$. *Therefore* $\cup_{n=1}^{\infty} E_n \in \Omega$.

*The above shows that* $\Omega$ *is a* $\sigma$-*algebra that contains all measurable rectangles; hence* $\Omega \supseteq \mathfrak{M} \otimes \mathfrak{N}$. *The proof that* $E^y \in \mathfrak{M}$ *for every* $y \in Y$ *is identical to the above case.* ∎

**Definition.** Let $f$ be a scalar function on $X \times Y$. For an element $x \in X$, the $x$-**section** of $f$ is the function $f_x : Y \to \mathbb{C}$ defined by $f_x(y) = f(x, y)$. Similarly, the $y$-**section** of $f$ is the function $f^y : X \to \mathbb{C}$ such that $f^y(x) = f(x, y)$.

**Proposition 8.8.3.** *If* $f : X \times Y \to \mathbb{C}$ *is* $\mathfrak{M} \otimes \mathfrak{N}$-*measurable, then, for every* $x \in X$, $f_x$ *is* $\mathfrak{N}$-*measurable, and, for every* $y \in Y$, $f^y$ *is* $\mathfrak{M}$-*measurable.*

*Proof.* Let $a \in \mathbb{R}$, and let $E = f^{-1}(a, \infty) = \{(x, y) \in X \times Y : f(x, y) > a\}$. Now $f_x^{-1}(a, \infty)$ is exactly the set $E_x$, which is measurable by the previous proposition. Thus $f_x$ is $\mathfrak{N}$-measurable. ∎

**Definition.** An **elementary set** in $X \times Y$ is a disjoint union of finitely many measurable rectangles. The collection of elementary sets will be given the symbol $\mathfrak{E}$.

It is clear that the collection of elementary sets also generates $\mathfrak{M} \otimes \mathfrak{N}$.

**Proposition 8.8.4.** *The collection* $\mathfrak{E}$ *of elementary sets is an algebra.*

*Proof.* It is clear that the intersection of two measurable rectangles is either empty or a measurable rectangle. Also $(A \times B)' = (A' \times Y) \cup (A \times B')$, so the complement of a measurable rectangle is an elementary set.

Let $E = \cup_{i=1}^{n} R_i$ and $F = \cup_{j=1}^{m} S_j$ be elementary sets, where each of $\{R_i\}$ and $\{S_j\}$ is a set of disjoint measurable rectangles. Now $E \cap F = \cup \{R_i \cap S_j : 1 \le i \le n, 1 \le j \le m\}$. This shows that $E \cap F \in \mathfrak{E}$, and that $\mathfrak{E}$ is closed under the formation of finite intersections

Now consider the complement of an elementary set $E = \cup_{i=1}^{n} R_i$. Since $E' = \cap_{i=1}^{n} R_i'$, $E' \in \mathfrak{E}$. Thus $\mathfrak{E}$ is closed under complementation. It is also clear that $\mathfrak{E}$ is closed under the formation of finite disjoint unions.

Now if $E_1, E_2 \in \mathfrak{E}$, then, by the above, $E_1' \cap E_2 \in \mathfrak{E}$; hence $E_1 \cup E_2 = E_1 \cup (E_1' \cap E_2) \in \mathfrak{E}$. ∎

**Definition.** Let $T$ be a nonempty set. A **monotone class in** $T$ is a collection $\mathfrak{C}$ of subsets of $T$ such that

(a) if $(E_n)$ is an ascending sequence in $\mathfrak{C}$, then $\cup_{n=1}^{\infty} E_n \in \mathfrak{C}$; and
(b) if $(E_n)$ is a descending sequence in $\mathfrak{C}$, then $\cap_{n=1}^{\infty} E_n \in \mathfrak{C}$.

**Proposition 8.8.5.** *Given an arbitrary collection $\mathcal{E}$ of subsets of a nonempty set $T$, there exists a (unique) smallest monotone class in $T$ that contains $\mathcal{E}$.*

*Proof.* The intersection of an arbitrary collection of monotone classes in $T$ is clearly a monotone class. The family of monotone classes containing $\mathcal{E}$ is nonempty since $\mathcal{P}(T)$ is such a monotone class. The intersection of the monotone classes in $T$ that contain $\mathcal{E}$ is the monotone class we seek. ∎

**Lemma 8.8.6 (the monotone class lemma).** *Let $\mathfrak{E}$ be an algebra of subsets in a nonempty set $T$. Then the smallest monotone class in $T$ containing $\mathfrak{E}$ is the $\sigma$-algebra generated by $\mathfrak{E}$. In particular, if an algebra in $T$ is a monotone class, then it is a $\sigma$-algebra.*

*Proof.* Let $\mathfrak{M}$ be the $\sigma$-algebra generated by $\mathfrak{E}$, and let $\mathfrak{M}_1$ be the smallest monotone class in $T$ containing $\mathfrak{E}$. Since $\mathfrak{M}$ is a monotone class containing $\mathfrak{E}$, $\mathfrak{M}_1 \subseteq \mathfrak{M}$. Thus we need to establish the reverse inclusion. It is clearly sufficient to show that $\mathfrak{M}_1$ is a $\sigma$-algebra.

We first show that $\mathfrak{M}_1$ is an algebra. Let $\mathfrak{M}_1' = \{E \subseteq T : E' \in \mathfrak{M}_1\}$. It is clear that $\mathfrak{M}_1'$ is a monotone class in $T$ and that $\mathfrak{E} \subseteq \mathfrak{M}_1'$. Thus $\mathfrak{M}_1 \subseteq \mathfrak{M}_1'$; hence $\mathfrak{M}_1$ is closed under complementation.

For a member $F \in \mathfrak{M}_1$, define $\Omega(F) = \{E \in \mathfrak{M}_1 : E \cup F \in \mathfrak{M}_1\}$. It is easy to verify that $\Omega(F)$ is a monotone class in $X$. Now if $G \in \mathfrak{E}$, then $\Omega(G)$ contains $\mathfrak{E}$, so $\Omega(G) = \mathfrak{M}_1$. Hence, for any $H \in \mathfrak{M}_1$, $H \in \Omega(G)$. By the very definition of $\Omega(H)$, $G \in \Omega(H)$, so $\mathfrak{E} \subseteq \Omega(H)$ for each $H \in \mathfrak{M}_1$. Because $\Omega(H)$ is a monotone class, $\mathfrak{M}_1 = \Omega(H)$, so $\mathfrak{M}_1$ is an algebra.

Now if $(E_n)$ is a sequence of members of $\mathfrak{M}_1$, let $B_n = \cup_{i=1}^{n} E_i$. Because $\mathfrak{M}_1$ is an algebra, each $B_n \in \mathfrak{M}_1$. Since $\mathfrak{M}_1$ is a monotone class, it follows that $\cup_{n=1}^{\infty} E_n = \cup_{n=1}^{\infty} B_n$ is in $\mathfrak{M}_1$. This shows that $\mathfrak{M}_1$ is a $\sigma$-algebra, and the proof is complete. ∎

The following result is immediate.

**Corollary 8.8.7.** *The $\sigma$-algebra $\mathfrak{M} \otimes \mathfrak{N}$ is the smallest monotone class that contains the algebra $\mathfrak{E}$ of elementary sets.* ∎

## Product Measures

**Theorem 8.8.8.** *Suppose $(X, \mathfrak{M}, \mu)$ and $(Y, \mathfrak{N}, \nu)$ are $\sigma$-finite measure spaces. For a subset $E \in \mathfrak{M} \otimes \mathfrak{N}$, and for $x \in X, y \in Y$, define*

$$\varphi(x) = \nu(E_x), \text{ and } \psi(y) = \mu(E^y).$$

*Then*

*(i) $\varphi$ is $\mathfrak{M}$-measurable,*
*(ii) $\psi$ is $\mathfrak{N}$-measurable, and*
*(iii) $\int_X \varphi d\mu = \int_Y \psi d\nu$.*

*Proof.* Let $\Omega$ be the collection of members of $\mathfrak{M} \otimes \mathfrak{N}$ for which all three conclusions of the theorem hold. We will show that $\Omega = \mathfrak{M} \otimes \mathfrak{N}$.

*First, we establish a number of facts.*

*(a) $\Omega$ contains all elementary sets.*
   If $E = A \times B$ is a measurable rectangle, then $\nu((A \times B)_x) = \chi_A(x)\nu(B)$, and $\mu((A \times B)^y) = \chi_B(y)\mu(A)$, are measurable;[14] hence $\int_X \nu(E_x)d\mu = \int_X \nu(B)\chi_A \, d\mu = \nu(B)\mu(A) = \int_Y \mu(A)\chi_B d\nu = \int_Y \mu(E^y)d\nu$. Now the result is true for elementary sets because of the additivity of measures and the linearity of integrals.

*(b) If $E_n \in \Omega$ and $E_1 \subseteq E_2 \subseteq \dots$, then $E = \cup_{n=1}^\infty E_n \in \Omega$.*
   Write $\varphi_n(x) = \nu((E_n)_x)$, $\psi_n(y) = \mu((E_n)^y)$, $\varphi(x) = \nu(E_x)$, and $\psi(y) = \mu(E^y)$. Now $\varphi_n(x)$ increases to $\varphi(x) = \nu(E_x)$, and $\psi_n(y)$ increases to $\psi(y) = \mu(E^y)$. By assumption, $\varphi_n$ and $\psi_n$ are measurable, so $\varphi$ and $\psi$ are measurable. Also by assumption, $\int_X \varphi_n d\mu = \int_X \psi_n d\nu$. By the monotone convergence theorem, conclusion (iii) holds for the set $E$.

*(c) If $E_1 \supseteq E_2 \supseteq \dots$ is a sequence in $\Omega$ and if $E_1 \subseteq A \times B$ for some measurable rectangle $A \times B$ with $\mu(A) < \infty$ and $\nu(B) < \infty$, then $E = \cap_{n=1}^\infty E_n \in \Omega$.*
   In the notation of the proof of fact (b), $\varphi_n$ decreases to $\varphi$, and $\psi_n$ decreases to $\psi$. Thus $\varphi$ and $\psi$ are measurable. Since $(E_1)_x \subseteq (A \times B)_x$, $\nu((E_1)_x) \leq \nu((A \times B)_x) = \nu(B)\chi_A(x)$. Therefore $\int_Y \varphi_1 d\mu = \int_X \nu((E_1)_x)d\mu \leq \int_X \nu(B)\chi_A d\mu = \mu(A)\nu(B) < \infty$. Similarly, $\int_Y \psi d\nu < \infty$. By assumption, $\int_X \varphi_n d\mu = \int_Y \psi_n d\nu$. Fact (c) now follows from the dominated convergence theorem.

---

[14] Recall that characteristic functions of measurable sets are measurable functions.

(d) If $(E_n)$ is a disjoint sequence in $\Omega$, then $E = \cup_{n=1}^{\infty} E_n \in \Omega$.

For each $n \in \mathbb{N}$, the set $\cup_{i=1}^{n} E_i$ is in $\Omega$ (see the proof of fact (a)). Now (d) follows from (b) applied to the ascending sequence $(\cup_{i=1}^{n} E_i)_{n=1}^{\infty}$.

Now we use the $\sigma$-finiteness assumption to write $X$ as the disjoint union of subsets $X_n$ of finite $\mu$-measure, and $Y$ as the disjoint union of subsets $Y_m$ of finite $\nu$ measure. For a member $E$ of $\mathfrak{M} \otimes \mathfrak{N}$, define $E_{m,n} = E \cap (X_n \times Y_m)$, and let $\Omega_1$ be the collection of all members $E$ of $\mathfrak{M} \otimes \mathfrak{N}$ such that, for all $m, n \in \mathbb{N}$, $E_{m,n} \in \Omega$. Facts (b) and (c) imply that $\Omega_1$ is a monotone class, and fact (a) implies that $\Omega_1$ contains all elementary sets. Thus $\Omega_1 = \mathfrak{M} \otimes \mathfrak{N}$ by corollary 8.8.7.

Thus $E_{m,n} \in \Omega$ for every $E \in \mathfrak{M} \otimes \mathfrak{N}$ and for all $m, n \in \mathbb{N}$. Since $E = \cup_{m,n} E_{m,n}$ and the sets $E_{m,n}$ are disjoint, fact (d) implies that $E \in \Omega$. ∎

Observe that conclusion (iii) of theorem 8.8.8 can be written as

$$\int_X \left\{ \int_Y \chi_E(x,y) d\nu(y) \right\} d\mu(x) = \int_Y \left\{ \int_X \chi_E(x,y) d\mu(x) \right\} d\nu(y).$$

Thus the order of integration can be switched in iterated integrals of characteristic functions of $\mathfrak{M} \otimes \mathfrak{N}$-measurable sets. This is clearly the first step to prove Fubini's theorem. First we need to define the product measure of two $\sigma$-finite measure spaces.

**Theorem 8.8.9.** Under the assumptions of theorem 8.8.8, the set function defined by

$$(\mu \otimes \nu)(E) = \int_X \varphi d\mu = \int_Y \psi d\nu$$

is the unique positive measure on $\mathfrak{M} \otimes \mathfrak{N}$ such that $(\mu \otimes \nu)(A \times B) = \mu(A)\nu(B)$ for all measurable rectangles $A \times B$. Furthermore, $\mu \otimes \nu$ is $\sigma$-finite. The measure $\mu \otimes \nu$ is the **product of the measures** $\mu$ and $\nu$.

Proof. Let $(E_n)$ be a disjoint sequence of $\mathfrak{M} \otimes \mathfrak{N}$-measurable subsets of $X \times Y$, and let $E = \cup_{n=1}^{\infty} E_n$. Since $E_x = \cup_{n=1}^{\infty} (E_n)_x$ and since the sequence $((E_n)_x)$ is disjoint, $\nu(E_x) = \sum_{n=1}^{\infty} \nu((E_n)_x)$. An application of the monotone convergence theorem yields

$$(\mu \otimes \nu)(E) = \int_X \sum_{n=1}^{\infty} \nu((E_n)_x) d\mu = \sum_{n=1}^{\infty} \int_X \nu((E_n)_x) d\mu = \sum_{n=1}^{\infty} (\mu \otimes \nu)(E_n).$$

The $\sigma$-finiteness of $\mu \otimes \nu$ is obvious. We leave the proof of the uniqueness part as an exercise. ∎

**Remark.** Both the existence and uniqueness of the product measure of $\sigma$-finite spaces can be based on the Hopf extension theorem. For a measurable rectangle $A \times B$ in $\mathfrak{M} \times \mathfrak{N}$, we define $\rho(A \times B) = \mu(A)\nu(B)$, and, for an elementary set $C = \cup_{i=1}^{n} A_i \times B_i$, we define $\rho(C) = \sum_{i=1}^{n} \mu(A_i)\nu(B_i)$. Then one can check that $\rho$ is countably additive on the algebra $\mathfrak{C}$ of elementary sets (it is not difficult). Now all the conditions of theorems 8.2.19 and 8.2.20 are met, and the (unique) Hopf extension of $\rho$ is the product measure $\mu \otimes \nu$. The approach we took to define the product measure has the slight advantage that it is better motivated by calculus concepts, as explained in the opening remarks of this section. In addition, Fubini's theorem follows without difficulty from the above results.

## Fubini's Theorem

**Theorem 8.8.10 (Tonelli's theorem).** *Suppose* $f : X \times Y \to \mathbb{C}$ *is an* $\mathfrak{M} \otimes \mathfrak{N}$-*measurable function.*

(a) *If $f$ is positive, let $\varphi(x) = \int_Y f_x d\nu$, and $\psi(y) = \int_X f^y d\mu$. Then $\varphi$ is $\mathfrak{M}$-measurable, $\psi$ is $\mathfrak{N}$-measurable, and*

$$\int_X \varphi d\mu = \int_{X \times Y} f d(\mu \otimes \nu) = \int_Y \psi d\nu.$$

(b) *In general, let $\varphi^*(x) = \int_Y |f|_x d\nu$, and $\psi^*(y) = \int_X |f|^y d\mu$. If $\varphi^* \in \mathfrak{L}^1(\mu)$ or if $\psi^* \in \mathfrak{L}^1(\nu)$, then $f \in \mathfrak{L}^1(\mu \otimes \nu)$.*

*Proof. Tonelli's theorem holds for the characteristic function of an $\mathfrak{M} \otimes \mathfrak{N}$-measurable set by the previous theorem. By the linearity of the integral, Tonelli's theorem holds for any $\mathfrak{M} \otimes \mathfrak{N}$-measurable simple function.*

*Now let $0 \le s_1 \le s_2 \le \dots$ be a sequence of $\mathfrak{M} \otimes \mathfrak{N}$-simple functions converging to $f(x, y)$ for every $(x, y) \in X \times Y$, and let $\varphi_n(x) = \int_Y (s_n)_x d\nu$. By the above paragraph, $\int_X \varphi_n d\mu = \int_{X \times Y} s_n d(\mu \otimes \nu)$. The monotone convergence theorem implies that $\int_X \varphi d\mu = \int_{X \times Y} f d(\mu \otimes \nu)$. The proof that $\int_Y \psi d\nu = \int_{X \times Y} f d(\mu \otimes \nu)$ is identical to the above.*

*Part (b) is obtained by applying part (a) to the function $|f|$.* ∎

**Theorem 8.8.11 (Fubini's theorem).** *If $f \in \mathfrak{L}^1(\mu \otimes \nu)$, then $f_x \in \mathfrak{L}^1(\nu)$ for a.e. $x \in X$, $f^y \in \mathfrak{L}^1(\mu)$ for a.e. $y \in Y$, the functions $\varphi(x) = \int_Y f_x d\nu$ and $\psi(y) = \int_X f^y d\mu$ are in $\mathfrak{L}^1(\mu)$ and $\mathfrak{L}^1(\nu)$, respectively, and*

$$\int_X \varphi d\mu = \int_{X \times Y} f d(\mu \otimes \nu) = \int_Y \psi d\nu. \tag{7}$$

*Proof.* It is clearly sufficient to prove the result when $f$ is a real function. Let $f^+$ and $f^-$ be the positive and negative parts of $f$, and write $\varphi_1(x) = \int_Y (f^+)_x d\nu$, and $\varphi_2(x) = \int_Y (f^-)_x d\nu$. Since $f^+ \leq |f|$, $f^+ \in \mathfrak{L}^1(\mu \otimes \nu)$, theorem 8.8.10 applies and $\int_X \varphi_1 d\mu = \int_{X \times Y} f^+ d(\mu \otimes \nu) < \infty$. Thus $\varphi_1 \in \mathfrak{L}^1(\mu)$ and example 1 in section 8.3 now implies that $\varphi_1(x)$ is finite for a.e. $x \in X$, that is, $(f^+)_x$ is integrable for a.e. $x \in X$. Similar results apply to $f^-$; $\varphi_2 \in \mathfrak{L}^1(\mu)$, and $\varphi_2$ is finite for a.e. $x \in X$. The function $\varphi = \varphi_1 - \varphi_2$ is defined for a.e. $x \in X$, and the identity $\int_X \varphi d\mu = \int_{X \times Y} f d(\mu \otimes \nu)$ follows from the fact that $f_x = (f^+)_x - (f^-)_x$ and the linearity of the integral. The remaining assertion of the theorem and the other identity in (7) are obtained by replicating the above proof for the function $f^y$. ∎

## Products of Lebesgue Measures

In the discussion below and until the end of the section, $k$ is a positive integer, and $\lambda_k$ denotes Lebesgue measure on the $\sigma$-algebra $\mathcal{L}^k$ of Lebesgue measurable subsets of $\mathbb{R}^k$. We also use the notation $\mathcal{B}^k$ to denote the $\sigma$-algebra of Borel subsets of $\mathbb{R}^k$.

In the following, we use the result of problem 10 on section 8.4 without explicit mention.

**Lemma 8.8.12.** *Let $r$ and $s$ be positive integers, and let $n = r + s$. If $Z$ is a set of Lebesgue measures 0 in $\mathbb{R}^r$ and $B \in \mathcal{L}^s$, then $Z \times B \in \mathcal{L}^n$, and $\lambda_n(Z \times B) = 0$.*

*Proof.* First assume that $B$ is bounded, and choose an open set $V$ of finite measure such that $B \subseteq V \subseteq \mathbb{R}^s$. Let $\epsilon > 0$. Choose an open set $U$ such that $Z \subseteq U \subseteq \mathbb{R}^r$ and $\lambda_r(U) < \epsilon$. Since we have not yet established the Lebesgue measurability of $Z \times B$, we estimate its outer measure: $m_n^*(Z \times B) \leq m_n^*(U \times V) = \lambda_n(U \times V) = \lambda_r(U)\lambda_s(V) < \epsilon\lambda_s(V)$. Since $\epsilon$ is arbitrary, $m_n^*(Z \times B) = 0$; hence $Z \times B$ is measurable of measure 0.

If $B$ is unbounded, consider the intersection $B_i$ of $B$ with the open ball in $\mathbb{R}^s$ of radius $i$ and centered at the origin. By what we just proved, for each $i \in \mathbb{N}$, $Z \times B_i \in \mathcal{L}^n$ has measure 0. Since $Z \times B = \bigcup_{i=1}^\infty (Z \times B_i)$, the proof is complete. ∎

**Proposition 8.8.13.** *Let $r$, $s$, and $n$ be as in lemma 8.8.12. Then*

(a) $\mathcal{B}^n \subseteq \mathcal{L}^r \otimes \mathcal{L}^s \subseteq \mathcal{L}^n$.
(b) *If $A \in \mathcal{L}^r$ and $B \in \mathcal{L}^s$, then $\lambda_n(A \times B) = \lambda_r(A)\lambda_s(B)$.*

*Proof.* (a) Every open cube in $\mathbb{R}^n$ is the product of two open cubes, one in $\mathbb{R}^r$ and one in $\mathbb{R}^s$. Thus $\mathcal{L}^r \otimes \mathcal{L}^s$ contains all open cubes in $\mathbb{R}^n$. Since every open subset of

$\mathbb{R}^n$ is a countable union of open cubes, $\mathcal{L}^r \otimes \mathcal{L}^s$ contains all open subsets of $\mathbb{R}^n$; hence $\mathcal{B}^n \subseteq \mathcal{L}^r \otimes \mathcal{L}^s$.

Suppose $A \in \mathcal{L}^r$ and $B \in \mathcal{L}^s$. By theorem 8.4.10(c), choose $F_\sigma$ sets $F \subseteq A$ and $K \subseteq B$ such that the sets $Z_1 = A - F$ and $Z_2 = B - K$ have measure 0. Now $A \times B = (F \times K) \cup Z$, where $Z = (F \times Z_2) \cup (Z_1 \times K) \cup (Z_1 \times Z_2)$. By the previous lemma, $\lambda_n(Z) = 0$. Since the product of $F_\sigma$ sets is an $F_\sigma$ set, $F \times K \in \mathcal{L}^n$. The $\mathcal{L}^n$-measurability of $A \times B$ is now immediate because it the union of two measurable sets. We have shown that $\mathcal{L}^n$ contains all measurable rectangles in $\mathcal{L}^r \otimes \mathcal{L}^s$. Hence $\mathcal{L}^r \otimes \mathcal{L}^s \subseteq \mathcal{L}^n$.

(b) First assume that $A$ and $B$ are bounded $G_\delta$ sets. Thus there exist descending sequences of bounded open sets $\{U_i\}$ in $\mathbb{R}^r$ and $\{V_j\}$ in $\mathbb{R}^s$ such that $A = \cap_{i=1}^\infty U_i$ and $B = \cap_{j=1}^\infty V_j$. Now

$$\lambda_n(A \times B) = \lambda_n(\cap_{i=1}^\infty \cap_{j=1}^\infty (U_i \times V_j)) = \lim_i \lambda_n(\cap_{j=1}^\infty (U_i \times V_j))$$
$$= \lim_i \lim_j \lambda_n(U_i \times V_j) = \lim_i \lim_j \lambda_r(U_i)\lambda_s(V_j) = \lambda_r(A)\lambda_s(B).$$

Now, for arbitrary (unbounded) $G_\delta$ sets $A$ and $B$, the result follows from the $\sigma$-finiteness of Lebesgue measure. We invite the reader to work out the details.

Finally, if $A$ and $B$ are Lebesgue measurable in their respective spaces, then, by theorem 8.4.10(c), choose $G_\delta$ sets $G$ in $\mathbb{R}^r$ and $H$ in $\mathbb{R}^s$ such that $A = G - Z_1$, $B = H - Z_2$, where $Z_1$ and $Z_2$ have measure 0. By lemma 8.8.12,

$$\lambda_n(G \times H) = \lambda_n(A \times B) + \lambda_n(A \times Z_2) + \lambda_n(Z_1 \times B) + \lambda_n(Z_1 \times Z_2) = \lambda_n(A \times B).$$

But $\lambda_n(G \times H) = \lambda_r(G)\lambda_s(H) = \lambda_r(A)\lambda_s(B)$, hence the result. ∎

Before we proceed to the next theorem, we will show that $\lambda_1 \otimes \lambda_1$ is not a complete measure. Let $E$ be a subset of $[0,1]$ that is not Lebesgue measurable. The set $A = E \times \{0\} \subseteq \mathbb{R}^2$ is contained in $B = [0,1] \times \{0\}$, which is in $\mathcal{L}^1 \otimes \mathcal{L}^1$. Clearly, $(\lambda_1 \otimes \lambda_1)(B) = 0$. However, $A$ is not in $\mathcal{L}^1 \otimes \mathcal{L}^1$ by proposition 8.8.2. As a by-product of this example, it follows that $\mathcal{L}^2$ is strictly larger than $\mathcal{L}^1 \otimes \mathcal{L}^1$.

**Theorem 8.8.14.** Let $r$ and $s$ be positive integers, and let $n = r + s$. Then $(\mathbb{R}^n, \mathcal{L}^n, \lambda_n)$ is the completion of $(\mathbb{R}^n, \mathcal{L}^r \otimes \mathcal{L}^s, \lambda_r \otimes \lambda_s)$.

Proof. By the above proposition, if $A \in \mathcal{L}^r$ and $B \in \mathcal{L}^s$, then $\lambda_n(A \times B) = \lambda_r(A)\lambda_s(B) = (\lambda_r \otimes \lambda_s)(A \times B)$. Thus $\lambda_n$ agrees with $\lambda_r \otimes \lambda_s$ on the set of measurable rectangles in $\mathcal{L}^r \otimes \mathcal{L}^s$. By the uniqueness of the product measure (theorem 8.8.9 and problem 5 at the end of this section), $\lambda_n$ extends $\lambda_r \otimes \lambda_s$. Since $(\mathbb{R}^n, \mathcal{L}^n, \lambda_n)$ is a complete measure space, it contains the completion of $(\mathbb{R}^n, \mathcal{L}^r \otimes \mathcal{L}^s, \lambda_r \otimes \lambda_s)$.

*The proof will be complete if we show that for a member $E$ of $\mathcal{L}^n$, there are members $A$ and $B$ of $\mathcal{L}^r \otimes \mathcal{L}^s$ such that $A \subseteq E \subseteq B$, and $(\lambda_r \otimes \lambda_s)(B - A) = 0$; see problem 4 on section 8.2. By theorem 8.4.10, there exists an $F_\sigma$ set $A \subseteq \mathbb{R}^n$ and a $G_\delta$ set $B \subseteq \mathbb{R}^n$ such that $A \subseteq E \subseteq B$, and $\lambda_n(B - A) = 0$. Since $A, B \in \mathcal{B}^n \subseteq \mathcal{L}^r \otimes \mathcal{L}^s$, the above paragraph implies that $(\lambda_r \otimes \lambda_s)(B - A) = \lambda_n(B - A) = 0$, as desired.* ∎

## Excursion: The Product of Finitely Many Measures

It is clear that the above definitions and constructions for the product of two measurable spaces can be extended to the product of any finite number of measurable spaces $\{(X_i, \mathfrak{M}_i), 1 \leq i \leq n\}$. A measurable rectangle is a set of the form $A_1 \times \ldots \times A_n$, $A_i \in \mathfrak{M}_i$, and an elementary set is a disjoint union of a finite number of measurable rectangles. It is easy to see that the collection, $\mathfrak{C}$, of elementary sets is an algebra. By definition, $\mathfrak{M}_1 \otimes \ldots \otimes \mathfrak{M}_n$ is the $\sigma$-algebra generated by the collection of measurable rectangle. Obviously, the algebra $\mathfrak{C}$ also generates $\mathfrak{M}_1 \otimes \ldots \otimes \mathfrak{M}_n$.

We first establish the following technical lemma

**Lemma 8.8.15.** *Let $(X_i, \mathfrak{M}_i), 1 \leq i \leq 3$ be measurable spaces. Then*

$$\mathfrak{M}_1 \otimes (\mathfrak{M}_2 \otimes \mathfrak{M}_3) = \mathfrak{M}_1 \otimes \mathfrak{M}_2 \otimes \mathfrak{M}_3.$$

*Proof. Recall that $\mathfrak{M}_1 \otimes \mathfrak{M}_2 \times \mathfrak{M}_3$ is generated by the set of all measurable rectangles $A_1 \times A_2 \times A_3$, where $A_i \in \mathfrak{M}_i$, while $\mathfrak{M}_1 \otimes (\mathfrak{M}_2 \otimes \mathfrak{M}_3)$ is generated by sets of the form $A_1 \otimes P$, where $A_1 \in \mathfrak{M}_1$ and $P \in \mathfrak{M}_2 \otimes \mathfrak{M}_3$. We show that $\mathfrak{M}_1 \otimes \mathfrak{M}_2 \otimes \mathfrak{M}_3 \subseteq \mathfrak{M}_1 \otimes (\mathfrak{M}_2 \otimes \mathfrak{M}_3)$. Every measurable rectangle $R = A_1 \times A_2 \times A_3$ in $X_1 \times X_2 \times X_3$ can be written as $R = A_1 \times P$, where $P = A_2 \times A_3$. Since $P \in \mathfrak{M}_2 \otimes \mathfrak{M}_3$, $\mathfrak{M}_1 \otimes (\mathfrak{M}_2 \otimes \mathfrak{M}_3)$ contains all measurable rectangles; hence $\mathfrak{M}_1 \otimes (\mathfrak{M}_2 \otimes \mathfrak{M}_3) \supseteq \mathfrak{M}_1 \otimes \mathfrak{M}_2 \otimes \mathfrak{M}_3$.*

*To prove the reverse containment, it is enough to show that $\mathfrak{M}_1 \otimes \mathfrak{M}_2 \otimes \mathfrak{M}_3$ contains every set of the form $A_1 \times P$, where $A_1 \in \mathfrak{M}_1$ and $P \in \mathfrak{M}_2 \otimes \mathfrak{M}_3$, which is a generating set for $\mathfrak{M}_1 \otimes (\mathfrak{M}_2 \otimes \mathfrak{M}_3)$.*

*Define a collection of subsets of $X_2 \times X_3$ as follows:*

$$\Omega = \{P \subseteq X_2 \times X_3 : X_1 \times P \in \mathfrak{M}_1 \otimes \mathfrak{M}_2 \otimes \mathfrak{M}_3\}.$$

*It is easy to see that $\Omega$ is a $\sigma$-algebra and that every measurable rectangle $A_2 \times A_3$ is in $\Omega$. Thus $\Omega$ contains the $\sigma$-algebra generated by all elementary rectangles in $X_2 \times X_3$; hence $\Omega \supseteq \mathfrak{M}_2 \otimes \mathfrak{M}_3$. It follows that, for every $P \in \mathfrak{M}_2 \otimes \mathfrak{M}_3$, $X_1 \otimes P \in \mathfrak{M}_1 \otimes \mathfrak{M}_2 \otimes \mathfrak{M}_3$. It is clear that, for every $A_1 \in \mathfrak{M}_1$, $A_1 \times X_2 \times X_3 \in \mathfrak{M}_1 \otimes$*

$\mathfrak{M}_2 \otimes \mathfrak{M}_3$; *hence the intersection of* $X_1 \times P$ *and* $A_1 \times X_2 \times X_3$ *is in* $\mathfrak{M}_1 \otimes \mathfrak{M}_2 \otimes$ $\mathfrak{M}_3$. *But the intersection of the latter two sets is exactly* $A_1 \times P$. *This concludes the proof.* ∎

By an argument almost identical to the above proof, it can be shown that $(\mathfrak{M}_1 \otimes \mathfrak{M}_2) \otimes \mathfrak{M}_3 = \mathfrak{M}_1 \otimes \mathfrak{M}_2 \otimes \mathfrak{M}_3$. Thus the formation of products of measurable spaces is associative.

It follows by induction that if $\{(X_i, \mathfrak{M}_i), 1 \leq i \leq n\}$ is a finite set of measurable spaces then

$$\mathfrak{M}_1 \otimes ... \otimes \mathfrak{M}_n = \mathfrak{M}_1 \otimes \mathfrak{R}_{n-1}, \text{ where } \mathfrak{R}_{n-1} = \mathfrak{M}_2 \otimes ... \otimes \mathfrak{M}_n.$$

This immediately suggests an inductive definition of the product of more than two measure spaces.

**Definition.** Let $\{(X_i, \mathfrak{M}_i, \mu_i), 1 \leq i \leq n\}$ be a set of $\sigma$-finite measure spaces. We define the product measure $\mu_1 \otimes ... \otimes \mu_n$ on $\mathfrak{M}_1 \otimes ... \otimes \mathfrak{M}_n = \mathfrak{M}_1 \otimes \mathfrak{R}_{n-1}$ by

$$\mu_1 \otimes ... \otimes \mu_n = \mu_1 \otimes \rho_{n-1}, \text{ where } \rho_{n-1} = \mu_2 \otimes ... \otimes \mu_n.$$

Theorem 8.8.9 and the inductive nature of the construction imply that $\mu_1 \otimes ... \otimes \mu_n$ is a $\sigma$-finite measure on $\mathfrak{M}_1 \otimes ... \otimes \mathfrak{M}_n$ and that, for a measurable rectangle $A_1 \times ... \times A_n$, we have $(\mu_1 \otimes ... \otimes \mu_n)(A_1 \times ... \times A_n) = \prod_{i=1}^{n} \mu_i(A_i)$.

**Theorem 8.8.16.** *Suppose that* $\{(X_i, \mathfrak{M}_i, \mu_i), 1 \leq i \leq n\}$ *is a set of $\sigma$-finite measure spaces. Then the product measure* $\mu_1 \otimes ... \otimes \mu_n$ *is the unique $\sigma$-finite measure on* $\mathfrak{M}_1 \otimes ... \otimes \mathfrak{M}_n$ *such that, for every measurable rectangle* $A_1 \times ... \times A_n$, $(\mu_1 \otimes ... \otimes \mu_n)(A_1 \times ... \times A_n) = \prod_{i=1}^{n} \mu_i(A_i)$. ∎

The existence of the product measure is by the inductive construction outlined before the statement of the theorem. The uniqueness of $\mu_1 \otimes ... \otimes \mu_n$ is by problem 5 at the end of the section.

**Fubini's theorem** (theorem 8.8.11) extends to the product of any finite number of measures in a straightforward manner. Using the notation we established earlier in this excursion, if $f \in \mathfrak{L}^1(\mu_1 \otimes ... \otimes \mu_n)$, then

$$\int_{X_1 \times ... \times X_n} f d(\mu_1 \otimes ... \otimes \mu_n) = \int_{X_1} \int_{X_2 \times ... \times X_n} f d\rho_{n-1} d\mu_1.$$

The repeated application of Fubini's theorem (induction) yields

$$\int_{X_1\times\ldots\times X_n} fd(\mu_1\otimes\ldots\otimes\mu_n) = \int_{X_1}\int_{X_2}\ldots\int_{X_n} fd\mu_n\ldots d\mu_2 d\mu_1.$$

## Exercises

1. Let $r$ and $s$ be positive integers, and let $n = r+s$. Prove that $\mathcal{B}^n = \mathcal{B}^r\otimes\mathcal{B}^s$.
2. Let $r$, $s$, and $n$ be as in problem 1. Prove that $\mathcal{L}^r\otimes\mathcal{L}^s$ is strictly contained in $\mathcal{L}^n$.
3. Let $T : \mathbb{R}^n \to \mathbb{R}^n$ be an invertible linear operator. Prove that, for a function $f$ that is either positive or integrable,

$$\int_{\mathbb{R}^n} fd\lambda = |det(T)| \int_{\mathbb{R}^n} (f\circ T)d\lambda. \qquad (*)$$

Hint: Let $A$ be the matrix of $T$ relative to the standard basis of $\mathbb{R}^n$. By theorem B.4 in Appendix B, $A$ is the product of elementary matrices. Prove that $(*)$ holds for linear mappings generated by elementary matrices. You need Fubini's theorem and a specialized version of problem 15 on section 8.4. Observe that a useful by-product of this exercise is that if $A$ is an orthogonal matrix, then, for all $E \in \mathcal{L}^n$, $\lambda(E) = \lambda(AE)$, where $AE = \{Ax : x \in E\}$. Thus, *the Lebesgue measure is rotation invariant*.
4. Prove that a proper subspace of $\mathbb{R}^n$ has Lebesgue measure 0. Hint: See problem 6 on section 8.4.
5. Complete the proof of theorem 8.8.9. Thus prove that if $\rho$ is a measure on $\mathfrak{M}\otimes\mathfrak{N}$ such that for $A \in \mathfrak{M}$, and $B \in \mathfrak{N}$, $\rho(A\times B) = \mu(A)\nu(B)$, then $\rho = \mu\otimes\nu$ on $\mathfrak{M}\otimes\mathfrak{N}$. The same result easily extends to the product of any finite number of measures.
6. Let $f : \mathbb{R}^2 \to \mathbb{R}^2$ be the function

$$f(x,y) = \begin{cases} 1 & \text{if } x \geq 0, x \leq y \leq x+1, \\ -1 & \text{if } x \geq 0, x+1 \leq y \leq x+2, \\ 0 & \text{otherwise.} \end{cases}$$

Prove that $\int_{-\infty}^{\infty}\int_{-\infty}^{\infty}f(x,y)dxdy \neq \int_{-\infty}^{\infty}\int_{-\infty}^{\infty}f(x,y)dydx$. This does not contradict theorem 8.8.11 because, clearly, $|f|$ is not integrable.
7. Let $X = [0,1]$, $\mathfrak{M}$ be $\mathcal{L}^1$-restricted to $[0,1]$, and $dx$ (or $dy$) denote the Lebesgue measure on $[0,1]$. Choose a sequence $\alpha_1 < \alpha_2 < \ldots$ in $(0,1)$ and,

for each $n \geq 1$, let $g_n$ be a continuous positive function that vanishes outside $[\alpha_n, \alpha_{n+1}]$ such that $\int_{\alpha_n}^{\alpha_{n+1}} g_n(x)dx = 1$. Define a function $f: \mathbb{R}^2 \to \mathbb{R}$ by $f(x,y) = \sum_{n=1}^{\infty} g_n(y)[g_n(x) - g_{n+1}(x)]$. Show that $\int_0^1 \int_0^1 f(x,y)dxdy \neq \int_0^1 \int_0^1 f(x,y)dydx$. Also prove directly that $|f|$ is not integrable on the unit square.

8. Show that $\int_0^1 \int_0^1 \frac{x^2-y^2}{(x^2+y^2)^2} dxdy = -\int_0^1 \int_0^1 \frac{x^2-y^2}{(x^2+y^2)^2} dydx = \frac{\pi}{4}$. By integrating the positive part of $f = \frac{x^2-y^2}{(x^2+y^2)^2}$ on the unit square, show directly that $f$ is not integrable on the unit square.

9. By integrating $e^{-y} \sin 2xy$ on the strip $[0,1] \times (0, \infty)$, show that $\int_0^\infty \frac{1}{y} e^{-y} \sin^2 ydy = \log(5)/4$.

10. Let $\{(X_i, \mathfrak{M}_i), 1 \leq i \leq n\}$ be a finite set of measurable spaces. Show that the complement of a measurable rectangle in $X_1 \times ... \times X_n$ is an elementary set. This fact is needed in the proof that the collection of elementary sets is an algebra.

## 8.9  A Glimpse of Fourier Analysis

This section has a number of axes. We extend the discussion of Fourier series of $2\pi$-periodic functions we started in section 4.10. We also study Fourier series of functions in $\mathfrak{L}^p(-\pi, \pi)$. Then we take a brief tour through the Fourier transform. Finally we take a last look at the orthogonal polynomials we encountered in section 4.10.

### Fourier Series of $2\pi$-Periodic Functions

In section 4.10, we looked at the sequence of partial sums $S_n f$ of a $2\pi$-periodic function $f$. The first tool we develop is an integral formula for $S_n f$. Using the notation of section 4.10,

$$S_n f(x) = \sum_{j=-n}^{n} \hat{f}(j)e^{ijx} = \sum_{j=-n}^{n} \frac{1}{2\pi} e^{ijx} \int_{-\pi}^{\pi} e^{-ijt} f(t)dt$$

$$= \frac{1}{2\pi} \int_{-\pi}^{\pi} \left( \sum_{j=-n}^{n} e^{ij(x-t)} \right) f(t)dt.$$

We define the **Dirichlet kernel** to be the sequence of functions

$$D_n(x) = \sum_{j=-n}^{n} e^{ijx}.$$

Then the above calculation yields

$$S_n f(x) = \frac{1}{2\pi} \int_{-\pi}^{\pi} f(t)D_n(x-t)dt = \frac{1}{2\pi}(f*D_n)(x).$$

Observe that $D_n(x) = 1 + \sum_{j=1}^{n} \left( e^{ijx} + e^{-ijx} \right) = 1 + 2\sum_{j=1}^{n} \cos(jx)$. Multiplying the two sides of last identity by $\sin(x/2)$ we obtain

$$\sin\left(\frac{x}{2}\right) D_n(x) = \sin\left(\frac{x}{2}\right) + 2\sum_{j=1}^{n} \sin\left(\frac{x}{2}\right)\cos(jx)$$

$$= \sin\left(\frac{x}{2}\right) + \sum_{j=1}^{n} \left[ \sin\left(j+\frac{1}{2}\right)x - \sin\left(j-\frac{1}{2}\right)x \right] = \sin\left(n+\frac{1}{2}\right)x,$$

from which we obtain the formula for $D_n(x)$ in closed form:

$$D_n(x) = \frac{\sin(n+\frac{1}{2})x}{\sin\frac{x}{2}}.$$

The Dirichlet kernel is clearly an even, $2\pi$-periodic function, and $D_n(0) = 2n+1$. Since $\sin(x/2) > 0$ on the interval $(0, \pi)$, $D_n(x)$ has simple roots at the roots of the function $\sin(n+1/2)x$, namely, $x_j = \frac{2\pi j}{2n+1}, j = \pm 1, \dots, n$.

The graph of $D_{10}$ appears in figure 8.7.

**Example 1.** We derive the following estimate of $\|D_n\|_1$:

$$\|D_n\|_1 = \frac{1}{2\pi} \int_{-\pi}^{\pi} |D_n(x)| dx > \frac{4}{\pi^2} \sum_{k=1}^{n} \frac{1}{k}.$$

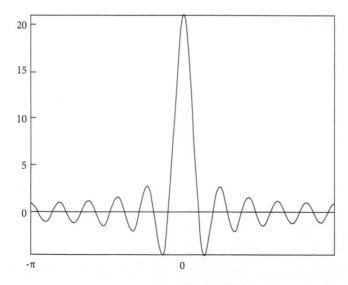

**Figure 8.7** $D_{10}$

Since $0 < sin(x/2) < x/2$ for $x \in (0, \pi)$,

$$\|D_n\|_1 = \frac{1}{\pi} \int_0^\pi \frac{|sin(n+1/2)x|}{sin\frac{x}{2}} dx \geq \frac{2}{\pi} \int_0^\pi \frac{|sin(n+1/2)x|}{x} dx$$

$$= \frac{2}{\pi} \int_0^{(n+1/2)\pi} \frac{|sin\, x|}{x} dx > \frac{2}{\pi} \int_0^{n\pi} \frac{|sin\, x|}{x} dx = \frac{2}{\pi} \sum_{k=1}^n \int_{(k-1)\pi}^{k\pi} \frac{|sin\, x|}{x} dx$$

$$> \frac{2}{\pi} \sum_{k=1}^n \frac{1}{k\pi} \int_{(k-1)\pi}^{k\pi} |sin\, x| dx = \frac{4}{\pi^2} \sum_{k=1}^n \frac{1}{k}. \blacklozenge$$

We are now ready to prove that the Fourier series of a continuous, $2\pi$-periodic function $f$ need not converge pointwise to $f$.

**Theorem 8.9.1.** *There exists a function $f \in C(S^1)$ such that $S_n f(0)$ does not converge to $f(0)$.*

*Proof.* We prove that, for some continuous, $2\pi$-periodic function $f$, the sequence $S_n f(0)$ is unbounded. For each $n \in \mathbb{N}$, define a functional $\lambda_n$ on $C(S^1)$ as follows: $\lambda_n(f) = S_n f(0)$. Then

$$|\lambda_n(f)| \leq \frac{1}{2\pi} \int_{-\pi}^\pi |f(t)| |D_n(t)| dt \leq \frac{\|f\|_\infty}{2\pi} \int_{-\pi}^\pi |D_n(t)| dt = \|D_n\|_1 \|f\|_\infty.$$

*It follows that $\left\|\lambda_n\right\| \leq \|D_n\|_1$. We show that $\|\lambda_n\| = \|D_n\|_1$. Let $\epsilon > 0$. Consider the function*

$$f(x) = \begin{cases} 1 & \text{if } D_n(x) \geq 0, \\ -1 & \text{if } D_n(x) < 0. \end{cases}$$

*Observe that $f(x)D_n(x) = |D_n(x)|$ for all $x \in [-\pi, \pi]$. By example 3 in section 8.7, there exists a function $g \in C(S^1)$ such that $\|f - g\|_1 < \epsilon$. Now*

$$\frac{1}{2\pi} \left| \int_{-\pi}^\pi D_n(x)g(x)dx - \int_{-\pi}^\pi |D_n(x)| dx \right| = \frac{1}{2\pi} \left| \int_{-\pi}^\pi D_n(x)g(x) - D_n(x)f(x)dx \right|$$

$$\leq \frac{1}{2\pi} \int_{-\pi}^\pi |D_n(x)| |f(x) - g(x)| dx \leq \|D_n\|_\infty \|f - g\|_1 < \epsilon \|D_n\|_\infty.$$

*It follows that $|\lambda_n(g)| = |\frac{1}{2\pi} \int_{-\pi}^\pi D_n(x)g(x)dx| > \frac{1}{2\pi} \int_{-\pi}^\pi |D_n(x)| dx - \epsilon \|D_n\|_\infty$. Since $\epsilon$ is arbitrary, $\|\lambda_n\| = \|D_n\|_1$.*

*In particular, the sequence of functionals $(\lambda_n)$ is not uniformly bounded, since $\|\lambda_n\| = \|D_n\|_1 \to \infty$ as $n \to \infty$. By the Banach-Steinhaus theorem, $\lambda_n$ cannot be pointwise bounded. Thus there exists a function $f \in \mathcal{C}(\mathcal{S}^1)$ such that $sup\{|S_n f(0)| : n \in \mathbb{N}\} = sup\{|\lambda_n(f)| : n \in \mathbb{N}\} = \infty$.* ∎

We now prove another classical theorem about the convergence of the means of the sequence of partial sums of the Fourier series of continuous, $2\pi$-periodic functions.

For a function $f \in \mathcal{C}(\mathcal{S}^1)$, consider the trigonometric polynomial

$$(\sigma_n f)(x) = \frac{S_0 f(x) + \ldots + S_n f(x)}{n+1}.$$

Since $S_j f(x) = \frac{1}{2\pi} f * D_j(x)$, $\sigma_n f(x) = \frac{1}{2\pi} f * K_n(x)$, where

$$K_n(x) = \frac{1}{n+1}(D_0 + \ldots + D_n).$$

The function $K_n$ is known as the **Feijer kernel**. We derive a formula for $K_n$ below. Form the formula for the Dirichlet kernel, we have

$$(n+1)\sin\left(\frac{x}{2}\right)K_n(x) = \sum_{j=0}^{n} \sin\left(j + \frac{1}{2}\right)x.$$

Thus

$$(n+1)\sin^2\left(\frac{x}{2}\right)K_n(x) = \sum_{j=0}^{n} \sin\left(\frac{x}{2}\right)\sin\left(j + \frac{1}{2}\right)x$$

$$= \frac{1}{2}\sum_{j=0}^{n}(\cos jx - \cos(j+1)x) = \frac{1}{2}(1 - \cos(n+1)x) = \sin^2\left(\frac{(n+1)x}{2}\right).$$

Hence

$$K_n(x) = \frac{\sin^2\left(\frac{(n+1)x}{2}\right)}{(n+1)\sin^2\left(\frac{x}{2}\right)}.$$

Clearly, $K_n$ is an even, positive, $2\pi$-periodic function, and since $\int_{-\pi}^{\pi} D_j(x)dx = 2\pi$ for all $j \in \mathbb{N}$, $\int_{-\pi}^{\pi} K_n(x)dx = 2\pi$ for all $n \in \mathbb{N}$.

The following property of $K_n$ is crucial for the next theorem: For $\delta \in (0, \pi), \lim_{n\to\infty} max\{K_n(x) : \delta \leq |x| \leq \pi\} = 0$. This is because if $\delta \leq |x| \leq \pi$, then $\sin^2(x/2) \geq \sin^2(\delta/2)$, and hence $0 \leq K_n(x) \leq \frac{1}{(n+1)\sin^2(\delta/2)} \to 0$ as $n \to \infty$.

**Theorem 8.9.2 (Feijer's theorem).** *For $f \in \mathcal{C}(\mathcal{S}^1)$, $\lim_n \|\sigma_n f - f\|_\infty = 0$.*

*Proof.* Since $f * K_n = K_n * f$, it is more convenient here to write $\sigma_n f(x) = \frac{1}{2\pi} \int_{-\pi}^{\pi} f(x - t)K_n(t)dt$. Let $\epsilon > 0$. By the uniform continuity of $f$, there exists a number $\delta > 0$ such that if $|t| < \delta$, then $|f(x - t) - f(x)| < \epsilon$ for all $x \in [-\pi, \pi]$. Choose a natural number $N$ such that, for $n > N$, $\max\{K_n(t) : \delta \le |t| \le \pi\} < \epsilon$. Recall that $\int_{-\pi}^{\pi} K_n(t)dt = 2\pi$. Now, for $n > N$,

$$|\sigma_n f(x) - f(x)| = \left| \frac{1}{2\pi} \int_{-\pi}^{\pi} (f(x - t) - f(x))K_n(t)dt \right|$$

$$\le \frac{1}{2\pi} \int_{-\pi}^{\pi} |f(x - t) - f(x)|K_n(t)dt$$

$$= \frac{1}{2\pi} \int_{|t|<\delta} |f(x - t) - f(x)|K_n(t)dt + \frac{1}{2\pi} \int_{\delta \le |t| \le \pi} |f(x - t) - f(x)|K_n(t)dt$$

$$\le \frac{\epsilon}{2\pi} \int_{|t|<\delta} K_n(t)dt + \frac{2\|f\|_\infty}{2\pi} \int_{\delta \le |t| \le \pi} \epsilon \, dt \le \epsilon + 2\epsilon\|f\|_\infty.$$

*Since $\epsilon$ is arbitrary, the proof is complete.* ∎

Observe that Feijer's theorem furnishes another proof that trigonometric polynomials are uniformly dense in $\mathcal{C}(\mathcal{S}^1)$.

## Fourier Series of $\mathfrak{L}^p$-Functions

Consider the Banach space $\mathfrak{L}^p(-\pi, \pi)$ with the norm

$$\|f\|_p = \left( \frac{1}{2\pi} \int_{-\pi}^{\pi} |f(x)|^p dx \right)^{1/p}, \ 1 \le p < \infty.$$

We are primarily interested in the cases $p = 1$ and $p = 2$, but a good number of the results in this subsection are valid for any $p \in [1, \infty)$. One can see directly that the Fourier coefficients $\hat{f}(n) = \frac{1}{2\pi} \int_{-\pi}^{\pi} f(t)e^{-int}dt$ of a function $f$ in $\mathfrak{L}^p(-\pi, \pi)$ are defined. This is because, for $p \ge 1$, $\mathfrak{L}^p(-\pi, \pi) \subseteq \mathfrak{L}^1(-\pi, \pi)$; hence

$$|\hat{f}(n)| \le \frac{1}{2\pi} \int_{-\pi}^{\pi} |f(t)|dt = \|f\|_1 < \infty.$$

It is convenient to refer to the set of Fourier coefficients $(\hat{f}(n))_{n \in \mathbb{Z}}$ of a function $f \in \mathfrak{L}^p(-\pi, \pi)$ by the notation $\mathfrak{F}(f)$. We think of $\mathfrak{F}$ as a linear transformation from $\mathfrak{L}^p(-\pi, \pi)$ to some suitable range space. For example, when $p = 2$, the range space of $\mathfrak{F}$ is $l^2(\mathbb{Z})$. We will show in example 2 below that, for all $p \ge 1$, and all $f \in \mathfrak{L}^p(-\pi,\pi)$, $\mathfrak{F}(f) \in c_0(\mathbb{Z})$. The norm on $c_0(\mathbb{Z})$ is the $\infty$-norm. Thus $\|\mathfrak{F}(f)\|_\infty = \sup\{|\hat{f}(n)| : n \in \mathbb{Z}\}$.

The following theorem is a special case of example 3 in section 8.7.

**Theorem 8.9.3.** *Trigonometric polynomials are dense in $\mathfrak{L}^2(-\pi,\pi)$.* ∎

We are now ready to extend the discussion of Fourier series we started in section 4.10 from $\mathcal{C}(\mathcal{S}^1)$ to $\mathfrak{L}^2(-\pi,\pi)$.

**Theorem 8.9.4.** *The set $\{u_n(t) = e^{int} : n \in \mathbb{Z}\}$ is an orthonormal basis for $\mathfrak{L}^2(-\pi,\pi)$.*

*Proof.* By theorem 8.9.3, $\mathrm{Span}(\{u_n : n \in \mathbb{Z}\})$ is dense in $\mathfrak{L}^2(-\pi,\pi)$. The assertion of this theorem follows directly from theorem 7.2.7. ∎

All the results we obtained in section 4.10 for Fourier series of continuous functions extend to $\mathfrak{L}^2(-\pi,\pi)$. The following theorem lists some of the properties. They follow directly from general Hilbert space theory.

**Theorem 8.9.5.** *The following are true for a function $f \in \mathfrak{L}^2(-\pi,\pi)$:*

(a) *The sequence $\left(\hat{f}(n)\right)_{n \in \mathbb{Z}}$ belongs to $l^2(\mathbb{Z})$.*
(b) *If $a = (a_n) \in l^2(\mathbb{Z})$, then the series $\sum_{-\infty}^{\infty} a_n u_n$ converges in $\mathfrak{L}^2(-\pi,\pi)$.*
(c) *$f(x) = \sum_{-\infty}^{\infty} \hat{f}(n)e^{inx}$, where convergence takes place in $\mathfrak{L}^2(-\pi,\pi)$.*
(d) *$\|f\|_2^2 = \sum_{-\infty}^{\infty} |\hat{f}(n)|^2$.*
(e) *If $\hat{f}(n) = 0$ for all $n \in \mathbb{Z}$, then $f = 0$ a.e.*
(f) *The mapping $f \to \mathfrak{F}(f)$ is a linear isometry from $\mathfrak{L}^2(-\pi,\pi)$ onto $l^2(\mathbb{Z})$.* ∎

The simplicity, elegance, and completeness of theorem 8.9.5 does not extend to functions in $\mathfrak{L}^1(-\pi,\pi)$. For example, the sequence of partial sums $S_n f(x) = \sum_{-n}^{n} \hat{f}(j)u_j(x)$ need not converge to $f$ in the 1-norm (see the section exercises), and $\mathfrak{F}$ does not map $\mathfrak{L}^1(-\pi,\pi)$ onto its range space, which we now describe.

**Example 2** (*the* Riemann-Lebesgue lemma). For $f \in \mathfrak{L}^1(-\pi,\pi)$, $\mathfrak{F}(f)$ is in $c_0(\mathbb{Z})$.

Observe that the assertion holds for trigonometric polynomials. Indeed, if $p(t) = \sum_{j=-N}^{N} a_j e^{ijt}$, then $\hat{p}(n) = 0$ whenever $|n| > N$.

To prove the general case, let $f \in \mathfrak{L}^1(-\pi,\pi)$, and let $\epsilon > 0$. Choose a trigonometric polynomial $p$ such that $\|f - p\|_1 < \epsilon$, and an integer $N$ such that $\hat{p}(n) = 0$ for $|n| > N$. Now if $|n| > N$,

$$|\hat{f}(n)| = |\hat{f}(n) - \hat{p}(n)| = \frac{1}{2\pi}\left|\int_{-\pi}^{\pi}(f(t) - p(t))e^{-int}dt\right|$$

$$\leq \frac{1}{2\pi}\int_{-\pi}^{\pi}|f(t) - p(t)|dt = \|f - p\|_1 < \epsilon. \blacklozenge$$

**Example 3.** If we view $\sigma_n$ as a linear operator on $\mathfrak{L}^1(-\pi,\pi)$, $\sigma_n$ is bounded, and $\|\sigma_n\| \leq 1$. For $f \in \mathfrak{L}^1(-\pi,\pi)$,

$$\|\sigma_n f\|_1 = \frac{1}{2\pi}\int_{-\pi}^{\pi}\frac{1}{2\pi}\left|\int_{-\pi}^{\pi}f(t)K_n(x-t)dt\right|dx \leq \frac{1}{4\pi^2}\int_{-\pi}^{\pi}\int_{-\pi}^{\pi}|f(t)|K_n(x-t)dtdx$$

$$= \frac{1}{4\pi^2}\int_{-\pi}^{\pi}|f(t)|\int_{-\pi}^{\pi}K_n(x-t)dxdt = \frac{1}{2\pi}\int_{-\pi}^{\pi}|f(t)|dt = \|f\|_1. \blacklozenge$$

**Example 4.** For $f \in \mathfrak{L}^1(-\pi,\pi)$, $\sigma_n f$ converges to $f$ in $\mathfrak{L}^1(-\pi,\pi)$.

Let $\epsilon > 0$, and choose $g \in \mathcal{C}(S^1)$ such that $\|f - g\|_1 < \epsilon$. By Feijer's theorem, $\sigma_n g$ converges to $g$ uniformly; hence $\sigma_n g$ converges to $g$ in $\mathfrak{L}^1(-\pi,\pi)$. Choose an integer $N$ such that, for $n > N$, $\|\sigma_n g - g\|_1 < \epsilon$. Using example 3, if $n > N$, then $\|\sigma_n f - f\|_1 \leq \|\sigma_n f - \sigma_n g\|_1 + \|\sigma_n g - g\|_1 + \|g - f\|_1 \leq 2\|f - g\|_1 + \|\sigma_n g - g\|_1 < 3\epsilon. \blacklozenge$

**Theorem 8.9.6 (the uniqueness theorem).** *If $f \in \mathfrak{L}^1(-\pi,\pi)$ and $\hat{f}(n) = 0$ for all $n \in \mathbb{Z}$, then $f = 0$ a.e. Consequently, the mapping $\mathfrak{F} : \mathfrak{L}^1(-\pi,\pi) \to c_0(\mathbb{Z})$ is injective.*

*Proof.* By assumption, $\sigma_n f = 0$ for all $n \in \mathbb{N}$. Since $\sigma_n f$ converges to $f$ by the previous example, $\|f\|_1 = 0$, and $f = 0$. a.e. $\blacksquare$

**Theorem 8.9.7.** *The linear mapping $\mathfrak{F} : \mathfrak{L}^1(-\pi,\pi) \to c_0(\mathbb{Z})$ is not onto.*

*Proof.* First observe that $\mathfrak{F}$ is bounded by virtue of the inequality $|\hat{f}(n)| \leq \|f\|_1$. If $\mathfrak{F}$ is surjective, then, by the open mapping theorem, $\mathfrak{F}$ would be invertible; hence, for every $f \in \mathfrak{L}^1(-\pi,\pi)$, $\|f\|_1 \leq M\|\mathfrak{F}(f)\|_\infty$, where $M = \|\mathfrak{F}^{-1}\|$. Now, for the sequence $(D_n)$ of Dirichlet kernels, $\|\mathfrak{F}(D_n)\|_\infty = 1$, while $\|D_n\|_1 \to \infty$ as $n \to \infty$. This contradiction delivers the result. $\blacksquare$

## The Fourier Transform

The Fourier transform of a function $f \in \mathfrak{L}^1(\mathbb{R})$ is, by definition, the function

$$\hat{f}(x) = \frac{1}{\sqrt{2\pi}}\int_{\mathbb{R}}f(t)e^{-ixt}dt.$$

The normalization constant $\frac{1}{\sqrt{2\pi}}$ is included only for the symmetry of the formulas and is not essential.

One can think of the Fourier transform as the continuous equivalent of Fourier series. Instead of using the discrete set of frequencies $\{e^{int}\}_{n\in\mathbb{Z}}$, the Fourier transform uses a continuum of frequencies $\{e^{ixt}\}_{t\in\mathbb{R}}$.

It is clear that $\hat{f}\in \mathfrak{L}^\infty(\mathbb{R})$ and that $\|\hat{f}\|_\infty \le \|f\|_1$. The following theorem narrows down the range space of the Fourier transform.

**Theorem 8.9.8.** *If $f\in \mathfrak{L}^1(\mathbb{R})$, then $\hat{f}\in \mathcal{C}_0(\mathbb{R})$.*

*Proof. First we prove that $\hat{f}$ is continuous. Suppose $(x_n)$ is a convergent sequence and that $\lim_n x_n = x_0$. Since $|f(t)e^{-ix_n t}| = |f(t)|$ and $f\in \mathfrak{L}^1(\mathbb{R})$, the dominated convergence theorem implies that*

$$\lim_n \hat{f}(x_n) = \lim_n \int_\mathbb{R} \frac{1}{\sqrt{2\pi}} f(t)e^{-ix_n t}\, dt = \int_\mathbb{R} \frac{1}{\sqrt{2\pi}} f(t)e^{-ix_0 t}\, dt = \hat{f}(x_0).$$

*To prove that $\hat{f}$ vanishes at $\infty$, write*

$$\hat{f}(x) = -\frac{1}{\sqrt{2\pi}} \int_\mathbb{R} f(t)e^{-ix(t+\pi/x)}\, dt = -\frac{1}{\sqrt{2\pi}} \int_\mathbb{R} (\tau_a f)(\xi)e^{-ix\xi}\, d\xi .(a = \pi/x).$$

*Thus*

$$2\hat{f}(x) = \frac{1}{\sqrt{2\pi}} \int_\mathbb{R} f(\xi)e^{-ix\xi}\, d\xi - \frac{1}{\sqrt{2\pi}} \int_\mathbb{R} (\tau_a f)(\xi)e^{-ix\xi}\, d\xi$$

$$= \frac{1}{\sqrt{2\pi}} \int_\mathbb{R} (f - \tau_a f)(\xi)e^{-ix\xi}\, d\xi.$$

*It follows that*

$$2|\hat{f}(x)| \le \frac{1}{\sqrt{2\pi}} \int_\mathbb{R} |f(\xi) - (\tau_a f)(\xi)e^{-i\xi x}|\, d\xi = \frac{1}{\sqrt{2\pi}} \|f - \tau_a f\|_1.$$

*As $|x| \to \infty$, $a \to 0$, and $\lim_{|x|\to\infty} \|f - \tau_a f\|_1 \to 0$, by theorem 8.7.9.* ∎

In this theorem, the fact that $\lim_{|x|\to\infty} \hat{f}(x) = 0$ is known as **the Riemann-Lebesgue lemma.**

The next goal is to prove the **inversion formula**. Guided by the inversion formula for a function $f \in \mathfrak{L}^1(-\pi, \pi)$ when $\mathfrak{F}(f) \in l^1(\mathbb{Z})$ (see problem 1 at the end of this section), one can reasonably conjecture that if $f$ and $\hat{f}$ are both in $\mathfrak{L}^1(\mathbb{R})$, then

$$f(x) = \frac{1}{\sqrt{2\pi}} \int_{\mathbb{R}} \hat{f}(t) e^{ixt} dt \text{ for almost every } x \in \mathbb{R}.$$

The proof bears some resemblance to that of Feijer's theorem in that we will find a family of functions $\{G_\sigma\}$ such that $\lim_{\sigma \downarrow 0} f * G_\sigma$ converges to $f$ in $\mathfrak{L}^1(\mathbb{R})$. Before we construct the family $G_\sigma$, it may be useful to find an even function that is equal to its own Fourier transform. One such function exists. The proof of the following proposition is left as an exercise.

**Proposition 8.9.9.** *For the function* $G_1(x) = \frac{1}{\sqrt{2\pi}} e^{-x^2/2}$, $\hat{G}_1 = G_1$. ∎

**Example 5.** The inversion formula holds for the function $G_1$.
  Because $G_1$ is even,

$$G_1(x) = G_1(-x) = \hat{G}_1(-x) = \frac{1}{\sqrt{2\pi}} \int_{\mathbb{R}} G_1(t) e^{ixt} dt = \frac{1}{\sqrt{2\pi}} \int_{\mathbb{R}} \hat{G}_1(t) e^{ixt} dt. \blacklozenge$$

**Definition. The Gauss kernel** For $\sigma > 0$, define

$$G_\sigma(x) = \frac{1}{\sigma\sqrt{2\pi}} exp\left\{\frac{-x^2}{2\sigma^2}\right\}.$$

Observe that $G_\sigma(x) = \frac{1}{\sigma} G_1(x/\sigma)$.

The family $\{G_\sigma : \sigma > 0\}$ is an **approximate identity** in the sense that

(a) $G_\sigma(x) \geq 0$ for all $x \in \mathbb{R}$ and all $\sigma > 0$,
(b) $\int_{\mathbb{R}} G_\sigma(x) dx = 1$ for all $\sigma > 0$, and
(c) For every $\delta > 0$, $\lim_{\sigma \downarrow 0} \int_{|x|>\delta} G_\sigma(x) dx = 0$.

**Example 6.** We prove property (c). For $|x| > \delta$, we have

$$\int_{|x|>\delta} G_\sigma(x) dx = \frac{2}{\sigma\delta\sqrt{2\pi}} \int_\delta^\infty \delta \, exp\left\{\frac{-x^2}{2\sigma^2}\right\} dx \leq \frac{2}{\sigma\delta\sqrt{2\pi}} \int_\delta^\infty x \, exp\left\{\frac{-x^2}{2\sigma^2}\right\} dx$$

$$= \frac{2\sigma}{\delta\sqrt{2\pi}} \int_{\delta/\sigma}^\infty y \, exp\left\{\frac{-y^2}{2}\right\} dy = \frac{2\sigma}{\delta\sqrt{2\pi}} exp\left\{\frac{-\delta^2}{2\sigma^2}\right\} \to 0 \text{ as } \sigma \downarrow 0. \blacklozenge$$

Other examples of approximate identities include the bump kernel we studied in section 8.7. Indeed, for a fixed $\eta > 0$, $\lim_{h \downarrow 0} \int_{|x| \geq \eta} \delta_h(x)dx = 0$, because, for every $h < \eta$, $\int_{|x| \geq \eta} \delta_h(x)dx = 0$.

**Example 7.** Let $g : \mathbb{R} \to \mathbb{R}$ be a bounded continuous function. Then $\lim_{\sigma \downarrow 0} \int_{\mathbb{R}} g(y)G_\sigma(y)dy = g(0)$.

Let $\epsilon > 0$, and choose $\delta > 0$ such that $|g(y) - g(0)| < \epsilon$ for all $y$ such that $|y| < \delta$. Also choose $\sigma_0 > 0$ such that $\int_{|y| > \delta} G_\sigma(y)dy < \epsilon$ whenever $0 < \sigma < \sigma_0$:

$$
\begin{aligned}
|g(y)G_\sigma(y) - g(0)| &= \left| \int_{\mathbb{R}} g(y)G_\sigma(y) - \int_{\mathbb{R}} g(0)G_\sigma(y)dy \right| \\
&\leq \int_{\mathbb{R}} |g(y) - g(0)|G_\sigma(y)dy \\
&= \int_{|y| < \delta} |g(y) - g(0)|G_\sigma(y)dy + \int_{|y| \geq \delta} |g(y) - g(0)|G_\sigma(y)dy \\
&\leq \epsilon \int_{|y| < \delta} G_\sigma(y)dy + 2\|g\|_\infty \int_{|y| > \delta} G_\sigma(y)dy < \epsilon(1 + 2\|g\|_\infty). \blacklozenge
\end{aligned}
$$

**Proposition 8.9.10.** *If* $1 \leq p < \infty$ *and* $f \in \mathfrak{L}^p(\mathbb{R})$, *then* $\lim_{\sigma \downarrow 0} \|f * G_\sigma - f\|_p = 0$.

*Proof.* Replicating the estimates in the proof of theorem 8.7.12, we obtain

$$
\|f * G_\sigma - f\|_p^p \leq \int_{\mathbb{R}} \|\tau_y f - f\|_p^p G_\sigma(y)dy.
$$

The function $g(y) = \|\tau_y f - f\|_p^p$ is continuous by problem 4 on section 8.7 and is bounded because $|g(y)| \leq (\|\tau_y f\|_p + \|f\|_p)^p = 2^p\|f\|_p^p$.

Applying the previous example, we have $\lim_{\sigma \to 0} \int_{\mathbb{R}} \|\tau_y f - f\|_p^p G_\sigma(y)dy = g(0) = 0$. $\blacksquare$

**Example 8.** We will later need the identity $G_\sigma(x) = \frac{1}{\sqrt{2\pi}} \int_{\mathbb{R}} G_1(\sigma t)e^{ixt}dt$:

$$
\begin{aligned}
G_\sigma(x) &= \frac{1}{\sigma}G_1(x/\sigma) = \frac{1}{\sigma}\hat{G}_1(-x/\sigma) = \frac{1}{\sigma\sqrt{2\pi}} \int_{\mathbb{R}} G_1(y) \, exp(iy\frac{x}{\sigma})dy \\
&= \frac{1}{\sqrt{2\pi}} \int_{\mathbb{R}} G_1(\sigma t)e^{ixt}dt. \blacklozenge
\end{aligned}
$$

**Theorem 8.9.11 (the inversion theorem).** *If $f$ and $\hat{f}$ are in $\mathfrak{L}^1(\mathbb{R})$, then for almost every $x \in \mathbb{R}$,*

$$f(x) = \frac{1}{\sqrt{2\pi}} \int_{\mathbb{R}} \hat{f}(t)e^{ixt}\,dt.$$

*Proof.* *Using the previous example, we have*

$$f * G_\sigma(x) = \int_{\mathbb{R}} f(x-t)G_\sigma(t)\,dt = \int_{\mathbb{R}} f(x-t)\frac{1}{\sqrt{2\pi}}\int_{\mathbb{R}} G_1(\sigma y)e^{iyt}\,dy\,dt$$

$$= \frac{1}{\sqrt{2\pi}} \int_{\mathbb{R}}\int_{\mathbb{R}} f(x-t)e^{iyt}\,dt\,G_1(\sigma y)\,dy$$

$$= \frac{1}{\sqrt{2\pi}} \int_{\mathbb{R}}\int_{\mathbb{R}} f(u)e^{iy(x-u)}\,du\,G_1(\sigma y)\,dy$$

$$= \int_{\mathbb{R}} \hat{f}(y)e^{ixy}G_1(\sigma y)\,dy.$$

*The summary of the above calculations is that*

$$f * G_\sigma(x) = \int_{\mathbb{R}} \hat{f}(y)e^{ixy}G_1(\sigma y)\,dy.$$

*Now consider the sequence $\sigma_n = 1/n$. By the identity we just established,*

$$f * G_{\sigma_n}(x) = \int_{\mathbb{R}} \hat{f}(y)e^{ixy}G_1(\sigma_n y)\,dy. \tag{8}$$

*On the one hand, $|\hat{f}(y)e^{ixy}G_1(\sigma_n y)| \le |\hat{f}(y)|/\sqrt{2\pi}, \hat{f} \in \mathfrak{L}^1(\mathbb{R})$, and $\lim_n G_1(\sigma_n x) = G_1(0) = \frac{1}{\sqrt{2\pi}}$. Thus, by the dominated convergence theorem, the right side of identity (8) converges to $\frac{1}{\sqrt{2\pi}}\int_{\mathbb{R}} \hat{f}(y)e^{ixy}\,dy$ for every $x \in \mathbb{R}$.*

*On the other hand, by proposition 8.9.10, the left side of identity (8) converges to $f$ in $\mathfrak{L}^1(\mathbb{R})$. By example 5 in section 8.3, the sequence $f * G_{\sigma_n}$ contains a subsequence that converges a.e. to $f$.*

*Putting the last two facts together, we arrive at the inversion formula.* ∎

Observe that the function $g(x) = \frac{1}{\sqrt{2\pi}}\int_{\mathbb{R}} \hat{f}(t)e^{ixt}\,dt$ is in $\mathcal{C}_0(\mathbb{R})$ by an argument identical to that in the proof of theorem 8.9.8. Thus the assumptions of the above theorem imply the $f$ is equal a.e. to a $\mathcal{C}_0(\mathbb{R})$ function.

**Corollary 8.9.12 (the uniqueness theorem).** *If $f \in \mathfrak{L}^1(\mathbb{R})$ and $\hat{f} = 0$, then $f(x) = 0$ for a.e. $x \in \mathbb{R}$.* ∎

## Orthogonal Polynomials: One More Time

In section 4.10, we studied the space $H$ of continuous, square integrable functions with respect to a weight function $\omega$. The inner product we used was given by the formula

$$\langle f, g \rangle = \int_a^b f(x)\overline{g(x)}\omega(x)dx.$$

We are now ready to settle a question that could not be answered completely in chapter 4. What is the smallest Hilbert space that contains the space $H$? The answer is now within our reach.

If we define a finite positive measure $\mu$ on the $\sigma$-algebra $\mathcal{L}$ of Lebesgue measurable sets by $d\mu = \omega d\lambda$, where $\lambda$ is Lebesgue measure on $\mathbb{R}$, then the Hilbert space we seek is (or should be, if there is justice) $\mathfrak{L}^2(\mathbb{R}, \mathcal{L}, \mu) = \mathfrak{L}^2(\mu)$.

In the case of Legendre polynomials, the situation is simple. Since $\omega(x) = 1$, the measure $\mu$ is nothing other than the Lebesgue measure on $(-1, 1)$, and $\mathfrak{L}^2(\mu) = \mathfrak{L}^2(-1, 1)$. As we observed in section 4.10, the space $H$ contains the space $\mathcal{C}[-1, 1]$. By theorem 4.10.8, the linear span of the sequence-normalized Legendre polynomials $(\tilde{P}_n)$ is dense in the space $\mathcal{C}[-1, 1]$. By example 1 in section 8.7, $\mathcal{C}[-1, 1]$ is dense in $\mathfrak{L}^2(-1, 1)$. It follows that the linear span of $(\tilde{P}_n)$ is dense in $\mathfrak{L}^2(-1, 1)$, and we have proved the following result.

**Theorem 8.9.13.** *The normalized Legendre polynomials $\tilde{P}_n$ form an orthonormal basis for $\mathfrak{L}^2(-1, 1)$.* ∎

The situation is far less obvious in the case of Hermite polynomials. In this case, $d\mu = e^{-x^2} d\lambda$, and it is true that the normalized Hermite polynomials $\tilde{H}_n = \frac{1}{\sqrt{n! 2^n \sqrt{\pi}}} H_n$ (see problem 15 on section 4.10) form an orthonormal basis for $\mathfrak{L}^2(\mu)$. Equivalently, we prove the following.

**Theorem 8.9.14.** *If $f \in \mathfrak{L}^2(\mu)$ and $\int_\mathbb{R} f(x)\tilde{H}_n(x)d\mu = 0$ for all $n \in \mathbb{N}$, then $f(x) = 0$ for a.e. $x \in \mathbb{R}$.*

*Proof.* Since $Span(\{\tilde{H}_0, \ldots, \tilde{H}_n\}) = Span(\{1, x, \ldots, x_n\})$, the assumption is equivalent to $\int_\mathbb{R} f(x)x^n e^{-x^2} dx = 0$ for all $n \in \mathbb{N}$.

*Because $\mu$ is a finite measure, $\mathfrak{L}^2(\mu) \subseteq \mathfrak{L}^1(\mu)$. In particular, $f \in \mathfrak{L}^1(\mu)$, and the function $g(t) = f(t)e^{-t^2} \in \mathfrak{L}^1(\lambda)$. The proof will be complete if we show that $\hat{g} = 0$.*

*We leave it to the reader to verify that, for a fixed $x \in \mathbb{R}$, the function $h(t) = e^{|xt|} \in \mathfrak{L}^2(\mu)$. It follows that the product $f(t)h(t) \in \mathfrak{L}^1(\mu)$; hence $f(t)e^{|xt|}e^{-t^2} \in \mathfrak{L}^1(\lambda)$. Now*

$$\sqrt{2\pi}\hat{g}(x) = \int_{\mathbb{R}} f(t)e^{-t^2}e^{-ixt}dt = \int_{\mathbb{R}} f(t)e^{-t^2}\sum_{n=0}^{\infty}\frac{(-ixt)^n}{n!}dt$$

$$= \sum_{n=0}^{\infty}\frac{(-ix)^n}{n!}\int_{\mathbb{R}} f(t)t^n e^{-t^2}dt = 0.$$

*The term-by-term integration of the series is justified by the dominated convergence theorem because the sequence of functions $f(t)e^{-t^2}\sum_{j=0}^{n}\frac{(-ixt)^j}{j!}$ is dominated by the integrable function $f(t)e^{|xt|}e^{-t^2}$.* ∎

## Exercises

1. Prove that if $f \in \mathfrak{L}^1(-\pi,\pi)$ and $\sum_{-\infty}^{\infty}|\hat{f}(n)| < \infty$, then $f(x) = \sum_{-\infty}^{\infty}\hat{f}(n)e^{inx}$ a.e. In particular, $f$ is equal a.e. to a continuous, $2\pi$-periodic function. See theorem 4.10.7.

2. Prove that there exists a function $f \in \mathfrak{L}^1(-\pi,\pi)$ such that $S_n f$ does not converge to $f$ in the 1-norm. Hint: View $\mathfrak{F}$ as a bounded linear operator on $\mathfrak{L}^1(-\pi,\pi)$, and use the Banach-Steinhaus theorem.

3. Prove that the family $h_\lambda(x) = \frac{\lambda}{\pi(x^2+\lambda^2)}$, $\lambda > 0$, is an approximate identity.

4. Show that the mapping $\mathfrak{F} : \mathfrak{L}^1(\mathbb{R}) \to \mathcal{C}_0(\mathbb{R})$ given by $f \mapsto \hat{f}$ is bounded and that $\|\mathfrak{F}\| = \frac{1}{\sqrt{2\pi}}$.

5. Suppose $f \in \mathfrak{L}^1(\mathbb{R})$, and let $a \in \mathbb{R}$. Prove that
   (a) if $g(x) = f(x)e^{iax}$, then $\hat{g}(x) = \hat{f}(x-a)$;
   (b) if $g(x) = f(x-a)$, then $\hat{g}(x) = \hat{f}(x)e^{-iax}$;
   (c) if $g(x) = f(-x)$, then $\hat{g}(x) = \overline{\hat{f}(x)}$; and
   (d) if $g(x) = f(x/a)$ and $a > 0$, then $\hat{g}(x) = a\hat{f}(ax)$.

6. Show that if $f,g \in \mathfrak{L}^1(\mathbb{R})$, then $f*g \in \mathfrak{L}^1(\mathbb{R})$, and $\widehat{(f*g)} = \hat{f}\hat{g}$.

7. Prove that if $f \in \mathfrak{L}^1(\mathbb{R})$ and the function $g(x) = -ixf(x) \in \mathfrak{L}^1(\mathbb{R})$, then $\hat{f}$ is differentiable and $\frac{d\hat{f}}{dx} = \hat{g}(x)$. Hint: Use the definition of derivative and the dominated convergence theorem.

8. Prove proposition 8.9.9. Hint: Apply the previous exercises to derive the differential equation $\frac{d\hat{G}_1(x)}{dx} = -x\hat{G}_1(x)$.

9. Let $g : \mathbb{R} \to \mathbb{R}$ be a bounded continuous function. Prove that, for every $x \in \mathbb{R}$, $\lim_{\sigma \downarrow 0}(g * G_\sigma)(x) = g(x)$. This result generalizes example 7.

10. Prove that there does not exist a function $\delta \in \mathfrak{L}^1(\mathbb{R})$ such that $\delta * f = f$ for all $f \in \mathfrak{L}^1(\mathbb{R})$.

11. Verify the details of the proof of theorem 8.9.14.

12. Prove that the sequence $\left( \dfrac{1}{\sqrt{n!2^n\sqrt{\pi}}} H_n(x)e^{-x^2/2} \right)$ is an orthonormal basis for $\mathfrak{L}^2(\lambda)$.

# The Equivalence of Zorn's Lemma, the Axiom of Choice, and the Well Ordering Principle

Before we embark on the task of proving theorem 2.2.1, we need to develop some background work.

**Notation.** If $S$ is a subset of a well-ordered set $A$, we use the notation $min\{S\}$ to denote the least element of $S$.

**Definition.** Let $A$ be a well-ordered set, and let $x \in A$. The initial **segment of A** determined by $x$ is the set

$$S(A,x) = \{y \in A : y < x\}.$$

Observe that $x = min\{A - S(A,x)\}$.

**Definition.** Let $(A_1, \leq_1)$ and $(A_2, \leq_2)$ be well-ordered subsets of a nonempty set $X$. No ordering of $X$ is assumed. We say that $(A_2, \leq_2)$ is a continuation of $(A_1, \leq_1)$ if $A_1 \subseteq A_2$, $A_1$ is a segment of $A_2$ and $\leq_1$ agrees with $\leq_2$ on $A_1$. Simply stated, $A_1$ is a segment of $A_2$, and $\leq_1$ is the ordering $A_1$ inherits from $(A_2, \leq_2)$. We use the notation $(A_1, \leq_1) \subseteq (A_2, \leq_2)$ to indicate that $(A_2, \leq_2)$ is a continuation of $(A_1, \leq_1)$ or that $(A_2, \leq_2) = (A_1, \leq_1)$.

A little reflection reveals that $\subseteq$ is a partial ordering of the collection $\mathfrak{W}$ of well-ordered subsets of $X$.

**Lemma A.1.** *In the notation of the above paragraph, let $\mathfrak{C} = \{(A_\alpha, \leq_\alpha)\}_{\alpha \in I}$ be a chain in $\mathfrak{W}$, and let $A = \cup_\alpha A_\alpha$. Then $A$ is well ordered.*

*Proof. Recall that to say that $\mathfrak{C}$ is a chain means that, for $\alpha, \beta \in I$, either $(A_\alpha, \leq_\alpha) \subseteq (A_\beta, \leq_\beta)$ or $(A_\beta, \leq_\beta) \subseteq (A_\alpha, \leq_\alpha)$. Here is an explicit definition of the ordering $\leq$ on $A$: for $a, b \in A$, let $\alpha, \beta \in I$ be such that $a \in A_\alpha, b \in A_\beta$. Since $\mathfrak{C}$ is a chain, say, $(A_\alpha, \leq_\alpha) \subseteq (A_\beta, \leq_\beta)$. Then $a, b \in A_\beta$. Define $a \leq b$ if $a \leq_\beta b$. The fact that $\leq$ is well defined follows from the fact that $\mathfrak{C}$ is a chain. It is a simple exercise to show that $\leq$ linearly orders $A$. We now show that $\leq$ is a well ordering on $A$. Let $S$ be a nonempty subset of $A$. Then $S \cap A_\alpha \neq \emptyset$ for some $\alpha$. Let $a$ be the least element of $S \cap A_\alpha$. We claim that $a$ is the least element of $S$. Let $b \in S$ be such that $b \leq a$, and assume that $b \in A_\beta$. If $(A_\beta, \leq_\beta) \subseteq (A_\alpha, \leq_\alpha)$, then $a, b \in S \cap A_\alpha$ and $b = a$, since $a$ is least in $S \cap A_\alpha$. If $(A_\alpha, \leq_\alpha) \subseteq (A_\beta, \leq_\beta)$, then $A_\alpha$ is a segment of $A_\beta$, and $b \leq a$; hence, $b \in A_\alpha$, and, as before, $b = a$. ∎*

Note that $(A, \leq)$ in the above lemma is an upper bound of the chain $\mathfrak{C}$. Thus if $A_\alpha \neq A$, then $(A, \leq)$ is a continuation of $(A_\alpha, \leq_\alpha)$. The reader is encouraged to work out the details. The crucial step to verify is the following: if $a \in A_\alpha, y \in A$, and $y < a$, then $y \in A_\alpha$. See lemma A.4.

**Theorem A.2.** *Zorn's lemma implies the well ordering principle.*

*Proof. Let $X$ be a nonempty set. We show that $X$ can be well ordered. Let $\mathfrak{W}$ be the collection of well-ordered subsets of $X$, and partially order $\mathfrak{W}$ by continuation. By lemma A.1, a chain in $\mathfrak{W}$ has an upper bound. By Zorn's lemma, $\mathfrak{W}$ has a maximal member $(A, \leq)$. We claim that $A = X$. If $A \neq X$, pick an element $z$ in $X - A$, and define an ordering $\leq_0$ on $Z = A \cup \{z\}$ as follows: retain the ordering $\leq$ on $A$, and define $a <_0 z$ for all $a \in A$. Now $(Z, \leq_0)$ is a strict continuation of $(A, \leq)$, which contradicts that maximality of $(A, \leq)$.* ∎

**Theorem A.3.** *The well ordering principle implies the axiom of choice.*

*Proof. Let $\{X_\alpha\}$ be a nonempty collection of nonempty sets. By assumption, each $X_\alpha$ can be well ordered. Let $x_\alpha$ be the least element of $X_\alpha$, and let $x = (x_\alpha)$. Clearly, $x$ is a choice function and $\prod_\alpha X_\alpha \neq \emptyset$.* ∎

We need a final set of details before we prove the last leg of theorem 2.2.1. The definition below makes sense for linearly ordered sets, but we limit the discussion to well-ordered sets because this is where our interest lies now.

**Definition.** A subset $B$ of a well-ordered set $A$ is said to be a **section of** $A$ if the conditions $a \in A, b \in B$ and $a < b$ imply that $a \in B$.

The following facts are obvious:

  (a) A segment of $A$ is a section of $A$.
  (b) $A$ is a section of itself.

The lemma below is crucial. It is a partial converse of fact (a).

**Lemma A.4.** *Every proper section $B$ of a well-ordered set $A$ is a segment of $A$.*

*Proof. By assumption, $A - B \neq \emptyset$. Let $x = min\{A - B\}$. We show that $B = S(A, x)$. If $y \in S(A, x)$, then $y < x$; hence $y \in B$ because otherwise $y$ would contradict the definition of $x$.*
*Conversely, suppose $y \in B$. Now $y \neq x$ since $y \in B$ and $x \notin B$. Also if $y > x$, then, by the definition of a section, $x \in B$, which is a contradiction. Thus $y < x$ and $y \in S(A, x)$.* ∎

We adopt the following assumptions and terminology for the remainder of this appendix.

Let $(X, \leq)$ be a partially ordered set such that every chain in $X$ has an upper bound but $X$ has no maximal element.

Given a proper chain $A$ in $X$, $A$ has an upper bound, $u$. Since $u$ is not maximal in $X$, there is $x \in X$ such that $x > u$. Clearly, $x \notin A$ because $x$ is a strict upper bound of $A$. Let $\mathfrak{A}$ be the collection of all chains in $X$. Invoking the axiom of choice, we can choose a strict upper bound of each chain in $X$. Thus we have a function $f : \mathfrak{A} \to X$ that assigns to each chain $A$ a strict upper bound, $f(A)$.

Fix an element $x_0 \in X$, and define a subset $A$ of $X$ to be **conforming** if

(a) $(A, \leq)$ is well ordered,

(b) $x_0$ is the least element in $A$, and

(c) For every $x \in A$, $f(S(A,x)) = x$.

**Lemma A.5.** *If $A$ and $B$ are conforming subsets of $X$ and $A - B \neq \varnothing$, then $B$ is a segment of $A$.*

*Proof.* Let $x = min\{A - B\}$, and define $C = S(A,x)$. It is easy to verify that $C \subseteq B$. We claim that $C = B$. Suppose, for a contradiction, that $B - C \neq \varnothing$, and let $y = min\{B - C\}$. We need three steps before we finalize the proof:

1. $S(B,y)$ *is a proper subset of* $C$: Suppose $u \in B, u < y$. If $u \notin C$, then $u \in B - C$, and $u < y$. This contradicts the definition of $y$. If $S(B,y) = C = S(A,x)$, then $y = f(S(B,y)) = f(S(A,x)) = x$, which is a contradiction because $y \in B$ and $x \notin B$. This proves our assertion that $S(B,y)$ is a proper subset of $C$.

2. $S(B,y)$ *is a section of* $C$: If $u \in C$, $v \in S(B,y)$, and $u < v$, we show that $u \in S(B,y)$. Since $u < v < y$, $u < y$. If $u \notin S(B,y)$, then $u \notin B$. Thus $u \in A - B$; hence $u \geq x$. But $u \in C = S(A,x)$; hence $u < x$. This contradiction proves that $u \in S(B,y)$; hence $S(B,y)$ is a section of $C$.

3. $S(B,y)$ *is a segment of* $C$; *thus* $S(B,y) = S(C,z)$, *where* $z \in C$: This follows directly from steps 1 and 2 and lemma A.4.

Now we conclude the proof. By step 3, $y = f(S(B,y)) = f(S(C,z)) = z$. This is a contradiction because $z \in C$, but $y \notin C$ by the definition of $y$. This contradiction proves that $B = C$. ∎

Let $U$ be the union of all the conforming subsets of $X$. The following is a direct result of lemma A.5.

Observation. If $A$ is a conforming subset of $X$, $a \in A$, $y \in U$ and $y < a$, then $y \in A$.

**Lemma A.6.** *$U$ is a conforming subset of $X$. Thus $U$ is the largest conforming subset of $X$.*

*Proof.* It is clear that $x_0$ is the least element of $U$.
The following facts follow directly from the above observation:

(a) If $T \subseteq U$ and $A$ is a conforming subset that intersects $T$, then the least element of $T \cap A$ is also the least element of $T$. Thus $U$ is well-ordered.

(b) If $x \in U$, and $A$ is a conforming subset that contains $x$, then $S(U,x) = S(A,x)$.

Thus $f(S(U,x)) = f(S(A,x)) = x$. ∎

Theorems A.2 and A.3 together with theorem A.7 below constitute the proof of theorem 2.2.1.

**Theorem A.7.** *The axiom of choice implies Zorn's lemma.*

*Proof.* Let $(X, \leq)$ be a partially ordered set such that each chain in $X$ has an upper bound. If $X$ is a chain, then it would have a maximal (in fact, a largest) element, and there is nothing more to prove. Therefore, we assume that $X$ is not a chain. We show that $X$ has a maximal element. Suppose, for a contradiction, that $X$ contains no maximal element.

We have shown in lemma A.6 that the set $U$ is the largest conforming subset of $X$. Since $U$ is well ordered and $X$ is not, $U \neq X$. Let $\omega = f(U)$. The set $U \cup \{\omega\}$ is clearly a conforming subset of $X$ that strictly contains $U$. This contradiction establishes the theorem. ∎

# Matrix Factorizations

The main purpose of this appendix is to prove a useful matrix factorization result: theorem B.4. Theorem B.3 is a useful by-product of this appendix.

**Definition.** Let $A$ be an $m \times n$ matrix. By an elementary row (column) operation on $A$, we mean one of the following operations:

(a) multiplying one row (column) by a nonzero scalar $s$

(b) interchanging two rows (columns)

(c) adding a multiple ($\mu$) of one row (column) to another row (column)

**Definition.** An elementary matrix is an matrix obtained by performing a single elementary row (or column) operation on the identity matrix.

Thus there are three types of elementary matrices:

(a) a scaling matrix (the entry $s \neq 0$ is the $i^{th}$ diagonal entry):

$$S(s, i) = \begin{pmatrix} 1 & & & & & & \\ & \ddots & & & & & \\ & & 1 & & & & \\ & & & s & & & \\ & & & & 1 & & \\ & & & & & \ddots & \\ & & & & & & 1 \end{pmatrix}$$

(b) an elementary permutation matrix (the off diagonal entries are $(i, j)$ and $(j, i)$):

$$P(i, j) = \begin{pmatrix} 1 & & & & & & \\ & \ddots & & & & & \\ & & 0 & \cdots & 1 & & \\ & & \vdots & \ddots & \vdots & & \\ & & 1 & \cdots & 0 & & \\ & & & & & \ddots & \\ & & & & & & 1 \end{pmatrix}$$

(c) a multiplier matrix (the entry $\mu$ is the $(j, i)$ entry):

$$M(\mu, i, j) = \begin{pmatrix} 1 & & & & & & \\ & 1 & & & & & \\ & & \ddots & & & & \\ & & & 1 & & & \\ & & & \vdots & \ddots & & \\ & & & \mu & & \ddots & \\ & & & \vdots & & & \\ & & & & & & 1 \end{pmatrix}$$

Elementary matrices are invertible since they have nonzero determinants: $det(S) = s, det(P) = -1$, and $det(M) = 1$.

The inverse of an elementary matrix is an elementary matrix of the same type. Clearly, a permutation matrix is equal to its own inverse. For a scaling matrix, $S(s, i)^{-1} = S(1/s, i)$.

Observe that a multiplier matrix can be written as $M(\mu, i, j) = I + \mu e_j e_i^T$. Using this, it is easy to verify that $(I + \mu e_j e_i^T)^{-1} = I - \mu e_j e_i^T$. Here $I$ is the identity matrix of the appropriate size.

**Lemma B.1.** *Let A be an $m \times n$ matrix.*

(a) *If E is an $m \times m$ elementary matrix obtained by performing a certain elementary row operation on $I_m$, then performing the same operation on A produces the matrix EA.*

(b) *If F is an $n \times n$ elementary matrix obtained by performing a certain elementary column operation on $I_n$, then performing the same operation on A produces the matrix AF.*

*Proof. Verifying the theorem when E or F is a scaling matrix or a permutation matrix is trivial.*

*If F is obtained from $I_n$ by adding $\mu$ times column j to column i, then $AF = A(I_n + \mu e_j e_i^T) = A + \mu(A e_j) e_i^T$. Now $A e_j$ is the $j^{th}$ column of A, and $(A e_j) e_i^T$ is a matrix whose only nonzero column is the $j^{th}$ column of A placed in the $i^{th}$ column. Hence the result.*

*Proving part (a) for the case of left multiplication by a multiplier matrix is similar and is left to the reader.* ∎

**Theorem B.2.** *Given a nonzero $m \times n$ matrix A, there exist elementary $m \times m$ matrices $E_1, E_2, \ldots, E_r$ and elementary $n \times n$ matrices $F_1, F_2, \ldots, F_s$ such that $E_r \ldots E_2 E_1 A F_1 F_2 \ldots F_s = D$, where D is a diagonal matrix of the form*

$$\begin{pmatrix} d_1 & & & \\ & \ddots & & \\ & & d_q & \\ & & & \end{pmatrix}, d_i \neq 0, 1 \leq i \leq q.$$

*Proof.* In light of lemma B.1, it is enough to prove that A can be reduced to a diagonal matrix through a sequence of elementary row and column operations. We proceed by induction on the number of rows, m. The result is true for a $1 \times n$ matrix and for any $n \in \mathbb{N}$. Consider the $1 \times n$ matrix $A = (a_1, \ldots, a_n)$. If $a_1 = 0$, we interchange the first entry with a later, nonzero entry. Once that is achieved, we subtract $a_i/a_1$ times entry 1 from entry $j$, $2 \leq j \leq n$. We obtain a matrix of the form $(a_1, 0, 0, \ldots, 0)$. This proves the base case of our inductive proof. Now we show the inductive step. Suppose the conclusion of the theorem holds for $k \times n$ matrices if $k < m$ and $n \in \mathbb{N}$. Let A be an $m \times n$ matrix. If $a_{1,1} = 0$, we can move a nonzero entry from a later row and/or column to the top left entry, so assume that $a_{1,1} \neq 0$. Subtracting $a_{i,1}/a_{1,1}$ times the top row from row $i$, $2 \leq i \leq m$, then subtracting $a_{1,j}/a_{1,1}$ times the first column from column $j$, $2 \leq j \leq n$, we obtain a matrix of the form

$$\left( \begin{array}{c|ccc} a_{11} & 0 & \cdots & \cdots & 0 \\ \hline 0 & & & \\ \vdots & & A' & \\ 0 & & & \end{array} \right). \tag{$*$}$$

Applying the inductive hypothesis to the sub-matrix $A'$, we can reduce $A'$ to a diagonal matrix through a combination of elementary row and column operations. Notice that operating on $A'$ does not perturb the top row or the first column of the matrix $(*)$. ∎

**Theorem B.3.** *Given an $m \times n$ matrix A, there exists an invertible $m \times m$ matrix Q and an invertible $n \times n$ matrix P such that $Q^{-1}AP$ is diagonal.*

*Proof.* Use the previous theorem and take $Q = (E_r \ldots E_1)^{-1}$ and $P = F_1 \ldots F_s$. ∎

**Theorem B.4.** *An invertible square matrix is the product of elementary matrices.*

*Proof.* Using theorem B.2, if A is invertible, so is D (recall that elementary matrices are invertible). In this case, still in the notation of theorem B.2, $q = n$ and $D = S_1 S_2 \ldots S_n$, where $S_i$ is the scaling matrix

$$S_i = \left( \begin{array}{ccccc} 1 & & & & \\ & \ddots & & & \\ & & d_i & & \\ & & & \ddots & \\ & & & & 1 \end{array} \right).$$

Thus $A = E_1^{-1} \ldots E_r^{-1} S_1 \ldots S_n \ldots F_s^{-1} \ldots F_1^{-1}$, as desired. ∎

# Bibliography

Bogachev, Vladimir I. *Measure Theory*. Vol. 1. Berlin, Heidelberg: Springer-Verlag, 2007.

Bowers, Adam and Nigel Kalton. *An Introductory Course in Functional Analysis*. New York: Springer, 2014.

Butcher, J. C. *Numerical Methods for Ordinary Differential Equations*. Chichester: Wiley, 2003.

Chartrand, Gary et al. *Mathematical Proofs: A Transition to Advanced Mathematics*. 3rd Edition. New York: Pearson, 2013.

Cohn, Donald. *Measure Theory*. 2nd Edition. Birkhäuser Advanced Texts. Basel: Springer, 2013.

Conte, S. D. and Carl De Boor. *Elementary Numerical Analysis: An Algorithmic Approach*. Tokyo: McGraw-Hill, 1980.

Crilly, Tony and Dale Johnson. *The emergence of topological dimension theory*, in I. M. James (ed.), *History of Topology*. New York: Elsevier, 1999, 1–24.

DeBarra, G. *Introduction to Measure Theory*. New York: Van Nostrand Reinhold Company, 1974.

Debnath, Lokenath and Piotr Mikusinski. *Introduction to Hilbert Spaces with Applications*. 3rd Edition. Amsterdam: Elsevier Academic Press, 2005.

Dorier, Jean-Luc. *A general outline of the genesis of vector space theory*. Historia mathematica, 22, no. 3, 1995, 227–61.

Fabian Marián et al. *Functional Analysis and Infinite-Dimensional Geometry*. CMS Books in Mathematics. New York: Springer, 2001.

Folland, Gerald. *Real Analysis*. 2nd Edition. Hoboken: John Wiley and Sons, 1999.

Garling, D. J. H. *A Course in Mathematical Analysis*. Cambridge: Cambridge University Press, 2013.

Griffel, D. H. *Applied Functional Analysis*. Ellis Horwood Sereis Mathematics and Its Applications. Chichester : Ellis Horwood, 1981.

Hewit, Edwin and Karl Stromberg. *Real and Abstract Analysis*. Vol. 25. Graduate Texts in Mathematics. Berlin, Heidelberg: Springer-Verlag, 1965.

Hillen, Thomas et al. *Partial Differential Equations*. Hoboken: Wiley and Sons, 2012.

Hönig, Chaim. *Proof of the well-ordering of cardinal numbers*. Proceedings of the American Mathematical Society 5 no. 2, 1954, 312.

Hutson, Vivian and John Sydney Pym. *Applications of Functional Analysis and Operator Theory*. Mathematics in Science and Engineering, Vol 146. London: Academic Press, 1980.

Johnson, L. W. and R. D. Riess. *Numerical Analysis*. Reading: Addison-Wesley, 1982.

Lang, Serge. *Real and Functional Analysis*. 3rd Edition. Graduate Texts in Mathematics. New York: Springer-Verlag, 1993.

Lewin, Jonathan. *A Simple Proof of Zorn's Lemma*. The American Mathematical Monthly 98 no. 4, 1991, 353.

MacCluer, Barbara. *Elementary Functional Analysis*. New York: Springer, 2009.

Mendelson, Bert. *Introduction to Topology*. 3rd Edition. New York: Dover Publications, 1990.

Mostow, George et al. *Fundamental Structures of Algebra*. New York: McGraw-Hill, 1975.

Munkres, James. *Topology: A First Course*. London: Prentice-Hall, 1975.

Muscat, Joseph. *Functional Analysis: An Introduction to Metric Spaces, Hilbert Spaces and Banach Algebras*. Cham: Springer, 2014.

O'Connor, J. J. and E. F. Robertson. *MacTutor History of Mathematics*. St. Andrews: University of St Andrews, 1998, http://www-history. mcs.st-andrews.ac.uk/index.html, accessed Nov. 6, 2020.

Oden, Tinsley and Leszek Demkowicz. *Applied Functional Analysis*. 2nd Edition. Boca Raton: CRC Press, 2010.

Pedersen, Gert K. *Analysis Now*. New York: Springer-Verlag, 1989.

Pinter, Charles. *A Book of Set Theory*. New York: Dover Publications, 2014.

Pitts, C. G. C. *Introduction to Metric Spaces*. University Mathematical Texts. Edinburgh: Oliver and Boyd, 1972.

Rudin, Walter. *Functional Analysis*. McGraw-Hill Series in Higher Mathematics. New York: McGraw-Hill, 1973.

Rudin, Walter. *Principles of Mathematical Analysis*. 3rd Edition. International Series in Pure and Applied Mathematics. New York: McGraw-Hill, 1976.

Rudin, Walter. *Real and Complex Analysis*. 2nd Edition. McGraw-Hill Series in Higher Mathematics. New York: McGraw-Hill, 1974.

Rynne, Bryan and Martin Youngson. *Linear Functional Analysis*. London: Springer-Verlag, 2008.

Searcóid, Micheál. *Metric Spaces. Springer Undergraduate Mathematics Series*. London: Springer, 2007.

Shalit, Orr Moshe. *A First Course in Functional Analysis*. Boca Raton: CRC Press, 2017.

Simmons, George. *Introduction to Topology and Modern Analysis*. New York: McGraw-Hill, 1963.

Smith Douglas et al. *A Transition to Advanced Mathematics*. 8th Edition. Andover: Cenage Learning, 2015.

Spence Lawrence et al. *Elementary Linear Algebra: A Matrix Approach*. 2nd Edition. Noida: Pearson, 2018.

Stakgold, Iver. *Green's Functions and Boundary Value Problems*. Pure and Applied Mathematics. New York: John Wiley and Sons, 1979.

Stoer, J. and R. Bulirsch. *Introduction to Numerical Analysis*. New York: Springer-Verlag, 1980.

Viro, O. Ya. et al. *Elementary Topology Problem Book*. Providence: American Mathematical Society, 2008.

Wade, William. *Introduction to Analysis*. Edinburgh: Pearson, 2014.

Young, Nicholas. *An Introduction to Hilbert Space*. Cambridge: Cambridge University Press, 1988.

# Glossary of Symbols

$(x_\alpha)_{\alpha \in I}$ Typical element in $\prod_{\alpha \in I} X_\alpha$   6

$A^B$ Set exponentiation   6

$I_X$ Identity function on $X$   4

$X = \prod_{\alpha \in I} X_\alpha$ Product of $\{X_\alpha\}$   6

$\chi_S$ Characteristic function of $S$   7

$\lim_n$ Limit as $n \to \infty$   11

$\liminf_n$ Limit inferior as $n \to \infty$   17

$\limsup_n$ Limit superior as $n \to \infty$   17

$\mathbb{N}$ Natural numbers   2

$\mathbb{Q}$ Rational numbers   2

$\mathbb{R}^n$ Euclidean $n$-space   6

$\mathbb{R}$ Real numbers   2

$\mathbb{Z}$ Integers   2

$\mathcal{P}(A)$ Power set of $A$   7

$\mathfrak{R}(f)$ The range of $f$   3

$\overline{\mathbb{R}}$ Extended real numbers   17

$\pi_\alpha : X \to X_\alpha$ Projection onto $X_\alpha$   7

$f(A)$ Direct image of $A$   3

$f : X \to Y$ Function from $X$ to $Y$   3

$f^{-1}(B)$ Inverse image of $B$   4

$g \circ f$ Composition of functions   4

$\inf(A)$ Infimum of a set   11

$\sup(A)$ Supremum of a set   11

$2^\mathbb{N}$ Binary sequences   31

$A - B$ Set difference   3

$A \approx B$ Set equivalence   26

$A'$ Derived set of $A$   112

$A^*$ Conjugate transpose of $A$   93

$B(x, r)$ Ball of radius $r$ centered at $x$   80

$Card(A)$ Cardinality of $A$   27

$Conv(A)$ Convex hull of $A$   82

$D_n$ Dirichlet kernel   430

$G_\sigma$ Gauss kernel   438

$H_n$ Hermite polynomials   189

$Ker(T)$ Kernel of $T$   62

$M^\perp$ Orthogonal complement of $M$   90

$P_M$ Projection operator on $M$   297

$P_n$ Legendre polynomials   184

$S^\Delta$ Upper Riemann sum   343

$S_n f$ Finite Fourier sum   176

$Span(A)$ Linear span of $A$   52

$T^*$ Adjoint of $T$   279

$T_n$ Tchebychev polynomials   187

$U \oplus V$ Direct sum   65

$U/V$ Quotient space   64

$X^*$ Dual space of $X$   256

$X^{**}$ Second dual space of $X$   270

$X_\infty$ One-point compactification   229

$\aleph_0$ Cardinality of $\mathbb{N}$   40

$\chi(x, y)$ Chordal distance between $x, y$   127

$\delta_h$ Bump kernel   415

$\hat{X}$ Image of $X$ in $X^{**}$   270

$\hat{f}_n$ Fourier coefficients of $f$   176

$\hat{f}$ Fourier transform   437

$\hat{x}_n$ Fourier coefficients of $x$   302

$\langle ., . \rangle$ Inner product   86

$\langle x, \lambda \rangle$ Duality bracket   279

$\mathbb{K}(I)$ Finitely supported functions $I \to \mathbb{K}$   50

$\mathbb{K}(\mathbb{N})$ Finitely supported seuences   50

$\mathbb{K}^n$ $n$-d space, $\mathbb{R}^n$ or $\mathbb{C}^n$   50, 433

$\mathbb{K}_{m \times n}$ Space of $m \times n$ matrices   50

$\mathbb{K}$ Base field, $\mathbb{R}$ or $\mathbb{C}$   50

$\mathbb{N}_n$ Natural numbers $\leq n$   27

$\mathbb{P}_n$ Polynomials of degree $\leq n$   50

$\mathbb{P}$ Space of polynomials   50

$\mathbb{R}_l^2$ Sorgenfrey plane   217

$\mathbb{R}_l$ Sorgenfrey line   199

$\mathcal{B}(X)$ Bounded functions on $X$   160

$\mathcal{B}[a, b]$ Space of bounded functions on $[a, b]$   50

$\mathcal{BC}(X)$ Continuous bounded functions on $X$   160

$\mathcal{C}(X)$ Continuous functions on $X$   160

$\mathcal{C}(S^1)$ $2\pi$-periodic functions   174

$\mathcal{C}[a, b]$ Space of continuous functions on $[a, b]$   51

$\mathcal{C}^\infty(\mathbb{R})$ Infinitely differentiable functions on $\mathbb{R}$   51

$\mathcal{C}_0(X)$ Continuous functions vanishing at $\infty$   237

$\mathcal{C}_c(X)$ Continuous functions of compact support   237

$\mathcal{H}$ Hilbert cube   118

$\mathcal{L}(X)$ Bounded operators on $X$   256

$\mathcal{L}(X,Y)$ Bounded transformations $X \to Y$   255

$\mathcal{N}(T)$ Null-space of $T$   62

$\mathcal{S}^1$ Unit circle   174

$\mathcal{S}^{n-1}_*$ Punctured unit sphere in $\mathbb{R}^n$   126

$\mathcal{S}^n$ Unit sphere in $\mathbb{R}^n$   126

$\mathfrak{L}^\infty$ Essentially bounded functions   403

$\mathfrak{L}^p$ Lebesgue spaces   403

$\mathfrak{M} \otimes \mathfrak{N}$ Product of $\sigma$-algebras   419

$\mathfrak{c}$ Cardinality of $\mathbb{R}$   40

$\mathcal{B}(X)$ Borel subsets of $X$   350

$\mathcal{B}^n$ Borel subsets of $\mathbb{R}^n$   350

$\mathcal{L}(\mathbb{R}^n)$ Lebesgue measurable subsets of $\mathbb{R}^n$   382

$\mathcal{L}^n$ Lebesgue measurable subsets of $\mathbb{R}^n$   382

$\mu \otimes \nu$ Product measure   423

$\nu \ll \mu$ Absolute continuity   398

$\nu^+$ Positive part of $\nu$   396

$\nu^-$ Negative part of $\nu$   396

$\overline{A}$ Closure of $A$   111

$\partial A$ Boundary of $A$   113

$\rho(T)$ Resolvent set of $T$   274

$\to^w$ Weak convergence   272

$\sigma(T)$ Spectrum of $T$   274

$\{e_1, e_2, ...\}$ Canonical vectors in $\mathbb{K}(\mathbb{N})$   52

$c_0$ Space of null sequences   51

$c$ Space of convergent sequences   51

$dim(U)$ Dimension of $U$   58

$dist(A, B)$ Distance between $A$ and $B$   113

$dist(x, A)$ Distance from $x$ to $A$   113

$f * g$ Convolution of functions   413

$f^+$ Positive part of $f$   367

$f^-$ Negative part of $f$   367

$int(A)$ Interior of $A$   110

$l^\infty$ Space of bounded sequences   51

$l^p$ Space of $p$-convergent series   77

$r(T)$ Spectral radius of $T$   276

$s_\Delta$ Lower Riemann sum   343

$sgn(g)$ Sign of g   405

$supp(f)$ Support of $f$   237

$vol(Q)$ Volume of box $Q$   343

$x \perp y$ $x$ is orthogonal to $y$   87

$|\nu|$ Total variation of $\nu$   396

# Index

Absolute continuity of measures 398
Absolutely convergent series 252
Adjoint operator 98, 279, 308
Alexandroff 231
Algebra of sets 349
Algebra over a field 68
Algebraic complement 65
Almost everywhere 369
Annihilator 281
Annihilator of a set 295
Antisymmetric relation 33
Apollonius identity 299
Approximate identity 438
Arzela-Ascoli theorem 164
Ascending sequence of sets 5
Ascoli's theorem 164, 170
Axiom of choice 35

Baire's theorem 139, 225
Banach algebra 273
Banach space 247
Banach, Stefan 245
Banach-Alaoglue theorem 287
Banach-Saks theorem 308
Banach-Steinhaus theorem 265
Barycentric coordinates 85
Basis 55
Bessel's inequality 302
Best approximation 91
Bicontinuous function 123, 203
Bijection 4
Binary sequences 6
Bolzano-Weierstrass property 15, 151
Bolzano-Weierstrass theorem 13
Borel algebra 350
Borel sets 350
Boundary of a set 113, 195
Boundary point 113, 195
Bounded away from zero 257
Bounded functions, space of 160
Bounded linear mapping 253
Bounded metric 114, 124
Bounded projection 272
Bounded sequence 12, 114
Bounded sequences, space of 51
Bounded set 11

Bounded subset 114
Box topoloogy 242
Bump function 414
Bump kernel 415

Canonical projections 260
Canonical vectors 52
Cantor intersection theorem 139
Cantor set 116
Cantor, Georg 25
Carathéodory condition 353, 382, 389
Carathéodory's theorem 83, 354
Cardinal arithmetic 41
Cardinality 27, 39
Cartesian product 3, 6
Cauchy criterion 15
Cauchy sequence 13, 136
Cauchy-Schwarz inequality 87, 293
Chain 33
Change of base 72
Characteristic function 7
Choice function 35
Chordal metric 126
Closed box 343
Closed graph 264
Closed graph theorem 264
Closed mapping 207
Closed set 107, 194
Closure of a set 111, 194
Closure point 111, 194
Co-dimension 67
Co-finite topology 193
Coarser metric 121
Coarser topology 199
Compact operator 320, 336
Compact space 149, 221
Compact subspace 150, 221
Compactification 231
Complemented subspace 271
Complement of a set 3
Complete measure 355
Complete metric space 137
Completeness of Lp 404
Completeness of R 11
Completeness of C 21
Completion of a measure 362

Completion of a metric  162
Completion of a norm  270
Completion of an inner product  298
Complex measure  397
Complex numbers  19
Componentwise convergence  131
Composition of functions  4
Conforming set  447
Conjugate Hölder exponents  78
Conjugate transpose  93
Connected components  211
Connected points  211
Connected space  208
Connected subset  210
Continuity at a point  120, 201
Continuity of inner products  121
Continuity of norms  121
Continuous bounded functions  160, 161, 202
Continuous function  120, 201, 202
Continuous functions, space of  160
Continuum hypothesis  45
Contraction  142
Contraction mapping theorem  141
Convergent sequence  11, 108
Convergent sequences, space of  51
Convergent series  252
Convex combination  82
Convex hull  82
Convex set  81
Convolution of functions  413
Coset of a subspace  64
Countable additivity  351
Countable intersection property  218
Countable set  29
Countably compact space  225
Counting measure  351

De Morgan's laws  3, 5
Decreasing sequence  12
Dedekind, Richard  1
Defining base  206
Defining subbase  206
Dense subset  133, 217
Dependent vectors  53
Derived set  112
Descending sequence of sets  5
Diagonalization  73
Diameter of a set  114
Dimension of a vector space  58, 59
Dini's theorem  171
Dirac measure  351
Direct sum of subspaces  65
Direct sums  65
Dirichlet kernel  430

Disconnected space  208
Discrete metric  106
Discrete topology  193
Disjoint family  5
Distance function  105
Distributive laws  3, 5
Dominated convergence theorem  372
Dual space  256
Duality bracket  278

Egoroff's theorem  409
Eigenspace  274
Eigenvalues  274
Eigenvectors  274
Elementary matrix  449
Elementary sets  420
Equicontinuity  163
Equivalence classes  7
Equivalence relation  7
Equivalent metrics  123
Equivalent sets  26
Essential upper bound  403
Essentially bounded functions  403
Euclidean metric  105
Euclidean $n$-space  6
Extended real line  17
Extreme point  83

$F_\sigma$ set  351
Fatou's theorem  371
Feijer kernel  433
Field  10
Finer metric  121
Finite intersection property  223
Finite measure  351
Finite rank operator  322
Finite sequence  4
Finite set  27
Finite-dimensional space  57
First countable space  220
Fixed point  142
Fourier coefficients  176, 302
Fourier series  176
Fourier transform  436
Fredholm alternative theorem  327
Fredholm integral equation  332
Fredholm theory  325
Fubini's theorem  424, 428
Functions of compact support  237, 374

$G_\delta$ set  351
Gauss kernel  438
Gelfand's theorem  277
Generalized continuum hypothesis  45

Gram-Schmidt process  92
Greatest element  33
Greatest lower bound  11
Grid  342

Hölder's inequality  78, 403
Hahn decomposition theorem  395
Hahn-Banach theorem  268
Half-spaces  90
Hamel basis  55
Hausdorff property  108
Hausdorff space  214
Heine-Borel theorem  154, 252
Hermite polynomials  188
Hermitian matrix  95
Hilbert cube  118, 250
Hilbert dimension  306
Hilbert space  293
Hilbert space isomorphism  304
Hilbert, David  291
Hilbert-Schmidt kernel  332
Hilbert-Schmidt theorem  329, 335
Homeomorphic spaces  125, 203
Homeomorphism  125, 203
Homomorphism  67
Hopf extension theorem  361
Hyperplane  90

Idempotent operator  69
Identity function  4
Image of a set  3
Increasing sequence  12
Independent vectors  54
Indexed sets  4
Indiscrete topology  193
Induced metric  115
Infimum  11
Infinite sequnce  4
Infinite set  27
Infinite-dimensional space  59
Infinity norm  76, 160
Initial segment  38, 445
Injective function  4
Inner product space  86
Inner regularity  387
Integrable function  368, 369
Integral of a function  365, 366, 368, 369
Interior of a set  110, 194
Interior point  110, 194
Intermediate value theorem  209
Interval  208
Invariance of dimension  59
Invariant subspace  69
Inverse functions  4

Inverse image  3
Inversion formula  438, 440
Invertible operator  273
Isolated point  112, 220
Isometric spaces  124
Isometry  124
Isomorphism  63

Jordan decomposition theorem  396

Kernel of a linear mapping  62
Krein-Millman theorem  84
Kronecker delta  52

Least upper bound  11, 34
Lebesgue measurable set  382
Lebesgue measure  382
Lebesgue number  152
Lebesgue outer measure  380
Lebesgue, Henri  341
Left shift operator  274
Legendre polynomials  92, 183
Limit inferior  17
Limit point  15, 20, 112, 197
Limit point of a sequence  17
Limit superior  17
Lindelöf space  134, 217
Linear basis  55
Linear combination  52
Linear functional  66
Linear mapping  61
Linear operator  68
Linear ordering  33
Liouville's theorem  276
Lipschitz function  142
Locally bounded function  202
Locally compact space  154, 227
Lower bound  11
Lower limit  17
Lower limit topology  199
Lower Riemann integral  344
Lower Riemann sum  343
Lower semicontinuous  203
$\mathcal{L}^p$ spaces  403
$l^p$ spaces  77
Luzin's theorem  411

Matrix of a linear mapping  70
Matrix representation  70
Maximal element  33
Maximal subspace  67
Mean square convergence  177
Measurable dissection  401
Measurable function  355, 359

Measurable rectangle 419
Measurable sets 351
Measurable space 351
Measure space 351
Metric 105
Metric space 105
Metrizable space 233
Metrization 233
Minimal spanning set 56
Minkowski's inequality 78, 404
Monotone class 421
Monotone class lemma 421
Monotone convergence theorem 371
Monotonic sequence 12
Monotonicity of measures 351
Mutually singular measures 396

Natural embedding 270
Negative set 394
Negative variation of a measure 396
Neighborhood of a point 108, 195
Neighborhood of a set 195
Norm 76
Norm of a bounded mapping 254
Norm topology 284
Normal matrix 95
Normal operator 98, 316
Normal space 215
Normed linear space 75
Nowhere dense set 116, 139, 197
Nowhere differentiable functions 144
Null sequences, space of 51
Nullity of a linear mapping 62
Null-space of a linear mapping 62

Obtuse angle criterion 156
One-point compactification 231
One-to-one correspondence 4
Open ball 80, 106
Open base 134, 198
Open cover 134, 149, 217, 221
Open mapping 207
Open mapping theorem 262
Open neighborhood 195
Open set 106, 193
Open subbase 199
Open subcover 134, 149, 217, 221
Operator algebra 273
Operator norm 256
Ordered pairs 3
Orthogonal complement 90, 295
Orthogonal matrix 93
Orthogonal polynomials 183
Orthogonal projection 91, 297

Orthogonal set 88
Orthogonal vectors 87
Orthonormal basis 300
Orthonormal set 88
Outer measure 352
Outer regularity 387

Parallelogram law 294
Parseval's identity 303
Partial ordering 33
Partition 342
Partition of unity 375
Path connected 212
Peano, Giuseppe 47
Perfect set 116
Piecewise linear function 61
Point spectrum 274
Pointwise boundedness 261
Polarization identity 294
Polytope 83
Positive measure 351
Positive operator 319
Positive set 394
Positive variation of a measure 396
Power set 7
Product measure 423
Product metric 130
Product spaces 130
Product topology 206, 239
Projection 7, 66
Projection operator 297
Projection theorem 296
Punctured circle 126
Punctured sphere 127
Pythagorean theorem 88, 294

Qoutient map 64
Quotient space 64, 281

Radon measurable set 389
Radon measure 389
Radon outer measure 389
Radon-Nikodym derivative 398
Radon-Nikodym theorem 400
Range of a function 3
Range of a linear mapping 62
Rank of a linear mapping 62
Real measure 394
Reflexive relation 7
Reflexive space 270
Regular measure 387
Regular space 215
Relation 7
Relative topology 196

Resolvent set 274
Restricted metric 115
Restricted topology 196
Riemann integrable function 344
Riemann integral 344
Riemann-Lebesgue lemma 435, 437
Riesz representation theorem 297
Riesz's lemma 251
Riesz-Fisher theorem 304
Riesz-Schauder theorem 323
Right shift operator 274

Schauder basis 253
Schröder-Bernstein theorem 37
Second countable space 134, 217
Second dual space 269
Section of a set 446
Sections of functions 420
Sections of sets 419
Segment of a set 445
Self-adjoint operator 98, 309
Separable space 133, 217
Sequential continuity 120
Sequentially compact space 151
Set exponentiation 6
$\sigma$-algebra 349
$\sigma$-compact 228
$\sigma$-finite 402
Simple function 365
Skew symmetric matrices 60
Sorgenfrey line 199
Sorgenfrey plane 217
Space of continuous functions 202
Space of bounded functions 160, 202
Space-filling curve 167
Span of a set 52
Spectral decomposition 97
Spectral radius 275
Spectral theorem 99, 329
Spectrum of an operator 274
Square integrable functions 183
Standard inner product 86
Standard matrix 61
Standard $n$-simplex 84
Step function 348
Stereographic projections 126
The Stone-Weierstrass Theorem 172
Strictly convex norm 307
Strong topology 284
Stronger metric 121
Stronger topology 199
Subadditivity of measure 352
Subcover 134, 149, 221
Subsequence 13

Subspace 51, 115
Subspace metric 115
Subspace topology 196
Sum of subspaces 65
Support of a function 237, 374
Supporting half-space 157
Supporting hyperplane theorem 158
Supremum 11
Supremum norm 76, 160
Surjective function 4
Symmetric relation 7

$T_1$ space 214
Tchebychev polynomials 186, 187
Three-term recurrence 184
Tietze extension theorem 409
Tonelli's theorem 424
Topological embedding 203
Topological space 193
Topology 193
Total ordering 33
Total variation of a measure 396, 400
Totally bounded space 151
Totally disconnected space 212
Transfinite induction 38
Transitive relation 7
Translation of a set 81
Triangle inequality 76, 105
Trigonometric polynomial 177
Tube lemma 224
Tychonoff's theorem 153, 224, 241

Uncountable set 29
Uniform boundedness 261
Uniform boundedness principle 262
Uniform continuity 14, 161
Uniform equicontinuity 163
Uniform norm 76, 160
Uniqueness theorem 181, 436, 440
Unitary matrix 93
Unitary operator 316
Upper bound 10, 34
Upper limit 17
Upper Riemann integral 344
Upper Riemann sum 343
Upper semicontinuous 203
Urysohn, Pavel 191
Urysohn metrization theorem 235
Urysohn's lemma 234, 373, 415
Usual metric 105
Usual topology 194

vector space 49
Vertices of a polytope 83
Volterra equation 143

Weak topology 284
Weak* topology 284
Weaker metric 121
Weaker topology 199

Weakly convergent sequence 272, 307
Weierstrass approximation theorem 165
Weierstrass M-test 140
Weierstrass, Karl 103
Weight function 182
Well-ordered set 34

Zorn's lemma 36